Organic Semiconductors for Optoelectronics

## Wiley Series in Materials for Electronic and Optoelectronic Applications

www.wiley.com/go/meoa



**Published Titles**

Bulk Crystal Growth of Electronic, Optical and Optoelectronic Materials, Edited by P. Capper Properties of Group-IV, III—V and II—VI Semiconductors, S. Adachi
Charge Transport in Disordered Solids with Applications in Electronics, Edited by S. Baranovski Optical Properties of Condensed Matter and Applications, Edited by J. Singh
Thin Film Solar Cells: Fabrication, Characterization, and Applications, Edited by J. Poortmans and V. Arkhipov
Dielectric Films for Advanced Microelectronics, Edited by M. R. Baklanov, M. Green, and K. Maex
Liquid Phase Epitaxy of Electronic, Optical and Optoelectronic Materials, Edited by P. Capper and M. Mauk Molecular Electronics: From Principles to Practice, M. Petty
Luminescent Materials and Applications, A. Kitai
CVD Diamond for Electronic Devices and Sensors, Edited by R. S. Sussmann
Properties of Semiconductor Alloys: Group-IV, III—V and II—VI Semiconductors, S. Adachi Mercury Cadmium Telluride, Edited by P. Capper and J. Garland
Zinc Oxide Materials for Electronic and Optoelectronic Device Applications, Edited by C. Litton, D. C. Reynolds, and T. C. Collins
Lead-Free Solders: Materials Reliability for Electronics, Edited by K. N. Subramunian Silicon Photonics: Fundamentals and Devices, M. Jamal Deen and P. K. Basu
Nanostructured and Subwavelength Waveguides: Fundamentals and Applications, M. Skorobogatiy Photovoltaic Materials: From Crystalline Silicon to Third-Generation Approaches, Edited by G. Conibeer and A. Willoughby
Glancing Angle Deposition of Thin Films: Engineering the Nanoscale, Matthew M. Hawkeye, Michael T. Taschuk, and Michael J. Brett
Physical Properties of High-Temperature Superconductors, R. Wesche
Spintronics for Next Generation Innovative Devices, Edited by Katsuaki Sato and Eiji Saitoh Inorganic Glasses for Photonics: Fundamentals, Engineering and Applications, Animesh Jha
Amorphous Semiconductors: Structural, Optical and Electronic Properties, Kazuo Morigaki, Sandor Kugler, and Koichi Shimakawa
Microwave Materials and Applications, Two volume set, Edited by Mailadil T. Sebastian, Rick Ubic, and Heli Jantunen
Molecular Beam Epitaxy: Materials and Applications for Electronics and Optoelectronics, Edited by Hajime Asahi and Yoshiji Korikoshi
Metalorganic Vapor Phase Epitaxy (MOVPE): Growth, Materials Properties, and Applications, Edited by Stuart Irvine and Peter Capper
Optical Properties of Materials and Their Applications, Second Edition, Edited by Jai Singh
Oxide Electronics, Edited by Asim Ray

# Organic Semiconductors for Optoelectronics

*Edited by*

*Hiroyoshi Naito*
Osaka Prefecture University
Osaka, Japan

WILEY

This edition first published 2021


*Registered Offices*
John Wiley & Sons, Inc., 111 River Street, Hoboken, NJ 07030, USA
John Wiley & Sons Ltd, The Atrium, Southern Gate, Chichester, West Sussex, PO19 8SQ, UK

*Editorial Office*
The Atrium, Southern Gate, Chichester, West Sussex, PO19 8SQ, UK

For details of our global editorial offices, customer services, and more information about Wiley products visit us at www.wiley.com.

Wiley also publishes its books in a variety of electronic formats and by print-on-demand. Some content that appears in standard print versions of this book may not be available in other formats.



*Library of Congress Cataloging-in-Publication Data*

Name: Naito, Hiroyoshi, editor.
Title: Organic semiconductors for optoelectronics / edited by Hiroyoshi Naito.
Description: First edition. | Hoboken, NJ : Wiley, 2021. | Series: Wiley series in materials for electronic and optoelectronic applications |
Includes bibliographical references and index.
Identifiers: LCCN 2020051135 (print) | LCCN 2020051136 (ebook) | ISBN 9781119146100 (hardback)

Subjects: LCSH: Organic semiconductors. | Optoelectronics.
Classification: LCC QC611.8.O7 O6967 2021 (print) | LCC QC611.8.O7 (ebook) | DDC 537.6/223–dc23
LC record available at https://lccn.loc.gov/2020051135
LC ebook record available at https://lccn.loc.gov/2020051136

Cover Design: Wiley
Cover Images: Courtesy and Copyright of NIPPON SHOKUBAI CO., LTD

Set in 10/12pt WarnockPro by Straive, Chennai, India

10 9 8 7 6 5 4 3 2 1

# Contents

# List of Contributors

**Kazuyoshi Tanaka** Fukui Institute for Fundamental Chemistry, Kyoto University, Kyoto, Japan

**Jai Singh** School of Engineering and Information Technology, Charles Darwin University, Australia

**Monishka Rita Narayan** School of Engineering and Information Technology, Charles Darwin University, Australia

**David Ompong** School of Engineering and Information Technology, Charles Darwin University, Australia

**Takashi Kobayashi** Department of Physics and Electronics, The Research Institute of Molecular Electronic Devices, Osaka Prefecture University, Sakai, Japan

**Takashi Nagase** Department of Physics and Electronics, The Research Institute of Molecular Electronic Devices, Osaka Prefecture University, Sakai, Japan

**Hiroyoshi Naito** Department of Physics and Electronics, The Research Institute of Molecular Electronic Devices, Osaka Prefecture University, Sakai, Japan

**Kenichiro Takagi** Department of Physics and Electronics, Osaka Prefecture University, Sakai, Japan

**Masahiro Funahashi** Department of Advanced Materials Science, Faculty of Engineering, Kagawa University, Takamatsu, Kagawa, Japan

**Akinori Saeki** Department of Applied Chemistry, Graduate School of Engineering, Osaka University, Suita, Osaka, Japan

**Andrey S. Mishchenko** RIKEN Center for emergent Matter Science (CEMS), Wako, Japan

**Takaaki Manaka** Tokyo Institute of Technology, O-okayama, Meguro-Ku, Tokyo, Japan

**Mitsumasa Iwamoto** Tokyo Institute of Technology, O-okayama, Meguro-ku, Tokyo, Japan

**Hiroyuki Matsui** Graduate School of Organic Materials Science, Yamagata University, Yamagata, Japan

**Yutaka Noguchi** Department of Electronics and Bioinformatics, Meiji University, Tokyo, Japan

**Hisao Ishii** Center for Frontier Science, Chiba University, Chiba, Japan

**Lars Jäger** Institute of Physics, University of Augsburg, Augsburg, Germany

**Tobias D. Schmidt** Institute of Physics, University of Augsburg, Augsburg, Germany

**Wolfgang Brutting** Institute of Physics, University of Augsburg, Augsburg, Germany

**Li Zhao** Center for Organic Photonics and Electronics Research, Kyushu University, Fukuoka, Japan

**DaeHyeon Kim** Center for Organic Photonics and Electronics Research, Kyushu University, Fukuoka, Japan

**Jean-Charles Ribierre** Center for Organic Photonics and Electronics Research, Kyushu University, Fukuoka, Japan

**Takeshi Komino** Center for Organic Photonics and Electronics Research, Kyushu University, Fukuoka, Japan

**Chihaya Adachi** Center for Organic Photonics and Electronics Research, Kyushu University, Fukuoka, Japan

**Mayumi Uno** Osaka Research Institue of Industrial Science and Technology (ORIST), Osaka, Japan

**Itaru Osaka** Graduate School of Advanced Science and Engineering, Hiroshima University, Hiroshima, Japan

**Kazuo Takimiya** RIKEN, Center for Emergent Matter Science, Saitama, Japan, and Graduate School of Science, Tohoku University, Sendai, Japan

**Koichi Okamoto** Department of Physics and Electronics, Osaka Prefecture University, Sakai, Japan

# Series Preface

## Wiley Series in Materials for Electronic and Optoelectronic Applications

This book series is devoted to the rapidly developing class of materials used for electronic and optoelectronic applications. It is designed to provide much-needed information on the fundamental scientific principles of these materials, together with how these are employed in technological applications. These books are aimed at (postgraduate) students, researchers, and technologists engaged in research, development, and the study of materials in electronics and photonics, and at industrial scientists developing new materials, devices, and circuits for the electronic, optoelectronic, and communications industries.

The development of new electronic and optoelectronic materials depends not only on materials engineering at a practical level, but also on a clear understanding of the properties of materials and the fundamental science behind these properties. It is the properties of a material that eventually determine its usefulness in an application. The series therefore also includes such titles as electrical conduction in solids, optical properties, thermal properties, and so on, all with applications and examples of materials in electronics and optoelectronics. The characterization of materials is also covered within the series as much as it is impossible to develop new materials without the proper characterization of their structure and properties. Structure–property relationships have always been fundamentally and intrinsically important to materials science and engineering.

Materials science is well known for being one of the most interdisciplinary sciences. It is the interdisciplinary aspect of materials science that has led to many exciting discoveries, new materials, and new applications. It is not unusual to find scientists with a chemical engineering background working on materials projects with applications in electronics. In selecting titles for the series, we have tried to maintain the interdisciplinary aspect of the field, and hence its excitement to researchers in this field.

*Arthur Willoughby*
*Peter Capper*
*Safa Kasap*

# Preface

The photoconductive and semiconducting properties of organic semiconductors were reported in 1906 and 1950, respectively, and since then, basic research has steadily continued. In 1980, molecularly dispersed polymers in which hole transport molecules were dispersed in insulating polymers were commercialized as photoreceptors for electrophotography. The manufacturing process for this organic photoreceptor was a coating process, which contributed to the low cost of the photoreceptor. Organic light-emitting diode (OLED) and organic solar cells were reported in 1987 and 1989, respectively. These devices were highly efficient at that time and showed the potential of the organic devices. OLEDs were commercialized as an automotive display in 1997, and are currently being used in high-definition OLED TVs and OLED lighting. In the future, it is expected that organic semiconductors will be successfully applied to flexible displays, biosensors, and other devices that could not be realized with conventional inorganic semiconductors. The development of future organic devices cannot be achieved without a proper understanding of the optoelectronic properties of organic semiconductors and how these properties influence the overall device performance. Therefore, it is intended here to have one single volume that covers fundamentals through to applications, with up-to-date advances in the field.

This book summarizes the basic concepts and also reviews some recent developments in the study of optoelectronic properties of organic semiconductors. It covers examples and applications in the field of electronic and optoelectronic organic materials. An attempt is made to cover both experimental and theoretical developments in each field presented in this book, which consists of 15 chapters contributed by experienced and well-known scientists on different aspects of optoelectronic properties of organic semiconducting materials. Most chapters are presented to be relatively independent with minimal cross-referencing, but chapters with complementary contents are arranged together to facilitate the reader with cross-referencing.

In Chapter 1 by Tanaka, the fundamental electronic properties of organic semiconducting materials are concisely reviewed and the chapter to provides basic concepts for understanding the electronic properties. In Chapter 2, Naito presents a review of electronic transport properties of organic semiconductors, and Chapter 3 by Singh *et al.* covers the theoretical concepts of optical properties of organic semiconductors. In Chapter 4, Kobayashi *et al.* have presented a comprehensive review of advanced, as well as standard experimental techniques, for the characterization of optical properties of organic semiconducting materials including fluorescent, phosphorescent and thermally assisted delayed fluorescent emitters. In Chapters 5 to 7, a comprehensive review of advanced and standard experimental techniques for the characterization of transport properties of organic semiconducting materials are presented. Naito reviews impedance spectroscopy, which is applicable to the measurement of drift mobility of thin organic semiconducting films in Chapter 5. Funahashi reviews standard time-of-flight measurements with different measurement configurations for drift mobility in organic liquid-crystalline semiconductors in Chapter 6, and Saeki reviews microwave and terahertz spectroscopy, which is a unique electrodeless technique, in organic and organic-inorganic perovskite solar cells in Chapter 7. Chapter 8, by Mishchenko, covers electron spin resonance study for the characterization of localized states. In Chapter 9, Manaka and Iwamoto present recent advances in second harmonic generation spectroscopy. In Chapters 10 to 12, reviews of device physics of key organic devices are presented. Matsui presents a comprehensive review of the device physics of organic field-effect transistors in Chapter 10 and, in Chapter 11 by Noguchi *et al.*, basic processes in OLEDs are reviewed. Zhao *et al.* discuss the relationship between out-coupling efficiency and molecular orientation in OLEDs in Chapter 12. Uno reviews the application of organic field-effect transistors to integrated circuits in Chapter 13 with Osaka and Takimiya reviewing high performance polymeric semiconductors for organic solar cells in Chapter 14. Finally, in Chapter 15, Okamoto covers plasmonics for the improvement of efficiencies of light-emitting and photovoltaic devices.

The aim of the book is to present its readers with recent developments in theoretical and experimental aspects of optoelectronic properties of organic semiconductors. Accomplishments and technical challenges in device applications are also discussed. The readership of the book is expected to be graduate students, as well as teaching and research professionals.

Finally, the Editor wishes to thank Jenny Cossham and Katrina Maceda for their help and encouragement in the editing and production processes.

Osaka, Japan *Hiroyoshi Naito*

# 1

# Electronic Structures of Organic Semiconductors

*Kazuyoshi Tanaka*

*Fukui Institute for Fundamental Chemistry, Kyoto University, Kyoto, Japan*

**CHAPTER MENU**



## 1.1 Introduction

Electric conductivities of organic materials are normally low and they are classified as insulators or semiconductors. In general, electric conductivity of the semiconductor is broadly considered to be in the range from $10^{-10}$ to $10^{2}$ $Scm^{-1}$ (Figure 1.1). Electric conductivity $\sigma$ is expressed by

$$\sigma = ne\mu \tag{1.1}$$

where $n$ is the number of charge carriers for electric transport, $e$ the elementary charge ($1.602 \times 10^{19}$ C), and $\mu$ the mobility of the carriers. Appearance of high conductivity in organic material per se is quite rare or completely absent. This is because organic materials do not have enough number of $n$ though they might have large $\mu$ in a potential sense embodied by, e. g., extended $\pi$-conjugation appropriate to the electric conduction throughout the material.

The above description means that organic materials can change into semiconductive or even metallic state in terms of appropriate injection of carriers if they are guaranteed to show appropriate $\mu$ values. From the latter half of the previous century, a great deal of attempts toward this direction have been piled up and nowadays organic semiconductors or organic metals have become quite common members in electronics materials such as organic field-effect transistor (OFET), organic light-emitting diode (OLED), organic photovoltaic (OPV) device, and so on. It is noted here that characteristic features of organic semiconductors or organic metals come from their structural low dimensionality. This is simultaneously accompanied with the fact that the direction of electric

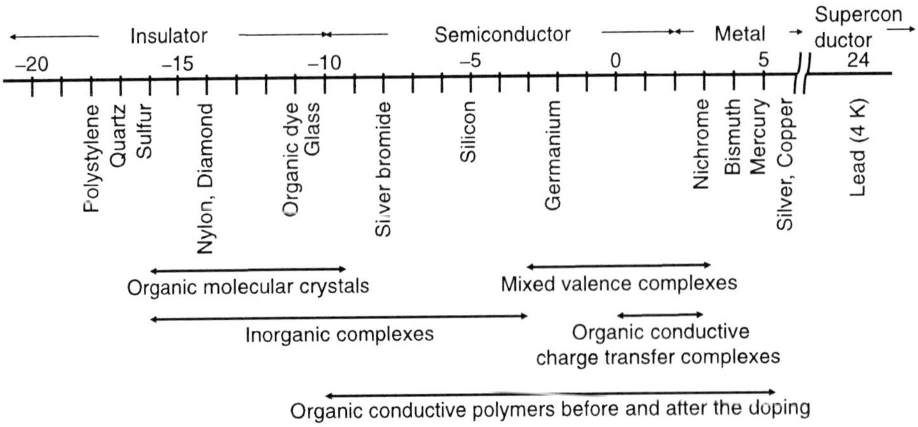


**Figure 1.1** Logarithmic representation of electric conductivity $\sigma$ (S/cm) of miscellaneous materials at room temperature.

transport is remarkably developed toward one or two directions in the material and, in this sense, these are called one-dimensional (1D) or two-dimensional (2D) materials. For example, polymer with rather rigid spine can be regarded as 1D material and graphene a complete 2D material. These low-dimensional materials often show peculiar behavior in relation to electronic properties when they are in the semiconductive or metallic state.

Analysis of the electronic structure is of primary importance in consideration of the semiconductive or metallic properties of organic materials. In this Chapter, we are to study (i) the ways of carrier injections and (ii) transition from the conductive state inherent in low-dimensional materials, with respect to organic semiconductors. Emphasis will also be put on understanding of the electronic properties of these materials based on their electronic structures. We first start from the electronic structures of organic materials with regular repetition of molecular unit, that is, crystalline structure, and then elucidate the electronic properties derived from the electronic structures. The prospects for typical conductive polymers and charge-transfer organic crystals are also to be afforded. In the last part, the electronic properties of organic amorphous material will also be dealt with.

## 1.2 Electronic Structures of Organic Crystalline Materials

In this Section, electronic structure and its related quantities of organic materials with crystalline structures are described with respect to the 1D system not only for the sake of simplicity but also due to being realistic in most of the organic semiconductors. Note that the 1D organic crystal has regular repetition of the unit cells as illustrated in Figure 1.2 being somewhat similar to the primary structure of ideal polymers. Extension to 2D or 3D crystal is quite straightforward. In order to describe the electronic structure of organic crystal, the orbital approximation occurring from one-electron picture is to be employed throughout this Section unless specially noted, since it allows us to

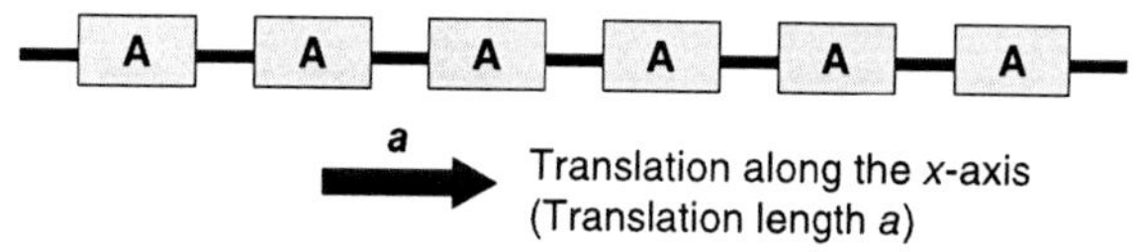


**Figure 1.2** Schematic drawing of 1D crystal.

have a simple but clear idea in the same spirit as the molecular orbital (MO) scheme for the ordinary organic molecules. In organic crystals, the wavefunction based on the one-electron picture is often mentioned as crystal orbital (CO) as is described later. We will try to figure out the electronic properties of organic crystals mainly derived from the COs.

### 1.2.1 Free-Electron Picture

First, we start from the simplest wavefunction of a free electron in 1D space, using the Schrödinger equation which is expressed as

$$-\frac{\hbar^2}{2m}\frac{d^2}{dx^2}\psi(x) = \varepsilon\psi(x) \tag{1.2}$$

without any potentials for a free electron. The wavefunction of a free electron at a point $x$ is accompanied with a variable $k$ as

$$\psi(x) = A\exp[ikx] + B\exp[-ikx] \tag{1.3}$$

where $i$ stands for the imaginary unit and $k$ is called wavevector (or wave number) being proportional to momentum $p$ of the electron, that is,

$$k = \frac{p}{\hbar} \tag{1.4}$$

with $\hbar = \frac{h}{2\pi}$, $h$ being the Planck's constant. As a matter of course, $k$ becomes vector $\mathbf{k}$ for 2D and 3D cases. Furthermore, $A$ and $B$ in Eq. (1.3) are the formal normalization constants. Each of the two terms in the right-hand side of Eq. (1.3) signifies the motion of a free electron to the $x$ and $-x$ directions.

The free-electron wavefunction basically describes the electron motion in a free space without any potentials as is mentioned above and, in this sense, is considered to describe the electrons inside the space of crystal as ideal gas. Note this wavefunction takes a complex value, which is natural in the picture of quantum mechanics. The energy of a free electron is a function of $k$ and is given by

$$\varepsilon_k = \frac{\hbar^2 k^2}{2m} \tag{1.5}$$

which has a continuous parabolic shape with a variable $k$ as shown in Figure 1.3. The plot of the energy value depending on $k$ is generally called energy band or band structure. According to the number of electrons, there appears the upper limit of energy levels filled with electrons called Fermi energy ($\varepsilon_F$) dividing both the valence and conduction bands. The wavevector at the position of $\varepsilon_F$ is called Fermi wavevector $k_F$. Note that $\pm k$ gives the same energy signifying the degeneracy according to inversion of the momentum, which is also mentioned as time-reversal symmetry due to the change of momentum direction.

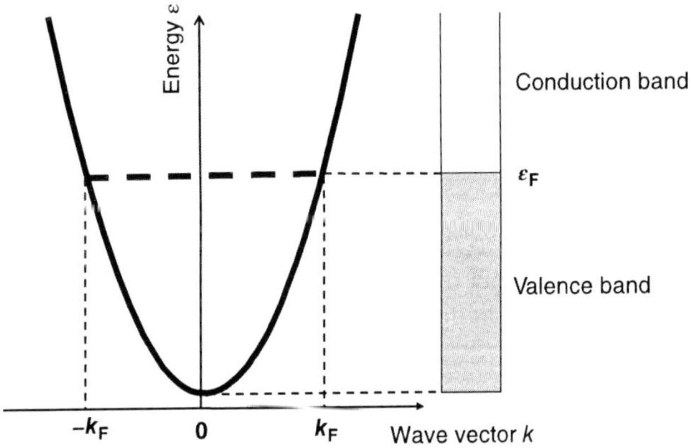


**Figure 1.3** Energy band of a free electron. $\varepsilon_F$ and $k_F$ signify Fermi energy and Fermi wave vector, respectively.

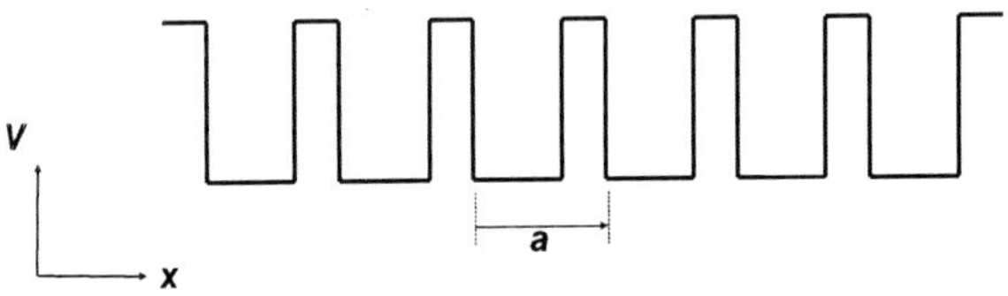


**Figure 1.4** Model potential of 1D crystal.

### 1.2.2 Tight-Binding Framework

#### 1.2.2.1 Formalism

The next step is to introduce an infinite array of the unit cells in the concerning organic 1D crystal structure already shown in Figure 1.2. This concept simultaneously brings about the spatial regular array of potentials $V(x)$ in Figure 1.4 into the Schrödinger equation as

$$\left\{-\frac{\hbar^2}{2m}\frac{d^2}{dx^2} + V(x)\right\}\psi(x) = \varepsilon\psi(x) \tag{1.6}$$

where

$$V(x + a) = V(x) \tag{1.7}$$

with $a$ being the translation length. Several examples of unit cells in the organic 1D and 2D crystals are given in Figure 1.5.

In order to obtain the plausible wavefunction for general 1D crystal with infinite repetition of the unit cells shown in Figure 1.2, periodic boundary condition (or Born-von Karman boundary condition) is introduced toward simple mathematical treatment as in the ordinary solid-state physics. This condition is embodied by considering a huge "ring" with an infinite diameter consisting of an infinite array of the unit cells as shown in Figure 1.6. This makes the 1D free-electron wavefunction in Eq. (1.3) change into the 1D Bloch function which satisfies the relationship

$$\psi(x + a) = \exp[ika]\psi(x) \tag{1.8}$$

where, again, $k$ signifies the wavevector and $a$ the translation length. The Bloch function is considered as deformation of the free-electron wavefunction into that modulated by

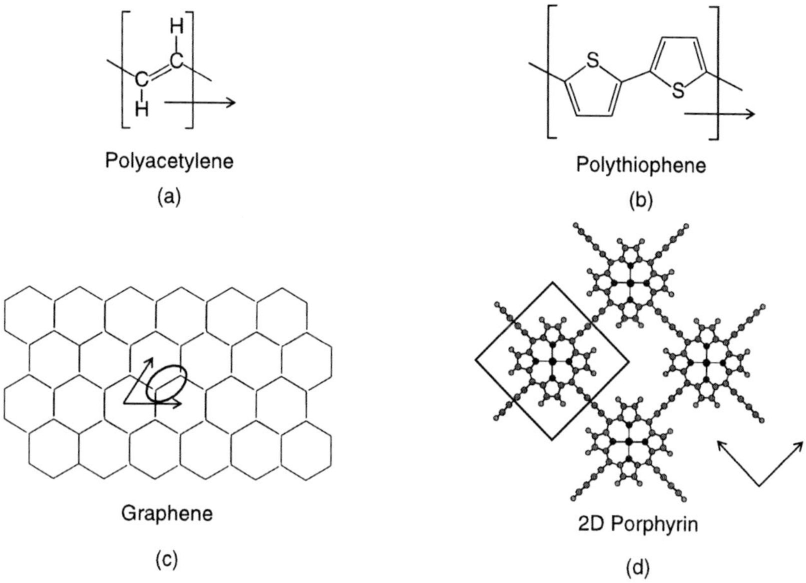


**Figure 1.5** Examples of 1D (a), (b) and 2D (c), (d) crystals and the unit cells (shown in parentheses, oval, or square). The arrows indicate the direction(s) of the translation.

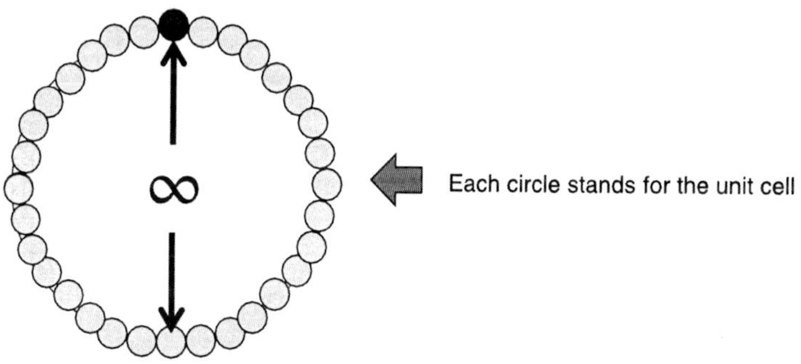


**Figure 1.6** Periodic boundary condition expressed by a ring with an infinitely large diameter. The first unit cell (black circle) becomes overlapped with the last unit cell after the infinite translation.

the array of the unit cells containing the atoms or molecules. Also note that Eq. (1.8) signifies that the translated wavefunction $\psi(x + a)$ is represented by multiplication of the phase factor $\exp[ika]$ to the original function $\psi(x)$. The value of $k$ ranges from $-\pi/a$ to $\pi/a$, which is called the first Brillouin zone or simply Brillouin zone.

Quantum chemical treatment of organic molecules are generally based on the linear combination of atomic orbitals (LCAO) framework, which can also be brought about into the Bloch function for organic crystal. [1] This is called CO after the conventional MO as has been already mentioned above. The CO is expressed by

$$\psi_s(k, x) = \frac{1}{\sqrt{N}} \sum_{j}^{N} \sum_{\mu}^{\text{cell}} \exp[ikja] C_{\mu,s}(k) \chi_\mu(x - ja) \tag{1.9}$$

**Table 1.1** Typical methods for crystal orbital (CO) calculation

| Method of calculation | Features |
|---|---|
| Hückel | for only $\pi$ electrons<br>total energy unhandled<br>spins unhandled |
| Extended Hückel | for all the valence electrons<br>band gap unreliable<br>total energy unhandled<br>spins unhandled |
| VEH (Valence effective Hamiltonian) | for all the valence electrons<br>seldom used recently<br>employs adjustable parameters |
| Semiempirical Hartree-Fock (CNDO, INDO, MINDO, MNDO, AM1, etc.) | for all the valence electrons<br>seldom used recently<br>employs adjustable parameters<br>total energy less reliable<br>band gap overestimated |
| Hartree-Fock | for all the electrons<br>total energy plausible<br>structural optimization possible<br>band gap overestimated |
| DFT (Density functional theory) | for all the electrons<br>total energy plausible<br>structural optimization possible<br>band gap plausible |

where $N$ formally stands for the total number of the unit cell numbered by $j$, $\chi_\mu$ $(x - ja)$ for the $\mu$-th atomic orbital (AO) involved in the $j$-th unit cell, $s$ the energy level of $\psi_s(k, x)$, and $C_{\mu,s}(k)$ the coefficient. Among these variables, the coefficients $C_{\mu,s}(k)$ are initially unknown and their values are to be variationally determined by solving the corresponding Schrödinger equation in terms of the secular equation. Note that $N$ is infinite in actuality, since there are an infinite number of the unit cells in Figure 1.6.

The concept employed above is often mentioned as the tight-binding method, since the wavefunction, based on the free-electron, is now modulated by AOs near the atomic region involved in each unit cell. In this sense, the COs still remain complex functions. It is straightforward to show that the tight-binding wavefunction in Eq. (1.9) satisfies the relationship in the 1D Bloch function of Eq. (1.8). There are several approximation methods for the actual calculation of COs, which are basically similar to those for the conventional MO calculations. Typical calculation methods are listed in Table 1.1. A few software packages are commercially available for the CO calculations.

The procedure to obtain the CO also gives the energy level $\varepsilon_s(k)$ of each $\psi_s(k, x)$. Since $\varepsilon_s(k)$ is obtained at each $k$ continuously existing in the Brillouin zone in the range $[-\pi/a, \pi/a]$ mentioned above, it constructs an energy-band structure for each $s$ at the same time, similar to that of a free electron in Figure 1.3. A simple image of two energy bands

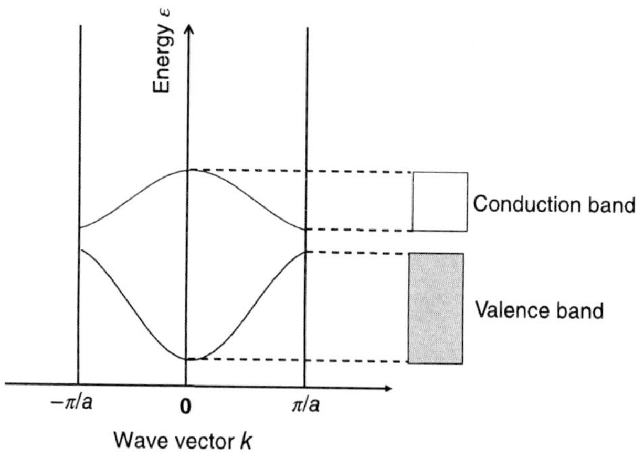


**Figure 1.7** Schematic drawing of the valence and the conduction bands.

obtained by the tight-binding scheme is illustrated in Figure 1.7, where two electrons are supplied per unit cell. Note that two electrons per unit cell occupy one band in total. In this case, the energetically lower part of the energy band is occupied by the electron and the upper part is unoccupied. The occupied branch is called the valence band and the unoccupied is called the conduction band.

#### 1.2.2.2 Simple Example

In order to understand the electronic structure of the organic 1D crystal, it will be appropriate to start from an explanation of the band-structure analysis of the simplest infinite 1D chain with iso-distant array of lattice; the unit cell of which consists of single atom A, the translation length being $a$ as in Figure 1.8a. This atom A can also be substituted by atomic group or so. For instance, when A is changed into a CH group, this chain can be considered as non-bond alternant polyacetylene in Figure 1.8b. Let us examine the electronic structure of this system within the framework of the simple Hückel approximation. The secular equation in this case is expressed by

$$\begin{vmatrix} \alpha - \varepsilon_s(k) & \beta(1 + \exp[-ika]) \\ \beta(1 + \exp[ika]) & \alpha - \varepsilon_s(k) \end{vmatrix} = 0 \tag{1.10}$$

where $\alpha$ denotes the Coulomb integral of the AO on A, $\beta$ the resonance integral between the adjacent AOs, and $a$ the translation length as denoted in Figure 1.8b. Note that both $\alpha$ and $\beta$ are of negative values. The resonance integral $\beta$ is also called the transfer integral $t$ ($t = -\beta$) in solid-state physics. The variable $k$ signifies the wavevector in the Brillouin zone $[-\pi/a, \pi/a]$ and the eigenvalue $\varepsilon_s$ is the function of $k$.

The band structure of the 1D iso-distant chain is then obtained as shown in Figure 1.9a from the eigenvalues $\varepsilon_1(k)$ and $\varepsilon_2(k)$ by solving Eq. (1.10)

$$\varepsilon_{1,2}(k) = \alpha \mp \sqrt{2\beta^2(1 + \cos ka)} \tag{1.11}$$

It is seen that at $k = \pm\frac{\pi}{a}$, the highest occupied (HO) band $\varepsilon_1(k)$, and the lowest unoccupied (LU) band $\varepsilon_2(k)$ stick together to give the zero-band gap. Hence, the 1D iso-distant chain should have the metallic property.

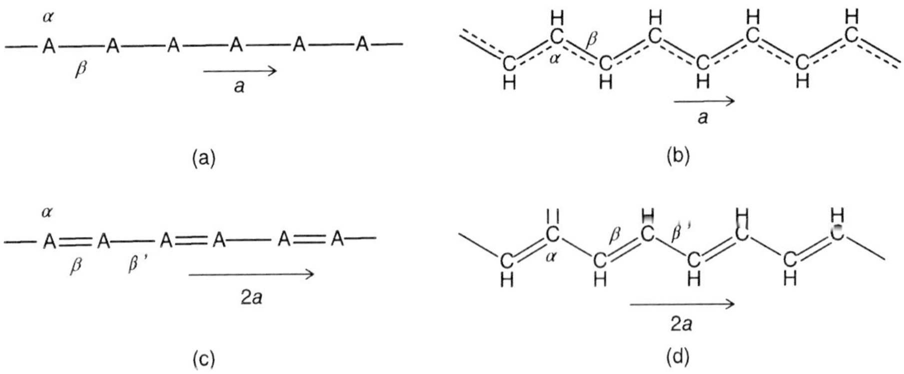


**Figure 1.8** Infinite repetition of atoms A and (b) polyacetylene with the iso-distant translation length. (c) and (d) represent the bond-alternant cases.

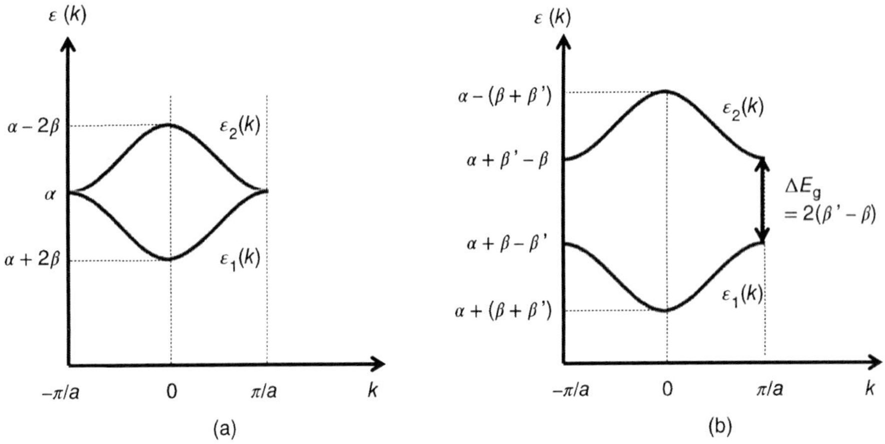


**Figure 1.9** Band structures corresponding to the 1D polymers in (a) Figure 1. 8(a), (b), and (b) Figure 1.8 (c), (d). Note that the translation length *a* in (b) is twice as long as that in (a) due to the dimerization in Figure 1.8.

Next, let us consider the case in which the above 1D chain is not iso-distant but with alternant distance (say, 1D alternant chain) as in Figure 1.8c. In this case, two kinds of resonance integrals $\beta$ and $\beta'$ exist corresponding to, e.g., A=A and A-A, bonds, respectively ($|\beta| > |\beta'|$). This chain can also be considered similar to the bond-alternant polyacetylene in Figure 1.8d. The secular equation for the 1D alternant chain is then given by

$$\begin{vmatrix} \alpha - \varepsilon_s(k) & \beta + \beta' \exp[-ika] \\ \beta + \beta' \exp[ika] & \alpha - \varepsilon_s(k) \end{vmatrix} = 0 \tag{1.12}$$

where the corresponding eigenvalues are obtained as

$$\varepsilon_{1,2}(k) = \alpha \mp \sqrt{\beta^2 + 2\beta\beta' \cos ka + \beta'^2} \tag{1.13}$$

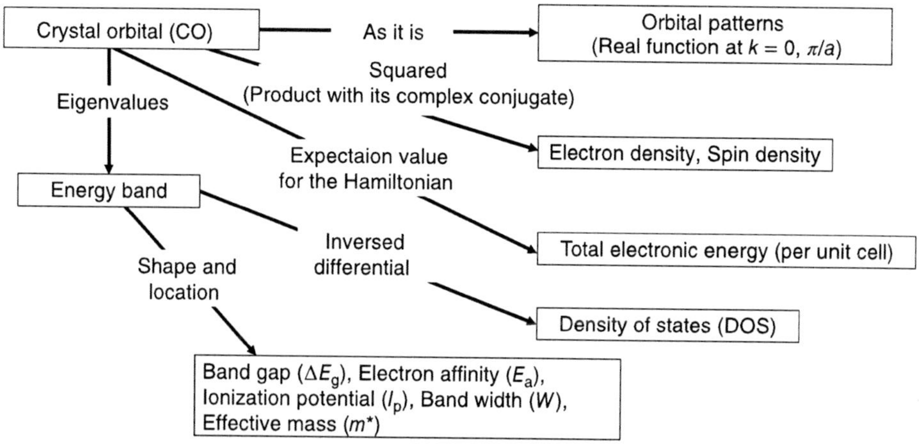


**Figure 1.10** Electronic properties derived from the crystal orbital (CO).

The band structure of the 1D alternant chain is illustrated in Figure 1.9b, where the band gap $\Delta E_g$ appears with the value

$$\Delta E_g = 2(\beta' - \beta) \tag{1.14}$$

This signifies that the 1D alternant chain should show semiconductive or insulating property depending on the value of $\Delta E_g$, which is also true for the bond-alternant polyacetylene.

It is an interesting problem to predict which system is the more energetically stable, the 1D iso-distant or the 1D alternant chain in the above. The answer to this question will be discussed in Section. 1.4.1.

### 1.2.3 Electronic Properties Based on the Electronic Structure

Several pieces of useful information on the electronic properties of organic 1D crystal can be obtained from the Bloch-type CO $\psi_s(k, x)$ and its energy level $\varepsilon_s(k)$, the diagram of which is shown in Figure 1.10. In the following, these will be described item by item.

#### 1.2.3.1 Characteristics of the Energy Band

The energy-band structure of organic crystals affords much information such as band gap, band width, ionization potential, electron affinity, and so on as seen in what follows. In particular, the highest occupied (HO) and the lowest unoccupied (LU) bands often play crucial roles not only in electronic property but also chemical reactivity. Though it is rather tiresome to examine all the energy bands of the organic 1D crystal, the analyses of the HO and the LU bands and their neighboring bands often give us sufficient information to consider the essential electronic properties.

It is of note that the classification of COs based on the symmetry such as, σ or π character, for example, reflects the corresponding energy band. The symmetry for the 1D crystal stems from the linear group in the space symmetry being a bit different from the point group to which the ordinary molecules belong. For instance, the energy-band structure of polythiophene with infinite chain length, as an example of organic 1D crystal, is shown in Figure 1.11a. Here, crossing of σ and π bands are seen. The fact that

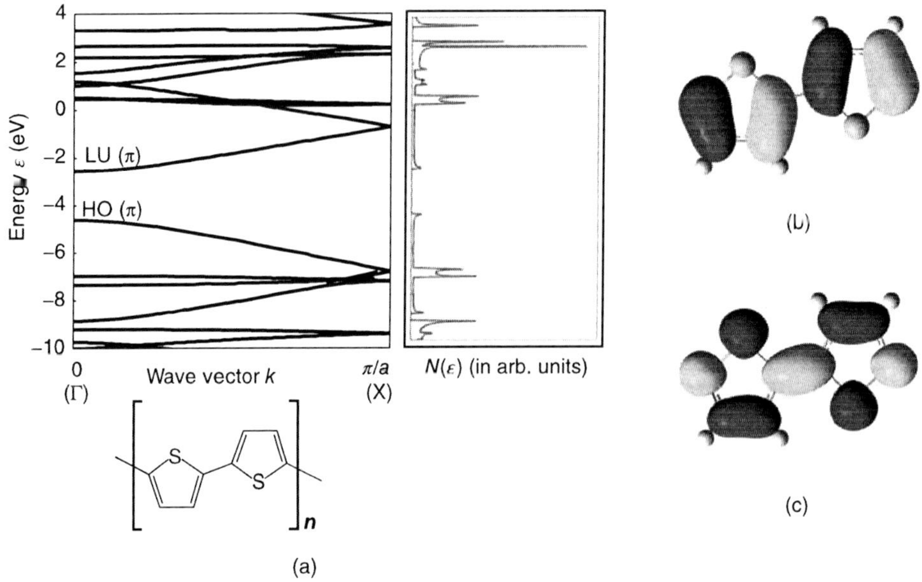


**Figure 1.11** (a) Band structure and DOS with the unit cell of polythiopehene, (b) the HOCO pattern, and (c) the LUCO pattern. Calculation was done by Crystal06 software with B3LYP/6-21G**. Both the HOCO and LUCO at $k = 0$ and $\pi/a$ are of $\pi$-types (Top view). Also note that the both HO and the LU bands are of $\pi$ type.

**Table 1.2** Electronic properties of polythiophene[a)]

| Quantity | Value |
|---|---|
| Band gap | 2.003 eV |
| HO band width | 4.292 eV |
| LU band width | 3.829 eV |
| Ionization potential | 4.496 eV |
| Electron affinity | 2.493 eV |
| Effective mass (Top of the HO band) | $-0.520\ m_0$ |
| Effective mass (Bottom of the LU band) | $0.560\ m_0$ |

a) Also see the caption of Figure 1.11.

these crossings can take place comes from the symmetry rule of the linear group, which has been discussed elsewhere. [2–4] Note that in Figure 1.11a the upper part of the HO bands and the whole of the LU bands are both of $\pi$-type leading to the extended $\pi$-conjugation toward the 1D direction or, in other words, throughout the polymer chain, which is appropriate to the intrachain conduction paths for both holes and electrons. Electronic properties accompanied with the band structure of polythiophene are listed in Table 1.2. The band structure can be experimentally obtained by the angle resolved photoelectron spectroscopy (ARPES) method.

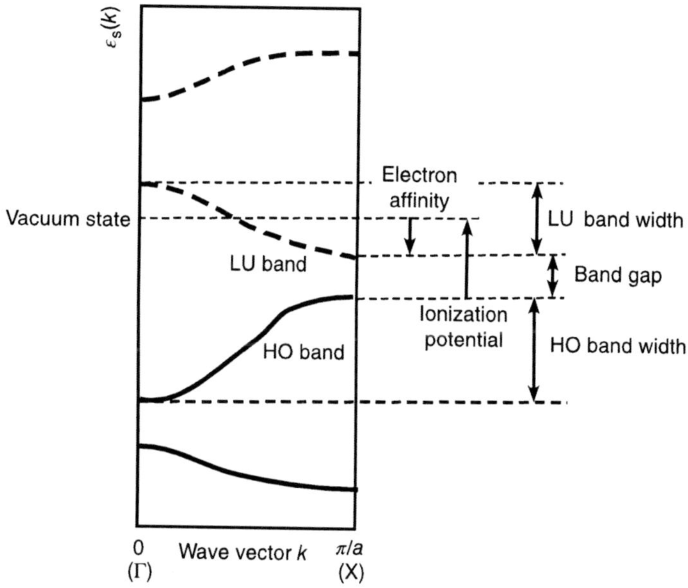


**Figure 1.12** Electronic properties derived from the energy band. Solid and dashed curves signify, respectively, the occupied and the unoccupied bands.

#### 1.2.3.2 Band Gap ($\Delta E_g$)

Band gap is defined as the energy difference between the top of the HO band and the bottom of the LU band as shown in Figure 1.12. This quantity is equal to the interband transition energy required for the excitation of the electron with the lowest energy in the 1D crystal. The band-gap value ($\Delta E_g$) thus gives a clue to the thermal activation energy for electric conduction and optical transition energy in both the semiconductor and insulator. Note that, as a matter of course, for the optical transition, the selection rule based on the symmetry should be satisfied. There can either be a direct or indirect band gap with respect to the change in the wavevector $k$ as is apparent from Figure 1.13. This characteristic is crucial to the optical transition, since those with indirect band-gap transition changing the $k$ value corresponds to the forbidden process.

#### 1.2.3.3 Fermi Energy ($\varepsilon_F$) and Fermi Level ($E_F$)

Fermi energy is denoted as $\varepsilon_F$ and has already appeared in, Figure 1.3 where the band-gap value is zero. In solid-state physics, this quantity is one of the most important variables since that indicates the highest energy of electrons in metals at absolute zero temperature. A surface consisting of the Fermi energy in the **k** space (wavevector space) is called the Fermi surface and exploration of its shape for many kinds of metals has been an interesting subject.

There is a similar but different concept, Fermi level $E_F$, which is also called chemical potential in the field of chemistry. At finite temperature, the Fermi level is occupied with one-half of the electron due to the Femi-Dirac distribution and, at zero temperature, $E_F$ becomes equal to $\varepsilon_F$. In a semiconductor without any impurities or defects (intrinsic semiconductor) or in an insulator, the position of $E_F$ is at about $\Delta E_g/2$, that is, in the middle of band gap. [5] In this sense, thermal activation energy $E_A$ for the electric

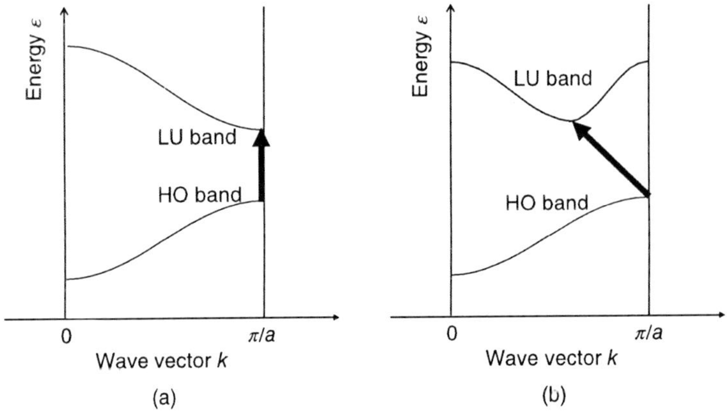


**Figure 1.13** (a) Direct band gap and (b) indirect band gap.

conduction in semiconductors is normally written as

$$E_A = \frac{\Delta E_g}{2} \tag{1.15}$$

and hence, the number of carriers $n$ in Eq. (1.1) in the semiconductor and insulator is expressed by

$$n \propto \exp\left[-\frac{\Delta E_g}{2k_B T}\right] \tag{1.16}$$

where $k_B$ is the Boltzmann constant considering the usual thermal activation process in these materials.

#### 1.2.3.4 Band Width (W)

The difference between the maximum and minimum energies of one energy band is called band width. For instance, the HO and LU band widths are illustrated in Figure 1.12. The larger the band width, the more the accompanying CO is delocalized over the entire 1D crystal. Hence, this is one of the important indices to consider the molecular design as to electric conduction path.

An energy band of π nature, for instance, tends to have a large band width indicating delocalized π conjugation over the whole crystal. On the other hand, the band width of the energy bands accompanied by the COs representing σ bond or lone pair are small as a whole, due to their rather flat bands. This signifies that electrons accommodated in these COs tend to localize at the relevant local atomic regions.

#### 1.2.3.5 Ionization Potential ($I_p$)

The ionization potential $I_p$ is defined as the energy required for extraction of an electron from the material (see Figure 1.12). In organic 1D crystal, this value is obtained from the $\varepsilon_{HO}$ being the energy of the highest occupied CO (HOCO) by changing the sign of its value so as to express the energetical depth of the top of the HO band from the vacuum state, that is

$$I_p = -\varepsilon_{HO} \tag{1.17}$$

In this sense, it is essentially similar to the work function of metals. This quantity measures the ease of oxidation of the concerning material. In other words, $I_p$ is equal to the energy paid to injection of a hole into the material. In order to make a p-type organic semiconductor small $I_p$ value, that is, the high-lying HO band is desirable for easy donation of electron or accommodation of a hole. Experimental value of $I_p$ can be obtained by ultraviolet photoelectron spectroscopy (UPS) measurement or the 1st oxidation potential measurement by the electrochemical manner. Though the material with small $I_p$ value is favorable for becoming p-type semiconductor, care should be taken to ensure it is easily oxidized under ambient condition.

#### 1.2.3.6 Electron Affinity ($E_a$)

The electron affinity $E_a$ is defined as the stabilization energy for injection of excessive electron to the material (see Figure 1.12) and is obtained by changing the sign of $\varepsilon_{LU}$ being the energy of the lowest unoccupied CO (LUCO) corresponding to the bottom of the LU band, that is

$$E_a = -\varepsilon_{LU} \tag{1.18}$$

This value signifies the energetical depth of the bottom of the LU band from the vacuum state. The larger $E_a$ value relates to that material which tends to accommodate the excessive electron more easily. Thus, this value measures ease of reduction of the material. Experimental value of $E_a$ can be obtained by the inverse photoelectron spectroscopy (IPES) based on the PES measurement for the electron-irradiated material. An electrochemical measurement of the 1st reduction potential also affords the value of $E_a$. In order to make an n-type organic semiconductor, large $E_a$ value, that is, the low-lying LU band is desirable for facile acceptance of electron.

#### 1.2.3.7 Density of States (DOS)

This quantity signifies the relative numbers of electron in the energy space and is obtained by

$$N(\varepsilon) = C\left(\frac{d\varepsilon(k)}{dk}\right)^{-1} \tag{1.19}$$

in the 1D crystal where $C$ is an appropriate normalization constant to give the number of electrons per 1 g, 1 cm$^3$, or unit cell and so on of the concerning material. In other words, the DOS represents the total number of the energy levels $\varepsilon_s(k)$ of the CO in an infinitesimally small energy width $d\varepsilon$ and hence becomes greater where the energy band is rather flat and small where the energy band is steep. For instance, the shape of the DOS is also shown in Figure 1.11a with respect to polythiophene. DOS at the Fermi level $E_F$ is often called Fermi density $N(E_F)$ and $N(E_F)d\varepsilon$ represents the number of charge carriers $n$ in Eq. (1.1).

Note that the DOS can be defined even if there are no energy bands as in amorphous materials (see Sec. 1.5). Hence, the DOS can be considered a more fundamental property than the energy band for the bulk material, irrespective of the fact that its structure is crystalline or amorphous.

#### 1.2.3.8 Effective Mass ($m^*$)

This quantity for 1D crystal is defined by

$$\frac{1}{m^*} = \frac{1}{\hbar^2}\frac{\mathrm{d}^2\varepsilon(k)}{\mathrm{d}k^2} \tag{1.20}$$

and is interpreted to signify the mass of an electron or a hole moving in the concerning CO. It is understood, from the mathematical aspect, that $m^*$ is proportional to the radius of curvature of the energy band at each point. This quantity is usually expressed by the ratio to the mass of a free electron $m_0$ such as $am_0$. The $m^*$ value can either be larger or smaller than $m_0$ depending on the band property, that is, the electron in the concerning CO behaves as "heavier" or "lighter" electrons. The negative value of $m^*$ at the top of the HO band is interpreted as that of a hole.

#### 1.2.3.9 CO Pattern

The orbital pattern which shows the shape of CO is sometimes of importance since it shows the nature of the chemical bond such as $\sigma$ type, $\pi$ type, lone pair, and so on, which is closely related to the structural characteristics of the organic material. Moreover, in the organometallic component, the d characters of the bond are exemplified by the orbital pattern. Note that the Bloch function is generally expressed by complex function as is seen in Eq. (1.9) at in between $k$ points in the Brillouin zone, and that only the COs at the Γ and X points become the real function.

Fortunately, almost all of the top of the HO and the bottom of the LU bands appear at the either Γ or X point as shown in Figure 1.11a. Hence, the highest occupied CO (HOCO) and the lowest unoccupied CO (LUCO) similar to the HOMO and the LUMO of the ordinary molecules become real functions in this case. Examples of the HOCO and the LUCO patterns of polythiophene is also given in Figures 1.11b and 11.c. The HOCO has in-phase overlap between the α-β and α'-β' carbons favoring the aromatic structure of the thiophene ring, whereas the LUCO between the β-β' and inter-ring carbon atoms favor the quinoid structure. These can be considered as suggesting that polythiophene favors the aromatic structure in the ground state and the quinoid structure in the excited state.

#### 1.2.3.10 Electron Density and Bond Order

The electron density of the 1D crystal is generally given by summation of the electron density derived from all occupied COs over the Brillouin zone as expressed by

$$P_{\mu\nu} = 2\sum_{s}^{occ}\left(\frac{Na}{2\pi}\right)\int_{-\frac{\pi}{a}}^{\frac{\pi}{a}} C_{\mu,s}{}^*(k)C_{\nu,s}(k)\mathrm{d}k \tag{1.21}$$

Note the summation of the off-diagonal value gives the bond order at the concerning AO pairs $\mu$ and $\nu$. The atomic net charge is given by the difference of the atomic electron density $\sum_{\mu}^{\text{Atom}} P_{\mu\mu}$ and the nuclear charge of that atom as in the MO calculation.

There can be several ways of the bond-order calculation such as Mulliken population or others as in the ordinary MO calculations. This reflects the somewhat arbitrary nature of the bond-order description dependent on the way of expression of accumulation of the electrons at the bond region. In the CO calculations, however, the Mulliken population method is generally employed for simplicity.

#### 1.2.3.11 Total Energy of 1D Crystal ($E_{tot}$)

This quantity $E_{tot}$ is given by the summation of the total electronic energy per unit cell $E_{elec}$ and the total inter-nuclear energy per unit cell $E_{nuc}$. [6] $E_{elec}$ is obtained by the expectation value toward the formal Hamiltonian of the 1D crystal with the Slater determinant consisting of all COs over the Brillouin zone. Hence, this Slater determinant is formally quite large and it is necessary to perform the calculation under some numerical algorithm. The $E_{tot}$ value allows us to consider the energetical stability of the concerning 1D crystal. In most of the software packages for CO calculation, there is attached, the function of the gradient optimization of $E_{tot}$ choosing the coordinates of all atoms in the unit cell and the translation length $a$ as the variables for this process.

#### 1.2.3.12 Mobility

Mobility $\mu$ in Eq. (1.1) is further decomposed to

$$\mu = \frac{e\tau}{m^*} \tag{1.22}$$

where $\tau$ is the relaxation time and $m^*$ effective mass as mentioned above. The relaxation time $\tau$ depends on miscellaneous factors in the material, such as defects, impurities, and phonon scattering. In this sense, though the both $\tau$ and $\mu$ ought to be the quantities depending on the actual material structures, there is the possibility to assess $\tau$ only by consideration of electron scattering by phonons of the 1D crystal. However, this examination process has not been well developed. So, the carrier number $n$ and effective mass $m^*$ can be the main factors to be considered for the band conduction in the organic crystal at present.

On the other hand, the intermolecular mobility of molecular crystals can somehow be dealt with based on the hopping concept as follows by estimating the diffusion coefficient $D$: [7]

$$\mu = \frac{e}{k_B T} D \tag{1.23}$$

where D is given by the expression

$$D = \frac{1}{2n} \sum_i r_i^2 W_i P_i \tag{1.24}$$

with

$$W_i = \frac{V_i(r)^2}{\hbar} \left( \frac{\pi}{\lambda k_B T} \right)^{\frac{1}{2}} \exp\left[ -\frac{\lambda}{4 k_B T} \right] \tag{1.25}$$

and

$$P_i = \frac{W_i}{\sum_i W_i} \tag{1.26}$$

The notations $r_i$, $V_i$, and $\lambda$ signify the following:

$r_i$: distance between the reference and the $i$-th neighboring molecular unit
$V_i$: electron-transfer coupling matrix element between the reference and the $i$-th neighboring adjacent molecular units

**Table 1.3** Calculated diffusion coefficient $D$ and mobility $\mu$

| | | Naphthalene | Anthracene | Tetracene | Pentacene |
|---|---|---|---|---|---|
| Diffusion coefficient $D$ ($cm^2/s$) | | 0.0342 | 0.0477 | 0.1096 | 0.1387 |
| Drift mobility | Calc. | 1.32 | 1.84 | 4.24 | 5.37 |
| $\mu$ ($cm^2/Vs$) | Exp. | 0.4–1 | 0.57–2.07 | 0.14, 0.4 | 3, 5–7 |

Data from Ref. 7 and references therein.

$\lambda$: reorganization energy related to the energy difference of a molecular unit in its neutral state and the cationic (or anionic) state depending the species of charge carriers.

Eq. (1.25) signifies that the larger $V_i$ and smaller $\lambda$ result in larger $\mu$ particularly in a molecular crystal. The value of $V_i$ is decided by the crystalline structure and the MO patterns of each molecular unit concerning the electron transfer. The value is closely related to the vibronic coupling constants. Hence, toward smaller $\lambda$, one should perform the appropriate molecular design to suppress the vibronic coupling as small as possible. [8] The calculated data of $D$ and $\mu$ for several oligoacenes are listed in Table 1.3. [7]

## 1.3 Injection of Charge Carriers

Appropriate injection of charge carriers into the originally insulating organic polymers and crystals with almost zero Fermi density (namely $N(E_F) = 0$) could change their intrinsic properties to show semiconductive or even metallic characteristics. There are several strategies for the injection:

1) By chemical or electrochemical dopings
2) By making charge-transfer (CT) complex
3) From the external circuit through the electrodes
4) Photoexcitation of electron from the HOCO to the LUCO (or from the HOMO and the LUMO in small molecule).

In this Section, we will mainly describe the above (1) and (2) which belong to typical chemical processes for organic materials. Processes (3) and (4) are briefly mentioned at the end of Section 1.3.2. By injection of charge carriers, organic materials can change into semiconductors with the electrical conductivity in the range from $10^{-10}$ $Scm^{-1}$ to sometimes $10^5$ $Scm^{-1}$. The highest value is about the same as those of actual metals. In some special cases, even superconductivity could appear.

For effective injection of charge carriers, the original organic materials should satisfy the conditions below:

1) The potential conduction path should exist to give good mobility $\mu$ for the charge carriers injected. For instance, the extended $\pi$-conjugation throughout the polymer chain should exist or a reasonably large overlap between, e.g., the p$\pi$ AOs in the organic molecular crystal should be constructed along the stacking direction of molecular planes after the carrier injection.

**Table 1.4** Miscellaneous dopant species

| | Acceptor |
|---|---|
| Halogen | $Cl_2$, $Br_2$, $I_2$, ICl, $ICl_3$, IBr, $IF_3$ |
| Lewis acid | $PF_5$, $AsF_5$, $SbF_5$, $BF_3$, $BCl_3$, $BBr_3$, $SO_3$, $GaCl_3$ |
| Protonic acid | HF, HCl, $HNO_3$, $H_2SO_4$, $HBF_4$, $HClO_4$, $FSO_3H$, $ClSO_3H$ |
| Transition metal halide | $NbF_5$, $TaF_5$, $MoF_5$, $WF_5$, $RuF_5$, $BiF_5$, $TiCl_4$, $ZrCl_4$, $HfCl_4$, $NbCl_5$, $TaCl_5$, $MoCl_5$, $MoCl_3$, $WCl_5$, $FeCl_3$, $TeCl_4$, $SnCl_4$, $SeCl_4$, $FeBr_3$, $TaBr_5$, $TeI_4$, $TaI_5$, $SnI_5$, $AuCl_3$ |
| Transition metal salt | $AgClO_4$, $AgBF_4$, $H_2IrCl_6$, $Ce(NO_3)_3$, $Dy(NO_3)_3$, $La(NO_3)_3$, $Pr(NO_3)_3$, $Sm(NO_3)_3$, $Yb(NO_3)_3$ |
| Organic molecule | TCNE, TCNQ, $F_4$TCNQ, DDQ, Chloranil |
| Others | $O_2$, $XeOF_4$, XeF, $NOSbF_6$, $NOSbCl_6$, $NOBF_4$, $NOPF_4$, $FSO_2OOSO_2F$, $(CH_3)_3OSbCl_6$ |
| Electrochemical doping | $ClO_4^-$, $BF_4^-$, $PF_6^-$, $AsF_6^-$, $SbF_6^-$ |
| | Donor |
| Alkali metal | Li, Na, K, Rb, Cs |
| Alkaline earth metal and others | Be, Mg, Ca, Sc, Ba, Ag, Eu, Yb |
| Electrochemical doping | $Li^+$, $Na^+$, $K^+$, $R_4N^+$, $R_4P^+$ (R=$CH_3$, $n$-$C_4H_9$, $C_6H_5$) |

2) Appropriate redox potentials or, in other words, small $I_p$ and large $E_a$, are desirable for satisfactory injection of holes and electrons, respectively.
3) In particular, there should be enough steric room for the dopants to enter inside the polymer structure.

### 1.3.1 Organic Conductive Polymers

In 1977, the doped polyacetylene film with $I_2$ recorded the increase in the electrical conductivity $\sigma$ up to $3.8 \times 10^2$ Scm$^{-1}$. [9] This can be regarded as the beginning for changing organic polymers to materials with semiconducting or even a metallic state, which are collectively called conductive polymers. Doping is a concept to introduce an oxidizing or reducing agent into organic polymers in order to inject holes or electrons, respectively. Dopants for oxidation are electron acceptors and for reduction electron donors. Miscellaneous dopant species are listed in Table 1.4. These dopants eventually produce finite numbers of charge carriers. Upon doping, the electrical conductivity $\sigma$ eventually increases to $10^0$-$10^5$ Scm$^{-1}$ as long as the original polymer satisfies the appropriate conditions mentioned above. Several members of organic conductive polymers are shown in Figure 1.14 and typical $\sigma$ values following doping of these are listed in Table 1.5.

Additional interesting phenomena in certain conductive polymers come from elementary excitations. For instance, *trans*-polyacetylene in the pristine state normally has a small number of radical spins (one spin per ca. 3000 carbon atoms) as the structural defect, which is detected by electron spin resonance (ESR) measurement. [10] Examination of this defect shows the spin is of $\pi$-type and moves in the polyacetyene chain with effective mass $6m_0$. This relatively heavy mass comes from the motion that this

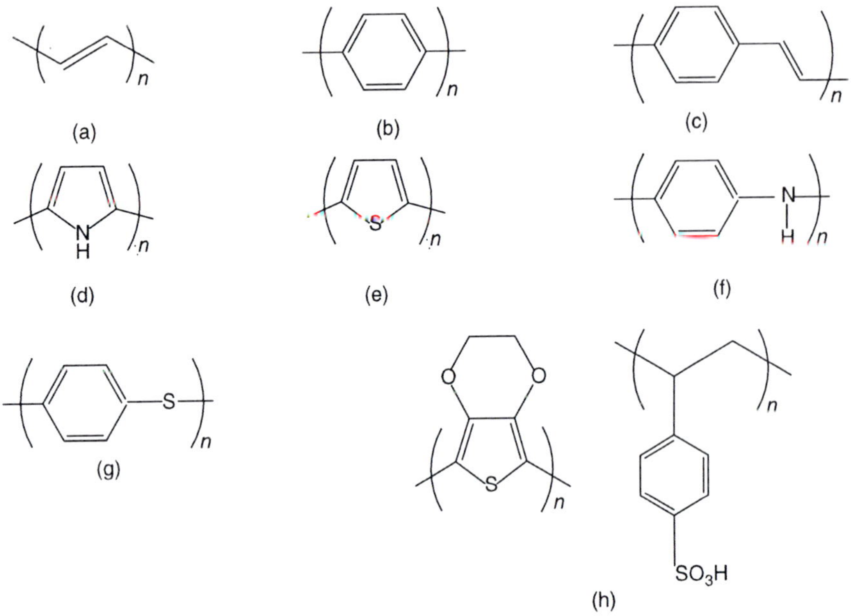


**Figure 1.14** Selected conductive polymers: (a) *trans*-polyacetylene (PA), (b) poly(*p*-phenylene) (PPP), (c) poly(*p*-phenylenevinylene) (PPV), (d) polyaniline (PAn), (e) polythiophene (PT), (f) polypyrrole (PPy), (g) poly(phenylene sulfide) (PPS), (h) poly(3,4-ethylenedioxythiophene) (PEDOT) and poly(stylene sulfonate) (PSS). PSS is used as the p-type dopant for PEDOT. PEDOT-PSS has been designed toward water solubility.

defect is accompanied by the change of the C-C bond distances as shown in Figure 1.15a. Research on this defect named it soliton or neutral soliton $S^0$. [11] This is a kind of defect of the bonding state, the energy of which pops up from the valence band into the band gap region as seen in Figure 1.15b. In this sense, a neutral soliton is an excited state and will only act as structural defect and not a carrier.

When the dopant is introduced into the polymer chain, the neutral soliton with odd spin will first disappear. This is because the dopant will extract or supply a spin depending on acceptor or donor to eventually extinguish the radical spin leading to a diamagnetic state or ESR-silent. Due to this scenario, the carbocation or carbanion is generated as in Figure 1.16. Neither of these show spin but have charges. Hence, these are called charged solitons, which can be a conductivity carrier. From the energetical viewpoints, they are elementary excitations.

Even after the whole neutral solitons are extinguished, the charged solitons continue to increase under the spinless state. In this state, the charge carrier is considered to be the charged solitons. At about 7 mol % of the dopant concentration is achieved, the Pauli paramagnetism suddenly rises up. From this doping region, energy levels of charged solitons plausibly grow to make a new metallic band. [12]

On the one hand, in the low concentration region of the dopant, the remaining neutral soliton and increasing charged soliton would coexist. They can combine to make cation radical or anion radical, which is called positive polaron or negative polaron,

**Table 1.5** Electronic properties and electric conductivities of selected conductive polymers

| | Ionization potential (eV)[a] | HO band width (eV)[a] | Band gap (eV)[a)b)] | Electric conductivity $\sigma$ after the doping (S/cm)[c)] |
|---|---|---|---|---|
| *trans*-Polyacetylene | 4.7 | 6.5 | 1.4 (1.4) | $1.7 \times 10^5$ ($I_2$) |
| Poly(*p*-phenylene) (22.7°)[d)] | 5.6 | 3.5 | 3.5 (3.4) | $5 \times 10^2$ ($AsF_5$) |
| Polythiophene[e)] | 5.0 | 2.6 | 1.6 (2.2) | $5.5 \times 10^3$ ($I_2$) |
| Polypyrrole | 3.9 | 3.8 | 3.6 (3.2) | $1.5 \times 10^3$ ($ClO_4^-$) |
| Polyaniline (30°)[d)] | 4.4 | 3.0 | 3.8 (3.3) | $1.5 \times 10^2$ (HCl) |
| Poly(*p*-phenylene vinylene) | 5.1 | 2.8 | 2.5 (ca. 3) | $1.4 \times 10^4$ ($H_2SO_4$) |
| Poly(*p*-phenylene sulfide) | 6.3 | 1.2 | - | $2.7 \times 10^0$ ($AsF_5$) |
| PEDOT-PSS[f)] | - | - | - | $9.3 \times 10^2$ |

a) Calculated values by the VEH method (see Table 1.1).
b) In parentheses is shown the experimental value.
c) In parentheses is shown the dopant.
d) In parentheses is shown the twisting angle between the phenyl rings employed for the calculation.
e) Poly[3-(2,5,8-trioxanonyl)thiophene]
f) PEDOT-PSS signifies poly(3,4-ethylenedioxylthiophene)-poly(stylenesulfonate), where PSS is a p-type dopant.

(i)T. A. Skotheim ed., "Handbook of Conducting Polymers", Marcel Dekker, New York (1986) and references therein.
(ii) K. Tanaka and K. Akagi, Development of Conductive Polymers in "Dr. Hideki Shirakawa and Conductive Polymers", K. Akagi and K. Tanaka eds., Kagaku-Dojin, Kyoto (2001), p. 48 and references therein (in Japanese).
(iii) Q. Zhang, Y. Sun, W. Xu and Daoben Zhu, *Adv. Mater.*, **26**, 6829 (2004) and references therein.

respectively (Figure 1.17). Polarons can also appear in poly(*para*-phenylene) (PPP), polythiophene (PT), and polypyrrole (PPy) (see, for example, Figure 1.18a). Polarons have radical spin and can be detected by the ESR or magnetic susceptibility measurement. Following increase in polaron concentration, two polarons combine to form two charged solitons becoming spinless. These two polarons can spatially couple to form bipolarons to act as bound-charge carriers throughout the polymer chain (Figure 1.18b) and the quinoidal structure appears inside a bipolaron. The length of a bipolaron ought to be decided by energetical balance and it has been estimated that bipolaron length is about five- or six-ring distance in case of PT. The positively or negatively charged polarons and bipolarons are also an elementary excitation found in the band-gap region (Figure 1.19).

All of these elementary excitations are summarized in Table 1.6. These show characteristic absorptions. When the amount of doping increases, new bipolaron bands grow, as in the heavily doped polyacetylene mentioned above, and then eventually turn out a metallic band as in Figure 1.20. [13]

### 1.3.2 Organic Charge-Transfer Crystals

In 1954, it was reported that a charge-transfer (CT) complex consisting of perylene (see Figure 1.21) and $Br_2$ showed considerably high electric conductivity of about 0.1 $Scm^{-1}$.

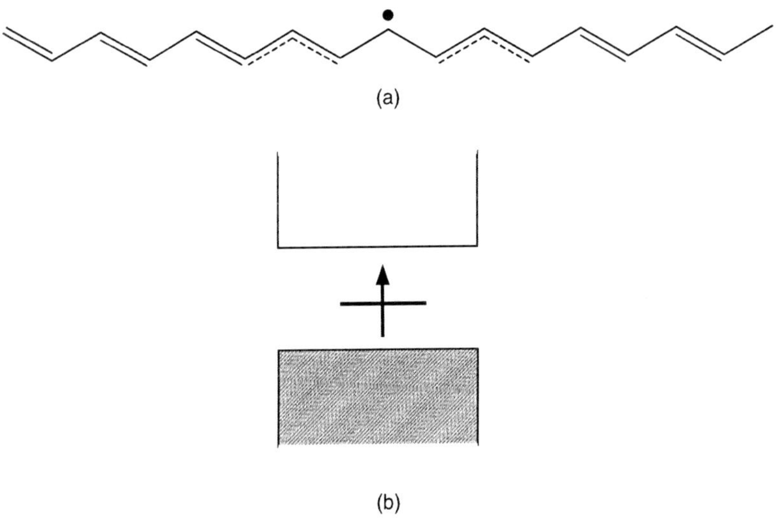

Figure 1.15 (a) A neutral soliton in polyacetylene and (b) its energy level.

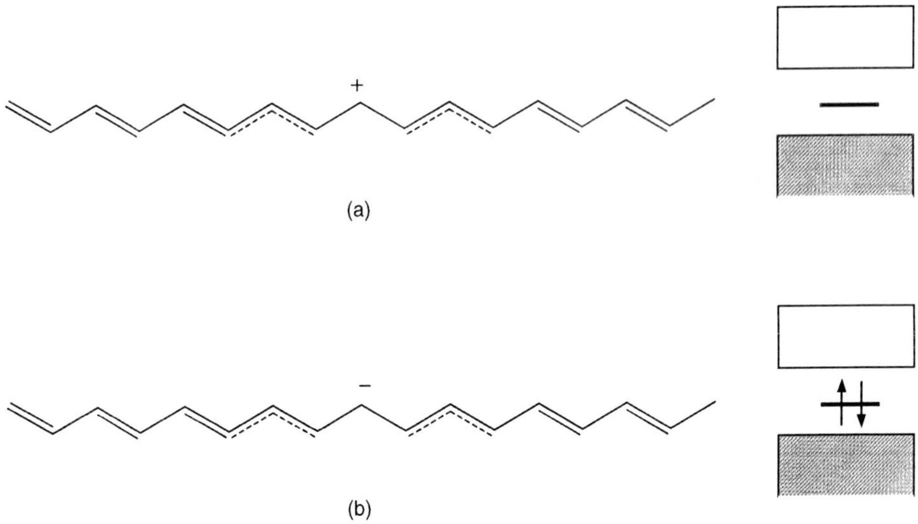

Figure 1.16 (a) A positively charged soliton with the energy level, and (b) a negatively charged soliton with its energy level in polyacetylene.

[14] This was the first CT complex as organic semiconductor, where perylene behaves as the donor (D) and $Br_2$ as the acceptor (A). The casting of D and A is rather relative depending on the combination of their $I_p$ and $E_a$ values.

TTF-TCNQ (tetracyanoquinodimethane-tetrathiafulvalene) (see Figures 1.21 and 1.22) is one of the most studied organic CT complexes showing interesting temperature dependency of electric conductivity in Figure 1.23 first found in 1973. [15] In this CT complex, TTF behaves as an electron donor (D) and TCNQ an acceptor (A) each making up a segregated column structure by individual stacking of D's and A's in

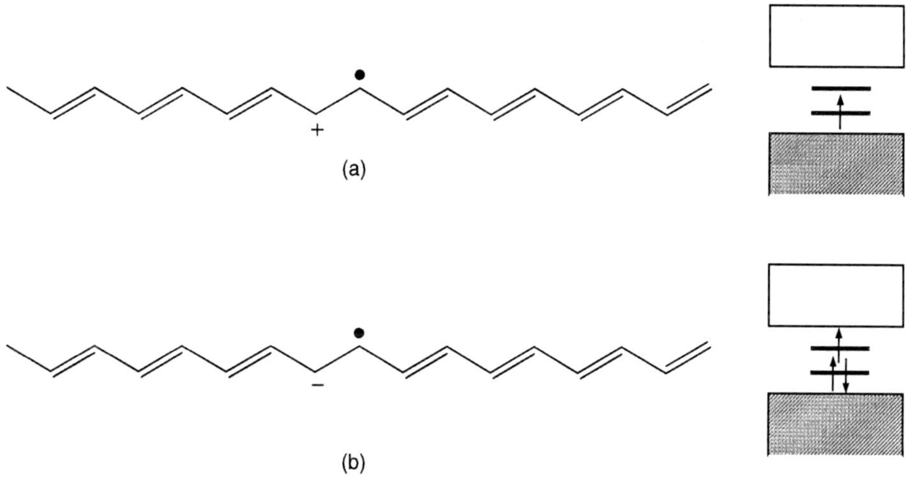

**Figure 1.17** (a) A positively charged polaron with the energy level, and (b) a negatively charged polaron with its energy level in polyacetylene.

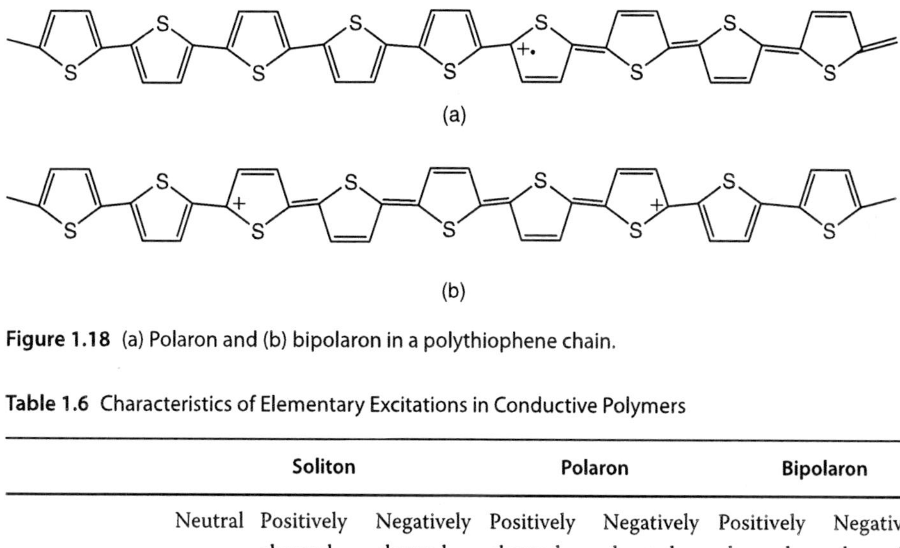

**Figure 1.18** (a) Polaron and (b) bipolaron in a polythiophene chain.

**Table 1.6** Characteristics of Elementary Excitations in Conductive Polymers

| | Soliton | | | Polaron | | Bipolaron | |
|---|---|---|---|---|---|---|---|
| | Neutral | Positively charged | Negatively charged | Positively charged | Negatively charged | Positively charged | Negatively charged |
| Charge number | 0 | +1 | −1 | +1 | −1 | +2 | −2 |
| Spin | 1/2 | 0 | 0 | 1/2 | 1/2 | 0 | 0 |

the crystalline state as in Figure 1.24. [16] It shows 1D electric conductivity along the column direction ($y$ direction) of about $6 \times 10^2 \mathrm{Scm}^{-1}$ at room temperature. This value is rather near to that of metal and the major charge carrier is estimated to be electron, that is, n-type carrier throughout the TCNQ column by thermopower measurement. [17] The amount of CT electron has been estimated as 0.59, based on the X-ray scattering observation, signifying $\mathrm{TTF}^{+0.59}$-$\mathrm{TCNQ}^{-0.59}$. [18] The electric conductivity shows large

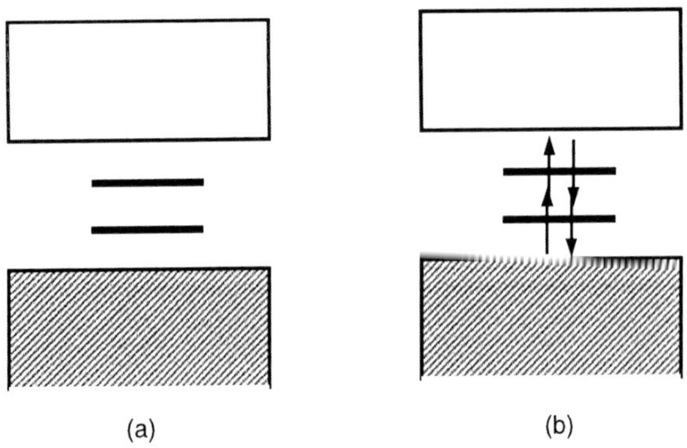


Figure 1.19 Energy levels of (a) a positively charged bipolaron and (b) a negatively charged bipolaron.

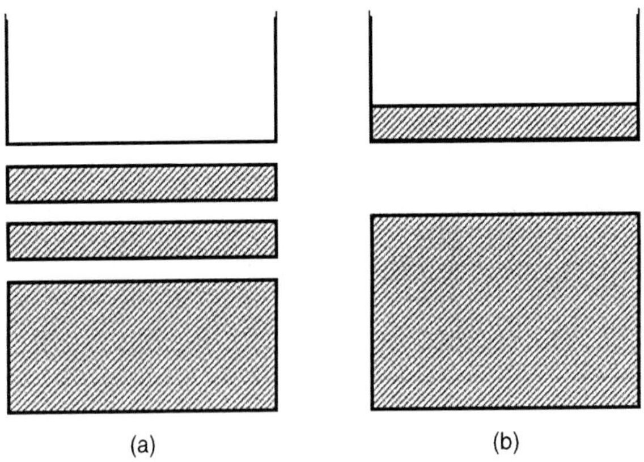


Figure 1.20 (a) Growth of bipolaron bands and (b) merging with other bands. Source: Modified from Ref. [13].

anisotropy along each crystallographic axis, that is, $\sigma_b$: $\sigma_c{}^*$: $\sigma_a$ = 500:3:1,[19] actually certifying the 1D conduction along the *b*-axis corresponding to the *y*-direction in Figure 1.24.

The formation of charge-transfer (CT) complex somewhat resembles the situation of doped organic conductive polymers, but these complexes, consisting of organic molecules with relatively low molecular weight, normally construct molecular assemblies or crystals. In this sense, we call these material organic CT crystal throughout this Chapter. Various component molecules behaving as D or A are indicated in Figures 1.21 and 1.22 and their $I_\mathrm{p}$ and $E_\mathrm{a}$ values in Table 1.7. From the standpoint of designing the donor with small $I_\mathrm{p}$ value, inclusion of the electron-donating groups such as $-NH_2$, -OH, -OMe, or –Me is favored. On the other hand, for the acceptor with large $E_\mathrm{a}$ value, the electron-accepting groups such as –CN, $-COCH_3$, -CHO, $-NO_2$, $-SO_3H$, -COOH, or –X (halogen) are favored.

The behavior of a simple CT complex is explained by the theory of Mulliken in which the wavefunction of the CT complex in the ground state $\psi$ is expressed by

$$\psi = c_0\psi(\mathrm{D}^0\mathrm{A}^0) + c_1\psi(\mathrm{D}^+\mathrm{A}^-) \tag{1.27}$$

where $\psi$ ($\mathrm{D}^0\mathrm{A}^0$) stand for the ground-state wavefunction consisting of neutral D and A, and $\psi$ ($\mathrm{D}^+\mathrm{A}^-$), a completely electron-transferred wavefunction. From the degree of

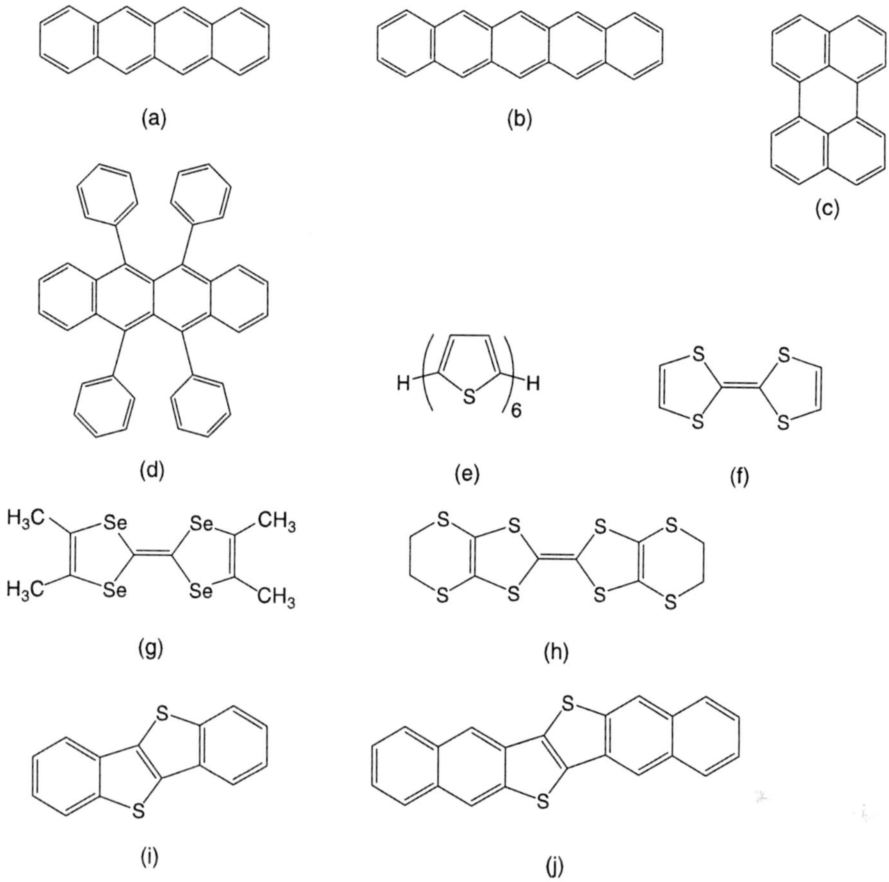

Figure 1.21 Selected organic donors: (a) tetracene, (b)pentacene, (c) perylene, (d) rubrene, (e) hexathiophene (sexithiophene), (f) tetrathiafulvalene (TTF), (g) tetramethyltetraselenafulvalene (TMTSF), (h) bis(ethylenedithio)-TTF (BEDT-TTF or ET), (i) [1]benzothieno[3,2-b][1]benzothiophene (BTBT), and (j) dinaphtho[2,3-b:2',3'-f]thieno[3,2-b]thiophene (DNTT).

mixing each wavefunction, it is roughly classified to represent an ionic state for $|c_0| \ll |c_1|$ and neutral state for $|c_0| \gg |c_1|$. Due to the mixing of $\psi(D^0A^0)$ and $\psi(D^+A^-)$, the total (actual) quantity of electron transferred $\delta$ from D to A becomes $0 < \delta < 1$. Note that $\delta > 1$ rarely occurs due to repulsion among the excessive electrons on A. The energy diagram of CT complex is simply illustrated in Figure 1.25, for instance, where the CT absorption is denoted by $h\nu_{CT}$.

Characteristic CT absorption of each type is expressed by

$$h\nu_{CT}{}^{N} = (I_p - E_a) - E_s \tag{1.28}$$

$$h\nu_{CT}{}^{I} = -(I_p - E_a) + (2\alpha - 1)E_s \tag{1.29}$$

where the superscripts N and I stand for neutral and ionic states, respectively, and $E_s$ and $\alpha$ signify the average electrostatic attractive energy and the Madelung constant, respectively, in the organic CT crystal. It was found that the relationship of

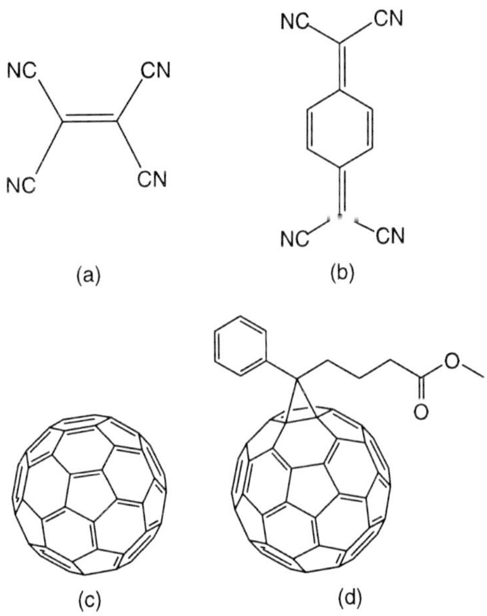


**Figure 1.22** Selected organic acceptors: (a) tetracyanoethylene (TCNE), (b) tetracyanoquinodimethane (TCNQ), (c) fullerene ($C_{60}$), and (d) phenyl-$C_{61}$-butyric acid methyl ester (PCBM). PCBM has been designed toward facile solubility in organic solvent.

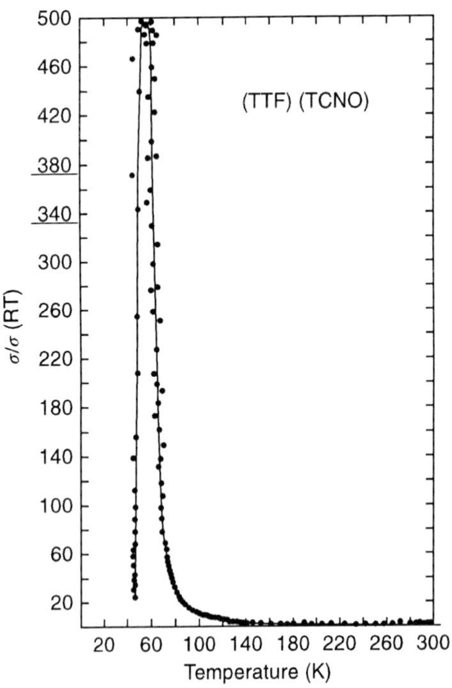


**Figure 1.23** Temperature dependence of electric conductivity of TTF-TCNQ. Ref [15]. Reproduce with permission of Elsevier.

$I_p - E_a$ (= ca. $\Delta E_{redox}$) vs $h\nu_{CT}$ can be plotted on a line as shown in Figure 1.26, [20] where the neutral-ionic boundary is obtained at $I_p - E_a = \alpha E_s$. Thus, referring to the plot of Figure 1.26, for unknown CT crystal, we can roughly predict whether it is of I or N type.

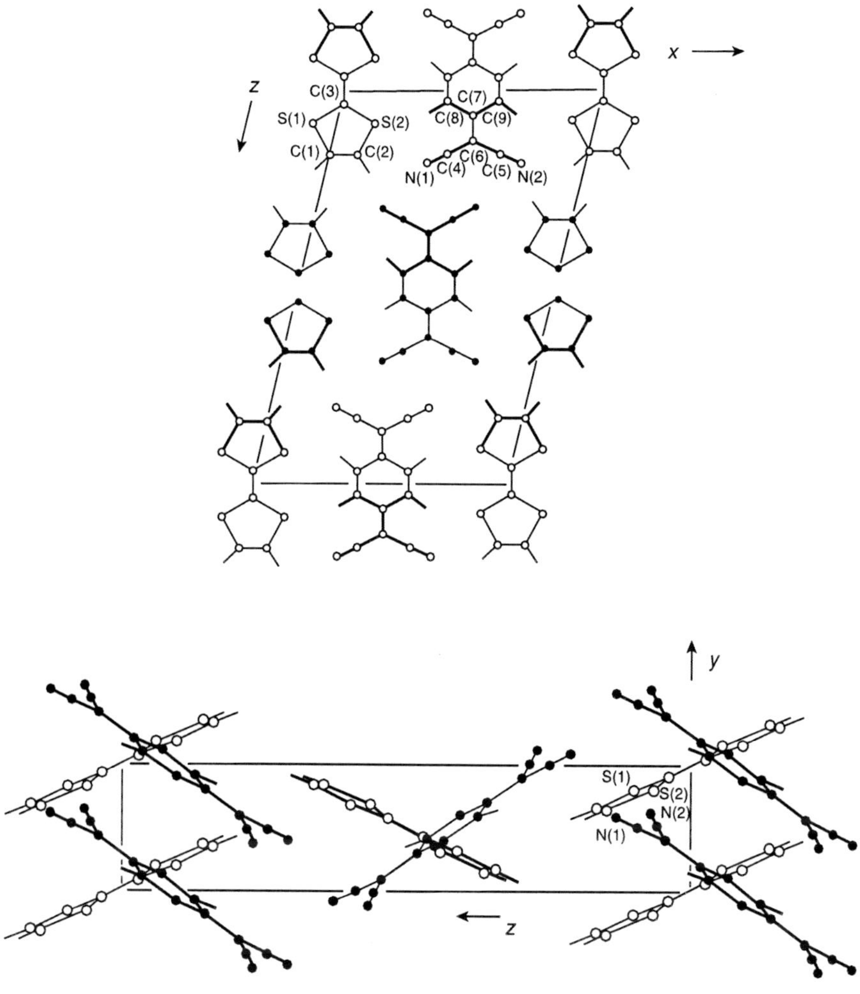


**Figure 1.24** Crystalline structure of TTF-TCNQ. Source: Ref. [16].

When donor and acceptor alternatively stack along the certain axis direction, it is called mixed column, and when they stack separately along the axis direction segregated column as schematically shown in Figure 1.27. The organic CT crystal with higher $\sigma$ value generally prefers the segregated-column structure since two kinds of carriers (electron and hole) can separately make motions through each column.

Fabrication of OFET often employs comparatively small-sized molecules of p-type such as pentacene, rubrene, BTBT or DNTT (see Figure 1.21). [21–23] These molecules are used under the charge-carrier injection from the external circuit. Since these molecules formally make crystals, the mobility can be analyzed with the use of Eqs. (1.23)–(1.26). A lesser number of organic electron acceptors are available compared with the donors except for fullerene ($C_{60}$) or its derivatives (see Figure 1.22). For the electron-collecting parts in OPV devices a PCBM molecule is often used, being a soluble $C_{60}$ derivative. [24]

**Table 1.7** Electronic properties of donor and acceptor molecules

| Donor | Ionization potential (eV)[a] | Oxidation potential(V)[b] |
|---|---|---|
| Tetracene | 6.89 | 0.77 |
| Pentacene | 6.58 | – |
| Perylene | 6.90 | 0.85 |
| Rubrene | 5.4[c] | 0.71 |
| TTF | 6.4 | 0.33 and 0.71 |
| BEDT-TTF | 6.21 | 0.55 and 0.85 |
| TMTSF | 6.27 | 0.44 and 0.72 |
| Hexathiophene | – | 0.46 |
| BTBT[d] | 5.6[c] | 1.29 |
| DNTT | 5.44[c] | 1.13 |
| Acceptor | Electron affinity (eV)[a] | Reduction potential (V)[b] |
| TCNE | 1.80 | 0.15 and −0.57 |
| TCNQ | 1.7 | 0.13 and −0.29 |
| Fullerene ($C_{60}$) | 2.68 | −0.60 |
| PCBM[e] | 2.63 | −0.77 |

a) Measured by photoelectron spectroscopy (PES) unless specially noted.
b) Indicated as the potentials vs SCE.
c) Estimated from the electrochemical data.
d) DPh-BTBT (2,7-diphenyl[1]benzothieno[3,2-b][1]-benzothiophene).
e) A derivative of $C_{60}$: Phenyl-$C_{61}$-butyric acid methyl ester.

(i)K. Seki, *Mol. Cryst. Liq. Cryst.*, **171**, 255 (1989).
(ii)G. Schukat, A. M. Richter and E. Ganghanel, *Sulfur Reports*, 7, 155 (1987).
(iii)P. Bäuerle, "Handbook of Oligo- and Polythiophenes", ed. D. Fichou, Wiley (1999), p. 114.
(iv)F. H. Herbstein, *Perspect. Struct. Chem.*, **4**, 166 (1971).
(v)F. Anger, T. Breuer, A. Ruff, M. Klues, A. Gerlach, R. Scholz, S. Ludwigs, G. Witte, and F. Schreiber, *J. Phys. Chem. C*, **120**, 5515 (2016).
(vi)K. Takimiya, H. Ebata, K. Sakamoto, T. Izawa, T. Otsubo and Y. Kunugi, *J. Am. Chem. Soc.*, **128**, 12604 (2006).
(vii)T. Yamamoto, S. Shinamura, E. Miyazaki and K. Takimiya, *Bull. Chem. Soc. Jpn.*, **83**, 120 (2010).
(viii)X.-B. Wang, H. –K. Woo and L. –S. Wang, *J. Chem. Phys.*, **123**, 051106 (2005).
(ix)Q. Xie, E. Perez-Cordero and L. Echegoyen, *J. Am. Chem. Soc.*, **224**, 3978 (1992).
(x)B. W. Larson, J. B. Whitaker, X.-B. Wang, A. A. Popov, G. Rumbles, N. Kopidakis, S. H. Strauss and O. V. Boltalina, *J. Phys. Chem. C*, **117**, 14958 (2013).

## 1.4 Transition from the Conductive State

### 1.4.1 Peierls Transition

Often, one might come across the description of Peierls transition for low-dimensional organic semiconductors, which causes the transition from the metallic to semiconductive or insulating state. This concept was originally developed by Peierls [25] purely for

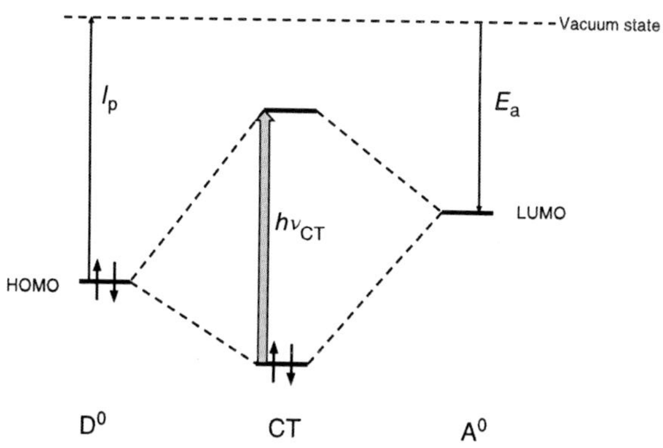


**Figure 1.25** Energy levels concerning the absorption of CT complex. $D^0$ and $A^0$ stand for the neutral donor and acceptor, respectively.

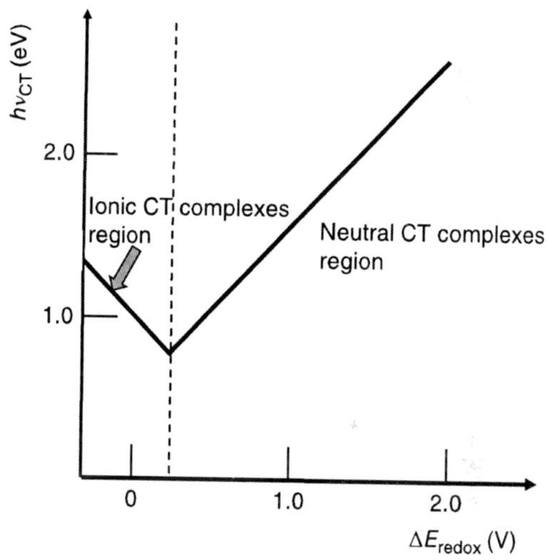


**Figure 1.26** Plot of the CT absorption energy vs. $\Delta E_{redox}$ for a number of CT complexes. Only the guide line for the eye is shown omitting each data point. Source: Modified from Ref. [20].

1D metallic chains shown in Figure 1.28 (left) which is inherently unstable to the metal-insulator transition at zero temperature in the presence of an electron-phonon interaction. Peierls' transition is generally brought about toward energetical stabilization by lowering the symmetry of the system, accompanied by the opening up of a band gap, as shown in Figure 1.28 (right). This situation is just applicable to what is described in Section. 1.2.2.

#### 1.4.1.1 Polyacetylene

Due to the above reason, polyacetylene is considered to have band structure with a band gap and the dimerized or bond-alternated structure after the Peierls transition has already occurred at room temperature. This situation is proven by existence of two kinds of vibrations corresponding to the C=C and C-C stretching modes. [26] The Peierls transition is essentially similar to the Jahn-Teller distortion in the ordinary molecule having the HOMO-LUMO degeneracy. For instance, a square cyclobutadiene normally causes the Jahn-Teller distortion to a rectangular structure with the opening the HOMO-LUMO gap while changing the magnetic state from triplet to singlet as shown in Figure 1.29.

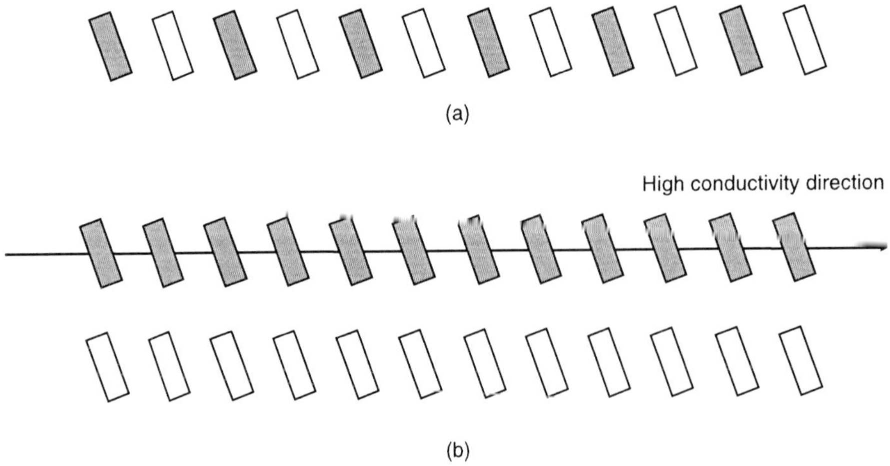


**Figure 1.27** Column array of organic CT crystals. (a) Mixed column and (b) segregated column. White and gray rectangles represent, e.g., donors and acceptors, respectively.

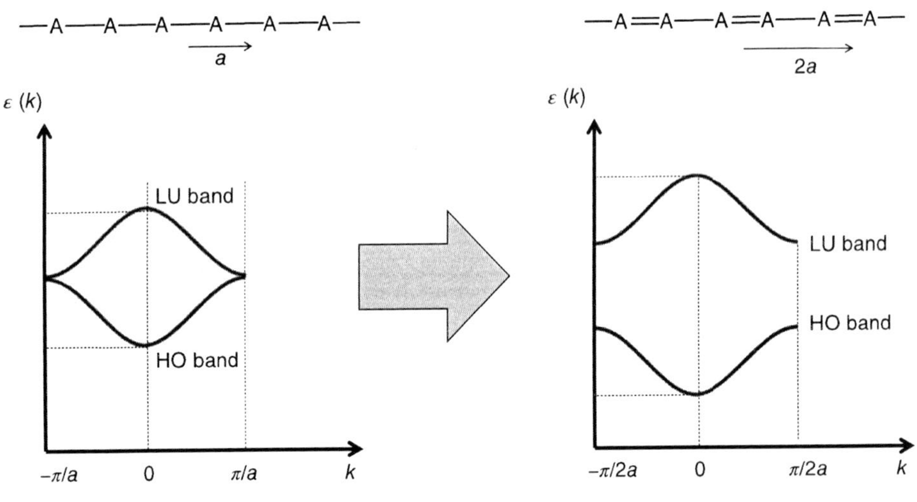


**Figure 1.28** A simple example of Peierls transition in the 1D metallic chain. Note that the translation length after the transition becomes 2*a*.

#### 1.4.1.2 TTF-TCNQ

The conductivity value of TTF-TCNQ described in Section 1.3.2 largely increases to nearly 58 K to ten times as large as that at room temperature and then, suddenly, a transition occurs to almost the insulating state at 53 K as seen in Figure 1.23. [15] Based on the X-ray scattering observation [18] and magnetic susceptibility measurement, [17] this conduction change at 58 K is also referred to as occurrence of Peierls transition such as in polyacetyene. In TTF-TCNQ, Peierls transition is often related to generation of charge density wave (CDW) with the wavevector $2k_F$ by an explicit consideration of electron-phonon interaction. Collective motion of this CDW seems to contribute to a

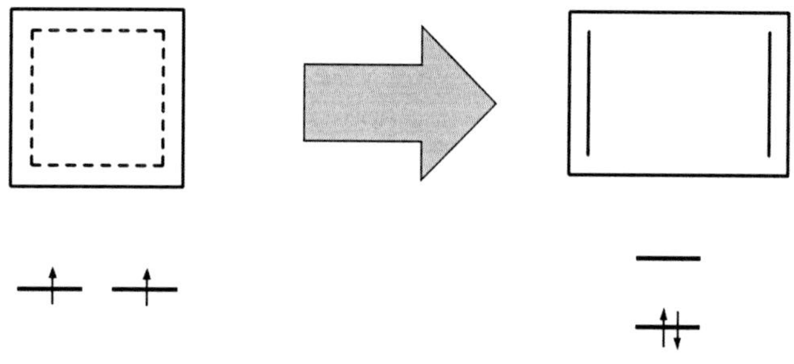

**Figure 1.29** Jahn-Teller distortion in cyclobutadiene with its spin change.

peculiar temperature dependence of metallic conductivity of TTF-TCNQ ($\sigma \propto T^{-2}$). [27]

In the field of organic CT crystal, strategies have been developed for suppression of the Peierls transition: (i) introduction of a larger atom, (ii) applying external pressure, and (iii) enhance the structural dimensionality of the CT complex, and so on. All of these relate to the introduction of an increase in intermolecular interactions leading to an increase of dimensionality. Attempts for this suppression have been motivated by expectation of the superconductivity upon decrease of temperature. [28] This is because, in higher dimensional systems such as 2D structures, occurrence of this transition tends to be suppressed by, what is called, imperfect nesting of the Fermi surface. [28]

There can be also other possibilities of metal-insulator transition than the Peierls transition such as formation of a Mott insulator in strongly electron-correlated materials. [29] In the system consisting of random potentials, a special kind of localization of electron wavefunction appears at quite unpredictable site and hence becomes insulating, which is called Anderson localization. [30]

### 1.4.2 Competition of Spin Density Wave and Superconductivity

The tetramethyltetraselenafulvalene (TMTSF) CT complex (Figure 1.21) designated by the formula $(TMTSF)_2PF_6$ is based on the designed molecule along the guiding principle toward raise of dimensionality. This complex crystal has a structure as shown in Figure 1.30, [31] where intermolecular Se—Se distance contributes to a slight enhancement of dimensionality probably brought about by larger AO lobes of selenium and the conduction occurs in the direction of *ab* plane. In this sense, such CT complex is mentioned as a pseudo-1D conductor with a slight contribution of 2D property, which is able to avoid the Peierls transition inherent in 1D conductors. Since the charge of $PF_6$ ought to be nominally -1, a TMTSF molecule is charged to be +0.5, hence this CT complex is rather called radical cation salt.

In $(TMTSF)_2PF_6$ the metal-semiconductor transition was actually suppressed down to 12 K as seen in Figure 1.31, [32] where the Pauli paramagnetism in the metallic phase, however, changes into the antiferromagnetic state. [33] In this sense, the transition in this material rather generates the spin density wave (SDW) accompanied with alternations of the spin directions as in Figure 1.32. It is of interest that this material has been found to show superconductivity at 0.9 K under application of the pressure 12 kbar.

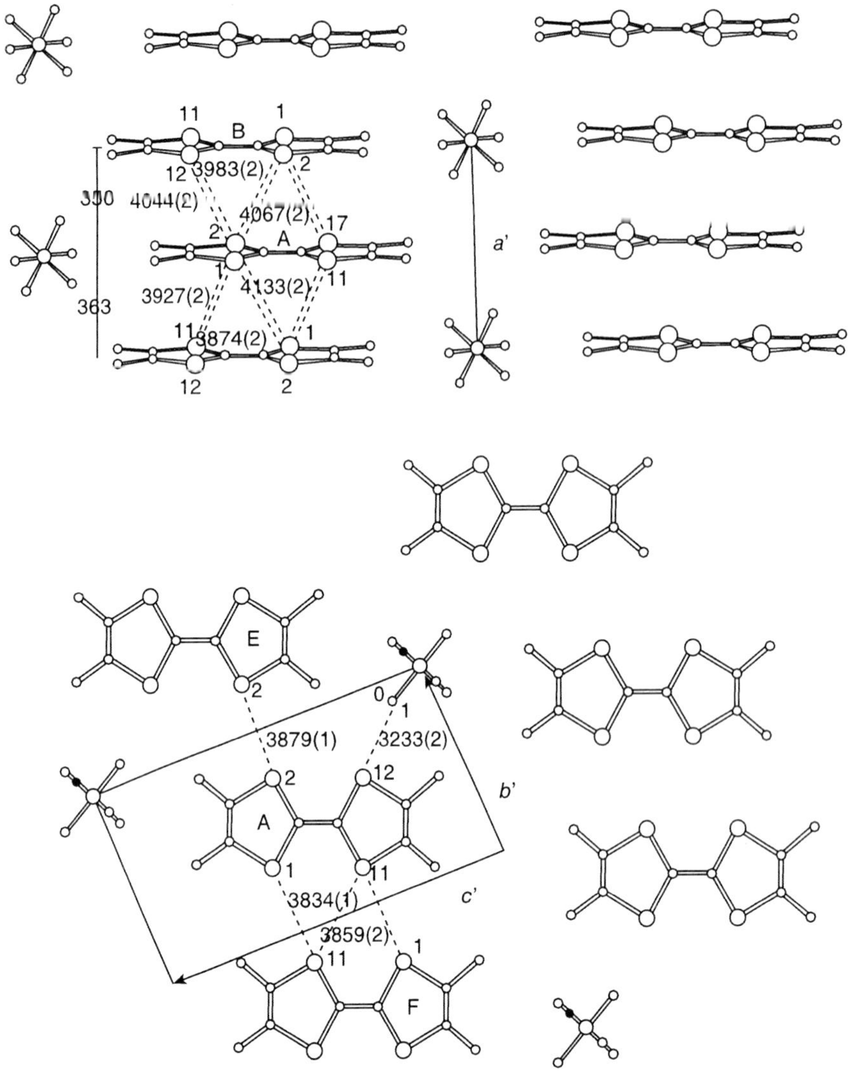


Figure 1.30 Crystalline structure of $(TMTSF)_2PF_6$, where a′, b′, and c′ are the projections of a, b, and c, respectively. Source: Ref [31]. Reproduce with permission of John Wiley & Sons.

[34] Moreover, $(TMTSF)_2ClO_4$ shows superconductivity at 1.4 K under ambient pressure. [35]

## 1.5 Electronic Structure of Organic Amorphous Solid

In the strictest sense, the actual materials including organic solid rarely have a complete crystalline structure but are more or less of one non-crystalline. There is always the

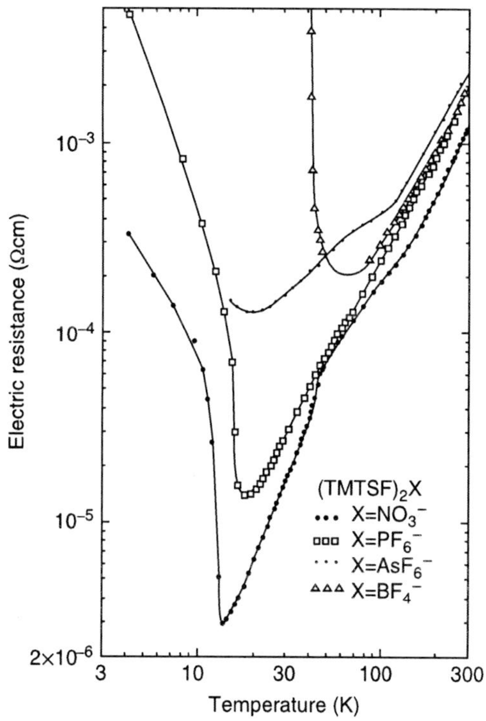


**Figure 1.31** Temperature dependence of electric resistance of $(TMTSF)_2X$. Source: Ref. [32].

**Figure 1.32** Antiferromagnetic alignment of $\alpha$ and $\beta$ spins.

possibility that some disordered structures caused by defects, dislocations, or impurities are included in organic crystals. In the amorphous solids, with breakdown of the crystalline structure, it should be considered that Bloch-type COs exist no more and the wavevector **k** loses its meaning. Hence, under these circumstances, energy bands cannot be defined but the DOS can still be considered instead of the energy band. The concept of the energy gap (no more band gap) is still valid since there are valence and conduction parts in the DOS as shown in Figure 1.33. The electronic structures of amorphous materials are rather complicated since these have to depend on the distribution of electron orbitals in each amorphous solid. In this Section, a couple of methods for approaching the electronic structure of amorphous solids is elucidated. Moreover, the hopping process related to the electric conduction mechanism is also mentioned.

### 1.5.1 Examination of Electronic Structures

Possible model structures for 1D organic amorphous solids, for example, would be expressed by (a) irregular substitution of the unit cell of different types, (b) irregular presence of a certain impurity or defect, and (c) miscellaneous fluctuation of the translation lengths of the unit cell. These are shown schematically in Figure 1.34.

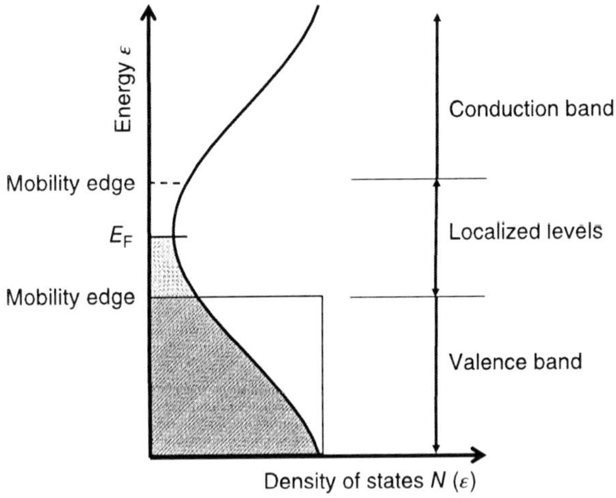


**Figure 1.33** Density of states (DOS) including localized levels.

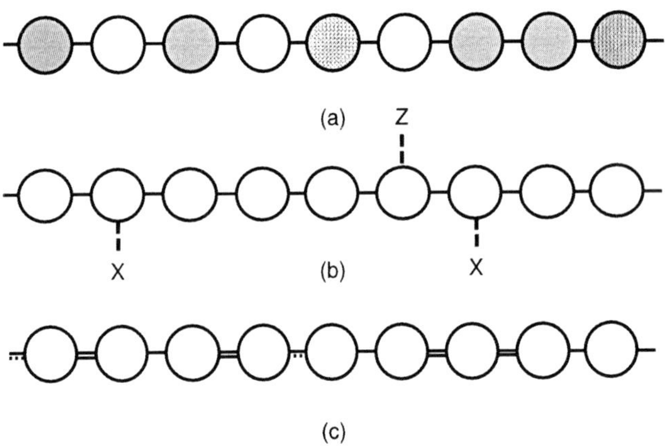


**Figure 1.34** Schematic representations of 1D organic amorphous solids: (a) irregular substitution of the unit cells, (b) irregular presence of impurities, and (c) fluctuation of the translation lengths of the unit cells.

Random copolymers would be included in the category of (a) in the above. There can be at least two ways to examine the electronic structures of amorphous solids as follows:

#### 1.5.1.1 Direct Calculation of the Local Structure

Approximate wavefunctions and their energy levels might be obtained by employing the appropriate fragment (or slab) representing characteristic local structure extracted from the concerning amorphous solid. Essentially, it is similar to the MO calculation for the ordinary molecules but the fragment in the above should be selected to reflect the property of the whole amorphous bulk as much as possible. Moreover, the boundary of the concerning fragment ought to be dealt with by putting the appropriate atom or functional group so as to compensate the rest of the amorphous bulk. It is naturally expected that the larger size of the fragment would give better results. This method is

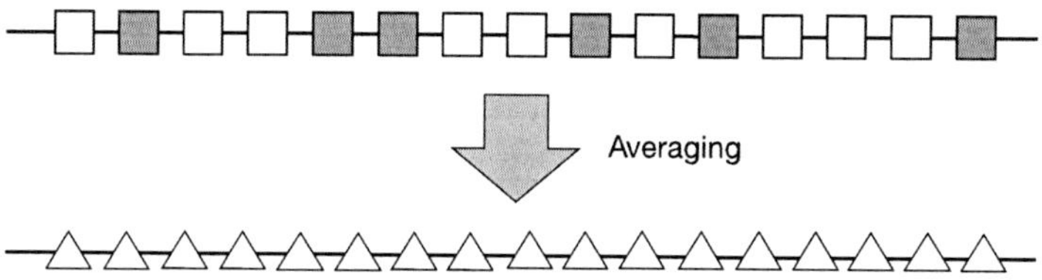


**Figure 1.35** Concept of effective mean approximation (EMA).

especially expected to be valid for van der Waals organic solid having randomness when the molecule unit therein is taken as the fragment.

#### 1.5.1.2 Effective-Medium Approximation

There is a possibility to replace the lattices of amorphous solid, for instance, with those in a virtual crystal having the uniform potential obtained by a kind of averaging process. This is called effective medium approximation (EMA) as illustrated in Figure 1.35. For instance, self-energy part in the Green function of the Hamiltonian for a random system can be set by using an appropriately obtained effective potential. [36] Examples of this method are shown in Figure 1.36 with respect to random inorganic copolymer of polythiazyl $(SN)_x$ whose unit cells are randomly mixed with (SNH) unit cells. [37] It is seen that the shape of the DOS clearly changes with an increase of the ratio of (SNH) unit, though polythiazyl itself is an electrically conductive polymer.

### 1.5.2 Localized Levels and Mobility Edge

In an amorphous solid there appear many localized energy levels in the energy gap and, collectively, these levels make certain zones in the DOS as shown in Figure 1.33. These zones, however, do not contribute to the electron delocalization or, in other words, electron transfer in the sense of the band picture. The electron transfer between such localized states is due to the hopping process, which will be described in the following Sections. The boundary line between the DOS dividing the delocalized and the localized levels is often called mobility edge in Figure 1.33, since this edge separates the characteristics of mobility of the electrons accommodated in those energy levels. The localized energy levels are often occupied by single electrons and hence accompanied by radical spins, which gives the paramagnetic behavior in total based on the Curie paramagnetism in the whole amorphous bulk.

### 1.5.3 Hopping Process

In general, in the amorphous materials, hopping process is dominant for electrical conduction mechanism. Hopping is a general name which expresses the whole process in which the conduction carrier moves by a through-space manner based on the spatial overlap between the wavefunctions located at each separated site, even if there are no chemical bonds. The hopping process thus signifies the transferring of the conduction carriers between the sites spatially separated, which simultaneously also signifies the transfer between the localized states in the energy space. These localized states normally exist out of the mobility edges in the DOS of amorphous solid as in Figure 1.33. The hopping process is of importance in polycrystalline or amorphous solids with conduction

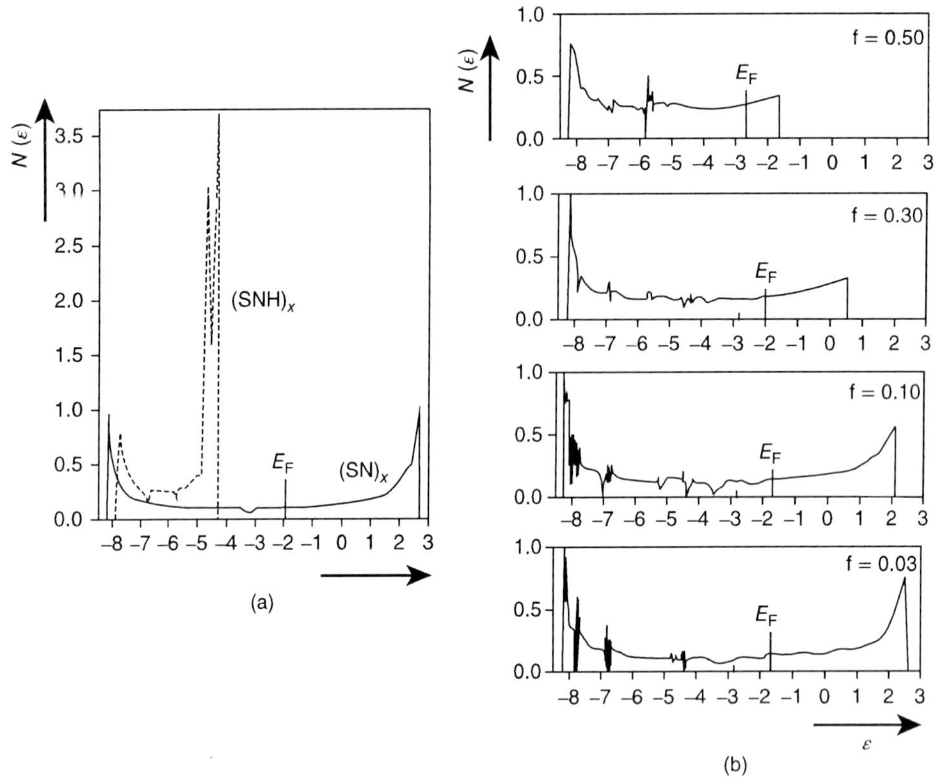


**Figure 1.36** (a) DOS of pure $(SN)_x$ and $(SNH)_x$. (b)DOS of random copolymer $[SN_{1-f}(SNH)_f]_x$ obtained by EMA (see text). Source: Ref. [37].

parts distributed in a random manner as in Figure 1.37. [38] For polymers, the interchain or interfibril hopping process is also of great importance.

#### 1.5.3.1 Hopping Process between the Nearest Neighbors

Let us consider the two sites A and B in Figure 1.38a separated with distance $R$, and the energies of the wavefunctions at those sites (namely, the energies of the localized levels) $E$(A) and $E$(B) with the energy difference $W$ as in Figure 1.38b. Then, the hopping probability $p$ for the hopping of A→B can be expressed by

$$p = (\mathrm{A})(\mathrm{B})(\mathrm{C}) \tag{1.30}$$

where (A), (B), and (C) are the Boltzmann factor, vibration factor, and overlapping factor, respectively, for the tails of the two wavefunctions as

$$\left.\begin{aligned} (\mathrm{A}) &= \exp\left[\frac{-W}{k_\mathrm{B}T}\right] \\ (\mathrm{B}) &= \nu_\mathrm{vib} \\ (\mathrm{C}) &= \exp[-2\alpha R] \end{aligned}\right\} \tag{1.31}$$

The variable $\alpha$ is a parameter with its dimension of the inversed distance and appears in the tail of the wavefunction, $\exp[-\alpha R]$, as in Figure 1.38c. For the strong localization

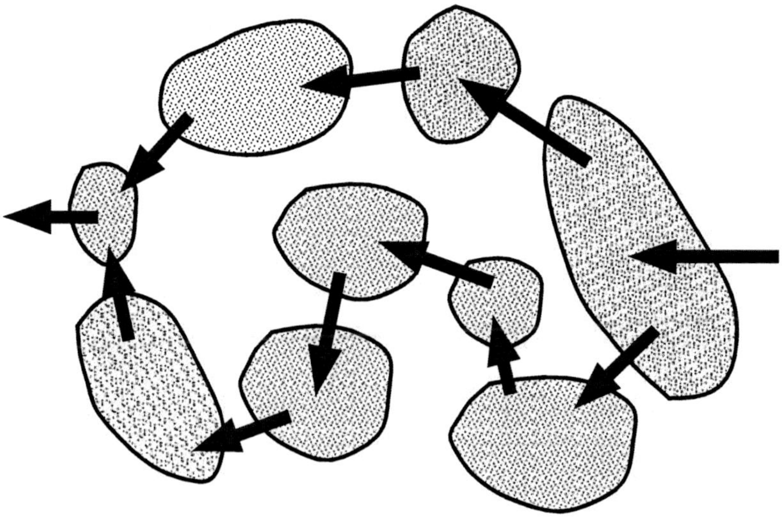

Figure 1.37 Hopping of the charge carriers between the metallic islands (dotted area). Source: Ref. [38].

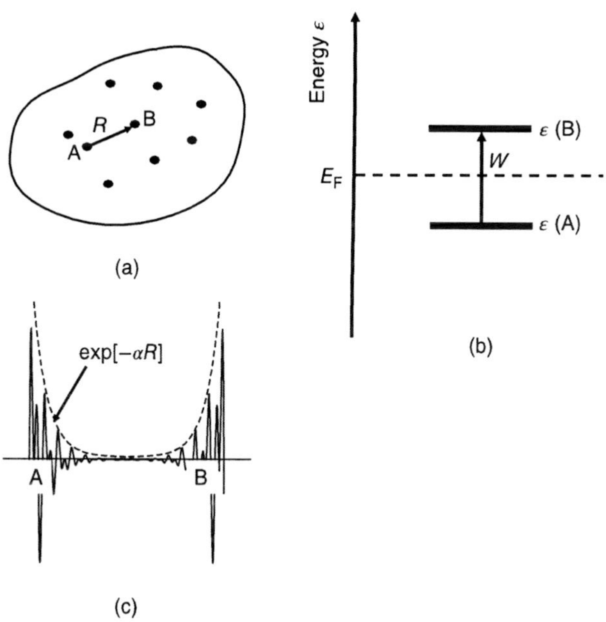


Figure 1.38 (a) Spatial hopping between sites A and B, (b) hopping between the energies of the sites A and B, and (c)envelope of the wavefunctions localized at the sites A and B.

phase embodied by $\alpha R_0 \gg 1$, with $R_0$ being the distance between the nearest neighboring sites, so as to give

$$\exp[-2\alpha R_0] \ll 1 \tag{1.32}$$

the hopping process between the nearest neighbors becomes dominant.

Supposing only the electrons near the Fermi level contribute to the above hopping process, the total number of such electrons per unit volume is given by $2N(\varepsilon_F)k_BT$ where $k_B$ is the Boltzmann constant and the Fermi DOS $N(\varepsilon_F)$ in the unit cm$^{-3}$. Under the DC electric field $F$ (Vcm$^{-1}$), along the direction B → A, the hopping probability for the direction A → B is expressed by

$$p = \nu_{vib}\exp[-2\alpha R]\left\{\exp\left[\frac{-W+eRF}{k_BT}\right] - \exp\left[\frac{-W-eRF}{k_BT}\right]\right\}$$
$$\cong \left(\frac{2eRF}{k_BT}\right)\nu_{vib}\exp\left[-2\alpha R - \frac{W}{k_BT}\right] \quad (1.33)$$

which gives the drift velocity of charge carriers $v_d$ (cm s$^{-1}$) as

$$v_d = pR = \left(\frac{2eR^2F}{k_BT}\right)\nu_{vib}\exp\left[-2\alpha R - \frac{W}{k_BT}\right] \quad (1.34)$$

By employing the current density $j$ (A cm$^{-2}$)

$$j = env_d = eN(\varepsilon_F)k_BTv_d \quad (1.35)$$

where $e$ (C) and $n$ (cm$^{-3}$) stand for elementary charge and carrier density at temperature $T$, respectively, then the electric conductivity becomes

$$\sigma = \frac{j}{F} = 2e^2R^2\nu_{vib}\exp\left[-2\alpha R - \frac{W}{k_BT}\right] \quad (1.36)$$

This signifies the activation process of Arrhenius type is introduced in $\sigma$.

#### 1.5.3.2 Variable Range Hopping (VRH)

When $\alpha R_0$ is nearly equal to unity, contribution from the hopping not only between the nearest neighboring sites but also from that to a greater number of sites becomes possible as illustrated in Figure 1.39. This is due to increase in the overlapping of the tails of the two wavefunctions between the many sites. At the low temperature region, the Boltzmann factor (A) becomes rather small in Eq. (1.30) and hence, the overlapping factor begins to rule the hopping probability $p$, as far as the vibration $\nu_{vib}$ is assumed to be independent of both temperature $T$ and radius $R$ of the sphere in Figure 1.39. This situation is called the variable range hopping (VRH). [38] Considering that hopping from the center to any sites in a 3D-sphere with the radius $R$, the maximum hopping probability per unit time at temperature $T$ is expressed by

$$p_{max} = \nu_{vib}\exp\left[-B_0\left\{\frac{\alpha^3}{k_BTN(\varepsilon_F)}\right\}^{\frac{1}{4}}\right] \quad (1.37)$$

where $B_0$ is a constant of about 2. Using Eqs. (1.35) and (1.36), the electric conductivity in the VRH mechanism becomes

$$\sigma = 2e^2R^2N(\varepsilon_F)\nu_{vib}\exp\left[-\left(\frac{B}{T}\right)^{\frac{1}{4}}\right] \quad (1.38)$$

This is called 3D-VRH since the 3D-sphere is considered here. For the 2D-sphere, namely within a circle, the electric conductivity is expressed by

$$\sigma \propto \exp\left[-\left(\frac{B}{T}\right)^{\frac{1}{3}}\right] \quad (1.39)$$

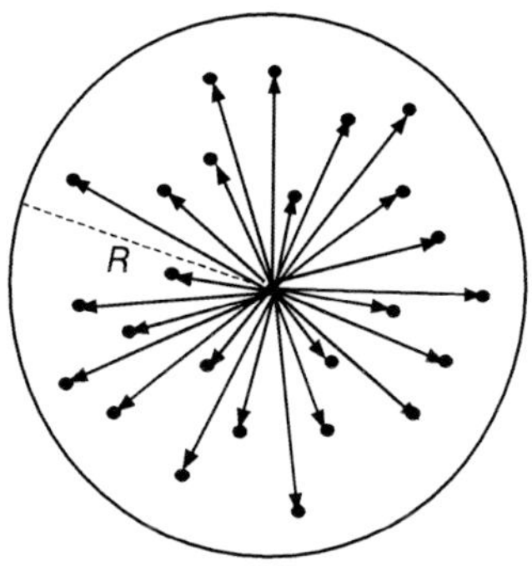


Figure 1.39 Concept of the 3D variable range hopping (VRH) in the sphere with radius R.

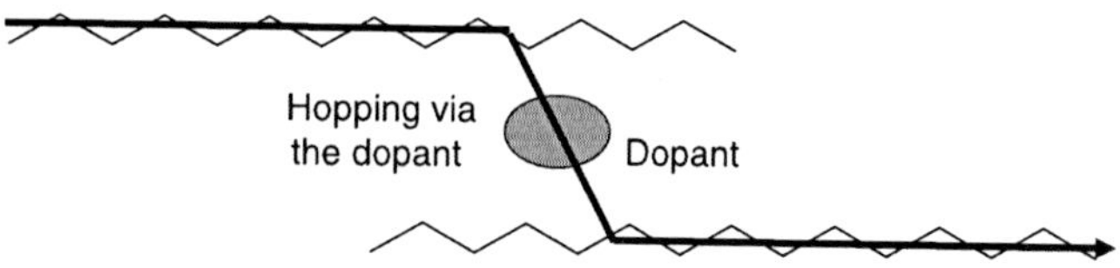


Figure 1.40 Interchain hopping utilizing the dopants as "stepping stones".

and, furthermore, temperature dependency of the VRH process can be described as

$$\sigma \propto \exp\left[-\left(\frac{B}{T}\right)^{\frac{1}{d+1}}\right] \tag{1.40}$$

where $d$ stands for the dimension number of the sphere considered. A great deal of amorphous solids, including organic materials, have been known to obey the 3D-VRH process.

#### 1.5.3.3 Hopping Process via the Dopants

An n-type polyacetylene doped with various alkaline metals exhibit the ESR spectra showing that the dopants influence the conduction electrons. [39] This suggests that the interchain hopping of conduction carriers utilizes the dopants themselves existing in the interchain region as if they are the stepping stones as shown in Figure 1.40. Similar ESR measurement results have been reported with respect to p-type doped amorphous carbon material obtained by pyrolysis of phenol-formaldehyde resin. [40]

## 1.6 Conclusion

In the last century, seeming organic semiconductors were studied mainly from scientific viewpoints. Nowadays, a wider function of these materials is being newly and steadily developed including more practical usage, mainly connecting with progress in printable processes and, in particular, toward fabrication of flexible electronic devices. In relation to this situation, understanding of the basic electronic structures of these materials should be of vital and essential importance. This Chapter has dealt with the subject of organic semiconductors from both the fundamental and historical aspects.

## Acknowledgment

This work was supported by a Grant-in-Aid for Scientific Research on Innovative Areas "New Polymeric Materials Based on Element-Blocks (No.2401)" (JSPS KAKENHI Grant Number JP24102014). The author would like to thank Dr. Hiroyuki Fueno for his assistance in the band-structure calculation.


## References


1 del Re, G., Ladik, J. and Biczó, G. (1967). *Phys. Rev.* 155: 997.
2 Heine, V. (1960). *Group Theory in Quantum Mechanics*. London: Pergamon, Chapter 6.
3 Tobin, M.C. (1960). *J. Molec. Spectrosc.* 4: 349.
4 McCubbin, W.L. (1971). *Chem. Phys. Lett.* 8: 507.
5 Kittel, C. (1996). *Introduction to Solid State Physics, 7e. New York: John Wiley & Sons*, Inc., Chap. 8.
6 Imamura, A., Tanaka, K., Yamabe T. and Fukui, K. (1979). Theoret. Chim. Acta (Berl.) 54: 1.
7 Deng, W.-Q. and Goddard III, W.A. (2004). *J. Phys. Chem. B* 108: 8614.
8 Sato, T., Tokunaga, K. and Tanaka, K. (2008). *J. Phys. Chem. A* 112: 758.
9 Shirakawa, H., Louis, E.J., MacDiarmid, A.G. et al. (1977). *J. C. S. Chem. Comun.* 16: 578.
10 Goldberg, I.B., Crowe, H.R., Newman, P. R. et al. (1979). *J. Chem. Phys.* 70: 1132.
11 Su, W.P., Schrieffer, J.R. and Heeger, A.J. (1980). *Phys. Rev. B* 22: 2099.
12 Tanaka, K. and Yamabe, T. (1985). Electronic Structure of Conductive Conjugated Systems and Their Physicochemical Properties. *Adv. Quantum Chem.* 17: 251.
13 Brédas, J.L., Thémans, B. and André, J.M. (1983). *Phys. Rev. B* 27: 7827.
14 Akamatsu, H., Inokuchi, H. and Matsunaga, Y. (1954). *Nature* 173: 168.
15 Coleman, L.B., Cohen, M.J., Sandman, D.J. et al. (1973). *Solid State Commun.* 12: 1125.
16 Kistenmacher, T.J., Phillips, T.E. and Cowan, D.O. (1974). *Acta Crystallogr. B* 30: 763.
17 Chaikin, P.M., Kwak, J.F., Jones, T.E. et al. (1973). *Phys. Rev. Lett.* 31: 601.
18 Kagoshima, S., Ishiguro, T. and Anzai, H. (1976). *J. Phys. Soc. Jpn.* 41: 2061.
19 Cohen, M.J., Coleman, L.B., Garito, A.F. and Heeger, A.J. (1974). *Phys. Rev. B* 10: 1298.
20 Torrance, J.B., Vazquez, J.E., Mayerle, J.J. and Lee, V.Y. (1981). *Phys. Rev. Lett.* 46: 253.
21 Anthony, J.E. (2006). *Chem. Rev.* 106: 5028.
22 Takeya, J., Yamagishi, M., Tominari, Y. et al. (2007). *Appl. Phys. Lett.* 90: 102120.
23 Yamamoto, T. and Takimiya, K. (2007). *J. Am. Chem. Soc.* 129: 2224.
24 Kim, K., Manoj, J.L., Namboothiry, A.G. and Carroll, D.L. (2007). *Appl. Phys. Lett.* 90: 163511.
25 Peierls, R.E. (1955). *Quantum Theory of Solids*. Oxford: Oxford University Press, Chapter 5.
26 Shirakawa, H. and Ikeda, S. (1971). *Polym. J.* 2: 231.

27 Andrieux, A., Schulz, H.J., Jerome, D. and Bechgaard, K. (1979). *Phys. Rev. Lett.* 43: 227.
28 Saito, G. and Yoshida, Y. (2007). *Bull. Chem. Soc. Jpn.* 80: 1.
29 Imada, M., Fujimori, A. and Tokura, Y. (1998). *Rev. Mod. Phys.* 70: 1039.
30 Anderson, P.W. (1978). *Rev. Mod. Phys.* 50: 191.
31 Thorup, N., Rindorf, G., Soling, H. and Bechgaard, K. (1981). *Acta Crystallogr. B* 37: 1236.
32 Bechgaard, K., Jacobsen, C.S., Mortensen, K. et al. (1980). *Solid State Commun.* 33: 1119.
33 Mortensen, K., Tomkiewicz, Y. and Bechgaard, K. (1982). *Phys. Rev. B* 25: 3319.
34 Jérome, D., Mazaud, A., Ribault, M. and Bechgaard, K. (1980). *J. Phys. Lett.(Fr.)* 41: 95.
35 Bechgaard, K., Carneiro, K., Rasmussen, F.B. et al. (1981). *J. Am. Chem. Soc.* 103: 2440.
36 Economou, E.N. (1979). *Green's Functions in Quantum Physics.* Berlin: Springer, Chapter 7.
37 Ladik, J.J. (1988). *Quantum Theory of Polymers as Solids.* New York: Plenum, Chapter 4.
38 Mott, N.F. and Davis, E.A. (1979). *Electronic Processes in Non-Crystalline Materials, Clarendon, 2e.* Oxford: Clarendon Press, Chapter 2.
39 Rachdi, F. and Bernier, P. (1986). *Phys. Rev. B* 11: 7817.
40 Tanaka, K., Koike, T., Nishino, H. et al. (1987). *Synth. Met.* 18: 521.

# 2

# Electronic Transport in Organic Semiconductors

*Hiroyoshi Naito*

*Department of Physics and Electronics, Osaka Prefecture University, Sakai, Japan,*

**CHAPTER MENU**



## 2.1 Introduction

Electronic transport properties of semiconductors are of fundamental importance, by which the performance of optoelectronic semiconductor devices such as bipolar transistors, field-effect transistors, solar cells and light-emitting diodes (LEDs) is governed. Physical quantities characterizing the electron- and hole-transport properties are mobilities, diffusion constants, deep-trapping lifetimes, bimolecular recombination constant and localized-state distributions [1], and several techniques such as time-of-flight (TOF) transient photocurrent, impedance spectroscopy, and thermally stimulated current have been employed for measuring these physical quantities. Once the physical quantities are known, the device performance can be predicted by means of device simulation using physical quantities as inputs to the simulators.

Amorphous, polycrystalline and single crystalline organic semiconductors have been successfully applied to organic light-emitting diodes (OLEDs), organic photovoltaics (OPVs), and organic field-effect transistors (OFETs). In this chapter, the transport properties of amorphous, polycrystalline, and crystalline semiconductors are described.

## 2.2 Amorphous Organic Semiconductors

The first successful application of amorphous organic semiconductors was organic photoreceptors for electrophotography [2]. Organic photoreceptors are generally double layer (charge generation and charge transport layers) structures and the layers are formed by molecularly doped polymers, which are charge generation molecules or charge carrier transporting molecules doped in electrically inactive polymers such as polycarbonate and polystyrene and which are dip-coated on Al drums. The photoreceptors are operated in air and hence air-stable organic materials for charge generation and carrier transport have been extensively developed.

Amorphous organic semiconductors have also been applied to OLED displays [3]. OLEDs can be classified into two categories; one is OLEDs with multi-layer devices, typical structures being anode/hole injection layer/hole transport layer/emissive layer/electron transport layer/cathode, and the other is polymer light-emitting diodes (PLEDs) with the device structure of anode/hole injection layer/polymer emissive layer/cathode. Multi-layered OLEDs are fabricated by means of vacuum deposition processes, while PLEDs are fabricated by means of printing processes such as inkjet printing, spin coating and gravure printing. In general, the performance of PLEDs is not better than that of multi-layered OLEDs because a polymer semiconductor for PLEDs should have the optoelectronic properties of high photoluminescence quantum efficiency and balanced electron and hole mobility (the polymer semiconductors are block copolymers and each monomer unit exhibits hole transport, electron transport, luminescence, solubility to organic solvent and so on) [4]. Crystallization of amorphous organic semiconductors in OLEDs causes the degradation of OLED performance and hence organic semiconductors with high-glass transition temperatures (>150°C) have been synthesized.

Electronic transport properties of amorphous organic semiconductors are caused by incoherent hopping of charge carriers in an exponential distribution or in a Gaussian distribution of localized tail states,

$$g(E) = \frac{N}{\sigma\sqrt{2\pi}} \exp\left(-\frac{E^2}{2\sigma^2}\right) \tag{2.1}$$

where $\sigma$ is the width of the Gaussian distribution (about 0.1 eV) [5] and $N$ is the total density of states ($10^{20}$~$10^{21}$ cm$^{-3}$). The transition rate of charge carrier from occupied site $i$ to unoccupied site $j$ (the distance between site i to site j is $r_{ij}$) is expressed as the Miller-Abrahams type of hopping rate [6],

$$v_{ij} = v_0 \exp\left(-\frac{2r_{ij}}{\alpha} - \frac{E_j - E_i + |E_i - E_j|}{2kT}\right), \tag{2.2}$$

where $\alpha$ is the extent of localized state wavefunction and is estimated to be $10^{-8}$ cm [7], $E_i$ and $E_j$ are the energies of site $i$ and site $j$ respectively, $k$ is the Boltzmann constant, $T$ is temperature, $v_0$ is the attempt-to-hop frequency whose order of magnitude is similar to that of phone frequency of $10^{12}$ s$^{-1}$ [8]. In the case of $E_i > E_j$, the transition emits phones whose energy is corresponding to the energy difference between site $i$ and site $j$, otherwise ($E_i < E_j$), the transition absorbs the phones.

Dependence of drift mobility on charge carriers of amorphous organic semiconductors on applied electric field and on temperature is described by the Gaussian Disorder Model (GDM) [5] when the charge carrier density is much smaller than $N$ and when no correlation exists between the energy of the localized state of a site and the spatial position of the site. Correlated Disorder Model (CDM) [9, 10] has been applied to the interpretation of the electric-field and temperature dependence of drift mobility when the correlation does exist.

When the charge carrier density is higher than $N/1000$, the mobility becomes dependent on the charge carrier density [11–13]. This is because the quasi-Fermi level approaches to the transport energy, which plays a role similar to the mobility edge [14] and will be described later, with increasing charge carrier density. Since such a condition is not fulfilled in OLEDs, we do not mention the charge-carrier density dependent mobility in this chapter. The density dependent mobility may be found in OFETs at high-gate voltages.

### 2.2.1 Measurements of Transport Properties

Several measurement methods have been proposed for studying transport properties of organic semiconductors. Reliable methods for the transport measurements are based on flight time measurements using samples with sandwiched electrode configuration. The flight time is the charge carrier transit time from one electrode to the other electrodes in the samples. Widely used methods are time-of-flight (TOF) transient photocurrent experiment [15], charge extraction by linear increasing voltage (CELIV) experiments [16], and impedance spectroscopy experiments [17]. The CELIV experiments cannot be directly applied to the mobility and charge collection experiments for amorphous semiconductors displaying dispersive transport and hence the TOF measurement is briefly described in this chapter and the transport measurements, including TOF, are fully discussed in this book.

#### 2.2.1.1 Time-of-Flight Transient Photocurrent Experiment

TOF experiments are suitable for low mobility semiconductors ($<1\ \text{cm}^2/\text{Vs}$). Hall effect measurements, which have been standard mobility measurements for crystalline semiconductors [18], cannot be carried out in such low mobility materials. The TOF experiments have been most extensively carried out for measuring drift mobility of organic and amorphous inorganic semiconductors. In addition to the mobility measurements, other transport properties such as photocarrier generation efficiency, deep-trapping lifetime, diffusion constants, and localized-state distributions can be determined by means of the TOF experiments [5, 19, 20].

The measurement system is shown in Figure 2.1. In the measurement, external voltage is applied to a sample with sandwiched electrode configuration (at least, one electrode is blocking). The sample thickness is thicker than 1 μm in order to define the flight distance of ~sample thickness. Pulsed strongly absorbed light is irradiated within the dielectric relaxation time of the sample to one electrode, and charge carrier sheet is created beneath the irradiated electrode. The injected charge carriers drift toward the counter electrode owing to the applied electric field, and the photocurrent flowing in the external circuit decays after the leading edge of the charge carrier sheet reaches the counter electrode. The charge carrier sheet moves at a constant speed (we call the transport non-dispersive transport), the photocurrent, $I(t)$, is constant, $I(t) = Q_0/t_r$, before the charge carrier transit time, $t_r$, and is zero when the sheet reaches the counter electrode, where $Q_o$ is the photoinjected charge density.

The time it takes for the charge carrier sheet to reach the counter electrode is called the transit time as mentioned above, and is given by $t_r = L/\mu F$, where $L$ is the sample

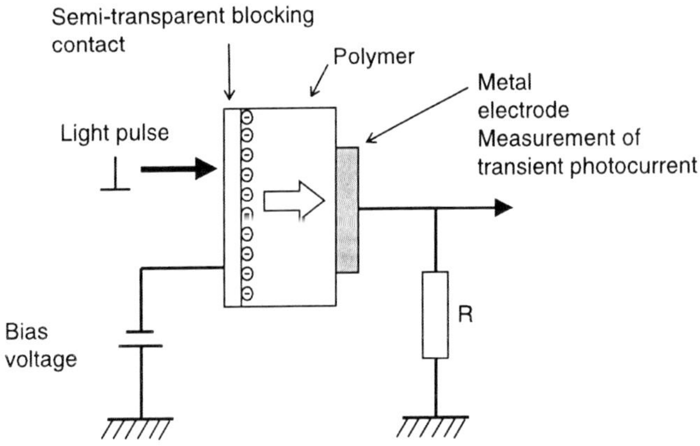


**Figure 2.1** Experimental setup of TOF transient photocurrent measurement.

film thickness and $\mu$ is the drift mobility. The drift mobility can thereby be determined by measuring the charge carrier transit time. As can be seen from Figure 2.1, the electron or hole mobility can be evaluated independently by changing the polarity of the irradiated electrode (the electron drift mobility can be evaluated in the electric field direction of Figure 2.1).

Representative photocurrent transients are shown in Figure 2.2 [21]. Figures 2.2(a)-(c) and (d) are called non-dispersion photocurrent transients because the transit time can be determined from the shoulder of the photocurrent transients. On the other hand, in Figure 2.2(d) and (e), the photocurrent transients do not clearly show such a shoulder and is called dispersive transport, in which charge carriers have distributed drift mobilities. The transit time is determined from a double logarithmic plot of the photocurrent transient as shown in the inset of Figure 2.2.

## 2.3 Experimental Features of Electronic Transport Properties

Experimental features of charge carrier transport of amorphous organic semiconductors revealed by the TOF method are electric-field and temperature dependent drift mobility. Figures 2.3 and 2.4 are electric field and temperature dependence of hole drift mobility of $N, N', N'', N''$-tetrakis(4-methylphenyl)-(1,1'-biphenyl)-4,4'-diamine (TTB) thin films respectively [22]. The TOF transient photocurrent as shown in Figures 2.3 and 2.4 have been observed in various amorphous organic semiconductors [2]. It can be seen from Figure 2.3 that the drift mobility increases with increasing square root of the electric field ($\log(\mu) \propto \sqrt{F}$), and from Figure 2.4 that the drift mobility shows the thermally activated behavior ($\log(\mu) \propto \exp(-\Delta E/kT)$).

## 2.4 Charge Carrier Transport Models

The experimental data shown in Figures 2.3 and 2.4 have been interpreted in terms of GDM or CDM. Before giving an overview of GDM and CDM, the multiple trapping

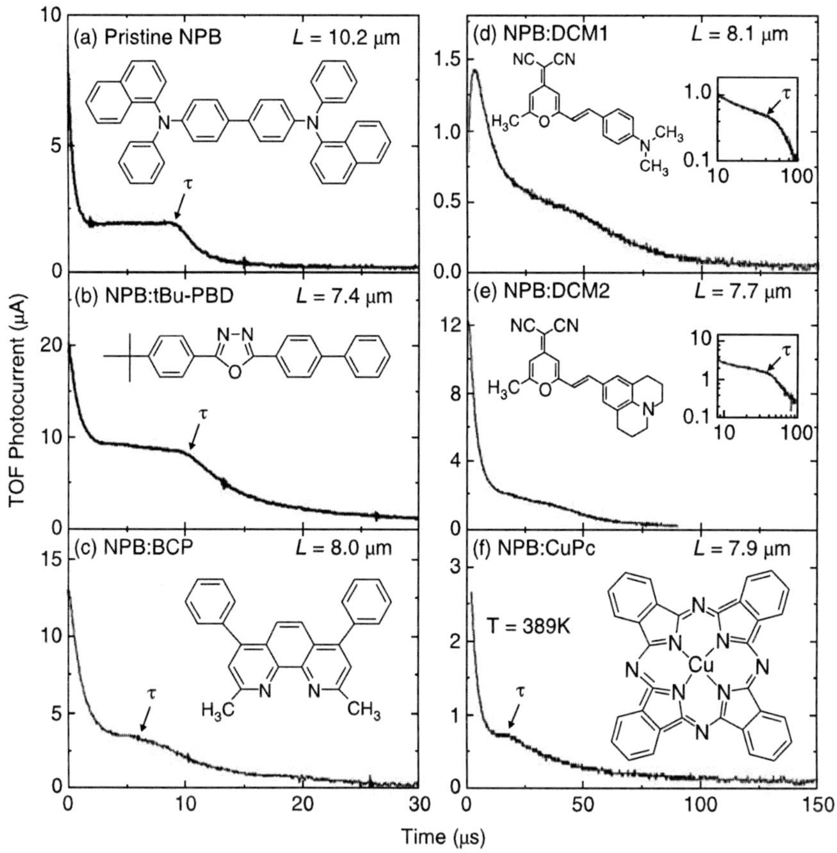


**Figure 2.2** Typical time of flight transients for (a) pristine NPB, (b) NPB:tBu-PBD, (c) NPB:BCP, (d) NPB:DCM1, (e) NPB:DCM2, and (f) NPB:CuPc under an applied field of 0.2 MV/cm at room temperature, while NPB:CuPc transient was taken at 389 K. In (d) and (e) the insets show the double logarithmic plots. The chemical structures with abbreviations and the film thicknesses *L* are also shown. Source: Based on Ref. [14].

model (MTM) proposed for the interpretation of charge transport in amorphous inorganic semiconductors is briefly described.

### 2.4.1 Multiple Trapping Model

In MTM, the band edge (mobility edge) is clearly defined, and the charge carrier transport process involves multiple trapping of free carriers at the band edge into the energetically distributed localized states generated by structural disorder. In the 1970s, the transient photocurrent of amorphous $As_2Se_3$, which exhibits excellent photoelectric characteristics as a photocopier in electrophotographic application, is similar to that in Figures 2.2(d) and 2.2(e), in which the transit times are not clearly seen. Several interpretations have been proposed to understand the dispersive transport, and now there is a consensus that MTM is the most appropriate interpretation of the dispersive transport in amorphous inorganic semiconductors [23]. As we will describe later, MTM is

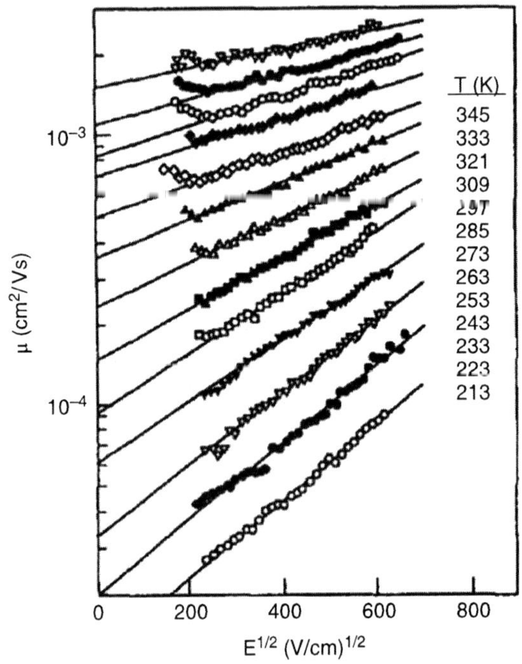


**Figure 2.3** Electric field dependences of hole drift mobilities in N,N′,N″,N″-tetrakis(4-methylphenyl)-(1,1′-biphenyl)-4,4′-diamine (TTB) at different temperatures. Source: Ref. [22].

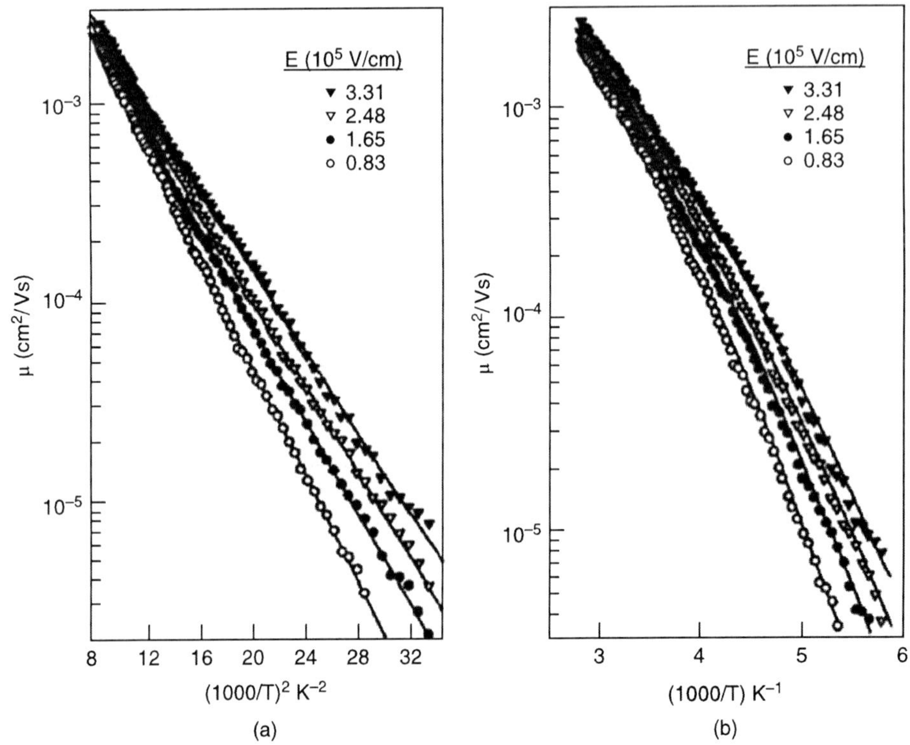


**Figure 2.4** Plots of hole drift mobilities versus $1/T^2$ at different electric field for TTB (a) and the same data versus 1/T (b). Source: Ref. [22].

still valid in amorphous organic semiconductors in the presence of the transport energy, known as trap-controlled hopping.

It is assumed exponentially distributed localized states,

$$g(E) = \frac{N}{E_0} \exp\left(\frac{E}{E_0}\right) \tag{2.3}$$

in MTM. Drift mobility is defined as

$$\mu = \mu_0 \frac{n_f}{n_f + n_t}, \tag{2.4}$$

where $n_f$ is free carrier density, $n_t$ is the trapped carrier density in the localized states, and $\mu_o$ is the microscopic mobility. After photoexcitation to samples in TOF experiments, photogenerated carriers are trapped in the localized states as time goes on. Most free carriers are trapped in the localized states at the energy level of

$$E_d(t) = -kT \ln(\nu t)$$

where $t$ is the time after the photoexcitation, as shown in Figure 2.5. The trapping of free carriers and detrapping of trapped carriers in localized states whose energy is ranging from the band edge to $E_d$ is balanced, while the trapping of free carriers into localized states whose energy is deeper than $E_d$ is dominant. The drift mobility is hence easily calculated to be

$$\mu \propto t^{-(1-\alpha_d)},$$

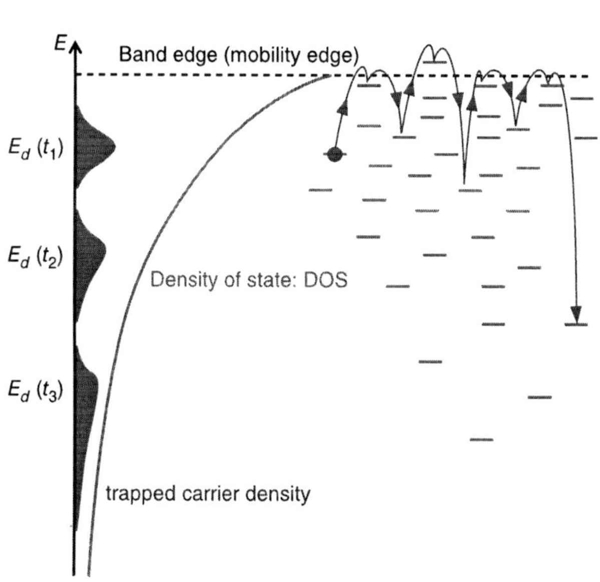


**Figure 2.5** Photocreated free carriers are trapped in an exponential distribution of localized states. Trapped carriers are distributed in a packet. The packet moves to a deeper energy region as time goes on ($t_1$<$t_2$<$t_3$).

and the transient photocurrent is proportional to the drift mobility and hence $I \propto t^{-(1-\alpha_d)}$, where $\alpha_d$ is the dispersion parameter, $\alpha_d = kT/E_0$. The transit time can be calculated from $\int_0^{t_r} F\mu(t)dt \cong L/2$ and is

$$t_r \propto (L/F)^{1/\alpha_d}.$$

The temperature dependence of the drift mobility can be expressed as

$$\mu \propto \exp(-\Delta E(F)/kT),$$

which is consistent with the thermally activated behavior observed in Figure 2.4(b). The empirical expression for electric-field and temperature dependence of drift mobility in Figures 2.3 and 2.4(b) has been known as Gill's empirical expression [7],

$$\mu = \mu_0 \exp\left(-\frac{E_0 - \beta\sqrt{F}}{kT}\right) \tag{2.5}$$

The MTM expression is also consistent with the Gill's empirical expression [24] in the experimental accessible electric field range.

The transient photocurrent after the transit time is governed by the thermal release of trapped carriers from the localized states (emission-limited current), and is expressed as $I \propto t^{-(1+\alpha_d)}$. The transit time is therefore determined from the inflection point in the double logarithmic plot as shown in Figures 2.2(d) and 2.2(e). The description mentioned above is intuitive but the same conclusions are obtained by solving the basic transport equations (the current continuity equation, the current equation and the rate equations of trapping and detrapping processes of charge carriers). The analytical solution (transient photocurrent) to the basic equations has been obtained in the case of $\alpha_d = 0.5$ [25].

### 2.4.2 Gaussian Disorder Model (GDM)

No well-defined band edge exists in amorphous organic semiconductors because the charge carrier transport is due to hopping between organic semiconducting molecules. It is rather difficult to analytically treat the hopping transport in amorphous organic semiconductors and so the hopping transport was treated in terms of Monte Carlo simulation instead. The empirical expressions for electric field and temperature dependence of drift mobility were derived from the simulation data, and were called GDM and CDM. GDM assumes the energetic disorder (diagonal disorder) on each lattice point of the simple cubic lattice and geometrical disorder (off-diagonal disorder) between the lattice points. The site energy and the distance between sites are assumed to vary statistically with Gaussian distribution functions. The energetic and the geometrical disorder is assumed to independently affect the charge carrier transport. Empirical expression was obtained from the transient photocurrent calculated in terms of Monte Carlo simulation based on Miller-Abraham hopping rate (Eq. [2.2]) [5],

$$\mu(\hat{\sigma}, \Sigma, F) = \begin{cases} \mu_0 \exp\left[-\left(\frac{2}{3}\hat{\sigma}\right)^2\right] \exp[C(\hat{\sigma}^2 - \Sigma^2)F^{1/2}] \ (\Sigma \geq 1.5) \\ \mu_0 \exp\left[-\left(\frac{2}{3}\hat{\sigma}\right)\right] \exp[C(\hat{\sigma}^2 - 2.55)F^{1/2}] \ (\Sigma < 1.5) \end{cases} \tag{2.6}$$

where $\hat{\sigma} = \frac{\sigma}{kT}$, $\sigma$ is the energetic disorder parameter, $\Sigma$ is the geometrical disorder parameter, $C$ is a constant and its typical value is $2.9 \times 10^{-4}$ $(\text{cm/V})^{1/2}$, $\mu_0$ is the hypothetical mobility without disorder.

The energetic disorder $\sigma$ is mainly due to the permanent dipoles of the charge-transporting molecules because the dipoles are randomly distributed, causing fluctuations in the electrostatic potential. In fact, it has been experimentally shown that the value of $\sigma$ is increased with increasing dipole moments of charge transporting molecules [26].

The origin of the geometrical disorder $\Sigma$ is simply the fluctuation of the distance between hopping sites. The geometrical disorder is also dependent on the fluctuation of the transfer integrals between charge-transporting molecules. The transfer integral changes significantly depending on the difference in orientation between the molecules because most of charge transporting molecules have planar molecular structures with a well-developed $\pi$-electron system [27]. In addition, the percolation of charge transporting path is the origin of $\Sigma$ [28].

### 2.4.3 Correlated Disorder Model (CDM)

In GDM, it was assumed that the energies of the hopping sites are uncorrelated. However, it is not likely because the interaction between the dipole moment of a charge-transporting molecule and a charge carrier on another charge-transporting molecule is a long-range interaction, meaning that the energies are not much different from those of neighboring molecules. The energy of the hopping site does not change spatially randomly but changes smoothly (there is a correlation between the energy of the hopping site and its spatial position). CDM takes the correlation into account, and Monte Carlo simulation showed that the incorporation of site-energy correlation in CDM leads to an extended electric field range within which the $\log(\mu) \propto \sqrt{F}$ dependence, Eq. (2.5), is obeyed [9]. The empirical formula in CDM gives a different formula from GDM [10].

$$\mu(\hat{\sigma}, F) = \mu_0 \exp\left(-\frac{9\hat{\sigma}^2}{25}\right) \exp\left[C_0'(\hat{\sigma}^{1.5} - \Sigma^2)\sqrt{\frac{eaF}{\sigma}}\right] \tag{2.7}$$

where $C_0' = 0.78$.

### 2.4.4 GDM vs. CDM

It is rather reasonable that there is a correlation between the energy of the hopping site and its spatial position as assumed in CDM. The electric field dependence of the mobility of $\log(\mu) \propto \sqrt{F}$ has been observed in amorphous organic semiconductors and, as mentioned above, such electric field dependence in an extended electric field range is reproduced by CDM. CDM has, therefore, been widely accepted as a charge carrier transport model for amorphous organic semiconductors [29–34].

On the other hand, it has been claimed that it is necessary to further examine the charge transport properties of molecularly doped polymers (MDPs) doped with a variety of charge transport materials to show the validity of CDM more strictly [35]. At present, it has been recognized that CDM can better describe the charge transport characteristics of amorphous organic semiconductors than GDM, but GDM is still employed to analyze the drift mobility data.

### 2.4.5 Polaronic Transport

The induced absorption band associated with polaron formation can be observed from photo-induced absorption and charge-induced absorption measurements in poly(3-hexylthiophene) (P3HT) and related polymers, indicating that hole polarons are mobile in the materials [36]. GDM and CDM do not take account of the polaronic transport. It has been known that the electric field and temperature dependences of the drift mobility observed in amorphous organic semiconductors cannot be explained by polaronic transport without taking account of the effect of disorder. The zero electric field mobility in GDM in which the polaronic transport with the polaron binding energy of $E_p$ (typical binding energy of polarons is about 0.15 eV or less) is taken into account and expressed as

$$\mu \propto \exp\left(-\frac{E_p}{2kT}\right)\exp\left[-A\left(\frac{\sigma}{kT}\right)^2\right],$$

where A=4/9 [37]. It has been shown that the electric field and temperature dependences of the drift mobility of amorphous organic semiconductors can be better explained by GDM incorporating polaronic transport [38]. However, it seems that there is no consensus that the polaronic transport should always be incorporated to GDM and CDM for the analysis of the drift mobility data.

### 2.4.6 Transport Energy

Transport energy is a very important concept for understanding hopping transport of amorphous organic semiconductors. Its existence has been discussed in the localized states with exponential [14, 39] and Gaussian distributions [40]. The transport energy in hopping transport system plays a role similar to the band edge (mobility edge) in MTM.

The transport energy is derived here in case of an exponential localized-state distribution, Eq. (2.3). The downward hopping rate from a hopping site with the energy $E_i$ to another hopping site with deeper energy is expressed as [41],

$$\nu_\downarrow = \nu_0 \exp\left[-\frac{2r(E_i)}{\alpha}\right],$$

where

$$r(E) = \left[\frac{4\pi}{3}\int_{-\infty}^{E} g(E)\left[1 - f\left(E', E_F\right)\right] dE'\right]^{-1/3}$$

$f(E', E_F)$ is the Fermi-Dirac distribution function, and $E_F$ is the Fermi energy. On the contrary, the upward hopping rate from a hopping site with $E_i$ to another hopping site with $E_x$($E_x$>$E_i$)is

$$\nu_\uparrow = \nu_0 \exp\left[-\frac{2r(E_x)}{\alpha} - \frac{E_x - E_i}{kT}\right],$$

where we assume that $\alpha$ is independent of the energy of localized states. The energy of the localized state at which the upward hopping rate is maximum is obtained from

$$\frac{\partial \nu_\uparrow(E_i, E_x)}{\partial E_x} = 0.$$

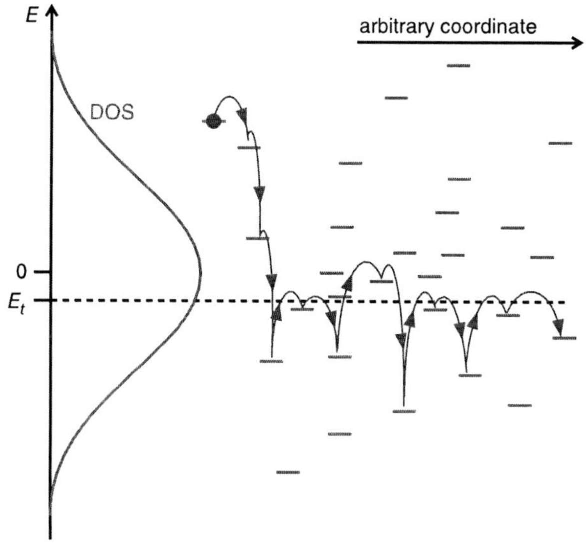


**Figure 2.6** Transport energy in a Gaussian distribution of localized states. Source: Modified from [42].

The energy $E_x$ obtained above is not dependent on $E_i$ and is the transport energy $E_t$. The transport energy in an exponential localized state distribution is

$$E_t = E_0 \ln\left(\frac{2}{9\pi\alpha^3 N}\right) - 3E_0 \ln\left(\frac{E_0}{kT}\right), \tag{2.8}$$

when $f(E', E_F) \ll 1$ [14]. The expression for $E_t$ tells us that the transport energy becomes deeper with a reducing temperature.

The transport energy in case of a Gaussian distribution of localized states, Eq. (2.1), has been derived in a way similar to the case of an exponential distribution mentioned above, and is located slightly below the peak energy ($E$=0) of the Gaussian distribution (Figure 2.6) [42].

The transport energy is a very important concept for understanding hopping transport of amorphous organic semiconductors as mentioned above. Computer simulation study of TOF transient photocurrent showed that the TOF transient photocurrents due to MTM and to multiple hopping mechanisms are indistinguishable in the case of exponential or Gaussian distributions of localized states [43]. This is an indication of the existence of a transport energy in multiple hopping mechanisms. The interpretation of optical and electronic experimental results of amorphous organic semiconductors has therefore been made on the basis of the basic transport equations with well-defined mobility edge.

### 2.4.7 Analytical Approach to Hopping Transport

The GDM and CDM expressions for electric field and temperature dependences of drift mobility obtained from Monte Carlo simulation data are empirical. It is important to note that the expression,

$$\mu(T) = \mu_0 \exp\left[-\left(C\frac{\sigma}{kT}\right)^2\right] \tag{2.9}$$

can be analytically derived using the transport energy concept [42]. The temperature dependence is exactly the same as that of GDM and CDM. The results show that the hopping transport of amorphous organic semiconductors, which could be handled only by empirical formulas based on Monte Carlo simulation, has become analytically solvable by using the concept of the transport energy.

### 2.4.8 Functional Forms of Localized State Distributions

A Gaussian distribution of localized states has been assumed in GDM and CDM, whereas an exponential distribution for localized states is also used for the analysis of transport properties of amorphous organic semiconductors. Functional forms of localized state distributions in amorphous organic semiconductors have thus still been a controversial issue. The reason for this is that no reliable spectroscopic methods for measuring localized state distribution is available. For instance, the thermal stimulation current (TSC) cannot accurately evaluate the energy distribution of the localized level because TSC probes the energy distribution of occupied localized states [44]. The steady-state space charge limiting current (SCLC) measurements can also probe localized state distributions but cannot probe the distributions over a wide energy range, and thus it is rather difficult to distinguish between Gaussian and the exponential distributions [45].

On the other hand, transient photocurrent [46] and impedance spectroscopies [17] can determine localized level distributions in a relatively wide energy range. In particular, it seems that the most suitable method for measuring localized-state distributions in amorphous organic semiconductors is impedance spectroscopy, which can measure localized states with extremely low densities [12].

At present, the temperature dependence of the drift mobility provides insight into the localized level distributions. For example, as shown in Figure 2.4, it is sometimes concluded that the localized-state distribution is Gaussian because the temperature dependence is better described by Eq. (2.9) rather than Eq. (2.5) [22]. However, even in a similar temperature range a clear difference, as shown in Figure 2.4 may not be observed in some cases. In addition, if we fit Eq. (2.5) to the temperature dependence of mobility in Figure 2.4, the extrapolated value of the mobility at $T \to \infty$ is 10 $cm^2/Vs$ or higher, which is too high for the mobility of an amorphous organic semiconductor. On the other hand, if we fit Eq. (2.9) to the temperature dependence of mobility in Figure 2.4, the extrapolated value of the mobility ranges from $10^{-2}$ to $10^{-1}$ $cm^2/Vs$. which is reasonable for the mobility of hopping conduction. For this reason, it was concluded that the Gaussian distribution is appropriate for the localized level distribution in amorphous organic semiconductors [5].

## 2.5 Prediction of Transport Properties in Amorphous Organic Semiconductors

In crystalline semiconductors, the electronic transport properties can be predicted by determining their crystal structures by X-ray diffraction, followed by band structure calculations [47]. In contrast, the prediction of the electronic transport properties of amorphous semiconductors is rather difficult. However, the recent development of computers greatly reduces the computation time and the prediction of the electronic transport properties of amorphous organic semiconductors becomes possible [48–50].

Molecular dynamics simulation has been used for preparing model molecular packing structures of amorphous organic semiconductors. Charge carrier transfer rates between a molecule and its surrounding molecules were evaluated using Marcus theory, from which local diffusion constants were obtained and then local mobilities were calculated from the Stokes-Einstein equation. The macroscopic mobility was estimated from the geometric average of the local mobilities. Hole mobilities whose chemical structures are shown in Figure 2.7 were calculated by means of the method mentioned above [50]. It is shown in Figure 2.8 that the calculated mobilities are in good agreement with the experimentally observed mobilities, indicating that the method can be used for the prediction of mobilities of organic molecular semiconductors.

Instead of the calculation of local mobilities, the macroscopic mobility in model molecular packing structures of amorphous organic semiconductors was directly calculated using a kinetic Monte Carlo simulation. The simulation is time-consuming but the quantitative agreement between the simulation and the experiment was obtained in *N*-*N*'-dicarbazole-3,5-benzene (mCP) [48] and 4,4'-di(*N*-carbazolyl)biphenyl (CBP) [49]. Further improvement of the computation speed of computers, as well as a calculation algorithm, will make quantitative prediction of the electronic transport properties of amorphous organic semiconductors possible in the very near future.

## 2.6 Polycrystalline Organic Semiconductors

### 2.6.1 Transport in Polycrystalline Semiconductors and Technological Importance of Polycrystalline Silicon

Electronic transport properties of polycrystalline semiconductors have been extensively studied because the material has the potential of low-cost, and high and stable device performance for thin-film transistors and photovoltaic application. In particular, polycrystalline silicon FETs, also known as thin-film transistors, were of great interest for applications in static random-access memories (RAMs), active-matrix liquid crystal displays and high-performance electrically erasable programmable read-only memories.

Polycrystalline semiconductors consist of small crystallites (grains) and boundaries of adjacent grains (grain boundaries), where a local potential barrier due to the high density of defects exists as illustrated in Figure 2.9, known as a polycrystalline model. It has been clear that the polycrystalline silicon FET performance is limited by the grain boundaries (GBs) in the channel [51, 52]. Consequently, passivation by hydrogen plasma was extensively used to reduce defect density at the grain boundaries and to improve intra-grain mobility [53]. Laser-annealing techniques were applied to improve polycrystalline silicon FETs characteristics through grain-size enlargement, and reduction of defect density in both grain boundaries and in the grain region [54].

A low-temperature process for the fabrication of polycrystalline silicon, known as low-temperature polycrystalline silicon (LTPS), has also been developed for the reduction of the process temperature below 600°C to avoid the distortion of glass substrates on which FETs are formed in display application. For instance, amorphous silicon FETs with the field-effect mobility of 0.23 $cm^2/Vs$ at room temperature has been transformed

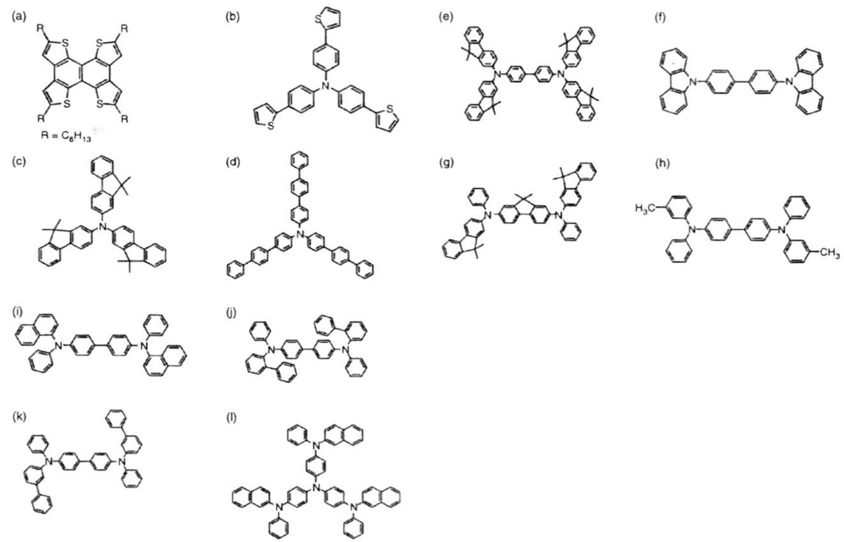


**Figure 2.7** Molecular structures. (a)C6-2TTN, (b)TTPA, (c)TFIA, (d)p-TTA, (e)FFD, (f)CBP, (g)PFFA, (h)TPD, (i)NPD, (j)o-BPD, (k)m-BPD, (l)2-TNATA. Source: Ref. [50].

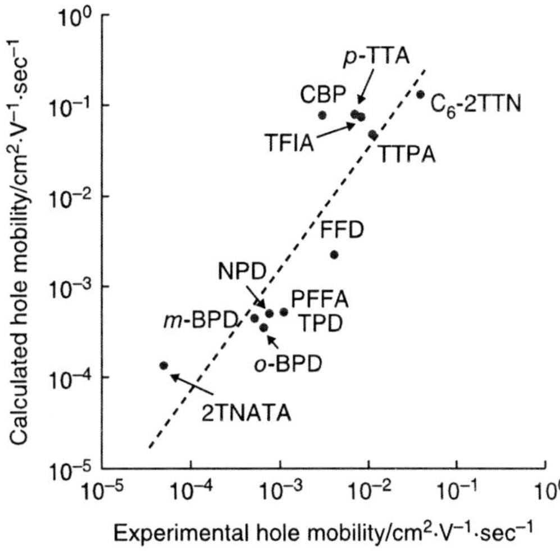


**Figure 2.8** Comparison of calculated and experimental hole mobilities. Source: Ref. [50].

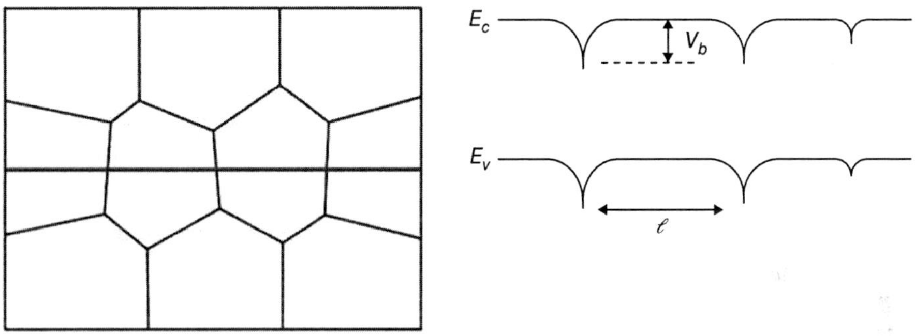


(a) (b)

**Figure 2.9** Illustration of polycrystal (a) and band diagram along the red line in (a) (b). Sources: Refs. [51, 52]

to polycrystalline silicon FETs with the field-effect mobility of 102 cm$^2$/Vs by means of a pulsed XeCl excimer laser irradiation [55].

### 2.6.2 Field-Effect Mobility in Organic Polycrystalline Semiconductors

Polycrystalline organic semiconductors have been used in organic field-effect transistors (OFETs) and organic photovoltaics (OPVs) for the same reason as in polycrystalline silicon. Polycrystalline OFETs play an essential role as key devices of logic circuits on flexible substrates. The performance of pentacene OFETs has been extensively studied in the 1990s–2000s, and the field-effect mobility of polycrystalline pentacene fabricated by vacuum evaporation was ranging from 3–6 cm$^2$/Vs [56, 57]. Potential distributions in polycrystalline pentacene OFETs were measured by means of atomic force microscope potentiometry [58]. The highest occupied molecular orbitals (HOMO)-band-edge fluctuations have been clearly observed as shown in Figure 2.10 and the crystallite size agree well with characteristic periods of HOMO-band-edge fluctuations in polycrystalline pentacene thin films, as illustrated in Figure 2.9.

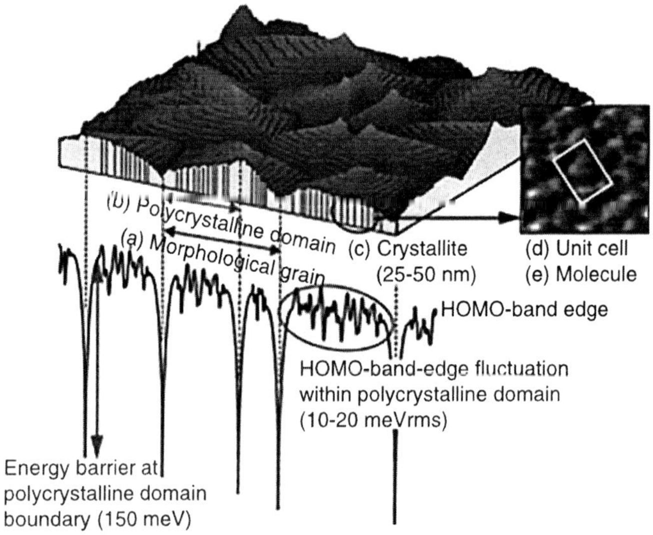


**Figure 2.10** Schematic illustration showing (top) crystallographic and (bottom) HOMO-band-edge structure in a pentacene polycrystalline film. A pentacene polycrystalline film consists of hierarchical structures, including (a) the micrometer-scale morphological grain, (b) the polycrystalline domain surrounded by large energy barriers for carrier transport, (c) the ten-nanometer-scale crystallite, (d) the unit cell of the crystal, and (e) the molecule. Source: Ref. [58].

Dangling bonds are formed at the grain boundaries of polycrystalline silicon and hence the termination of dangling bonds with hydrogen atoms was effective in the improvement of the charge carrier transport. On the other hand, such treatment has not been usually carried out in polycrystalline organic semiconductors because almost no dangling bonds exist at grain boundaries of organic semiconductors that consist of closed-shell molecules with van der Waals interactions (generally, chain ends of polymeric semiconductors are chemically modified to remove localized electronic states). However, even at the grain boundaries of polycrystalline organic semiconductors, localized electronic states acting as charge carrier traps are formed, for instance, because of accumulated impurities at the boundaries or charge carrier localization due to weak electronic overlap.

The relationship between the field-effect mobility and the grain size of polycrystalline organic semiconductors was discussed in [59]. The field-effect mobility, $\mu_{FE}$, of polycrystalline semiconductors in the case of a series connection of grains and grain boundaries is expressed as

$$\frac{1}{\mu_{FE}} = \frac{1}{\mu_G} + \frac{1}{\mu_{GB}}$$

where $\mu_G$ and $\mu_{GB}$ are the mobilities in both the grains and grain boundaries, respectively. Since $\mu_G$ is generally much higher than $\mu_{GB}$, the $\mu_{FE}$ of polycrystalline semiconductors is limited by charge carrier transfer process across the grain boundaries. Hence, the $\mu_{FE}$ of polycrystalline semiconductors is generally raised with the increasing size of the grains, and indeed the increase in the $\mu_{FE}$ with grain size in polycrystalline octithiophene OFETs (Figure 2.11[a]) was observed and the expression relating the $\mu_{FE}$ and the

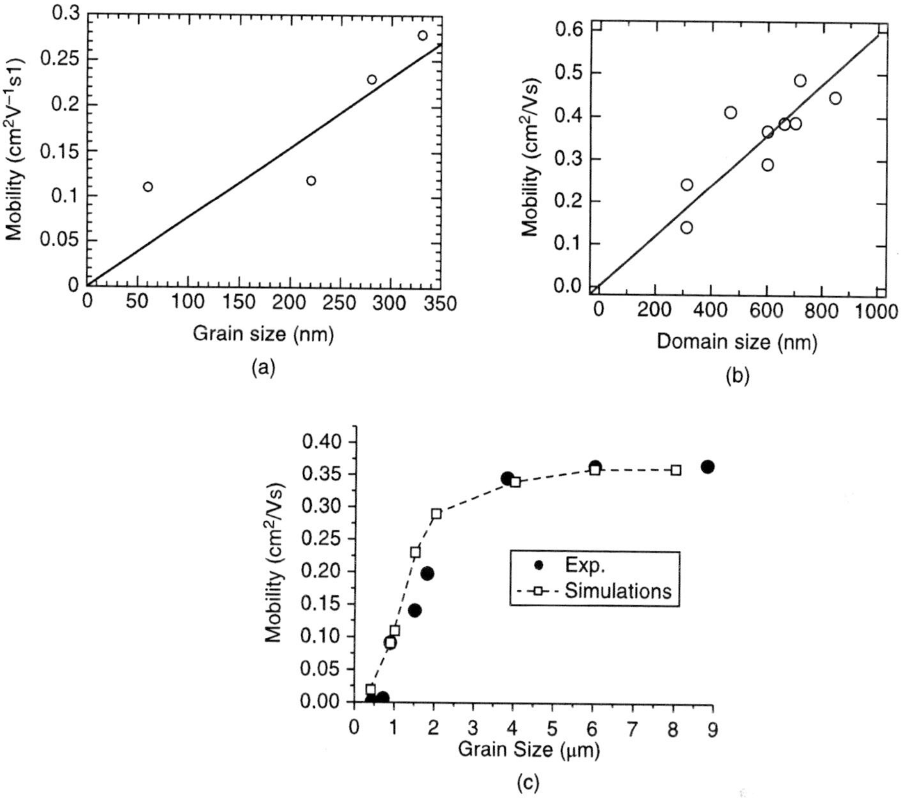


**Figure 2.11** Variation of the mobility with grain size in octithiophene (a) and in pentacene (b). Comparison between experiment and calculation (Eq. 2.12) of mobility of pentacene OFETs as a function of grain size (c). Sources: (a) Ref. [59], (b) Ref. [60], and (c) Ref. [62].

grain size $\ell$ ($\mu_{FE}$ is proportional to $\ell$),

$$\mu_{FE} = \frac{q\langle v\rangle \ell}{8kT} \exp\left(-\frac{E_b}{kT}\right) \tag{2.10}$$

have been derived, where $\langle v\rangle$ is the electron mean velocity and $E_b$ is the barrier height between adjacent grains. The expression was derived using a polycrystalline model in Figure 2.9, where charge carriers pass over back-to-back Schottky barriers generated by defect states at grain boundaries via thermionic emission. The experimental observation (Figure 2.11) is consistent with the prediction of Eq. (2.10), in the case where the grain sizes of octithiophene OFETs were below 350 nm.

The grain size dependence of $\mu_{FE}$ was also studied in polycrystalline pentacene FETs [60] (Figure 2.11(b)). The $\mu_{FE}$ is proportional to the grain size as in the case of octithiophene OFETs, mentioned above. However, theoretical treatment in [59] seems to be unsatisfactory because the calculated $\langle v\rangle$ was too small as the thermal velocity. To remove the problem, the experimental results were analyzed using a polycrystalline model with the diffusion theory which is used when the mean free path of charge carriers is smaller than the width of barriers (the mean free path is almost equal to the molecular distance of pentacene). The $\mu_{FE}$ is given by

$$\mu_{FE} = \frac{q\ell}{2kT}\mu_G\sqrt{\frac{N_A E_b}{2\epsilon_s}}\exp\left(-\frac{E_b}{kT}\right), \tag{2.11}$$

where $N_A$ is the acceptor density and $\epsilon_s$ the dielectric constant of pentacene. The grain size dependence of $\mu_{FE}$ in Figure 2.11(b) is well explained with slightly lower $\mu_G$ reported in single crystalline pentacene thin films [61].

Such dependences were also examined in polycrystalline pentacene OFETs with larger grain sizes up to 8 μm. The $\mu_{FE}$ in pentacene OFETs is increased linearly with grain size below 2 μm, as observed in Figure 2.11(c), and saturates above 6 μm. The different expression relating the $\mu_{FE}$ and the grain size $\ell$,

$$\mu_{FE} = \frac{1}{1 + n_{GB}\left(\frac{\ell}{L}\right)\exp\left(\frac{E_b}{kT}\right)}\mu_G \exp(\sqrt{F/F_0}) \tag{2.12}$$

was proposed to explain the experimental results in Figure 2.11(c) [62], where $n_{GB}$ is the number of grain boundaries along the channel of the OFETs, $L$ is the channel length, $F$ is the electric field and $F_0 = 3 \times 10^5$ $V/cm$. The calculation results of the expression are in good agreement with the experimental results in Figure 2.11(c).

We showed three different expressions (Eqs. [2.10–2.12]) for relating grain size and barrier height to the field-effect mobility in OFETs, and see that $E_b$ is one of the limiting factors of the field-effect mobility in polycrystalline OFETs. An electron spin resonance (ESR) study showed that the charge transport is limited by $E_b$ both in representative small-molecule and polymer-based polycrystalline OFETs [63]. Two types of motional narrowing in ESR spectra were observed in dinaphtho(2,3-b:2',3'-f)thieno(3,2-b)thiophene (DNTT) and poly(2,5-bis[3-hexadecylthiophene-2-yl] thieno[3,2-b]thiophene) (PB16TTT) OFETs. Both types of motional narrowing are due to charge carrier motion in grains and across grain boundaries, from which the activation energies of charge carrier transport in grain and across grain boundaries were estimated. The activation energies of charge carrier motion across grain boundaries coincide with those for field-effect mobilities of DNTT and PB16TTT OFETs. The result is an indication that the charge transport is limited by $E_b$. The thermally activated behavior of charge carrier transport in grains shows that the charge carrier transport in grains is due to a trap-controlled band transport, which is also known as MTM.

### 2.6.3 Performance of Field-Effect Transistors with Polycrystalline Organic Semiconductors

Grain boundaries in polycrystalline organic semiconductors can limit the charge carrier transport in OFETs as mentioned above. However, trapping at the interface between organic semiconductors and gate dielectrics [64] and the morphology (surface roughness) of gate dielectrics also impact the performance of OFETs [65]. When $SiO_2$ is used as gate dielectrics, $SiO_2$ surface is treated by self-assembled monolayers (SAM) such as hexamethyldisilazane (HMDS) and octadecyltrichlorosilane (PDTS) to remove silanol on the $SiO_2$ surface that act as trapping centers [66]. After the SAM treatment, $SiO_2$ surfaces are hydrophobic and the formation of organic semiconductors by printing processes for OFETs with bottom gate/top electrodes configuration is rather difficult.

In contrast to the fabrication of OFETs with bottom gate/top electrodes configuration, that of OFETs with top gate/bottom electrodes configuration is relatively easier by means of printing processes because organic semiconductors are formed by a printing process from the solutions onto wettable substrates. In addition, the surface of the printed organic semiconductor thin films is highly ordered because of the hydrophobic effect of air [67, 68].

Several solution-processable organic semiconductor materials exhibiting high field-effect mobility, including polymer [69, 70] and small-molecule semiconductors [71–74], have been synthesized. Especially, the small-molecule semiconductors containing benzothienobenzothiophene (BTBT) core units with alkyl side chains are prototypical materials because of high crystallinity and high chemical stability in ambient air. OFETs based on dioctyl BTBT (2,7-dioctyl[1]benzothieno[3,2-b][1]benzothiophene (C8-BTBT) have achieved high $\mu_{FE}$ exceeding 1 cm$^2$/Vs, even in polycrystalline thin films formed using the conventional spin coating technique [72]. The ordering and packing of C8-BTBT films can be significantly improved using edge cast [75, 76], solvent annealing [77], double-shot inkjet printing [78], and off-center spin coating [79]. These methods enable us to obtain single-crystalline C8-BTBT films that exhibit a high average $\mu_{FE}$ of greater than 10 cm$^2$/Vs.

Figure 2.12 shows an example for high OFET characteristics of polycrystalline OFETs with top gate/bottom electrodes configuration [80]. The OFET structure was schematically shown in Figure 2.14. Polycrystalline didodecyl BTBT (C12-BTBT) thin films were prepared from toluene – mesitylene binary solution of C12-BTBT and the maximum field-effect mobility of C12-BTBT is 10.6 cm$^2$/Vs. The uniformity of the field-effect mobility is rather good and is 7.8∓1.0 cm$^2$/Vs (62 OFETs). In addition, The OFET is electronically stable; the gate bias stress test showed that almost no changes in the transfer characteristics were observed after the application of the gate voltage of -30 V for $10^4$ s.

## 2.7 Single-Crystalline Organic Semiconductors

Single-crystalline organic semiconductors have been extensively studied. One of the reasons for this is that relatively high mobility can be obtained, which leads to the improvement of the cut-off frequency of OFETs. High-speed OFETs will be successfully applied to medium-scale integrated circuits. Another reason is that single-crystalline thin films for OFETs can be fabricated by means of printing processes.

C8-BTBT and C12-BTBT exhibiting excellent $\mu_{FE,}$ even in the polycrysatalline phase, are rod-like molecules with alkyl chains "symmetrically" linked to the BTBT core in Figure 2.12(a) [72]. However, the BTBT molecules have exhibited phase transitions from crystalline to liquid crystalline or other crystal phases at temperatures around 90–130 °C [72], leading to the degradation of transistor performance at elevated temperatures during device fabrication. It has been reported that asymmetrical substituted BTBT cores aggregate to form a highly crystalline liquid crystalline phase [74]. Rod-like molecules with different substituents such as alkyl and phenyl groups connected to the electronic skeleton in an "asymmetric" manner have recently been attracting attention in order to control heat resistance, solubility, and layered crystallinity of the asymmetric rod-like molecules (Figure 2.13a).

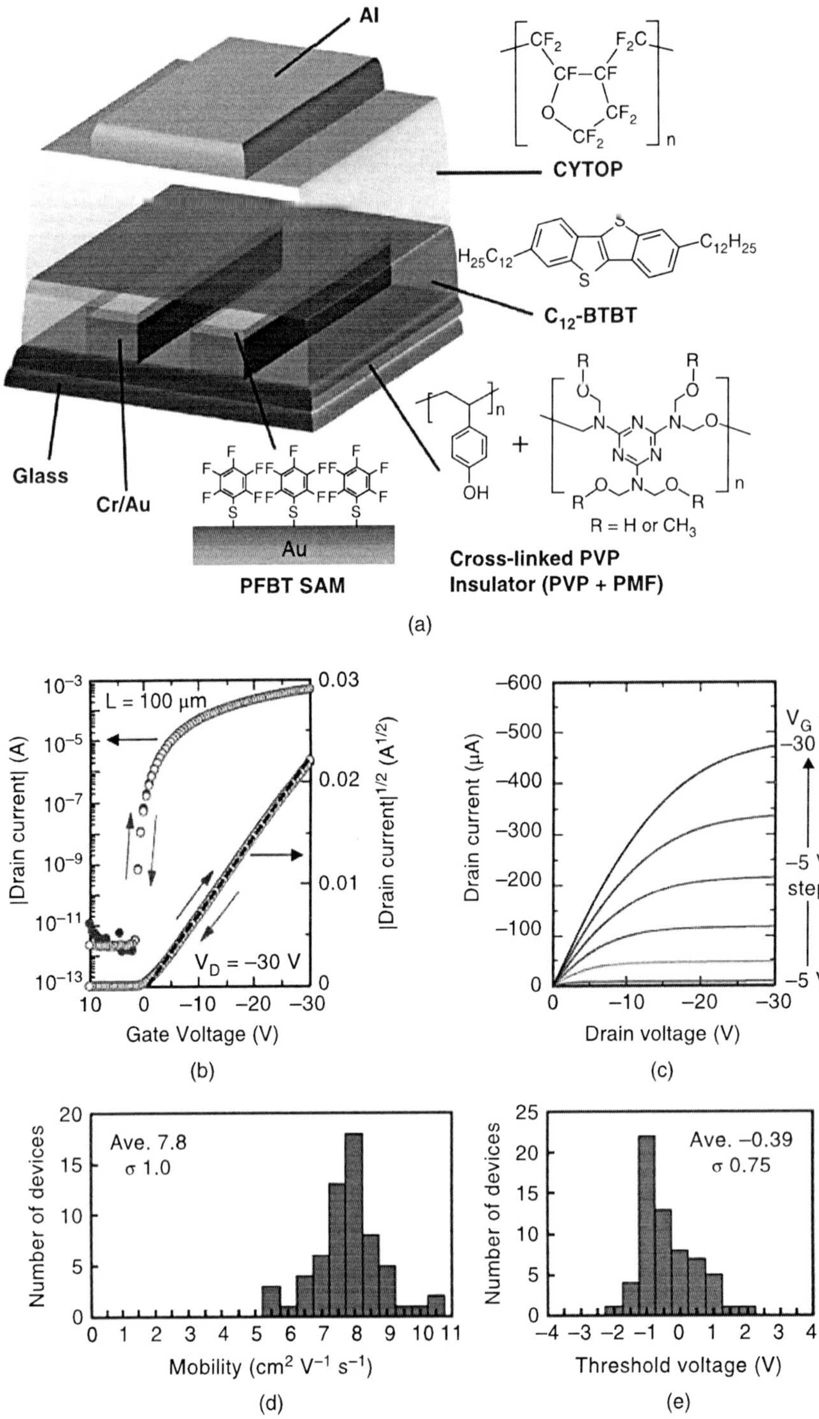


**Figure 2.12** Schematic of a top-gate/bottom-contact C12-BTBT FET, along with the chemical structures of C12-BTBT, cross-linked PVP, CYTOP, and PFBT SAM were deposited using spin coating technique (a), transfer (b) and output (c) characteristics, and histograms of field-effect mobility (d) and threshold voltage (e) of the C12-BTBT FETs (L=100 μm). C12-BTBT thin films were spin-coated using 0.5 wt% binary solvent of toluene:mesitylene (8:2). Source: Modified from Ref. [80].

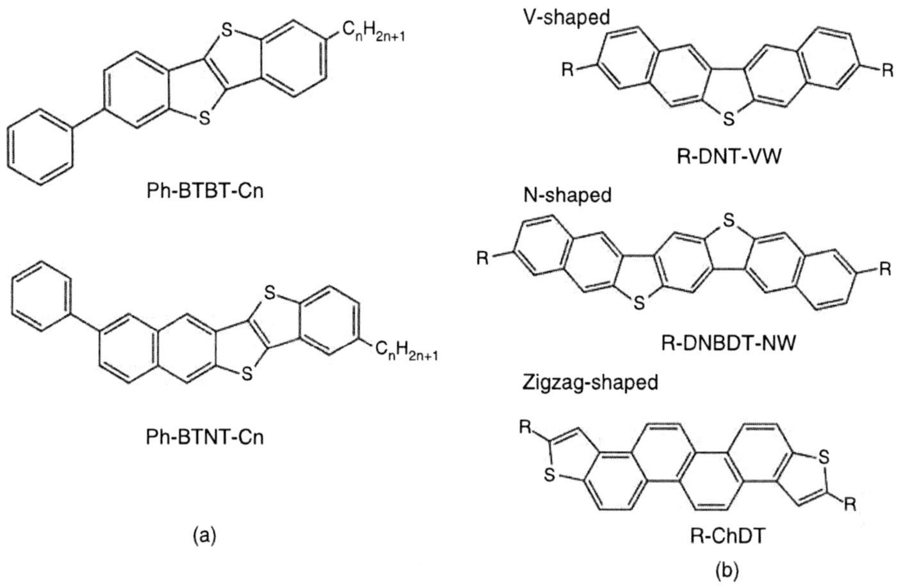


**Figure 2.13** Chemical structures of phenyl/alkyl-substituted benzothieno[3,2-b]benzothiophene (Ph-BTBT-Cn) and phenyl/alkyl-substituted benzothieno[3,2-b]naphtho[2,3-b]thiophene (Ph-BTNT-Cn) (a), and V-shaped (dinaphtho[2,3-b:2′,3′-d]-thiophene (DNT-V)), N-shaped (dinaphtho[2,3-d:2′,3′-d′]benzo-[1,2-b:4,5-b′]dithiophene (DNBDT-N)) and zigzag shaped molecules (chryseno[2,1-b:8,7-b′]dithiophene (ChDT)) (b). Source: Modified from Ref. [84].

Recently, it has been reported that the detrimental effects of molecular motions on the performance of organic semiconductors such as pentacene and BTBT derivatives [81–84]. In addition, it has been suggested that the dynamical disorder of organic semiconductors in the solid state might have a strong impact on the electronic transport [84]. Then, new organic semiconductors with V-shaped, N-shaped and zigzag-shaped molecular structures have been synthesized to reduce unfavorable molecular motions (Figure 2.13(b)). The organic semiconductors with those specific shapes also exhibit high thermal stability, high solubility, and high crystallinity [85].

### 2.7.1 Band Conduction in Single-Crystalline Organic Semiconductors

Band conduction is not always observed in single-crystalline organic semiconductors and the observation of band conduction is dependent on the bandwidths (transfer integrals). Band conduction takes place when the bandwidths are in the order of about 0.1–0.2 eV or more [86].

It was found that the $\mu_{FE}$ of rubrene single-crystal FETs is increased with decreasing temperature below room temperature, and hence the temperature dependence of the $\mu_{FE}$ is likely to be due to band conduction [87]. Indeed, the bandwidth of rubrene single crystals was found to be 0.4 eV along the well-stacked direction, studied by means of angle-resolved ultraviolet photoelectron spectroscopy [88]. Hall effect measurements in rubrene single-crystal OFETs have been carried out [89]. The reciprocal of the Hall coefficient $R_H$ is plotted against the gate voltage in Figure 2.14, and the charge per unit area $Q$ determined from the capacitance of the gate insulator and the gate voltage is also

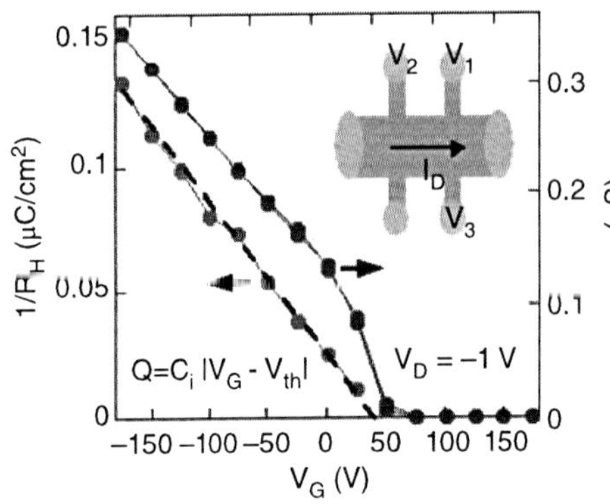


**Figure 2.14** Results of Hall measurements for a rubrene single-crystal transistor. The reciprocal of the Hall coefficient is in good agreement with the charge per unit area Q (broken line) determined from the capacitance of the gate insulator and the gate voltage. The four-terminal conductivity is also shown. Source: Ref. [89].

shown. The charge carrier density determined from $R_H$ is in good agreement with that from $Q$ and rises with increasing absolute value of the gate voltage. The result implies that holes injected by the field effect are not localized in the molecules but are widely distributed among the molecules. The mobility of rubrene was determined to be about 8 $cm^2/Vs$ from the dependence of the four-terminal conductivity on the gate voltage shown in Figure 2.14. The high degree of carrier delocalization means that the conduction in the OFET is band-like at room temperature. A result similar to Figure 2.14 has been observed in other high mobility crystalline semiconductors such as DNTT and C8BTBT [89].

In the case of band conduction, a standard approach for the prediction of the electronic transport properties [47] can be employed. The organic semiconductor molecules for high-performance OFETs as shown in Figure 2.13 form a lamella-like layered structure with an edge-on molecular orientation on the substrate. Figure 2.15 shows packing structures and transfer integrals for DNBDT-N and C10-DNBDT-NW [85]. DNBDT-N forms a 1D $\pi$–$\pi$ stacking structure, while C10-DNBDT-NW forms herringbone packing. The 2D cofacial molecular arrangement of C10-DNBDT-NW

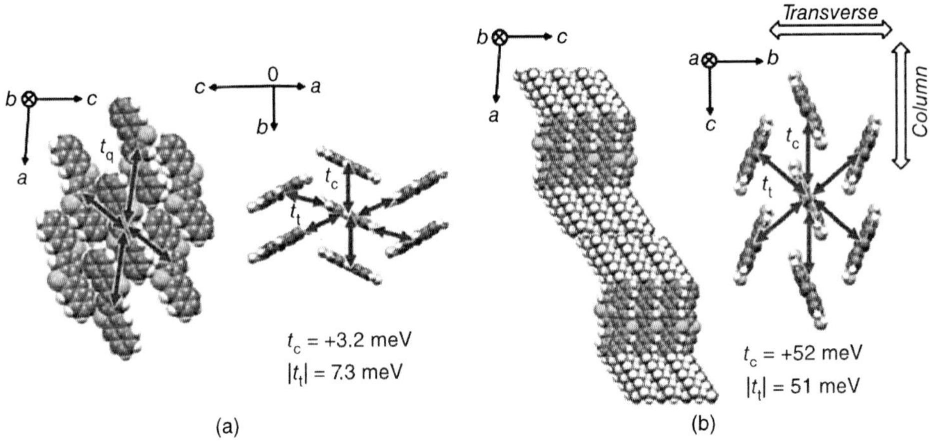


**Figure 2.15** Packing structures and transfer integrals for (a) dinaphtho[2,3-d:2′,3′-d′]benzo-[1,2-b:4,5-b′]dithiophene (DNBDT-N) and (b) decyl-substituted DNBDT-NW in Figure 2.13(b) (C10-DNBDT-NW). Source: Ref. [89].

through strong π–π and C–H···π interactions result in almost equal transfer integrals in the column and transverse directions ($t_c$ =+52 meV and $t_t$ =+51 meV), which suggest that C10-DNBDT-NW OFETs exhibit relatively high $\mu_{FE}$ and that the $\mu_{FE}$ is not strongly dependent on the direction of the crystal axis in Figure 2.15.

Mobility is expressed as

$$\mu = \frac{q\tau}{m^*},$$

obtained from the equation of the motion of charge carrier in crystalline solids, where $\tau$ is the average scattering time (average time between scattering events caused by defects in the solid and/or lattice vibration) and $m^*$ is the effective mass of the charge carrier. The effective mass of holes is calculated from the valence band structures, $E(k)$, of crystalline organic semiconductors and is given by

$$\frac{1}{m^*} = \frac{1}{\hbar^2}\frac{d^2E}{dk^2}.$$

where $\hbar$ is the Planck constant divided by 2π and $k$ is the wavevector. The effective mass of hole in the column and transverse directions of C10-DNBDT-NW is relatively small ($1.86m_0$ and $5.85m_0$) [85], where $m_0$ is the mass of free electrons. Such theoretical calculations of transfer integrals and effective masses are useful for predicting the charge transport properties of newly designed molecules.

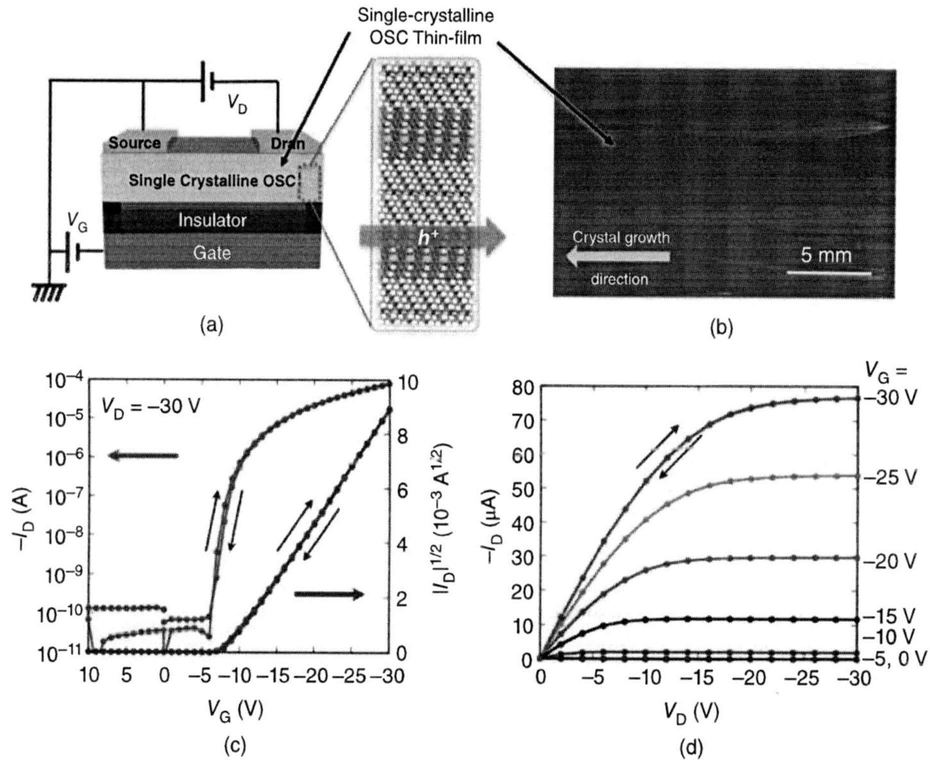


**Figure 2.16** Bottom-gate–top-contact OFET structure and packing structure of C10-DNBDT-NW on the substrate (a) and optical microscopy image (b). Transfer (c) and output (d) characteristics of the OFET based on a C10-DNBDT-NW single-crystalline thin film. Source: Ref. [89].

### 2.7.2 Performance of Field-Effect Transistors with Single Crystalline Organic Semiconductors

C10-DNBDT-NW formed excellent single-crystalline thin films using an edge-casting technique. Figure 2.16 shows the structure of C10-DNBDT-NW OFET, optical microscope image of the single-crystalline thin film, and the transfer and output characteristics of the OFETs [85]. We see uniform film formation and excellent OFET performance with the maximum $\mu_{FE}$ of 16 cm$^2$/Vs.

Single crystalline Ph-BTNT-Cn (Figure 2.13) thin films were formed on a perfluoropolymer, CYTOP, known as the most lyophobic insulating amorphous polymer using an extended meniscus-guided coating technique [90]. The single-crystalline thin films were formed from chlorobenzene solutions of Ph-BTNT-C10 and Ph-BTNT-C12, and the mixture is effective for the formation of uniform bilayer of the molecules on CYTOP [90]. Figures 2.17(a) and (b) show the OFET structure and the transfer curves of the OFETs, respectively. The OFETs exhibit excellent transfer characteristics (56 OFETs) and the distribution of the subthreshold swing is shown in Figure 2.17(c) (the average value of 67 mV/dec). The values are close to the theoretical limit (60 mV/dec at 300 K) mainly because of extremely low interface state density between CYTOP and Ph-BTNT-Cn. The maximum $\mu_{FE}$ is 5.5 cm$^2$/Vs. The distribution of $V_{on}$, defined as the voltage at $I_d = 10^{-12}$ A, is shown in Figure 2.17(d), and the values of $V_{on}$ are close to 0 V.

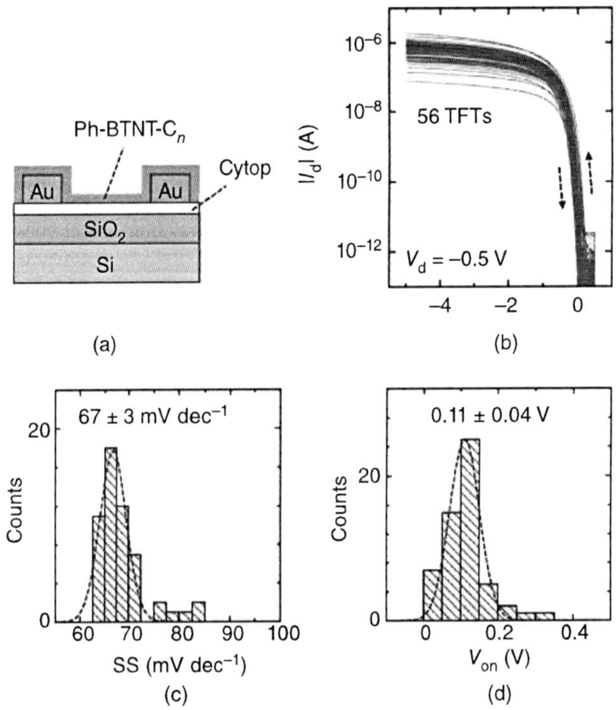


**Figure 2.17** Bottom-gate–top-contact OFET structure of printed Ph-BTNT-Cn single crystal (a). Transfer curves of 56 OFETs on the same substrate (b). Statistical distribution of the subthreshold swing (SS) (c) and Von (d) for 56 OFETs. Channel length of 100 μm, channel width of 800 μm, and gate capacitance of 23 nF/cm$^2$ 90. Source: Ref. [90].

The reproducibility of the OFET characteristics in Figure 2.17 is much better than that of the polycrystalline OFET characteristics in Figure 2.12. The low-cost printing processes for the fabrication of single-crystalline organic semiconductor thin films and the reproducible OFET characteristics as shown in Figure 2.17 demonstrate the potential usefulness of single crystalline OFETs for medium scale integrated circuits.

## 2.8 Concluding Remarks

The charge transport mechanisms of amorphous, polycrystalline and single-crystalline organic amorphous semiconductors were outlined. The basic concepts in this chapter might be useful in characterizing new organic semiconducting materials, in understanding new phenomena of electronic transport properties observed in organic semiconductors, and in studying device physics of key organic semiconducting devices, OLEDs, OPVs and OFETs.

## Acknowledgment

The author would like to thank the Japan Society for the Promotion of Science (a Grant-in-Aid for Scientific Research, (A) JP17H01265, (B) JP19H02599, (B) JP20H02716 and JP20K21007) for their financial support.

## References

1 Kao, K.C. (2004). *Dielectric Phenomena in Solids.* London: Elsevier.
2 Borsenberger, P.M. and Weiss, D.S. (1993). *Organic Photoreceptors for Imaging System.* New York: Mercel Dekker.
3 Yersin, H. (ed.) (2018). *Highly Efficient OLEDs: Materials Based on Thermally Activated Delayed Fluorescence.* Weinheim: Wiley-VCH.
4 Blom, P.W.M. (2020). *Adv. Mater. Technol.* 5: 2000144.
5 Bässler, H. (1993). *Phys. Status Solidi B* 175: 15.
6 Miller, A. and Abrahams, E. (1960). *Phys. Rev.* 120: 745.
7 Gill, W. (1972). *J. Appl. Phys.* 43: 5033.
8 Bässler, H. (1981). *Phys. Status Solidi B* 107: 9.
9 Gartstein, Yu.N. and Conwell, E.M. (1995). *Chem. Phys. Lett.* 245: 351.
10 Novikov, S.V., Dunlap, D.H., Kenkre, V.M. et al. (1998). *Phys. Rev. Lett.* 81: 4472.
11 Maennig, B., Pfeiffer, M., Nollau, A. et al. (2001). *Phys. Rev. B* 64: 195208.
12 Tanase, C., Meijer, E., Blom, P. and de Leeuw, D. (2003). *Phys. Rev. Lett.* 91: 216601.
13 Tanase, C., Blom, P. and de Leeuw, D. (2004). *Phys. Rev. B* 70: 193202.
14 Monroe, D. (1985). *Phys. Rev. Lett.* 54: 146.
15 Mort, J. and Pai, D.M. (eds.) (1976). *Photoconductivity and related phenomena.* Amsterdam: Elsevier.
16 Juška, G., Arlauskas, K., Viliūnas, M. et al. (2000). *Phys. Rev. B* 62: R16235.
17 Takagi, K., Abe, S., Nagase, T. et al. (2015). *J. Mater. Sci. Mater. Electron.* 26: 4463.

18 Schroder, D.K. (2006). *Semiconductor Material and Device Characterization, 3e*. New Jersey: Wiley Interscience.
19 Nagase, T., Kishimoto, K., and Naito, H. (1999). *J. Appl. Phys.* 86: 5026.
20 Kasap, S.O., Polischuk, B., and Dodds, D. (1990). *Rev. Sci. Inst.* 61: 2080.
21 Tsung, K.K. and So, S.K. (2008). *Appl. Phys. Lett.* 92: 103315.
22 Heun, S. and Borsenberger, P.M. (1995). *Chem. Phys.* 200: 245.
23 Tiedje, T. and Rose, A. (1980). *Solid State Commun.* 37: 49.
24 Naito, H. and Kanemitsu, Y. (1994). *Phys. Rev. B* 49: 10131.
25 Tiedje, T. (1984). Time-resolved charge transport in hydrogenated amorphous silicon. In: *The Physics of Hydrogenated Amorphous Silicon II Electronic and Vibrational Properties* (eds. J.D. Joannopoulos and G. Lucovsky), 261–300. Heidelberg: Springer-Verlag.
26 Hirao, A. and Nishizawa, H. (1996). *Phys. Rev. B* 54: 4755.
27 Slowik, J.H. and Chen, I. (1983). *J. Appl. Phys.* 54: 4467.
28 Ogawa, N. and Naito, H. (2002). *Electr. Eng. Jpn.* 140: 1.
29 Borsenberger, P.M., Gruenbaum, W.T., and Magin, E.H. (1996). *Physica B* 228: 226.
30 Borsenberger, P.M. and Bässler, H. (1991). *J. Chem. Phys.* 95: 5327.
31 Borsenberger, P.M. and Bässler, H. (1992). *Phys. Status Solidi B* 170: 291.
32 Borsenberger, P.M. and O'Regan, M.B. (1995). *Chem. Phys.* 200: 257.
33 Borsenberger, P.M., Gruenbaum, W.T., and Magin, E.H. (1996). *Physica B* 228: 233.
34 Shen, Y., Diest, K., Wong, M.H. et al. (2003). *Phys. Rev. B* 68 (081204).
35 Schein, L.B. and Tyutnev, A. (2008). *J. Phys. Chem. C* 112: 7295.
36 Kobayashi, T., Kinoshita, K., Nagase, T., and Naito, H. (2011). *Phys. Rev. B* 83 035305.
37 Bässler, H., Borsenberger, P.M., and Perry, R.J. (1994). *J. Polym. Sci. B Polym. Phys.* 32: 1677.
38 Fishchuk, I.I., Kadashchuk, A., Bässler, H., and Nešpůrek, S. (2003). *Phys. Rev. B* 67: 224303.
39 Baranovskii, S.D., Thomas, P., and Adriaenssens, G.J. (1995). *J. Non-Cryst. Solids* 190: 283.
40 Baranovskii, S.D., Faber, T., Hensel, F., and Thomas, P. (1997). *J. Phys. C* 9: 2699.
41 Oelerich, J.O., Huemmer, D., Weseloh, M., and Baranovskii, S.D. (2010). *Appl. Phys. Lett.* 97: 143302.
42 Baranovskii, S.D., Cordes, H., Hensel, F., and Leising, G. (2000). *Phys. Rev. B* 62: 7934.
43 Hartenstein, B., Bassler, H., Jakobs, A., and Kehr, K.W. (1996). *Phys. Rev. B* 54: 8574.
44 Baranovskii, S.D., Zhu, M., Faber, T. et al. (1997). *Phys. Rev. B* 55: 16226.
45 Lanyon, H.P.D. (1963). *Phys. Rev.* 130: 134.
46 Nagase, T. and Naito, H. (2000). *J. Appl. Phys.* 88: 252.
47 Yu, P.Y. and Cardona, M. (2010). *Fundamentals of Semiconductors: Physics and Materials Properties, 4e*. Berlin: Springer-Verlag.
48 Suzuki, F., Shizu, K., Kawaguchi, H. et al. (2015). *J. Mater. Chem. C* 3: 5549.
49 Uratani, H., Kubo, S., Shizu, K. et al. (2016). *Sci. Rep.* 6: 39128.
50 Asada, T. and Koseki, S. (2018). *Org. Electron.* 53: 141.
51 Levinson, J., Shepherd, F.R., Scanlon, P.J. et al. (1982). *J. Appl. Phys.* 53: 1193.
52 Fossum, J.G., Ortiz-Conde, A., Schichijo, H., and Banerjee, S.K. (1983). *IEEE Trans Electron. Dev.* 30: 933.

53 Kamins, T.I. and Marcoux, D.J. (1980). *IEEE Electron Dev. Lett.* 1: 159.
54 Angelis, C.T., Dimitriadis, C.A., Miyasaka, M. et al. (1999). *J. Appl. Phys.* 86: 4600.
55 Sera, K., Okumura, F., Uchida, H. et al. (1989). *IEEE Trans. Electron. Dev.* 36: 2868.
56 Yang, H., Shin, T.J., Ling, M.-M. et al. (2005). *J. Am. Chem. Soc.* 127: 11542.
57 Zhang, X.-H., Tiwari, S.P., Kim, S.-J., and Kippelen, B. (2009). *Appl. Phys. Lett.* 95: 223302.
58 Matsubara, R., Sakai, M., Kudo, K. et al. (2011). *Org. Electron.* 12: 195.
59 Horowitz, G. and Hajlaoui, M.E. (2001). *Synth. Met.* 122: 185.
60 Matsubara, R., Ohashi, N., Sakai, M. et al. (2008). *Appl. Phys. Lett.* 92 (242108).
61 Jurchescu, O.D., Popinciuc, M., van Weeks, B.J. and Palstra, T.T.M. (2007). *Adv. Mater.* 19: 688.
62 Di Carlo, A., Piacenza, F., Bolognesi, A. et al. (2005). *Appl. Phys. Lett.* 86 (263501).
63 Matsui, H., Kumaki, D., Takahashi, E. et al. (2012). *Phys. Rev. B* 85 (035308).
64 Veres, J., Ogier, S.D., Leeming, S.W. et al. (2003). *Adv. Func. Mat.* 13: 199.
65 Jung, Y., Kline, R.J., Fischer, D.A. et al. (2008). *Adv. Funct. Mater.* 18: 742.
66 Veres, J., Ogier, S., Lloyd, G., and de Leeuw, D. (2004). *Chem. Mater.* 16: 4543.
67 Kushida, T., Nagase, T., and Naito, H. (2011). *Appl. Phys. Lett.* 98 063304.
68 Takagi, K., Nagase, T., Kobayashi, T., and Naito, H. (2014). *Org. Electron.* 15: 372.
69 McCulloch, I., Heeney, M., Bailey, C. et al. (2006). *Nat. Mater.* 5: 328.
70 Li, J., Zhao, Y., Tan, H.S. et al. (2012). *Sci. Rep.* 2: 754.
71 Payne, M.M., Parkin, S.R., Anthony, J.E. et al. (2005). *J. Am. Chem. Soc.* 127: 4986.
72 Ebata, H., Izawa, T., Miyazaki, E. et al. (2007). *J. Am. Chem. Soc.* 129: 15732.
73 Okamoto, T., Mitsui, C., Yamagishi, M. et al. (2013). *Adv. Mater.* 25: 6392.
74 Iino, H., Usui, T., and Hanna, J. (2015). *Nat. Commun.* 6: 6828.
75 Uemura, T., Hirose, Y., Uno, M. et al. (2009). *Appl. Phys. Express* 2 (111501).
76 Nakayama, K., Hirose, Y., Soeda, J. et al. (2011). *Adv. Mater.* 23: 1626.
77 Liu, C., Minari, T., Lu, X. et al. (2011). *Adv. Mater.* 23: 523.
78 Minemawari, H., Yamada, T., Matsui, H. et al. (2011). *Nature* 475: 364.
79 Yuan, Y., Giri, G., Ayzner, A.L. et al. (2014). *Nat. Commun.* 5: 3005.
80 Sanda, S., Nagase, T., and Kobayashi, T. (2018). *Org. Electron.* 58: 306.
81 Fratini, S., Ciuchi, S., Mayou, D. et al. (2017). *Nat. Mater.* 16: 998.
82 Troisi, A., Orlandi, G., and Anthony, J.E. (2005). *Chem. Mater.* 17: 5024.
83 Ciuchi, S. and Fratini, S. (2012). *Phys. Rev. B* 86: 245201.
84 Illig, S., Eggeman, A.S., Troisi, A. et al. (2016). *Nat. Commun.* 7: 10736.
85 Okamoto, T., Yu, C.P., Mitsui, C. et al. (2020). *J. Am. Chem. Soc.* 142: 9083.
86 Brédas, J.L., Calbert, J.P., da Silva Filho, D.A., and Cornil, J. (2002). *PNAS* 99: 5804.
87 Podzorov, V., Menard, E., Borissov, A. et al. (2004). *Phys. Rev. Lett.* 93 (086602).
88 Machida, S., Nakayama, Y., Duhm, S. et al. (2010). *Phys. Rev. Lett.* 104 (156401).
89 Takeya, J. (2014). *J. Imaging Soc. Jpn.* 53: 492.
90 Kitahara, G., Inoue, S., Higashino, T. et al. (2020). *Sci. Adv.* In: *6: eabc8847.*

# 3

# Theory of Optical Properties of Organic Semiconductors

*Jai Singh, Monishka Rita Narayan and David Ompong*

*School of Engineering and Information Technology, B-purple 12, Faculty of EHSE, Charles Darwin University, Darwin, NT 0909, Australia*

**CHAPTER MENU**



## 3.1 Introduction

Organic semiconductors consist of organic molecules, which are usually hydrocarbons. A hydrocarbon molecule usually consists of hydrogen and carbon atoms bonding, sometimes with other atoms of oxygen and nitrogen, depending on the structures. Organic solids are formed by the weak Van der Waals forces, leading to frail bonding caused by the weak overlap of the electronic wave functions between neighbouring molecules. The result of this weak bonding is that the inter-molecular separation in organic solids is usually much greater than those in inorganic solids leading to far narrower electronic bands than in inorganic solids and hence the energies of the valence and conduction bands of solids can be well approximated by those of the highest occupied molecular orbital (HOMO) and lowest unoccupied molecular orbitals (LUMO) of individual molecules, respectively. In earlier studies on excitons in organic solids these were referred to as molecular excitons because both the excited electron and hole are excited on the same molecule to form Frenkel excitons [1, 2]. Although approximate, it is now fully accepted that HOMO and LUMO of individual molecules represent the valence and conduction bands and the corresponding energy gap represents the energy gap of an organic semiconductor. As described below, if not all, most optical, electronic and opto-electronic properties of organic semiconductors are studied using the HOMO and LUMO energies for the energy band energies.

There are two classes of organic semiconductors: 1) low molecular weight compounds and 2) polymers. The common feature of both groups is that the constituent molecules are planar and their carbon atoms are bonded through $sp^2$ hybridization. In such a bonding, the $\sigma$ bonds are formed between the planar s, $p_x$ and $p_y$ orbitals and these bonds are much stronger, contributing to the stability of the planar structure. The $p_z$ components of the p-bonds, known as the $\pi$-bonds, are loosely bonded and form delocalized conjugated bonds which contribute to the electronic and optical properties of molecules. As a result, the low energetic electron excitations of conjugated molecules occur through $\pi$-$\pi^*$ type transitions from the HOMO to the LUMO energy levels. As described above, the energy gap between the LUMO and HOMO is identified as the energy band gap in organic semiconductors, which is usually between 1.5 and 3 eV. Such transitions cause the absorption or emission in the visible region of solar radiation. The energy gap may be controlled by the degree of conjugation of the individual systems, which opens various possibilities for the modification of opto-electronic properties of organic semiconductors [3, 4].

Another common aspect in the electronic and opto-electronic properties of organic semiconductors is that, as soon as electronic excitation occurs from HOMO to LUMO energy levels, either through optical excitations or charge carrier injections, the excited electron and hole pairs in organic semiconductors immediately form Frenkel excitons. This is because the dielectric constant of most organic solids is about 3-4, which is far lower than that of inorganic semiconductors, e.g., it is 12 in Si. This means that the binding energy $E_B = \frac{\kappa e^2}{\varepsilon r}$ between the excited pair of electrons (e) and hole (h) is at least four times larger than that in Si. (Here, $k = (4\pi\varepsilon_o)^{-1} = 9\times10^9$, $e$ is electronic charge, $\varepsilon$ is the dielectric constant of the organic material and $r$ is the average separation between the excited e-h pair). This is the reason organic semiconductors are also known as excitonic semiconductors and this is the key difference between inorganic semiconductors like Si and Ge and organic semiconductors like poly(thiophene) (P3HT) and poly(fluorene-alt-pyridine) (PFPy).

In this chapter, a theoretical study of the optical properties of organic semiconductors is reviewed with a view to its application in devices such as organic solar cells (OSCs) and organic light-emitting devices (OLEDs). From the theoretical point of view, the operation of OSCs is opposite to that of OLEDs based on electroluminescence: one shines light on OSCs to generate current and one passes current in OLEDs to generate light. In practice, however, it is not the same device on which, if you first shine light will generate current and, if you pass current first it will generate light. The device structure of the organic layers in OSCs and OLEDs is very different from each other. To understand the operation of OSCs and OLEDs, it is very important to understand the opto-electronic properties, such as photon absorption, photoluminescence, electroluminescence and charge-carrier generation in organic semiconductors. In Section 3.2, the theory of optical absorption is presented and exciton generation, photoluminescence and electroluminescence are presented in Section 3.3, followed by the exciton dissociation in Section 3.4.

## 3.2 Photoexcitation and Formation of Excitons

Light absorption of energy larger than or equal to the band gap of an organic solid results in the production of excited electron ($e$) and hole ($h$) pairs. An electron is excited to the LUMO leaving a hole in the HOMO, which instantly form an exciton due to the strong Coulomb interaction between $e$ and $h$ caused by the low dielectric constant ($\varepsilon = 3$–$4$)

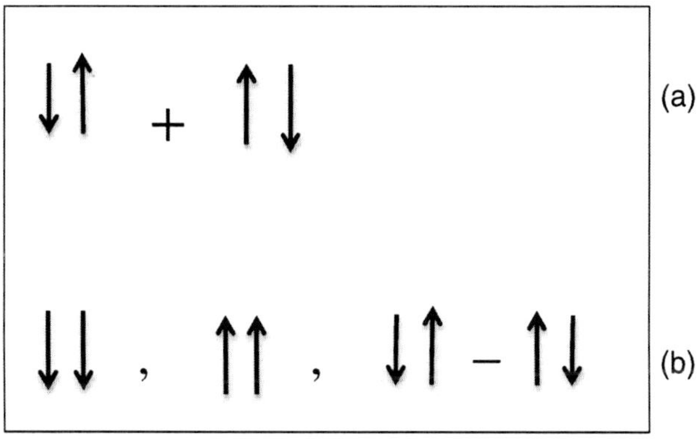


**Figure 3.1** Schematic diagrams of (a) one spin configuration for the singlet and (b) three spin configurations for the triplet excitons.

of organic solids. In organic solids, as stated above, the energy band gap is equal to the energy difference between the LUMO and HOMO energy levels. As the electronic intermolecular interaction is weak in organics, the formation of such excitons is usually of Frenkel type [5–8]. There are two possible spin configurations for the formation of excitons: 1) singlet excitons and 2) triplet excitons as shown in Figure 3.1.

Absorption of a photon in any semiconductor can excite an electron from the HOMO valence to the LUMO conduction bands leaving a hole in the valence band. Such an excited state will usually be a singlet because of the spin conservation. In organic semiconductors, a triplet state usually lies at a lower energy than a singlet state and hence it can be excited from the singlet-excited state through intersystem crossing [9]. The mechanism of intersystem will be described below. The triplet state can also be excited directly by photons of energy larger than or equal to the triplet mediated by the exciton-spin-orbit-photon interaction operator as described below. The mechanism of absorption of a photon and formation of a singlet exciton due to exciton-photon interaction is presented in Section 3.2. 1 and for a triplet exciton due to the exciton-spin-orbit-photon interaction in Section 3.2.2.1.

### 3.2.1 Photoexcitation of Singlet Excitons due to Exciton-photon Interaction

We consider an organic solid in which an electron is excited directly from HOMO to LUMO upon photon absorption without any change in its spin configuration. The interaction operator, $\hat{H}_{Is}$, between radiation and a pair of electron and hole is given by [10]:

$$H_{Is} = -\left(\frac{e}{m_e^*}\mathbf{p}_e - \frac{e}{m_h^*}\mathbf{p}_h\right) \cdot \mathbf{A} \tag{3.1}$$

where the subscript $I$ and $s$ denote interaction and singlet, $m_e^*$ and $m_h^*$ are the effective masses and $\mathbf{p}_e$ and $\mathbf{p}_h$ are the linear momenta of the excited electron and hole, respectively, $e$ is the electronic charge and $\mathbf{A}$ is the vector potential of photons given by:

$$\mathbf{A} = \sum_{\lambda} A_{0s}(\hat{\varepsilon}_\lambda c_\lambda^+ e^{-i\omega_\lambda t} + \varepsilon_\lambda c_\lambda e^{-i\omega_\lambda} t) \tag{3.2}$$

where $A_{os} = \left(\frac{\hbar}{2\varepsilon_0 n^2 V\omega_\lambda}\right)^{1/2}$ and $n$ is the refractive index, $V$ is the illuminated volume of the material, $\omega_\lambda$ is the photon frequency, $\varepsilon_o$ is the vacuum permittivity, $\hbar$ is the reduced Planck's constant and $\hat{\varepsilon}_\lambda$ is the unit polarization vector of photons. $c_\lambda^+(c_\lambda)$ is the creation (annihilation) operator of a photon in mode $\lambda$ . The first term of A represents the emission while the second term relates to absorption of a photon. In this chapter, only the absorption term is considered from here onwards.

Using the centre of mass coordinate, $\mathbf{R}_x = (m_e^*\mathbf{r}_e + m_h^*\mathbf{r}_h)/M$ where $M = m_e^* + m_h^*$ and relative coordinate, $\mathbf{r} = \mathbf{r}_e - \mathbf{r}_h$ , the absorption interaction operator, $\hat{H}_{as}$ for singlet, can be written as [11, 12]:

$$H_{as} = -\frac{e}{\mu_x}\sum_\lambda \left(\frac{\hbar}{2\varepsilon_0 n^2 V\omega_\lambda}\right)^{1/2} e^{i\omega_\lambda t}(\varepsilon_\lambda \cdot \mathbf{p})C_\lambda \tag{3.3}$$

where the subscript $a$ denotes absorption, $\mu_x$ is the reduced mass of exciton: $\mu_x^{-1} = m_e^{*-1} + m_h^{*-1}$ and $\mathbf{p} = -i\hbar\nabla_r$ is the relative momentum between $e$ and $h$ .

The field operator of an electron in the LUMO level $(L)$,$\hat{\Psi}_L(r_e)$ can be expressed as:

$$\hat{\Psi}_L(r_e) = N^{-1/2}\sum_\ell \phi_\ell^L(r_e)a_{L\ell} \tag{3.4}$$

where $N$ is the number of molecules in the sample, $\phi_\ell^L(r_e)$ is the wave function of an electron at the excited molecular site $\ell$ and $a_{L\ell}$ is the annihilation operator of an electron in LUMO of site $\ell$ .

Likewise, the field operator of a hole, $\hat{\Psi}_H(r_h)$, in the HOMO level $(H)$ , can be expressed as:

$$\hat{\Psi}_H(r_h) = N^{-1/2}\sum_m \phi_m^H(r_h)d_{Hm} \tag{3.5}$$

where $\phi_m^H(r_h)$ is the wave function of a hole at the excited molecular site $m$ and $d_{Hm}$ is the annihilation operator of a hole in HOMO of site $m$ .

Now using Eqs. (3.4)–(3.5), the interaction operator in Eq. (3.3) can be written in the second quantized form as:

$$H_{as} = -\frac{e}{\mu_x}\sum_\lambda \left(\frac{\hbar}{2\varepsilon_o n^2 V\omega_\lambda}\right)^{1/2} e^{i\omega_\lambda t}Z_{\ell m\lambda}B_{LH\ell m}^+ c_\lambda \tag{3.6}$$

where $B_{LH\ell m}^+ = \frac{1}{\sqrt{2}}[a_{L\ell}^+(+1/2)d_{Hm}^+(-1/2) + a_{L\ell}^+(-1/2)d_{Hm}^+(+1/2)]$ [13] is the creation operator of a singlet exciton by exciting an electron in LUMO at site $\ell$ and a hole in HOMO at site $mm$ and $Z_{lm\lambda} = \int \phi_\ell^L(r_e)\varepsilon_\lambda.\mathbf{p}\phi_m^H(r_h)d\mathbf{r}_e d\mathbf{r}_h$ .

For deriving the rate of absorption, we consider the initial state $|i\rangle$ consisting of a photon in mode $\lambda$ and the final state $|f\rangle$ has a singlet exciton. Such initial and final states can be respectively expressed as:

$$|i>= c_\lambda^+|0>|0_p> \tag{3.7}$$

And

$$|f>=|\frac{1}{\sqrt{2}}[a_L^+(+1/2)d_H^+(-1/2) + a_L^+(-1/2)d_H^+(+1/2)]|0>|0_p> \tag{3.8}$$

where |0> represents the vacuum states of the solid where HOMO is completely occupied and LUMO is empty and $|0_p>$ is the vacuum state of photons (no photons).

Using these, the transition matrix element for transition from the initial to the final state is obtained as [11, 12]

$$\langle f|\hat{H}_{as}|i\rangle = -\frac{e}{\mu_x}\sum_{\lambda}\left(\frac{\hbar}{2\varepsilon_o n^2 V\omega_\lambda}\right)^{1/2} e^{i\omega_\lambda t} p_{LH} \tag{3.9}$$

where

$$p_{LH} = N^{-1}\sum_{l}\sum_{M} z_{lm\lambda}\delta_{\ell,m} \tag{3.10}$$

and $\delta_{\ell,m}$ represents that the matrix element is non-zero only when $e$ and $h$ are created on the same site.

Now applying Fermi's golden rule, the rate of absorption, $R_a$, of a photon to excite an exciton can be written as:

$$R_a = \frac{2\pi}{\hbar}\sum_{\lambda}|<f|\hat{H}_a|i>|^2\delta(E_f - E_i) \tag{3.11}$$

where $E_i$ and $E_f$ are the initial and final state energies given by:

$$E_i = E_{HOMO} + \hbar\omega_\lambda \tag{3.12a}$$

where $\hbar\omega_\lambda$ is the energy of the absorbed photon in mode $\lambda$, and

$$E_f = E_{LUMO} \tag{3.12b}$$

where $E_{LUMO}$ is the LUMO energy and $E_{HOMO}$ is the HOMO energy.

Substituting Eqs. (3.9) and (3.12) into Eq. (3.11), we obtain the rate of formation of singlet exciton, $R_{as}$, by absorption of a photon as:

$$R_{as} = \frac{2\pi e^2}{\mu_x^2}\left(\frac{1}{2\varepsilon_o n^2 V\omega_\lambda}\right)\sum_{E_{LUMO}E_{HOMO}}|P_{LH}|^2\delta(E_{LUMO} - E_{HOMO} - \hbar\omega\lambda) \tag{3.13}$$

Applying the two-level approximation, the rate of singlet excitation in organic solids can be written as:

$$R_{as} = \frac{\pi e^2}{\varepsilon_o n^2\omega\mu_x^2}|P_{LH}|^2\delta(E_{LUMO} - E_{HOMO} - \hbar\omega_\lambda) \tag{3.14}$$

Now, using the dipole approximation, the transition matrix element, $p_{LH}$, is obtained as:

$$p_{LH} = i\omega\mu_x\langle\varepsilon_\lambda.\mathbf{r}\rangle \tag{3.15}$$

where $\mathbf{r}$ is the dipole length and $\langle\varepsilon_\lambda.\,\mathbf{r}\rangle$ denotes integration over the photon wave vector $\mathbf{k}$ for all photon modes $\lambda$, which can be evaluated as follows: Considering a wave vector $\mathbf{k}$ can be associated with every photon mode, one can write: $\sum_{\lambda} = \frac{2V}{(2\pi)^3}\int d^3\mathbf{k}$ with $k = \omega_\lambda/c$, where $k = (4\pi\varepsilon_o)^{-1} = 9\times10^9$ , $c$ is the speed of light and $k^2dk = \frac{\omega_\lambda^2}{c^3\hbar}d(\hbar\omega_\lambda)$, we get:

$$\sum_{\lambda} = \frac{2V}{(2\pi)^3}\int d^3\mathbf{k} = \frac{2V}{(2\pi)^3}\int_0^{E_{LUMO}-E_{HOMO}}\int_0^{\pi}\int_0^{2\pi}\frac{\omega_\lambda^2}{c^2\hbar}\sin\theta\sin^2\phi_\lambda d(\hbar\omega_\lambda)d\theta d\phi_\lambda \tag{3.16}$$

Replacing Eq. (3.16) into Eq. (3.14) and substituting $\varepsilon = n^2$, we get $R_{as}$ as:

$$R_{as} = \frac{4ke^2\sqrt{\varepsilon}|r_s|^2(E_{LUMO} - E_{HOMO})^3}{3\hbar^4 c^3} \tag{3.17}$$

### 3.2.2 Excitation of Triplet Excitons

The triplet excitons can be excited in two ways in organic semiconductors; directly through strong exciton-spin-orbit-photon interactions and indirectly through the intersystem crossing as described below.

#### 3.2.2.1 Direct Excitation to Triplet States Through Exciton-Spin-Orbit-Photon Interaction

By absorption of a photon of adequate energy, an electron can be excited to a triplet spin configuration through the electron-spin-orbit-photon interaction, $\hat{H}_{It}$, given by [14]:

$$\hat{H}_{It} = -\frac{eg}{2m_e^2c^2}\mathbf{s}.\left(\mathbf{p} \times \sum_{n=1}^{N} \mathbf{E}_n\right) \tag{3.18}$$

where the subscript $t$ denotes triplet, $g = 2$ is the gyromagnetic ratio, $\mathbf{E}_n = -\nabla V$ is the electric field of the electron generated by the $n^{th}$ nucleus in the molecule, where $\nabla V_n$ is the gradient of the scalar nuclear potential and **s** and **p** are the spin angular and orbital momenta of the electron, respectively. Once light interacts with the sample, the interaction operator in Eq. (3.18) modifies to:

$$\hat{H}_{It} = -\frac{eg}{2m_e^2c^2}\mathbf{s}.\left(\mathbf{p} + \frac{e}{c}\mathbf{A}\right) \times \left(-\frac{1}{c}\frac{\partial \mathbf{A}}{\partial \mathbf{t}} - \sum_{n=1}^{N} \nabla V_n\right) + \frac{eg}{m_e c}\mathbf{s}.\mathbf{H} \tag{3.19}$$

where $\mathbf{H} = \nabla \times \mathbf{A}$ is the magnetic field of the electromagnetic radiation and $\nabla V_n = -(Z_n ek/r_{en}^3)\mathbf{r}_{en}$ where $\mathbf{r}_{en}$ is the position vector of the electron from the nucleus and $|\mathbf{r}_{en}| = \mathrm{r}_{en}$. For $Z_n >> 1$, the interaction between the excited electron and other valence electrons in the atom is considered to be negligible. **A** is the vector potential of photons suitable for the exciton-spin-orbit-photon interaction through the triplet excitation given by:

$$\mathbf{A} = \sum_{\lambda} A_{ot}(\hat{\varepsilon}_\lambda c_\lambda^+ e^{-i\omega_\lambda t} + \varepsilon_\lambda c_\lambda e^{i\omega_\lambda t} \tag{3.20}$$

where $A_{ot} = [2\pi c^2\hbar/\varepsilon_0\varepsilon\omega_\lambda V]^{1/2}$. Like in Eq. (3.2), here the first term of A corresponds to the emission while the second term is the absorption of a photon.

The operator in Eq. (3.19) can be further simplified by taking into account that within the dipole approximation we get $\nabla \times \mathbf{A} = \mathbf{0}$, which eliminates a few terms as shown below:

$$\frac{e}{c^2}\mathbf{s}.\left(\mathbf{A} \times \frac{\partial \mathbf{A}}{\partial t}\right) = 0, \tag{3.21a}$$

$$\frac{1}{c}\mathbf{s}.\mathbf{p} \times \frac{\partial \mathbf{A}}{\partial \mathbf{t}} = -\frac{i\hbar}{c}\mathbf{s}.\nabla \times \frac{\partial \mathbf{A}}{\partial \mathbf{t}} = -\frac{i\hbar}{c}\mathbf{s}.\frac{\partial}{\partial t}(\nabla \times \mathbf{A}) = 0 \tag{3.21b}$$

Substituting Eq. (3.21) into Eq. (3.19), we get the interaction operator as:

$$\hat{H}_{It} = -\frac{eg}{2m_e^2c^2}\left(-\sum_{n=1}^{N} \frac{Z_n eks.\mathbf{L}_n}{r_{en}^3} - \frac{e}{c}\mathbf{s}.\left(\mathbf{A} \times \sum_{n=1}^{N} \nabla V_n\right)\right) \tag{3.22}$$

where $\mathbf{L}_n = \mathbf{r}_{en} \times \mathbf{p}$ is the orbital angular momentum of electron. The first term of Eq. (3.22) is the well-known stationary spin-orbit interaction operator [14] and is obtained in the absence of radiation. Its inclusion in the Hamiltonian as a perturbation can only split the degeneracy of a triplet state. As this term is a stationary operator, it cannot cause any transition. Only the last term, which depends on spin, radiation and time can be considered as the time-dependent perturbation operator and hence can cause transition.

Substituting the absorption term for **A** from Eq. (3.20) into Eq. (3.20), the time-dependent interaction operator for triplet absorption, $\hat{H}_{at}$, becomes:

$$\hat{H}_{at} = -\frac{e^3 gk}{2m_e^2 c^2} \sum_{\lambda,n} \frac{Z_n}{r_{en}^2} \left( \frac{2\pi\hbar}{\varepsilon_0 \varepsilon \omega_\lambda V} \right)^{1/2} e^{i\omega_\lambda t} \mathbf{s}.(\varepsilon_\lambda \times \mathbf{r}_{en}) c_\lambda \tag{3.23}$$

where $\mathbf{r}_{en} = \mathbf{r}_{en}/r_{en}$ is a unit vector. For evaluating the triple scalar product of three vectors, without the loss of any generality, we may assume that vectors $\varepsilon_\lambda$ and $\mathrm{r}_{en}$ are in the xy-plane at an angle $\phi_\lambda$, then we get $\varepsilon_\lambda \times \mathbf{r}_{en} = \sin\phi_{\lambda n}\hat{\eta}$, $\hat{\eta}$ being a unit vector perpendicular to the xy-plane. This gives $\mathbf{s}.(\varepsilon_\lambda \times \mathbf{r}_{en}) = \mathbf{s}.\hat{\eta} \sin\phi_{\lambda n} = s_z \sin\phi_{\lambda n}$, which simplifies Eq. (3.23) to:

$$\hat{H}_{at} = -\frac{e^3 gk}{2m_e^2 c^2} \sum_{\lambda,n} \frac{Z_n}{r_{en}^2} \left( \frac{2\pi\hbar}{\varepsilon_0 \varepsilon \omega_\lambda V} \right)^{1/2} e^{i\omega_\lambda t} \sin\phi_{\lambda n} s_z c_\lambda \tag{3.24}$$

The electron spin-orbit-photon interaction operator in Eq. (3.24) represents interaction for an electron with spin-orbit. For exciting an electron and hole pair, this interaction can be written as [15]:

$$\hat{H}_{at} = -\frac{e^3 gk}{2\mu_x^2 \varepsilon c^2} \sum_{\lambda,n} \left[ \frac{Z_n}{r_{en}^2} \left( \frac{2\pi\hbar}{\varepsilon_0 \varepsilon \omega_\lambda V} \right)^{1/2} \sin\phi_{\lambda en} s_{ez} + \sum_{\lambda,n} \frac{Z_n}{r_{hn}^2} \left( \frac{2\pi\hbar}{\varepsilon_0 \varepsilon \omega_\lambda V} \right)^{1/2} \sin\phi_{\lambda hn} s_{hz} \right] e^{i\omega_\lambda t} c_\lambda \tag{3.25}$$

where $r_{en}$ and $r_{hn}$ are the electron and hole distances from their nuclear site $n$ and $s_{ez}$ and $s_{hz}$ are the spin projections along the z-axis of the electron and hole, respectively. It may be noted that the interaction operator in Eq. (3.25) remains the same for an electron-hole pair and an exciton, excited in a triplet state.

Using the field operators in Eqs. (3.4) and (3.5) in Eq. (3.25), the time-dependent operator of spin-orbit-exciton-photon interaction is obtained in second quantization as:

$$\hat{H}_{at} \approx -\frac{e^3 gZk}{\mu_x^2 c^2 \varepsilon r_t^2} \left( \frac{2\pi\hbar}{\varepsilon_0 \varepsilon V} \right)^{1/2} \sum_{\lambda} \sum_{\sigma_e,\sigma_h} \frac{\sin\phi_\lambda}{\sqrt{\omega_\lambda}} e^{i\omega_\lambda t} \times (s_{ez} + s_{hz}) a_L^{+}(\sigma_e) d_H^{+}(\sigma_h) \delta_{\sigma_e,\sigma_h} c_\lambda \tag{3.26}$$

where $r_t$ is the average separation between electron and hole in a triplet exciton and is approximated by $< \mathrm{HOMO}|\mathrm{r}_{en}^{-2}|\mathrm{LUMO} > \approx < \mathrm{HOMO}|\mathrm{r}_{\mathrm{hn}}^{-2}|\mathrm{LUMO} > \approx (r_t/2)^{-2}$.

It may be noted that the sum over sites $n$ does not appear in Eq. (3.26). This is because the interaction operator depends on the atomic number, $Z_n$, and the inverse square of the distance between an electron and nucleus and a hole and nucleus. Therefore, only the contribution of the nearest heaviest atom is significant while contributions from the other atomic sites become negligible. Hence, the summation over $n$ may be removed as an approximation.

For simplifying the sum over spins in Eq. (3.26), one needs to consider the three spin configurations of a triplet exciton state which are expressed as [13]:

$$a_L^+(+1/2)d_H^+(+1/2) = a_L^+(+1/2)a_H(-1/2) \tag{3.27a}$$

$$\begin{aligned}&\frac{1}{\sqrt{2}}[a_L^+(+1/2)d_H^+(-1/2) - a_L^+(-1/2)d_H^+(+1/2)]\\ &\quad = \frac{1}{\sqrt{2}}[a_L^+(+1/2)a_H(+1/2) - a_L^+(-1/2)a_H(-1/2)]\end{aligned} \tag{3.27b}$$

$$a_L^+(-1/2)d_H^+(-1/2) = a_L^+(-1/2)a_H(+1/2) \tag{3.27c}$$

Using the property of $s_{ez}$ and $s_{hz}$ operators as $s_{ez}a_L^+\left(\pm\frac{1}{2}\right) = \pm\frac{1}{2}\hbar a_L^+\left(\pm\frac{1}{2}\right)$ and $s_{hz}d_H^+\left(\pm\frac{1}{2}\right) = \mp\frac{1}{2}\hbar d_H^+\left(\pm\frac{1}{2}\right)$, we find that only the contribution from Eq. (3.27b) is non-zero. This reduces the operator in Eq. (3.26) to the following term:

$$\begin{aligned}\hat{H}_{at} \approx &-\frac{2\hbar e^3 gZk}{\mu_x^2 c^2 \varepsilon r_t^2}\left(\frac{2\pi\hbar}{\varepsilon_0 \varepsilon V}\right)^{1/2} \sum_\lambda \frac{\sin\phi_\lambda}{\sqrt{\omega_\lambda}} e^{i\omega_\lambda t}\\ &\times \left[\frac{1}{\sqrt{2}}(a_L^+(+1/2)d_H^+(-1/2) + a_L^+(-1/2)d_H^+(+1/2))\right] c_\lambda\end{aligned} \tag{3.28}$$

It may be noted in Eq. (3.28) that the operator $s_z$ has flipped the spin from the triplet configuration to singlet configuration. Thus, in such case, the exciton-spin-photon interaction plays two roles:

a. Enhances the rate of triplet excitation due to incorporation of heavy metal atoms and
b. Flips the spin from triplet to singlet configuration which facilitates dissociation. This mechanism is schematically presented in Figure 3.2.

We now consider a transition from an initial state $|i>$ given in Eq. (3.7) to the final state $|f>$ in Eq. (3.8), whose spin has been flipped by the spin-orbit interaction. Then the transition matrix element is obtained as:

$$<f|\hat{H}_{at}|i> = -\frac{2\hbar e^3 gZk}{\mu_x^2 c^2 \varepsilon r_t^2}\left(\frac{2\pi\hbar}{\varepsilon_0 \varepsilon V}\right)^{1/2} \frac{\sin\phi_\lambda}{\sqrt{\omega_\lambda}} e^{i\omega_\lambda t} \tag{3.29}$$

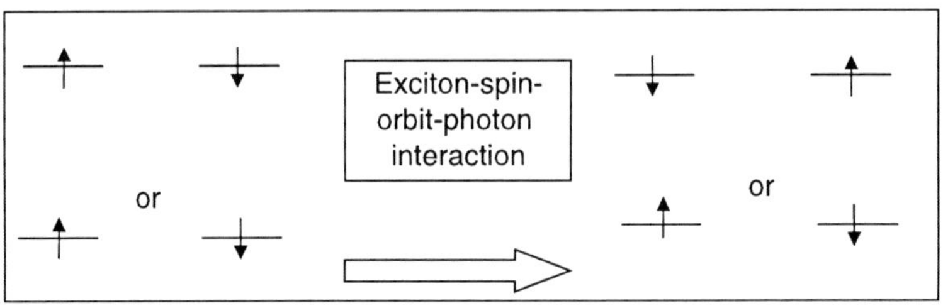


**Figure 3.2** Schematic diagram of the effect of exciton-spin-orbit-photon interaction in directly exciting a triplet exciton state by absorption of a photon.

Applying Fermi's golden rule as in Eqs. (3.11) and (3.12) and simplifying the sum over $\lambda$ as carried out in Eq. (3.16) with $g = 2$, $R_{at}$ is obtained as:

$$R_{at} = \frac{32e^6 Z^2 k^2 (E_{LUMO} - E_{HOMO})}{\mu_x^4 c^7 \varepsilon_0 \varepsilon^3 r_t^4} \tag{3.30}$$

It is desirable to express the rates of absorption of singlet and triplet excitons derived in Eqs. (3.17) and (3.30), respectively, in terms of their excitonic Bohr radii, which are known parameters in many organic materials. Using $r_s = a_{xs}/\varepsilon$ and $r_t = a_{xt}/\varepsilon$, where $a_{xs}$ and $a_{xt}$ are the excitonic Bohr radii of singlet and triplet excitons, respectively, and substituting these in Eqs. (3.17) and (3.30), we get the rate of a photon absorption for a singlet excitation due to exciton-photon interaction and triplet excitation due to exciton-spin-orbit-photon interaction, respectively, as:

$$R_{as} = \frac{4ke^2 (E_{LUMO} - E_{HOMO})^3 a_{xs}^2}{3c^3 \varepsilon^{1.5} \hbar^4} \tag{3.31}$$

$$R_{at} = \frac{32e^6 Z^2 k^2 \varepsilon (E_{LUMO} - E_{HOMO})}{c^7 \varepsilon_o \mu_x^4 a_{xt}^4} \tag{3.32}$$

where, $a_{xs} = \frac{\alpha^2}{(\alpha-1)^2} a_{xt}$, $a_{xt} = \frac{\mu\varepsilon}{\mu_x} a_o$ where $\alpha$ is a material-dependent constant representing the ratio of the magnitude of the Coulomb and exchange interactions between the electron and hole in an exciton, $a_o$ is the Bohr radius and $\mu$ is the reduced mass of electron in a hydrogen atom [13].

The rates of singlet and triplet absorption derived in Eqs. (3.31) and (3.32) are sensitive to the energy gap, $(E_{LUMO} - E_{HOMO})$., of the material. Hence, for various donor organic materials used in OSCs, the rates of formation of singlet and triplet excitons are calculated as a function of $(E_{LUMO} - E_{HOMO})$ obtained from the experimental studies and presented in Table 3.1. The inverse of these rates gives the time of formation of the corresponding exciton, ($\tau_I = R_{ai}^{-1} i = s$ or $t$).

According to Eqs. (3.31) and (3.32), $R_{as}$ does not depend on $Z$ but $R_{at}$ depends on $Z^2$. Thus, in the case of triplet excitation, the higher the $Z$ value, the faster will be $R_{at}$. Normally, organic materials consist of hydrocarbons, where carbon, C, has the highest atomic number. In experimental studies [13, 16–18], doping with the heavy metal atoms such as Ir in the donor organic materials is observed to enhance the photon-electron conversion efficiency of OSCs. Therefore, we have calculated the rates and corresponding formation times of singlet and triplet excitons with Z=6 for C and Z=77 for Ir, for several organic materials used as donors in OSCs and the results are presented in Table 3.1.

For all organic materials used as donors and listed in Table 3.1, the rate of excitation of singlet excitons is five orders of magnitude greater than that of triplet excitons. This is because $R_{as}$ is highly sensitive to the absorption energy, $(E_{LUMO} - E_{HOMO})^3$ while $R_{at}$ is only linearly dependent on $(E_{LUMO} - E_{HOMO})$. According to Table 3.1, TFB has the highest absorption energy of 3.59 eV and provides the fastest rate of absorption of $6.36 \times 10^{10} S^{-1}$ for singlet excitons and $5.28 \times 10^5 S^{-1}$ for triplet excitons as compared to other organic materials. Therefore, utilizing wide-band gap materials can be expected to greatly enhance the singlet and triplet absorption rates in OSCs.

The rates are also sensitive to the singlet and triplet excitonic Bohr radii as shown in Eqs. (3.31) and (3.32). In the case of a singlet exciton, the rate is proportional to $a_{xs}^2$ while

**Table 3.1** Rates of absorption of a photon to form a singlet exciton due to exciton-photon interaction and triplet exciton due to exciton-photon-spin-orbit interaction calculated using Eqs. (3.31) and (3.32), respectively, as a function of energy gap of various donor organic materials.

| Organic Material | $E_{LUMO}$ $-E_{HOMO}$ [eV] | $R_{as}$ $10^{10}$ [$s^{-1}$] | $\tau_{as}$ $10^{-11}$ [s] | $R_{at}$ (C) $10^3$ [$s^{-1}$] | $\tau_{at}$ (C) $10^{-4}$ [s] | $R_{at}$ (Ir) $10^5$ [$s^{-1}$] | $\tau_{at}$ (Ir) $10^{-6}$ [s] | Ref. |
|---|---|---|---|---|---|---|---|---|
| PCBM | 2.40 | 1.90 | 5.26 | 2.14 | 4.67 | 3.53 | 2.83 | [18] |
| PFPy | 2.94 | 3.49 | 2.86 | 2.62 | 3.81 | 4.32 | 2.31 | |
| P3HT | 2.10 | 1.27 | 7.85 | 1.87 | 5.34 | 3.09 | 3.24 | [7] |
| α-NPD | 3.10 | 4.10 | 2.44 | 2.77 | 3.61 | 4.56 | 2.19 | [20] |
| P3OT | 1.83 | 0.84 | 11.9 | 1.63 | 6.12 | 2.69 | 3.72 | [21] |
| Pt(OEP) | 1.91 | 0.96 | 10.4 | 1.70 | 5.87 | 2.81 | 3.56 | [22] |
| MEV-PPV | 2.17 | 1.41 | 7.12 | 1.94 | 5.16 | 3.19 | 3.13 | [23] |
| PPV | 2.80 | 3.02 | 3.31 | 2.50 | 4.00 | 4.12 | 2.43 | [24] |
| MDMO-PPV | 2.20 | 1.46 | 6.83 | 1.96 | 5.09 | 3.23 | 3.09 | [25] |
| PEOPT | 1.75 | 0.74 | 13.6 | 1.56 | 6.40 | 2.57 | 3.89 | [26] |
| PTPTB | 1.70 | 0.68 | 14.8 | 1.52 | 6.59 | 2.50 | 4.00 | |
| BBL | 1.90 | 0.94 | 10.6 | 1.70 | 5.90 | 2.79 | 3.58 | [27] |
| F8BT | 1.80 | 0.80 | 12.5 | 1.61 | 6.22 | 2.65 | 3.78 | |
| PFB | 2.81 | 3.05 | 3.28 | 2.51 | 3.99 | 4.13 | 2.42 | |
| TFB | 3.59 | 6.36 | 1.57 | 3.20 | 3.12 | 5.28 | 1.89 | |

where,

PCBM = [6,6]-phenyl-$C_{61}$-butyric acid methyl ester.
PFPy = poly(fluorene-alt-pyridine)
P3HT = poly(thiophene)
α-NPD = N,N′-diphenyl-N,N′-bis(1-naphthyl)-1-1′biphenyl-4,4″diamine
P3OT = poly(3-octylthiophene-2,5-diyl)
Pt(OEP) = platinum octaethyl porphyrin
MEV-PPV= poly(2-methoxy-5-(2′-ethylhexyloxy)-1,4-phenylenevinylene)
MDMO-PPV = poly[2-methoxy-5-(3′,7′-dimethyloctyloxy)-1,4-phenylenevinylene]
PEOPT = poly(3-(4′-(1″,4″,7″-trioxaoctyl)phenyl)thiophene)
PTPTB = poly-(N -dodecyl-2,5-bis(2′-thienyl)pyrrole-(2,1,3-benzothiadiazole))
BBL = poly(benzimidazobenzophenanthroline ladder)
F8BT = poly(9,9-dioctylfluorene-cobenzothiadiazole)
PFB = poly(9,9-dioctylfluorene-co-bis-N,N-(4 -butylphenyl)-bis-N,N-phenyl-1,4-phenylenediamine)
TFB = poly(9,9-dioctylfluoreneco-N-(4-butylphenyl) diphenylamine)

for a triplet exciton, it is proportional to $a_{xt}^{-4}$. Hence, the larger the exciton Bohr radius of a singlet exciton, the faster will be the rate of formation of singlet excitons by photon absorption. However, in the case of triplet excitation, the smaller exciton Bohr radius gives faster triplet absorption rate. In addition, the dielectric constant of the material also plays a key role in the absorption rates. Organic solids with low $\varepsilon$ favours singlet excitation while a larger value leads to faster triplet excitation as obtained in Eqs. (3.31) and (3.32), respectively.

According to Table 3.2, the rate of triplet excitation by incorporating Ir atom (Z=77) increases by two orders of magnitude in all donors than without it. This clearly shows

**Table 3.2** The calculated intersystem crossing rate ($k_{isc}$) from Eq. (3.49) and experimental rates ($k_{isc}^{exp}$), for some OSC materials along with their highest atomic number ($Z$) and singlet-triplet energy difference ($\Delta E$). For these calculations we have used $\varepsilon = 3$, $a_x = 4.352\,nm$, and $\omega_v = 8 \times 10^{14}\,s^{-1}$.

| Organic material | Z | $\Delta E(eV)$ | $k_{isc}(s^{-1})$ | $k_{isc}^{exp}(s^{-1})$ | Ref. |
|---|---|---|---|---|---|
| NPD (Ir doped) | 77 | 0.90 | $1.1\text{x}10^{11}$ | | [33] |
| CBP (Ir doped) | 77 | 0.90 | $1.1\text{x}10^{11}$ | | [33] |
| P3HT | 16 | 0.80 | $3.7\text{x}10^{9}$ | | [34] |
| SubPc | 9 | 0.71 | $9.2\text{x}10^{8}$ | $9.1\text{x}10^{8}$ | [35] |
| F8BT | 16 | 0.70 | $2.8\text{x}10^{9}$ | $1.2\text{x}10^{7}$ | [36] |
| Toluene | 6 | 0.70 | $4.0\text{x}10^{8}$ | $8.5\text{x}10^{6}$ | [37] |
| Naphthalene | 6 | 1.47 | $1.8\text{x}10^{9}$ | $5.0\text{x}10^{6}$ | [38] |
| 1-Bromonaphthalene | 35 | 1.30 | $4.7\text{x}10^{10}$ | $\approx\text{x}10^{9}$ | [39] |
| Benzophenon | 16 | 0.30 | $5.2\text{x}10^{8}$ | $\approx\text{x}10^{10}$ | [37] |
| Platinum-acetylide | 78 | 0.80 | $8.8\text{x}10^{10}$ | $> \text{x}\ 10^{11}$ | [40] |

where;

NPD=N,N′-bis (naphthalen-1-yl)-N, N′-bis(phenyl)-benzidine
CBP= 4,4′-bis(9-carbazolyl)-1,1′-biphenyl
P3HT = poly(3-hexylthiophene)
SubPc= boron subphthalocyanine chloride
F8BT = poly(9,9-dioctylfluorene-cobenzothiadiazole)

that the incorporation of heavy metal atom enhances the rate of triplet excitation in OSCs due to enhanced exciton-spin-orbit-photon interaction. As described above, this exciton-spin-orbit-photon interaction also flips the spin to singlet configuration to facilitate the absorption by spin conservation. At the same time, flipped spins to singlet configuration make it easier to dissociate the exciton into free-charge carriers, which is essential for the operation of a solar cell. Assuming that faster dissociation means faster collection of the charge carriers at the opposite electrodes, this can lead to higher power conversion efficiencies. Such an enhancement in the photon-electron conversion efficiency of OSCs incorporated with Ir complexes is in agreement with the experimental results [20, 29, 30].

#### 3.2.2.2 Indirect Excitation of Triplet Excitons Through Intersystem Crossing and Exciton-Spin-Orbit-Phonon Interaction

In this case, an exciton is already excited in the singlet exciton state in an organic molecule of an organic solid and the molecule also has a triplet state at a lower energy by $\Delta E$ such that the molecular vibrational energies of singlet and triplet overlap as shown in Figure 3.3.

If the singlet exciton is excited ata higher molecular virational energy state of a molecule overlapping with a molecular vibrational energy state of the triplet state, then singlet can get transferred to the triplet, however, the spin configuration will not flip to a triplet configuration. The flipping of the spin can only occur through the exciton-spin-orbit interaction. Accordingly, the mechanism of intersystem crossing can occur through the combination of excitons, molecular vibrations and spin-orbit

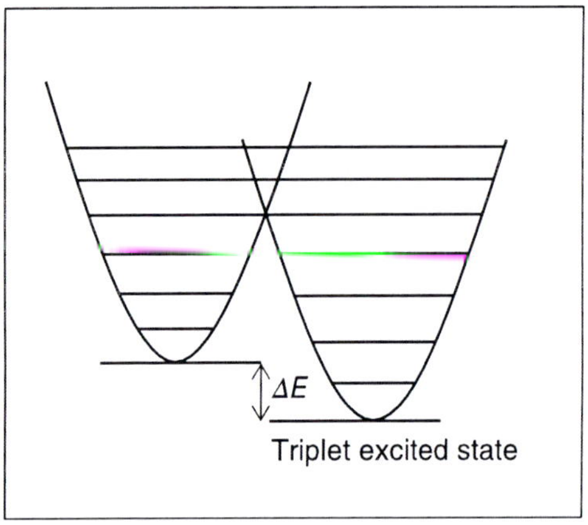


**Figure 3.3** Schematic illustration of a singlet excited state whose molecular vibrational energy states overlap with those of a triplet excited state which is at a lower electronic energy by $\Delta E$.

interactions. It is interesting to note that although the intersystem crossing has been known for very long, until recently, no such interaction operator is known to exist in the literature. We have derived a new exciton-spin-orbit-phonon interaction operator suitable for ISC in organic solids [9] as described below.

The stationary part of the spin-orbit interaction for an exciton in a molecule consisting of N atoms can be written as [9, 14]:

$$H_{so} = -\frac{e^2 gk}{2\mu_x^2 c^2} \sum_n^N \left( \frac{z_n}{r_{en}^3} \mathbf{s}_e \bullet \mathbf{l}_{en} + \frac{z_n}{r_{hn}^3} \mathbf{s}_h \bullet \mathbf{l}_{hn} \right) \tag{3.33}$$

where $e$ is the electron charge, $g = 2$ is the gyromagnetic ratio, $k = 1/4\pi\varepsilon_o$ is the Coulomb constant, $\mu_x^{-1} = m_e^{-1} + m_h^{-1}$ is the reduced mass of exciton, $c$ is the speed of light, $\mathbf{s}_e(\mathbf{s}_h)$ is the electron (hole) spin, $\mathbf{l}_{en} = \mathbf{r}_{en} \times \mathbf{p}_e$ is electron angular momentum and $\mathbf{r}_{en}(\mathbf{p}_e)$ is the position vector (orbital momentum) of the electron from the nth nucleus. Similarly, $\mathbf{l}_{hn} = \mathbf{r}_{hn} \times \mathbf{p}_h$ is the hole angular momentum and $\mathbf{r}_{hn}(\mathbf{p}_h)$ is the position vector (orbital momentum) of the hole from the nth nucleus. For a non-rigid structure, Eq. (3.31), can be expanded in Taylor series about the equilibrium positions of molecules. Terminating the expansion at the first order, we get:

$$H_{so} = H_{so}^0 + H_{sov} \tag{3.34}$$

where $H_{so}^0$ is the zeroth order term and represents the interaction in a rigid structure and $H_{sov}$ is the first order term which gives the interaction between exciton-spin-orbit interactions and molecular vibrations and it is obtained as:

$$H_{sov} = -\frac{3e^2 gkz}{2\mu_x^2 c^2} \sum_{n,v} \left( \frac{\mathbf{s}_e \bullet \mathbf{l}_e}{r_e^{\,4}} R_{nv} + \frac{\mathbf{s}_h \bullet \mathbf{l}_h}{r_h^{\,4}} R_{nv} \right) \tag{3.35}$$

where $R_{nv}$ is the molecular displacement from the equilibrium position due to the intramolecular vibrations. In Eq. (3.35), the subscript $n$ on $r_{en}$ and $r_{hn}$ is dropped using the approximation that the quantity within parentheses in Eq. (3.35) depends on $r_{en}^{-4}$ and $r_{hn}^{-4}$ thus the nearest nucleus to the electron and hole is expected to play the dominant influence and, as such, the presence of other nuclei may be neglected as an approximation. This approximation helps in reducing the summation to only one nucleus for each electron and hole and hence the subscript $n$ is dropped. In carrying out the Taylor series expansion, it is further assumed that the distances $r_{en}$ and $r_{hn}$ of the electron and hole with reference to the individual nuclei of a molecule can be replaced by their distances $r_e$ and $r_h$, respectively, with reference to the equilibrium position of the individual molecules. This approximation may be regarded to be quite justified within the Born-Oppenheimer approximation regime.

In the second quantization, $R_{nv}$, can be expressed as [31]:

$$R_{nv} = (q_{vo} - q_{oo})(b_{nv}^{+} + b_{nv}) \tag{3.36}$$

where $b_{nv}^{+}(b_{nv})$ is the vibrational creation (annihilation) operator in vibrational mode $v$.

For expressing the operator in Eq. (3.35) in the second quantization, we can write the field operator for an electron in the LUMO and that of a hole in the HOMO, respectively, as:

$$\hat{\psi}_e = \sum_{\sigma_e} \varphi_{LUMO}\, a_L(\sigma_e)$$

$$\hat{\psi}_h = \sum_{\sigma_h} \varphi_{HOMO}\, d_H(\sigma_h),$$

$$d_H(\sigma_h) = a_H^{+}(-\sigma_h) \tag{3.37}$$

where $\varphi_{LUMO}$ and $\varphi_{HOMO}$ are the wavefunctions of the electron in the LUMO and hole in the HOMO, respectively. It may be clarified here that we are dealing with molecules, hence, the valence and conduction bands wavefunctions are those of HOMO and LUMO, respectively. Using Eq. (3.37) the interaction operator in Eq. (3.35) can be expressed in the second quantization as:

$$\hat{H}_I \approx -\frac{24e^2gkz}{\mu_x^2c^2r_x^4} \sum_{v,\sigma_e,\sigma_h} (q_{vo} - q_{oo}) a_L(\sigma_e) d_H(\sigma_h) \delta_{\sigma_e,\sigma_h},$$

$$\times (\mathbf{s}_e \bullet \mathbf{l}_e + \mathbf{s}_h \bullet \mathbf{l}_h)(b_{nv}^{+} + b_{nv}) \tag{3.38a}$$

where $r_x$ is the average separation between the electron and hole in the exciton and it is approximated as:

$$\langle \phi_{HOMO} || r_{en}^{-4} || \phi_{LUMO} \rangle \approx \langle \phi_{HOMO} || r_{hn}^{-4} || \phi_{LUMO} \rangle \approx (r_x/2)^{-4} \tag{3.38b}$$

To evaluate the spin and orbital angular momentum operators in the interaction operator Eq. (3.35), we can use [32]:

$$s \bullet l = \frac{J^2 - l^2 - s^2}{2} \tag{3.39}$$

where $J$ is the total angular momentum. Eq. (3.39) can be rearranged as:

$$J^2 = l^2 + s^2 + 2(s_x l_x + s_y l_y + s_z l_z) \tag{3.40}$$

Defining $s_i$ and $l_i$ ($i = x, y$) in Eq. (3.40), in terms of orbital angular momentum raising (lowering) operators as: $L_+ = l_x + il_y$ ($L_- = l_x - il_y$) and spin angular momentum raising (lowering) operator as $S_+ = s_x + is_y$ ($S_- = s_x - is_y$), respectively, we get:

$$J^2 = l^2 + s^2 + 2\left(\frac{S_+L_+ + S_-L_+ + S_+L_- + S_-L_-}{4}\right)$$

$$-2\left(\frac{S_+L_+ - S_-L_+ - S_+L_- + S_-L_-}{4}\right) + 2s_z l_z \tag{3.41}$$

Using Eq. (3.41) in Eq. (3.39) we obtain:

$$s \bullet l = s_z l_z + \frac{S_- L_+}{2} + \frac{S_+ L_-}{2} \tag{3.42}$$

It is this term in the interaction operator in Eq. (3.38) that flips the spin of the exciton from the singlet to triplet configuration. Using Eq. (3.42) in Eq. (3.38a) we get:

$$\hat{H}_I \approx \frac{12\hbar^2 gkze^2}{\mu_x^2 c^2 r_x^4} \sum_{\nu,\sigma_e,\sigma_h} (q_{vo} - q_{oo}) \times a_L(\sigma_e) d_H(\sigma_h) \delta_{\sigma_e,\sigma_h} (b^{\dagger}_{n\nu} + b_{n\nu}) \tag{3.43}$$

here, we assume $l_{ez} \approx l_{hz} = l_z = \hbar$, which is the angular momentum associated with the first excited state with the magnetic quantum number =1.

It may be of interest to note that both the interaction operators derived in Eq. (3.35) and (3.43) are time-independent which means these are a part of any organic solids and these will cause an intersystem crossing to occur whenever vibrational overlapping appears between singlet and triplet excited states. Contrary to this, the interaction operator in Eq. (3.28) is a time-dependent operator which will only become active when a light is shone on the material causing direct transitions to triplet states.

The rate of transitions through the intersystem crossing can now be calculated as follows. Assuming that the initial state $|i\rangle$ consists of an exciton in the singlet spin configuration and molecular vibrations and the final state $|f\rangle$ consists of a triplet exciton and molecular vibrations. Using the occupation number representations, the initial state can be written as:

$$|i\rangle = \frac{N^{-3/2}}{\sqrt{2}} \sum_{n_1,m_1,\nu_1} \sum_{\sigma_e,\sigma_h} [a^{+}_{n_1 L_1}(+\sigma_e) d^{+}_{m_1 H_1}(-\sigma_h) + a^{+}_{n_1 L_1}(-\sigma_e) d^{+}_{m_1 H_1}(+\sigma_h)] b^{+}_{n_1 \nu_1} |0\rangle |\nu_1\rangle \tag{3.44}$$

where the electron is created in the LUMO at site $n_1$ and a hole in the HOMO at site $m_1$, $|0\rangle$ represents the electronic vacuum state and $|\nu_1\rangle$ is the initial molecular vibrational occupation state. Likewise, the final state can be expressed as:

$$|f\rangle = \frac{N^{-3/2}}{\sqrt{2}} \sum_{n_2,m_2,\nu_2} \sum_{\sigma^{\prime}_e,\sigma^{\prime}_h} [a^{+}_{n_2 L_2}(+\sigma^{\prime}_e) d^{+}_{m_2 H_2}(-\sigma^{\prime}_h) - a^{+}_{n_2 L_2}(-\sigma^{\prime}_e) d^{+}_{m_2 H_2}(+\sigma^{\prime}_h)] |0\rangle |\nu_2\rangle \tag{3.45}$$

where the electron is located in the LUMO at site $n_2$ and a hole in the HOMO at site $m_2$, and $|\nu_2\rangle$ is the final molecular vibrational occupation state after the spin flip into the triplet excited state.

Using the usual anticommutation relations for fermions and commutation relations for boson operators, the transition matrix element is obtained from Eqs. (3.43)–(3.45) as:

$$\langle f|\hat{H}_I|i\rangle = -\frac{12gkze^2\hbar^2 n_\nu}{\mu_x^2 c^2 r_x^4}(q_{vo} - q_{oo}) \tag{3.46}$$

where $n_\nu$ is the effective number of vibrational levels taking part in the transition process.

Using Fermi's golden rule then the rate of intersystem crossing $k_{isc}$ can be written as:

$$K_{isc} = \frac{2\pi}{\hbar} \sum_f |\langle f||\hat{H}_I||i\rangle|^2 \delta(E_f - E_i) \tag{3.47}$$

Here, $E_f = E_T + n_2\hbar\omega_v$ is the final triplet state energy and $E_i = E_S + n_1\hbar\omega_v$ is the initial singlet state energy including the energy of corresponding vibrational energies. $E_S$ and $E_T$ are the singlet and triplet exciton energies, respectively.

Substituting Eq. (3.46) into Eq. (3.47), we get $k_{isc}$ as:

$$K_{isc} = \sum_v \left( \frac{12g\varepsilon^4 kze^2\hbar^2 n_v}{\mu_x^2 c^2 a_x^4}(q_{vo} - q_{oo}) \right)^2 \delta(-\Delta E + n_v\hbar\omega_v) \tag{3.48}$$

where $r_x = \frac{a_x}{\varepsilon}$ is used to express the rate in terms of the excitonic Bohr radius $a_x$ and $\varepsilon$ is the dielectric constant. Expressing $E_f - E_i = -\Delta E + n_v\hbar\omega_v$, where $\Delta E = E_S - E_T$, $n_v = n_2 - n_1$ and the square of the molecular displacement due to excitation as: $(q_{vo} - q_{oo})^2 = \frac{8\pi r_x^2}{3}$ [32], $k_{isc}$ is obtained as:

$$K_{isc} = \frac{3072\pi^2\varepsilon^6 k^2 Z^2 e^4 \hbar^3}{\mu_x^4 c^4 a_x^6 (\hbar\omega_v)^3}(\Delta E)^2 \quad \text{s}^{-1}. \tag{3.49}$$

The rate derived in Eq. (3.49) depends on excitonic Bohr radius, molecular vibrational energy, the atomic number $Z$ of the heaviest atom and $\nabla E$, the exchange energy between the singlet and triplet states. The derivation of the rate in Eq. (3.49) also clarifies how the phenomenon of ISC occurs. An exciton is first excited to the singlet exciton state which is higher in energy than the triplet state. The higher energy of the singlet excited state provides the required exciton-spin-orbit – molecular vibration interaction energy to flip the spin to triplet state before the transfer can take place. This is the reason that $k_{isc}$ in Eq. (3.49) vanishes if $\Delta E = 0$. It is unlikely that the singlet and triplet states in any molecule can become isoenergetic and if that is the case then the singlet absorption will dominate.

According to Table 3.2, the calculated rates are found to be in reasonable agreement with experimental results and minor discrepancies may be attributed to the approximations used in deriving Eq. (3.49). The rate in Eq. (3.49) can be applied to calculate the ISC rate in any molecular solids.

Comparing the rates of absorption derived in Eqs. (3.31), (3.32) and (3.49) as listed in Tables 3.1 and 3.2, we find that in most organic solids the rate of singlet absorption $R_{as}$ is of the order of $10^{11}$ s$^{-1}$ and formation time $\tau_{as} \geq 10$ ps, and the rate of the direct triplet absorption is $R_{at} \approx 10^3$ s$^{-1}$ and $\tau_{at} \approx 10^{-4}$ s without the involvement of any heavy metal atoms but with the incorporation of Ir as the heavy metal atom we get $R_{at} \approx 10^5$ s$^{-1}$ and $\tau_{at} \approx 1$ μs. In comparison, the rate of intersystem crossing $K_{isc} \geq 10^8$ s$^{-1}$ ($\tau_{isc} \approx 1$ ns) without the metal atom and $K_{isc} \geq 10^{11}$ s$^{-1}$ ($\tau_{isc} \approx 1$ ps) with Ir atoms involved. It is interesting to note that the rate of intersystem crossing to form triplet excitons is as fast as the rate of formation of singlet excitons. If the organic solids do not have any heavy metal compounds, then the rate of formation of triplet excitons via direct transitions is three orders of magnitude lower than that via the intersystem crossing. However, it may be noted that there is some energy loss in exciting triplet excitons through the intersystem crossing which is lost as thermal vibrational energy may heat a device.

## 3.3 Exciton up Conversion

In Section 3.2, we described the theory of intersystem crossing where an exciton in singlet state converts into a triplet state through the exciton-spin-orbit interaction which flips the spin and vibrational energy overlap that moves exciton from singlet to triplet state. The reverse process of converting a triplet exciton into a singlet exciton is also possible and is desirable in the operation of organic light emitting devices (OLEDs), where electrons and holes are injected from the opposite electrodes. As shown in Figure 3.1, statistically, there is only one possible spin configuration for forming singlet excitons and three spin configurations for forming triplet excitons. However, the radiative recombination of triplet excitons is not spin allowed unless one has strong spin-orbit interaction, which is usually the case in organic solids. In such a situation, only singlet excitons emit light and that limits the internal quantum efficiency to 25% because triplets cannot recombine and hence emit radiation. If the triplets could easily be converted into singlets then the internal quantum efficiency of 100% can be achieved. This is the principle of operation of OLEDs based on thermally activated delayed fluorescence (TADF) [41]. In organic solids, it is relatively easy to substitute another molecule (a donor) whose singlet energy may be close to the triplet state of the host solid (acceptor) and also to introduce heavy metal atoms to enhance the exciton-spin-orbit interaction. If the energy difference is very small, the up conversion can occur even at room temperature. The thermal energy will be adequate to raise the vibrational energy of the triplet exciton to overlap with the singlet state of the donor to enable the transfer through the exciton-spin-orbit interaction as schematically shown in Figure 3.4.

The rate of up conversion to the same energy should be the same as derived in Eq. (3.49) for the intersystem crossing but for thermally activated up conversion, to a singlet exciton state at energy $\Delta E$ higher than that of the triplet exciton state (see Figure 3.4) one needs to multiply the rate of intersystem crossing in Eq. (3.49) by the Boltzmann factor of activation energy $\Delta E$, then the thermally activated rate of up conversion $K_{uc}$ becomes [41b]:

$$K_{uc} = K_{isc} exp(-\Delta E/\kappa T) \tag{3.50}$$

where $\kappa = 8.6 \times 10^{-5}$ eV/K is the Boltzmann constant and $T$ is temperature in Kelvin. The rate of thermally activated up conversion increases as $(\Delta E)^2$ and it becomes zero as $\Delta E = 0$, which does not look realistic. Why should the rate of up conversion become zero if the singlet and triplet exciton states are iso-energetics, i.e., $\Delta E = 0$? Actually, as in the case of intersystem crossing, if $\Delta E = 0$, then no up conversion occurs only the singlet exciton state will be excited directly.

An example of thermally activated up conversion material is 2-biphenyl-4,6-bis (12-phenylindolo [2,3-a]carbazole-11-yl)-1,3,5-triazine (PIC-TRZ) containing an indolocarbazole donor unit and a triazine acceptor unit [41]. Following the up conversion

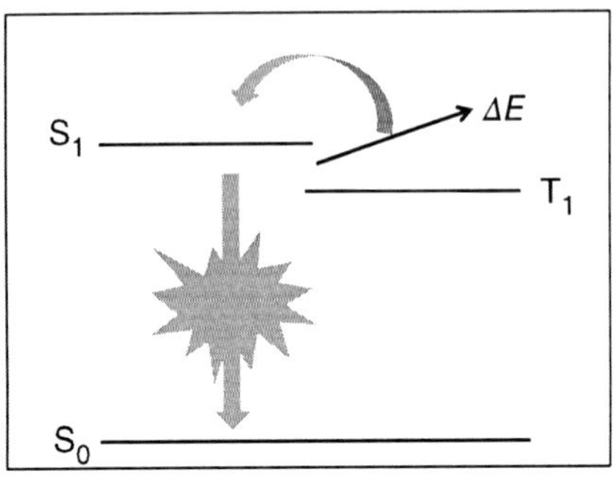


**Figure 3.4** Schematic illustration of thermally activated up conversion from a triplet exciton to singlet exciton state higher in energy by $\Delta E$.

to a singlet state, fluorescence occurs due to the radiative recombination of thus up converted singlet excitons, known as the thermally activated delayed fluorescence (TADF). Using TADF in an OLED, the generated triplet excitons, which cannot contribute to electroluminescence due to unfavourable spin configurations, can be made to recombine radiatively and thus the internal quantum efficiency can be enhanced through the up conversion. The synthesized molecule of PIC-TRZ exhibits a very small $\Delta E$ providing both efficient up conversion from T1 to S1 levels and intense fluorescence that leads high electroluminescence (EL) efficiency.

The invention of OLEDs, based on TADF, has become a very active field of research in organic electronics.

## 3.4 Exciton Dissociation

As described in Section 2, photo-excitations in organic semiconductors form excitons instantly and excitons are the excited pairs of $e$ and $h$ bound in hydrogen like electronic states. These are not free excited pairs of charge carriers and therefore the formation of excitons in devices like (OSCs) is not good for their operation because one needs to dissociate excitons into free $e$ and $h$ excited pairs which can be transported to the opposite electrodes of OSCs to produce photocurrents. On the contrary, formation of excitons is good for OLEDs because it enables radiative recombination and emission of light. In this section, the dissociation of excitons in OSCs is described in detail.

In the development of OSCs, first a single layer of organic semiconductor sandwiched between anode and cathode was prepared. In such a structure, there is no other force than the electrical force available due to the difference in work functions of the two electrodes to dissociate the excitons excited within the organic layer. The built-in electric field due to the difference in the work functions of the electrodes is given by:

$$F = \frac{\phi_e - \phi_a}{r} \tag{3.51}$$

where $\phi_e$ and $\phi_a$ are the work functions of cathode and anode, respectively, and $r$ is the separation between them, which is about the total thickness of an OSC. However, as the exciton is an electrically neutral entity, the built-in-electric field is expected to have little contribution in the exciton dissociation. This built-in electric field can only contribute in charge separation and collection after the excitons have dissociated. As a result, the single layered OSCs have very poor power conversion efficiency (PCE). Then, in 1986, the bilayer concept was invented by Tang [42] with one layer of a donor material, which can easily give away its electrons, and the second layer of an acceptor material, which can easily accept electrons. The two layers have an interface between them and are sandwiched between two electrodes. Many combinations of donor and acceptor materials have been invented and tried for OSCs. The important point in selecting the donor and acceptor materials is that the LUMO and HOMO of a donor molecule is at higher energies than those of an acceptor molecule as shown in Figure 3.5.

It is generally accepted that the generation of photo charge carriers in a bilayer OSC occurs through the following five processes in sequence [31, 43]: 1) photon absorption from the sun in the donor and/or acceptor excites electron-hole pairs which instantly form neutral Frenkel excitons, 2) diffusion of the excited excitons to the donor-acceptor

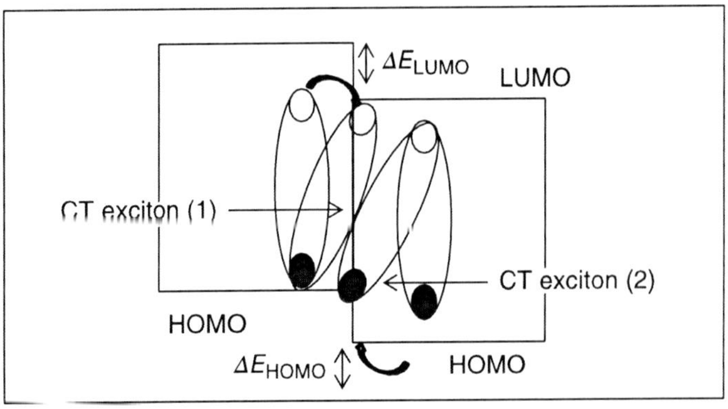


**Figure 3.5** Schematic illustration of the formation of CT excitons at the D-A interface in a BHJ OSC. CT exciton (1) is formed when Frenkel exciton is excited in the donor and CT exciton (2) is formed when Frenkel exciton is excited in the acceptor.

(D–A) interface, 3) formation of charge transfer (CT) excitons at the D–A interface by transferring the electron to the acceptor, from excitons excited in the donor and/or acceptor by transferring the hole to the donor from excitons excited in the acceptor [44], 4) dissociation of the CT excitons at the D–A interface and 5) transport and collection of the dissociated free charge carriers at their respective electrodes to generate photocurrent, which is the main purpose of any solar cell. In a bilayer structure, excitons are required to diffuse to the D–A interface to form CT excitons leading to the subsequent dissociation. How exactly the dissociation of CT excitons takes place, will be discussed below but it may be noted that an exciton, being electrically neutral, cannot be directed to move in any particular direction by any external or built-in electric field. Therefore, exciton can only diffuse from one point to another in a random motion and when it reaches the D–A interface, it will form CT exciton. This means that the exciton diffusion length $L_D$ must be larger than the thickness of the donor or acceptor layers. In organic semiconductors, $L_D$ is short ~10 nm and hence the bilayer structured OSCs also have very poor PCE. One way forward is to make the active organic layer from a blend of the donor and acceptor materials and the structure thus obtained is called bulk heterojunction (BHJ) OSC. The BHJ OSCs are one of the most promising alternative photovoltaic technologies because of the advantages of high absorption coefficient, light weight, flexibility, and the potential of low-cost solution process capability, etc. Also, in a BHJ OSC, the photogeneration of charge carriers occurs through the above five processes but excitons do not have to diffuse to a D–A interface at a fixed distance. The BHJ OSCs have reached PCE nearly 10% [45–48] with an expectation of achieving 15% in the very near future [47].

The processes 1–3 and 5, listed above for the operation of BHJ OSCs, have been extensively investigated and understood. The heterojunction structure is the only possible way of allowing excitons to diffuse and change into CT excitons at a nearby D–A interface. However, the process (4) of the dissociation of CT excitons at the D–A interface assisted only by the built-in electric field can be very inefficient [49], as discussed below hence the mechanism of dissociation has not been fully understood. Therefore, for an efficient dissociation of CT excitons into free electron and hole pairs, one has to consider other possibilities.

As both donor and acceptor materials are organic semiconductors with similar dielectric constants, the formation of CT excitons from Frenkel excitons excited either in the donor or acceptor material does neither make the CT excitons loosely bound nor the electron and hole in a CT exciton become separated farther. This is because of the following: excitons excited in the donor molecules form CT excitons (1) by electron transfer from the LUMO of the donor to the LUMO of the neighbour acceptor molecules, having a lower energy by $\Delta E_{LUMO}$, across the D–A interface, and those excited in the acceptor molecules form CT excitons (2) by hole transfer from the HOMO of the acceptor to the HOMO of the donor molecules [47], having a lower energy by $\Delta E_{HOMO}$, as clearly illustrated in Figure 3.5. For an efficient formation of CT excitons, both HOMO and LUMO levels of the donor must be at higher energies than those of the acceptor. As the CT excitons are created by transferring the electrons and holes to lower energy states, it may expected that the CT exciton states are even more stable than their predecessor Frenkel excitons and hence cannot be dissociated in any way easier than the corresponding Frenkel excitons.

The binding energy of singlet excitons is about 0.06 eV and that of a triplet is 0.7 eV in most organic solids. Therefore, the binding energy of the corresponding CT excitons is expected to be at least the same, if not greater. As the formation of a CT exciton involves two molecules, electron excited on an acceptor molecule and hole on a donor molecule, it is speculated that the electron and hole become farther apart in a CT exciton and hence can easily be dissociated due to the built-in electric field generated by the difference in the electrode work functions. A study on the dissociation of CT excitons by Devizis et al. [43], reveals that only charge pairs with an effective electron-hole separation distance less than 4 nm are created during the dissociation of Frenkel excitons, which is about the same as the excitonic Bohr radius of singlet Frenkel excitons in organic solids [31]. Therefore, an exciton with a separation of 4 nm between their charge carriers is not dissociated yet but may dissociate after the formation of a CT exciton. As already described above, due to the difference in work functions of electrodes, the built-in electric field cannot act on an electrically neutral particle, similar to a CT exciton efficiently, and hence cannot dissociate it. Therefore, the cause of dissociation of a CT exciton is puzzling and not fully explored.

In view of the above, although the formation of a CT exciton is a prerequisite intermediate state for its dissociation, as was identified earlier [31, 50, 51], its dissociation requires some excess energy to overcome its binding, which is about the same as that of an exciton binding energy as explained above. It is usually assumed that if the built-in electric field given in Eq. (3.51) is adequate to offer an energy greater than or equal to the binding energy of a CT exciton, it may be expected to dissociate. However, if CT excitons can be dissociated by this built-in field, then Frenkel excitons can also be dissociated, whether excited in the donor or acceptor because, as discussed above, they have similar or even lower binding energy. However, that would mean the formation of CT excitons at the D–A interface in BHJ OSCs plays no role, which is contrary to the observed higher efficiency in BHJ OSCs, and hence cannot be accepted. It has also been suggested that a CT exciton will subsequently dissociate following Onsager's theory [52]. However, a CT exciton is a discrete quantum state holding a quantum of solar energy, it is not exactly a geminate pair of varying distance between the two charge carriers ($e$ and $h$). In this situation, whether Onsager's theory can be applied for the dissociation of an exciton becomes, on its own, a topic of debate and controversy.

Although, there is significant amount of work done in studying the morphology of the BHJ OSCs for efficient charge generation [53], not much on the mechanism of dissociation of CT excitons has been discussed. We have recently proposed [31, 54] that a CT exciton can be dissociated only if it is given an excess energy equivalent to or greater than its binding energy from an external source. The only possible source of such energy available in OSCs is due to the formation of CT excitons, which releases excess energy from the conversion of a Frenkel exciton (at higher energy) to a CT exciton (lower energy), as it has been modelled recently [31].

### 3.4.1 Process of Conversion from Frenkel to CT Excitons

As an exciton is electrically neutral, it cannot be directed to move in any particular direction to reach the D–A interface. So, while it is excited within a material sandwiched between an anode and cathode, it can diffuse in any direction randomly. To direct an exciton to move in any particular direction, one has to introduce an interface with an acceptor (another organic material with a lower LUMO energy) in that direction. In this way, a cascade type of structure of LUMO and HOMO of donor and acceptor with their interfaces parallel to the electrodes can direct an exciton to move to the nearest D–A interface by forming a CT and then to the next D–A interface and thus forming and remaining in a CT exciton state at each interface, as shown in Figure 3.6 for a cascade of three materials. In this case, in a blended structure of ternary materials, the excitons can be excited in each of the three materials.

If the exciton is excited in the first donor, then the CT exciton can be formed at the first interface by transferring the electron to the LUMO of the first acceptor thus releasing an energy $\Delta E^1_{\mathrm{LUMO}}$. This CT exciton can then change to another CT exciton such that the

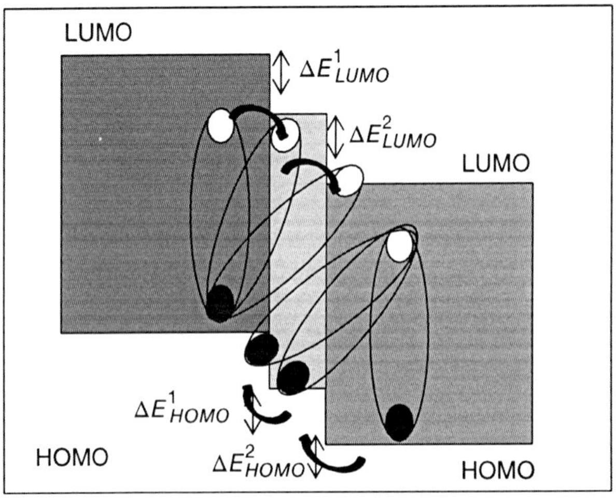


**Figure 3.6** Schematic illustration of the formation of CT excitons in a ternary BHJ OSC, where $\Delta E^1_{\mathrm{LUMO}}$ and $\Delta E^2_{\mathrm{LUMO}}$ represent the LUMO energy off-sets and $\Delta E^1_{\mathrm{HOMO}}$ and $\Delta E^2_{\mathrm{HOMO}}$ HOMO energy offsets at interfaces 1 and 2, respectively, counted from left to right.

electron is transferred to the second acceptor material by releasing an energy, $\Delta E^2_{\text{LUMO}}$ but the hole is still in the first donor material. In this case, the exciton was excited in the first acceptor material (middle donor/acceptor), then it has the possibility of forming a CT exciton in two ways: (1) electron goes down to the third acceptor's LUMO by releasing an energy $\Delta E^2_{\text{LUMO}}$ and/or (2) the hole goes up to the HOMO of the first donor by releasing an energy $\Delta E^1_{\text{HOMO}}$.

Figure 3.6 Schematic illustration of the formation of CT excitons in a ternary BHJ OSC, where $\Delta E^1_{\text{LUMO}}$and $\Delta E^2_{\text{LUMO}}$ represent the LUMO energy off-sets and $\Delta E^1_{\text{HOMO}}$ and $\Delta E^2_{\text{HOMO}}$ HOMO energy offsets at interfaces 1 and 2, respectively, counted from left to right.

Finally, if the exciton is excited in the second acceptor (third material), it has only one way of forming a CT exciton at the second D–A interface by transferring its hole to the acceptor of HOMO of the middle material and releasing an energy $\Delta E^2_{\text{HOMO}}$, which may then move to the first interface and form a CT exciton by transferring itself to the first donor and releasing an energy $\Delta E^1_{HOMO}$. However, none of these CT excitons can be considered dissociated as discussed above.

### 3.4.2 Dissociation of CT Excitons

The dissociation of a CT exciton can only occur due to the mechanism of its formation as discussed in our earlier work [31, 54]. Accordingly, as the formation of a CT exciton involves release of energy due either to transfer of an electron to a lower energy state of LUMO of the acceptor at a D–A interface or transfer of a hole to a higher HOMO level of the donor. If this released energy, usually in the form of molecular vibrational energy, impacts back on the CT exciton it may dissociate it if this energy is greater or at least equal to the binding energy of CT excitons. Thus, the formation of a CT exciton also involves its dissociation if the energy offsets are greater than or equal to the binding energy of CT excitons. This may be the reason for assuming that the formation of a CT exciton leads to automatic dissociation of excitons and hence the formation of CT state is a prerequisite state for the dissociation [50, 51]. Accordingly, the condition of dissociation of CT excitons can be given by [31]:

$$\Delta E_{\text{LUMO}} \text{ or } \Delta E_{\text{HOMO}} \geq E_B, \tag{3.52}$$

where $\Delta E_{\text{LUMO}} = E^D_{\text{LUMO}} - E^A_{\text{LUMO}}$ is the difference between the energy of donor LUMO, $E^D_{\text{LUMO}}$, and that of acceptor LUMO, $E^A_{\text{LUMO}}$, $\Delta E_{\text{HOMO}} = E^D_{\text{HOMO}} - E^A_{\text{HOMO}}$ is the difference between the energy of donor HOMO, $E^D_{\text{HOMO}}$, and that of acceptor HOMO, $E^A_{\text{HOMO}}$, and $E_B$ is the binding energy of excitons. Based on the condition in Eq. (3.1) and using the newly derived exciton-molecular vibration interaction operator as a perturbation, the rate of dissociation of CT excitons is obtained as [31]:

$$R^{i,j}_D = \frac{1}{96\pi^2\hbar^3\varepsilon^2 E^i_B}\left[(E^D_j\text{-}E^A_j)\text{-}E^i_B\right]^2 (\hbar\omega_v)\mu_x a^2_{xi}, \tag{3.53}$$

where $i$ denotes parameters associated with S (singlet) and T (triplet) excitons and $j$ denotes LUMO when CT excitons are formed from excitons excited in the donor and HOMO when CT excitons are formed from excitons excited in the acceptor material. The rate given in Eq. (3.53) is applicable to the dissociation of both singlet and triplet excitons but one has to use the corresponding parameters. In deriving, the dissociation rate in Eq. (3.53), it is assumed that a singlet Frenkel exciton changes to a singlet CT exciton and a triplet Frenkel exciton changes to a triplet CT exciton at the D–A interface.

Thus, the exciton dissociation at a D–A interface may be regarded as a two-step process: (1) The formation of CT excitons at the interface and (2) if the condition in Eq. (3.52) is met they may dissociate efficiently. Thus, the CT formation at D-A interface is a prerequisite for its dissociation but it is not dissociated because, as stated above, a CT exciton is no different in binding energy than a Frenkel exciton and like a Frenkel exciton, it is also an electrically neutral entity and hence cannot be influenced by the built-in electric field caused by the electrodes. Therefore, for an efficient dissociation of a CT exciton at the D–A interface, it is important that the condition in Eq. (3.52) is met. This is also supported by the experimental result by He et al. [55], who have fabricated a BHJ OSC using PTB7 as donor and $PC_{71}BM$ as acceptor and reported a PCE of 9.2% for an inverted structure. We will not discuss the effect of inverted structure on the PCE here but it is to be noted that the LUMO energy of PTB7 as donor is at -3.31 eV and that of $PC_{71}BM$ as acceptor is at -4.3 eV giving $\Delta E_{\mathrm{LUMO}}$= 0.99 eV, which is much larger than the binding energy of both singlet and triplet CT excitons and hence satisfies the condition in Eq. (3.52) very well for dissociating excitons excited in the donor material. Likewise, for the dissociation of excitons excited in the acceptor $PC_{71}BM$ with HOMO energy at -5.15 eV and HOMO of donor PTB7 at -6.1 eV, gives $\Delta E_{\mathrm{HOMO}}$= 0.95 eV, which is also more than the binding energy of both singlet and triplet excitons and satisfies the condition in Eq. (3.52). The dissociation rates calculated from Eq. (53) for excitons excited in the donor are $1.89\times10^{14}$ $s^{-1}$ for singlet and $4.82\times10^{9}$ $s^{-1}$ for triplet excitons. The rates of dissociation for excitons excited in the acceptor are $1.64\times10^{14}$ $s^{-1}$ for singlet excitons and $3.16\times10^{9}$ $s^{-1}$ for triplet excitons. These high rates imply faster dissociation of singlet than triplet excitons in PTB7: $PC_{71}BM$ BHJ OSCs, leading to faster free charge carriers generation resulting in enhanced photocurrent and hence higher power conversion efficiency. The concept of having$\Delta E_j^i \geq E_B^i$, $i$ = S or T and $j$ = LUMO or HOMO for an efficient dissociation of CT excitons is very important, although it has not yet been fully realized.

**Note: This chapter was submitted for publication in 2018.**

## References


1 Singh, J. (1994). *Excitation energy transfer processes in condensed matter: theory and applications*. New York: Plenum Press.
2 Davydov, A.S. (1971). *Theory of Molecular Excitons (trans. S.B. Dressner)*. New York: Springer.
3 Brütting, W. (2005). *Physics of Organic Semiconductors*. Weinheim: Wiley-VCH.
4 Swist, A. and Soloducho, J. (2012). *Chemik* 66: 289.
5 Narayan, M.R. and Singh, J. (2012). *Phys. Status Solidi (c)* 9: 2386.
6 Roncali, J. (2009). *Acc. Chem. Res.* 42: 1719.
7 Kippelen, B. and Brédas, J.L. (2009). *Energy Environ. Sci.* 2: 251.
8 Brédas, J.L., Norton, J.E., Cornil, J., and Coropceanu, V. (2009). *Acc. Chem. Res.* 42: 1691.
9 Ompong, D. and Singh, J. (2016). *Phys. Status Solidi C* 13: 89.
10 Singh, J. and Oh, I.-K. (2005). *J. Appl. Phys.* 97 (063516).

11 Narayan, M. and Singh, J. (2013). *J. Appl. Phys.* 114: 154515.
12 Singh, J. and Williams, R.T. (eds.) (2015). *Excitonic and Photonic Processes in Materials. Singapore: Springer Chapter* 8.
13 Singh, J. and Shimakawa, K. (2003). *Advances in amorphous semiconductors.* London: Taylor & Francis.
14 Singh, J. (2007). *Phys. Rev. B* 76 085205.
15 Singh, J. (2011). *Phys. Status Solidi A* 208: 1809.
16 Xu, Z., Hu, B., and Howe, J. (2008). *J. Appl. Phys.* 103 043909.
17 Yang, C.-M., Wu, C.-H., Liao, H.-H. et al. (2007). *Appl. Phys. Lett.* 90 (133509).
18 Schulz, G.L. and Holdcroft, S. (2008). *Chem. Mater.* 20: 5351.
19 Huang, J., Yu, J., Guan, Z., and Jiang, Y. (2010). *Appl. Phys. Lett.* 97: 143301.
20 Singh, J. (2010). *Phys. Status Solidi C* 7: 984.
21 Shafiee, A., Salleh, M.M., and Yahaya, M. (2011). *Sains Malays.* 40: 173.
22 Singh, J., Baessler, H., and Kugler, S. (2008). *J. Chem. Phys.* 129 041103.
23 Li, Y., Cao, Y., Gao, J. et al. (1999). *Synth. Met.* 99: 243.
24 Da Costa, P.G. and Conwell, E. (1993). *Phys. Rev. B* 48: 1993.
25 Thompson, B.C. and Fréchet, J.M. (2007). *Angew. Chem. Int.* 47: 58.
26 Winder, C. and Sariciftci, N.S. (2004). *J. Mater. Chem.* 14: 1077.
27 Bittner, E.R., Ramon, J.G.S., and Karabunarliev, S. (2005). *J. Chem. Phys.* 122: 214719.
28 Saleh, B.E. and Teich, M.C. (1991). *Fundamentals of photonics*, vol. 22. New York: Wiley.
29 Singh, J. and Oh, I.-K. (2005). *J. Appl. Phys.* 97 (063516).
30 Singh, J. (ed.) (2012). *Organic Light Emitting Devices.* New York: InTech.
31 Narayan, M.R. and Singh, J. (2013). *J. Appl. Phys.* 114: 73510.
32 Gasiorowicz, S. (1996). *Quantum Physics.* New York: Wiley.
33 Luhman, W.A. and Holmes, R.J. (2009). *Appl. Phys. Lett.* 94: 153304.
34 Gautam, B.R. (2013). *Magnetic field effect in organic films and devices.* Ann Arbor: The University of Utah.
35 Medina, A., Claessens, C.G., Rahman, G.M.A. et al. (2008). *Chem. Commun.* 1759: 61.
36 Ford, T.A., Avilov, I., Beljonne, D., and Greenham, N.C. (2005). *Phys. Rev. B* 71: 125212.
37 Cogan, S., Haas, Y., and Zilberg, S. (2007). *J. Photochem. Photobiol. A* 190: 200.
38 Donald, L.P., Gary, M.L., George, S.K., and Engel, R.G. (2011). *A Small Scale Approach to Organic Laboratory Techniques.* Belmont: Cengage Learning.
39 Klessinger, M. and Michl, J. (1995). *Excited states and photochemistry of organic molecules.* New York: Wiley.
40 Guo, F., Kim, Y.G., Reynolds, J.R., and Schanze, K.S. (2006). *Chem. Commun.* 17: 1887.
41 (a) Endo, A., Sato, K., Yoshimura, K. et al. (2011). *Appl. Phys. Lett.* 98 (093302); (b) Shakeel, U. and Singh, J. (2018). *Org. Electr.* 59: 121.
42 Tang, C.W. (1986). *Appl. Phys. Lett.* 48: 183.
43 Devizis, A., Jonghe-Risse, J.D., Hang, R. et al. (2015). *J. Am. Chem. Soc.* 137: 8192.
44 Wright, M. and Uddin, A. (2012). *Sol. Energy Mat. & Solar. Cells* 107: 87.
45 He, Z., Zhong, C., Su, S. et al. (2012). *Nat. Photon.* 6: 591.
46 Lu, L., Zheng, T., Wu, Q. et al. (2015). *Chem. Rev.* 115: 12666.

**47** Service, R.F (2011). *Science* 332 (6027): 293.

**48** Green, M.A., Emery, K., Hishikawa, Y. et al. (2014). *Prog. Photovoltaics. Res. Appl.* 22: 701.

**49** Peumas, P., Yakimov, A., and Forest, S.R. (2003). *J. Appl. Phys.* 93: 3393.

**50** Deibel, C., Strobel, T., and Dyakonov, V. (2010). *Adv. Mat.* 22: 4097.

**51** Brabec, C.J., Zerja, G., Cerullo, G. et al. (2001). *Chem. Phys. Lett.* 340: 232.

**52** Mihailetchi, V.D., Koster, L.J.A., Hummelen, J.C., and Blom, P.W.M. (2004). *Phys. Rev. Lett.* 93: 216601.

**53** Scharbar, M.C. and Sariciftci, N.C. (2013). *Prog. Polymer Sci.* 38: 1929.

**54** Singh, J., Narayan, M., Ompomg, D., and Zhu, F. (2017). *J. Mater. Sci. Mater. Electron.*: 149.

**55** He, Z., Zhong, C., Su, S. et al. (2012). *Nat. Photon.* 6: 591.

# 4

# Light Absorption and Emission Properties of Organic Semiconductors

*Takashi Kobayashi, Takashi Nagase and Hiroyoshi Naito*

*Department of Physics and Electronics, The Research Institute of Molecular Electronic Devices, Osaka Prefecture University, Sakai, Japan*

**CHAPTER MENU**



## 4.1 Introduction

Organic semiconductors have been extensively studied in many research fields associated with light. For instance, organic light-emitting diodes (OLEDs) have already been used in commercially available smart phones, digital audio players, and high-definition monitors, but the fundamental research for development of new emitters as well as for enlargement of device size, improvement of efficiency, and cost reduction still continues in many countries [1]. Organic photovoltaics (OPVs) have also attracted considerable attention because they have a strong potential to reduce the power generation cost. The best power conversion efficiency of OPV exceeded 10% several years ago [2]. To achieve such high efficiency, it was necessary to develop low-band gap polymers that can absorb a wider spectral range of the sunlight [3]. In those research fields, it is essential to have an understanding of light absorption and the emission properties of organic semiconductors. Also, in other research fields including organic thin-film transistors, optical spectroscopy has been utilized as a tool to obtain various information regarding excited states in target compounds [4, 5]. Thus, basics of optical spectroscopy and a background to understand the experimental results will be valuable for the reader in any research area of organic electronics. Although discussion in this chapter is limited to the optical properties in a visible range, a lot of information concerning the electronic states and excited-state structures of organic semiconductors can be obtained even from conventional absorption measurements. Furthermore, photoluminescence (PL) spectroscopy allows us to focus on the lowest accessible excited state(s) of all the excited states. With PL spectroscopy, decay channels of the excited state can also be investigated.

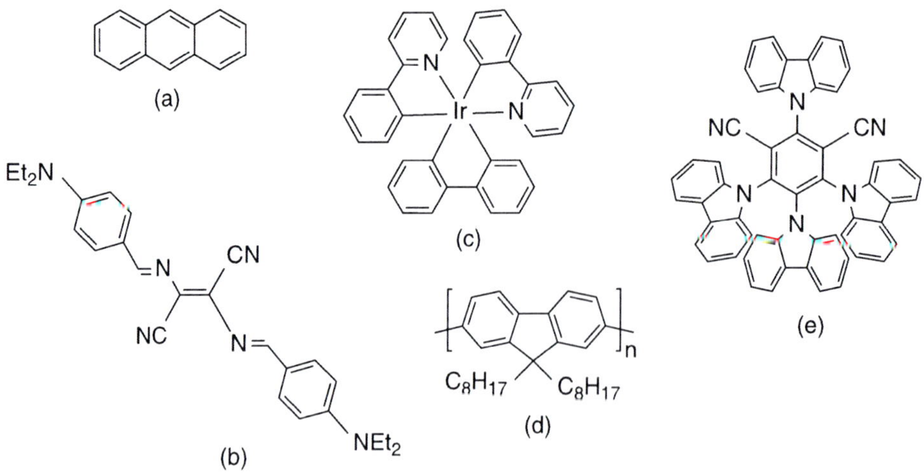


**Figure 4.1** Chemical structures of some organic semiconductors: (a) anthracene, (b) N,NV-bis[4-(N,N-diethylamino)benzylidene]diaminomaleonitrile (DE2), (c) *fac*-tris(2-phenylpyridine)iridium (III) [$Ir(ppy)_3$], (d) poly(9,9-dioctylfluorene) (PFO), and (e) 1,2,3,5-tetrakis(carbazol-9-yl)-4,6-dicyanobenzene (4CzIPN). In panel (b), $Et_2$ means two ethyl groups.

Organic semiconductors include a variety of organic compounds. They are classified into fluorescence, phosphorescence, and thermally activated delayed fluorescence (TADF) emitters in the field of OLEDs [6–9]: fluorescence emitters only consist of elements with low atomic numbers, such as carbon, hydrogen, nitrogen, and oxygen atoms, whereas phosphorescence emitters include heavy atoms such as Ir, Os, and Pt; phosphorescence emitters may be categorized to metal complexes in some research fields. TADF emitters do not contain any heavy atoms but have intermolecular charge transfer (CT) excited states so that their optical properties are different from those of conventional fluorescence emitters. Although $\pi$ conjugated polymers are usually classified into fluorescence emitters, they also show unique optical properties because of reasons originating from their huge molecular weight. In this chapter, therefore, organic semiconductors are classified into fluorescence, phosphorescence, and TADF emitters and $\pi$ conjugated polymers. In Figure 4.1, we show the chemical structures of some organic semiconductors. Following the classification mentioned above, molecules (a) and (b) are fluorescence emitters, and ones (c), (d), and (e) are, respectively, classified to phosphorescence emitters, $\pi$ conjugated polymers, and TADF emitters. In the following section, their electronic states and absorption properties are explained. After that, we will describe two advanced topics, i.e., determination of excited-state structure with nonlinear optical spectroscopy, and the decay mechanism of excited states.

## 4.2 Electronic States in Organic Semiconductors

Semiconducting properties of organic semiconductors are determined, or at least largely influenced, by $\pi$ electrons in $\pi$ bonds. Since $\pi$ bonds are weaker than $\sigma$ bonds and, furthermore, $\pi$ electrons are delocalized throughout a network of connected p-orbitals, it is much easier to add an electron to and extract one from $\pi$ electrons as well as to excite π electrons with a visible light. In many organic semiconductors, the highest occupied molecular orbital (HOMO) and the lowest unoccupied molecular orbital (LUMO) correspond to π bonding and π antibonding orbitals, respectively. Thus, the optical transition between those levels is often referred to as a π-π* transition; * indicates an unoccupied orbital. However, actual optical properties of organic semiconductors cannot be understood only in terms of π-π* transitions. In this section, in the order of fluorescence, phosphorescence, and TADF emitters and π conjugated polymers, we will show what else is also important in order to understand their actual optical properties.

### 4.2.1 Fluorescence Emitters

In general, in an absorption spectrum with a scale in energy, the area of an absorption band is proportional to the oscillator strength, which is determined by the square of the transition dipole moment between the associated initial ($i$) and final ($f$) states $\boldsymbol{m}_{if}$:

$$\boldsymbol{m}_{if} = \int \Psi_f(-e\boldsymbol{r})\Psi_i d^3\boldsymbol{r}, \tag{4.1}$$

where $\Psi_i$ and $\Psi_f$ are the wavefunctions (or molecular orbitals) of $i$ and $f$ states, $e$ is the elementary charge, and $\boldsymbol{r}$ is the position vector. Recently, the oscillator strength, as well as the excitation energy of each transition, can be simulated by a quantum chemical calculation. In the case of anthracene, which is one of the typical fluorescence emitters, it is expected that only two electronic transitions appear in a spectral range below 4.5 eV: at 3.28 eV with a moderate oscillator strength and 3.90 eV with a very weak oscillator strength. However, as shown in Figure 4.2, the actual absorption spectrum of anthracene in ethanol has more than two peaks. Such an oscillator structure (vibronic structure) is often observed in fluorescence emitters, and is due to a strong coupling between electrons and molecular vibrations (vibronic coupling). In the case of anthracene, the 0-0 transition at 3.3 eV in the absorption spectrum corresponds to the *pure* electronic transition, and the others, i.e. 0-1, 0-2, 0-3, etc. transitions, are electronic transitions with phonon emission.

In Figure 4.3, we show the potential energy curves of a molecule in the electronic ground and excited states. In this figure, $Q_G$ represents the nucleus arrangement where

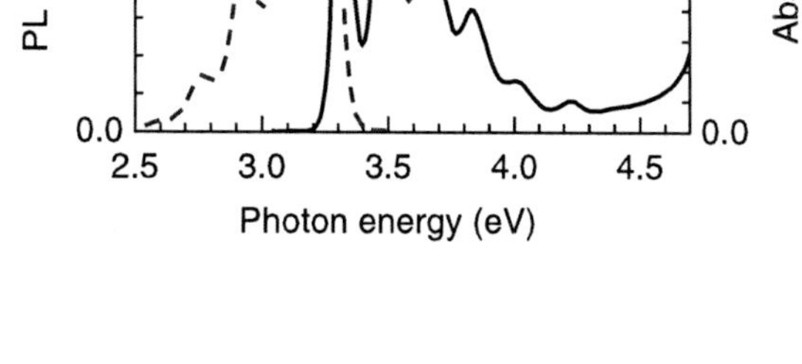


**Figure 4.2** Absorption and PL spectra of anthracene in methanol.

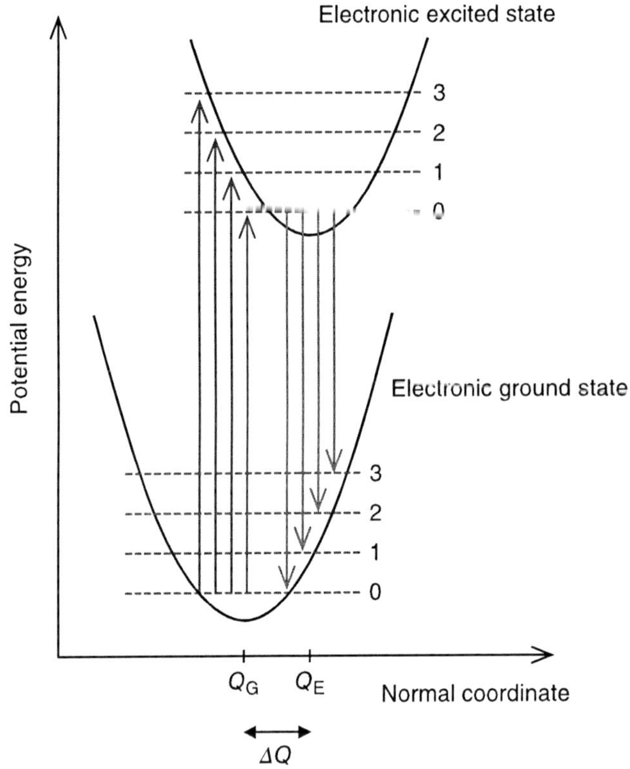


**Figure 4.3** Potential energy curves of a molecule in the electronic ground and excited states along with several molecular vibrational levels. The upward and downward arrows indicate absorption and emission processes, respectively. The potential minima of the ground and excited states are represented by $Q_G$ and $Q_E$, respectively.

the total energy is minimized in the electronic ground state, and the horizontal axis can be understood as a degree of discrepancy in nucleus arrangements from $Q_G$. In many organic materials, an electronic excited state is created by an electronic transition from a bonding orbital to an antibonding one. Thus, the potential minimum $Q_E$ in the electronic excited state is different from $Q_G$. In the case of $Q_G \neq Q_E$, the upward transitions into higher vibrational levels can take place in addition to the transitions into the 0th vibrational level. Since the vibrational levels are not continuous but discrete, a vibronic structure consisting of several peaks with the energy interval corresponding to the associated phonon energy appears. In the case of anthracene, the associated phonon energy is estimated to be 0.17~0.18 eV. In organic semiconductors, a vibronic structure due to a C=C double bond stretching mode with a phonon energy around 0.18 eV (or around 1,400 $cm^{-1}$) is frequently observed. Note that here, we assume no population at the higher vibrational level in the ground state (this is not the case if an associated phonon energy is very small, or the sample temperature is high).

A vibronic structure is also observed in the PL spectrum of many fluorescence emitters. After photoexcitation process is completed, excited states immediately relax to the 0th vibrational level in the electronic excited state. Thus, the downward transitions can occur only from the 0th level. Also, in the downward transitions, the final states are limited to the discrete vibrational levels in the electronic ground state. This explains the vibronic structure observed in the PL spectrum. If we assume that the potential curves are parabolic (a harmonic oscillator approximation), the associated phonon energy and the discrepancy between the potential minima $\Delta Q = Q_G - Q_E$ are common to the downward and upward transitions. The only difference is that the upward transitions with photon emission appear in the high energy side with respect to the 0-0 transition while the downward transitions with phonon emission appear in the low energy side. Consequently, the absorption and PL spectra are in a mirror image symmetry.

The relative intensity of the 0-$n$ transition $F_{0n}$ is determined by a transition probability between the 0th and $n$th vibrational levels. Using the vibrational wavefunctions at the 0th and $n$th levels, $\xi_0(Q)$ and $\xi_n(Q)$, $F_{n0}$ can be expressed as

$$F_{0n} = \left| \int \xi_n^*(Q)\xi_0(Q - \Delta Q)dQ \right|^2. \tag{4.2}$$

This $F_{0n}$ is called the Franck-Condon factor. It is known that Eq. (4.2) can be simplified into a Poisson distribution function under the assumption of the harmonic potential [10]. $F_{0n}$ does not alter the oscillator strength, i.e. $\boldsymbol{m}_{if}$, and only divides it into several transitions. The origin of the vibronic structure is the discrepancy between $Q_G$ and $Q_E$. This can be confirmed that if $\Delta Q = 0$ in Eq. (4.2), $F_{0n}$ would be zero expect for $n = 0$ because of the orthogonality of vibrational wavefunctions. As $\Delta Q$ increases from zero, transitions into higher vibrational levels gain the oscillator strength whereas transitions into lower vibrational levels are gradually reduced. Finally, $\Delta Q$ becomes large so that a harmonic oscillator approximation is no longer valid. In such a case, the following is expected [10]: the center of vibration shifts as the vibrational level increases, and energy gaps between adjacent vibrational levels are not constant. In such a case, instead of vibronic structure, a broad and featureless spectral shape is experimentally observed.

The absorption and PL spectra shown in Figure 4.2 were measured with anthracene in a dilute solution, where intermolecular coupling is negligible. In solid state, e.g. in crystals or neat thin films, intermolecular coupling may remarkably change optical properties of molecules. In a special case (J-aggregates), a redshifted and narrower absorption and PL bands with respect to those in solution are observed. One example will be shown later (Section 4.3.3.1). Another special case (H-aggregates) is characterized by the gradual onset of the absorption and its blueshifted peak with respect to that in solution. It is also known that in J-aggregates, PL quantum efficiency (PLQE), which is defined as the number of emitted photons per absorbed photon, is largely suppressed even if the molecule has high PLQE in solution. Another case that is frequently encountered in the studies of OLEDs is that PLQE is significantly reduced in spite of the small change in the absorption and PL spectra. These phenomena are called concentration quenching. To avoid this, emitters are often lightly doped in a host matrix in OLEDs. For the reader who is interested in intermolecular coupling [11–14].

### 4.2.2 Phosphorescence Emitters

In fluorescence emitters, triplet excited states generated after recombination of injected electrons and holes have to decay nonradiatively into the ground state because of

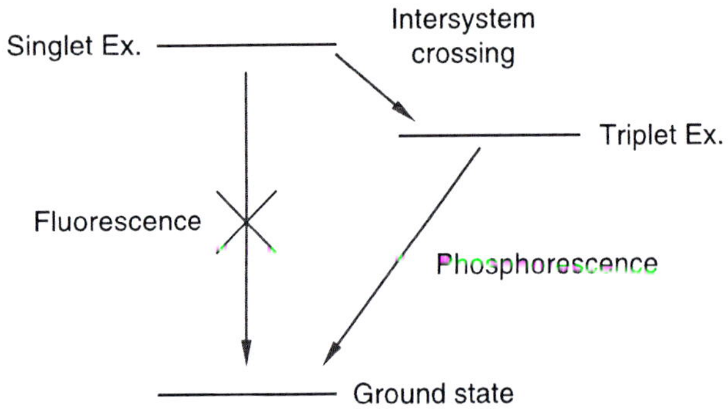


**Figure 4.4** Simplified energy diagram and a main decay channel of excited states in phosphorescence emitters. In many phosphorescence emitters, intersystem crossing is so efficient that fluorescence is completely quenched.

the extremely low transition dipole moment between the triplet excited states and the ground state. On the other hand, in phosphorescence emitters, singlet and triplet excited states are slightly mixed by the spin-orbit coupling so that triplet excitons can decay into the ground state with phosphorescence emission (Figure 4.4). Intersystem crossing from singlet excited states into triplet excited states is also accelerated. Therefore, phosphorescence emitters can potentially convert all of the created excited states into photons [6, 7, 15]. In efficient phosphorescence emitters, electronic excitations always involve 5d-orbitals of the metal, such as $Ir^{3+}$, $Os^{2+}$, and $Pt^{2+}$, and the spin-orbit coupling is carried by those 5d-orbitals. The electronic states in phosphorescence emitters are entirely different from those in fluorescence emitters. In the case of $Ir(ppy)_3$, which is a well-known efficient phosphorescence emitter [6], the LUMO corresponds to an unoccupied $\pi$ orbital ($\pi^*$) of ligands, and the HOMO mainly consists of an occupied d-orbital. The electronic transition from the HOMO to the LUMO induces charge transfer (CT) from the metal to the ligand(s). So, this kind of transition is referred to as a metal-to-ligand charge transfer (MLCT) transition. Note that some metal complexes have ligand center (LC or $\pi$-$\pi^*$) or metal center (MC) transitions. Our discussion is, however, limited to Ir, Os, or Pt complexes with MLCT excitations, which are suitable for OLED applications [7].

In general, CT transitions, including MLCT transitions, exhibit a broad and featureless absorption band with a smaller magnitude compared to that due to $\pi$-$\pi^*$ transitions. In CT molecules, the wavefunctions of the HOMO and the LUMO are distributed on different units of a molecule. Resultant small spatial overlap between the two wavefunctions leads to suppression of the transition dipole moment [see Eq. (4.1)]. Long transfer distance also induces large molecular deformation, which is expressed as a large $\Delta Q$ in Figure 4.3 and is the origin of a broad and featureless absorption spectrum. In Figure 4.5, as an example, we show the absorption and PL spectra of neat thin films of $Ir(ppy)_3$. The absorption band observed around 3.2 eV is due to MLCT transitions, and is indeed smaller than the $\pi$-$\pi^*$ band observed around 4.3 eV. The MLCT band seems to have several shoulders but those are not due to vibronic coupling. The actual excitations near the HOMO-LUMO gap in phosphorescence emitters are very complicated [7, 16]. In the case of $Ir(ppy)_3$, there is a doubly degenerated 5d-orbaital (HOMO-1) right below the HOMO, and similarly there is a doubly degenerated $\pi^*$-orbital (LUMO+1) right above

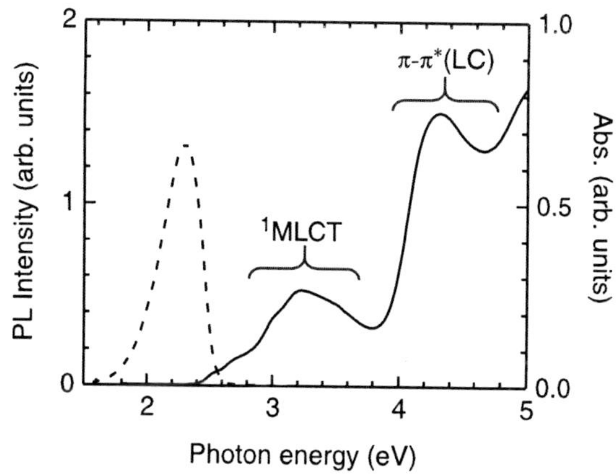


**Figure 4.5** Absorption and PL (phosphorescence) spectra of Ir(ppy)$_3$ neat thin film. Absorption bands around 3.2 and 4.3 eV are due to singlet MLCT and π-π* (or LC) transitions, respectively. PLQE of neat thin films is significantly reduced by concentration quenching but if the purity of the sample is sufficiently high, the PL spectrum can be measured.

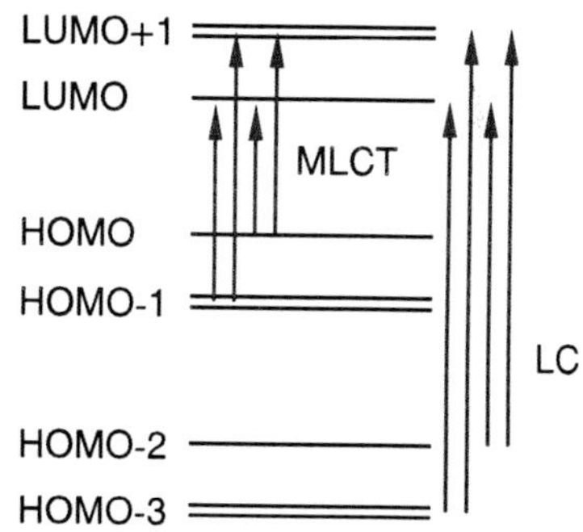


**Figure 4.6** Simplified energy diagram of Ir(ppy)$_3$ near the HOMO-LUMO gap. Whether some levels are degenerated or not, depends on the molecular structure and the surrounding circumstance.

the LUMO (Figure 4.6). Therefore, Ir(ppy)$_3$ possesses four possible MLCT transitions near the HOMO-LUMO gap. In addition, an undegenerated and a doubly degenerated π-orbital exists below the HOMO-1, and thus four LC (or π-π*) transitions may occur. Furthermore, some of these levels are quantum mechanically mixed by configuration interaction, which is a consequence of electron-electron interaction. From a quantum chemical calculation [16], it has been confirmed that the highest three occupied molecular orbitals have both 5d and ligand π characters whereas the LUMO is the lowest π* orbital. Theoretical work has also revealed that the absorption bands below 4.0 eV consist of more than ten transitions and that among them, the lowest singlet transition, i.e., the transition into the singlet MLCT state, appears at 2.8 eV. This means that the lowest absorption band observed below 2.8 eV is not due to an excitation into singlet MLCT states but into triplet MLCT states. In phosphorescence emitters, singlet-triplet mixing is induced by spin-orbit coupling. The resultant singlet admixtures to triplet MLCT states are responsible for the lowest absorption band. The observed PL (phosphorescence) originates from the same triplet MLCT states.

### 4.2.3 TADF Emitters

Although OLEDs, based on phosphorescence emitters, exhibit excellent external quantum efficiency (EQE), the use of Ir, Pt, or Os leads to an increase in material cost. One strategy that has been undertaken to reduce the cost of OLEDs without reduction of EQE is to up-convert generated triplet excited states into singlet excited states by thermal energy [17]. To realize such up-conversion, the energy gap between the lowest singlet and triplet excited states $\Delta E_{st}$ should be small so that triplet excited states can overcome the gap by thermal energy [8, 9, 17]. In 2012, Uoyama et al. reported that an excellent EQE comparable to that achievable with phosphorescence emitters is indeed recorded in OLEDs using a TADF emitter, i.e. 4CzIPN [8].

The singlet-triplet splitting is caused by a quantum mechanical effect, which takes the spin correlation into account, and $\Delta E_{st}$ is determined by the exchange interaction. To reduce the $\Delta E_{st}$ value, the spatial overlap between the wavefunctions of the HOMO and LUMO should be decreased. It is CT molecules that fulfill this condition as mentioned above. In fact, many CT molecules exhibit TADF. Figure 4.7 shows an absorption spectrum of neat thin films of 4CzIPN, where broad absorption bands characteristic to CT transitions are observed above 2.6 eV. In the case of 4CzIPN, the PL observed at room temperature consists of fluorescence and delayed fluorescence from the lowest CT excited state. Thus, its spectral shape is also broad and featureless compared to that of fluorescence emitters.

CT molecules also have excited states whose wavefunctions are localized on the same unit as the HOMO. Such excited states, which are called locally-excited (LE) states, induce neither charge transfer nor large molecular deformation. As a result, transitions into LE states do not induce significant change in the dipole moment, and PL from the lowest LE state exhibits a clear vibronic structure. Because of the former feature, LE states are not stabilized as much as CT excited states in a polar solvent. In some cases, the order of the lowest LE state and CT excited state is reversed depending on the polarity of the solvent, and a resultant drastic change of PL properties is observed. This solvent dependence of PL properties has been used to study the nature of the lowest excited states in molecules with CT excited states including TADF emitters [18–20].

### 4.2.4 π Conjugated Polymers

Electronic states of π conjugated polymers are basically the same as those in fluorescence emitters: the absorption bands observed in a range from near infrared to near ultraviolet is due to π-π* transitions. The differences between π conjugated polymers and small molecules (fluorescence emitters) essentially come from huge molecular weight

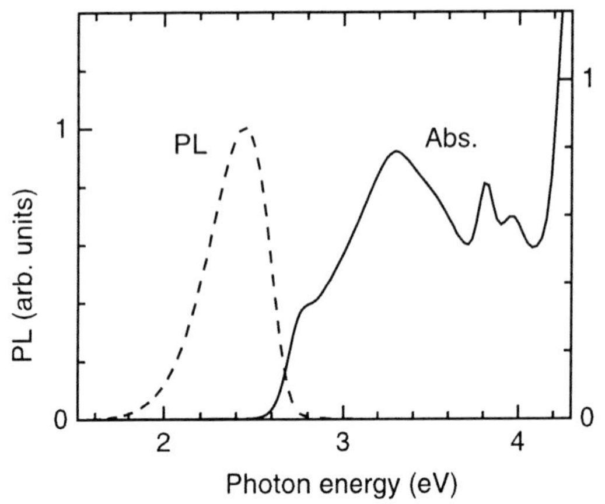


**Figure 4.7** Absorption and PL spectra of neat thin films of 4CzIPN.

of the polymers. It is technically impossible to eliminate molecular weight distribution. In addition, the polymer chains can be bent or twisted more easily compared to small molecules. Ideally, $\pi$ conjugated polymers adopt a fully planar conformation, and $\pi$ electrons are delocalized throughout the polymer chain. In actual polymers, however, the delocalization is limited by bends and twists. As a result, a single polymer chain is split into several segments with various delocalization lengths. Such a limited delocalization length is called effective conjugation length. Optical properties of a segment are dependent on its effective conjugation length $L$ [21–23]. For instance, between $L$ and the lowest excitation energy $E$, there is a following relationship:

$$E = A + \frac{B}{L}, \tag{4.3}$$

where $A$ and $B$ are constants depending on the repeating unit of a polymer [23–25]. Differentiating Eq. (4.3) yields

$$\Delta E = -\frac{B}{L^2}\Delta L. \tag{4.4}$$

This equation indicates that distribution of excitation energy, that is, inhomogeneous broadening, is expanded when an average of $L$'s become short. Such a situation is realized in a solution sample or in glassy thin films, where polymer chains adopt a random-coil conformation. Figure 4.8 shows absorption and PL spectra of two types of

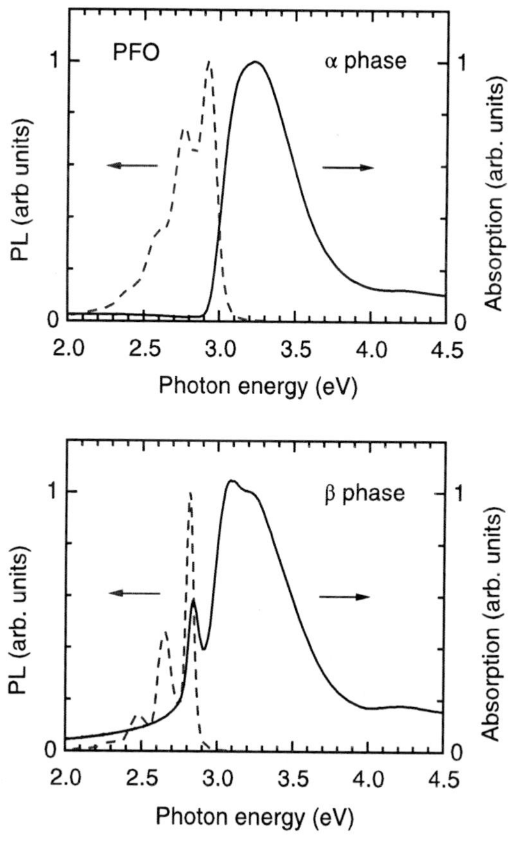


**Figure 4.8** Absorption and PL spectra of thin films of α- and β-phase PFO.

thin films of PFO, which is a well-known soluble polymer with blue fluorescence. Glassy PFO ($\alpha$ phase) can be fabricated by spin-coating using a solvent with a low boiling point such as toluene. The absorption spectrum of $\alpha$ phase is broad and featureless, and is almost the same as that of PFO in solution [26]. On the other hand, the PL spectrum of $\alpha$ phase is relatively narrow and a clear vibronic structure can be seen. This is because PL preferentially originates from segments with longer $L$'s within a sample, as will be discussed later (Section 4.4.3.1).

Optical properties of PFO are largely dependent on the fabrication conditions [26–28]. For instance, when PFO thin films are fabricated by drop-casting instead of spin-coating, polymer chains adopt more planar conformations ($\beta$ phase). In such conformations, bends and twists are well suppressed, and $L$'s are elongated. From Eqs. (4.3) and (4.4), it is expected that the elongation of $L$ results in the redshift and narrowing of the absorption bands. As shown in Figure 4.8, such redshifted and narrower absorption and PL bands are observed. The additional small absorption band at 3.2 eV is attributed to the 0-0 transition of $\beta$-phase PFO, and its 0-1 transition superimposed to the broad and large absorption band due to the residual $\alpha$ phase can be seen around 3.05 eV. In the PL spectrum of $\beta$-phase PFO, further redshifted and narrow peaks with respect to those in $\alpha$ phase can be found. The difference in the PL spectral shape can also be understood in terms of the elongation of $L$'s [26].

## 4.3 Determination of Excited-state Structure Using Nonlinear Spectroscopy

Excited electronic states in an isolated organic molecule in a vacuum can be revealed to a great extent by quantum chemical calculations. In many cases, the results with time-dependent density functional theory are sufficient for understanding of the measured absorption spectrum. However, it is still impossible to obtain sufficient results for some cases. For example, in the case of $\pi$ conjugated polymers, the conformational dependence of the electronic states, as well as the huge number of the constituent atoms, make it difficult to calculate the absorption spectrum that can be compared with the experimental results. There are also some reverse cases; although the theoretical data are available, the corresponding experimental data cannot be obtained with conventional absorption and PL spectroscopy. An example of the latter cases is the ratio of the $\pi$ and 5d orbital components in the HOMO of a phosphorescence emitter. The ratio can be calculated [16] but cannot be determined with those spectroscopies. It is also impossible to experimentally find excited states that are silent in the absorption spectrum because of the symmetry. Therefore, there are needs for experimental techniques that can be used in any situations to gain a deeper understanding of the excited states in organic semiconductors. A solution for the needs is nonlinear optical spectroscopy, which offers a means for investigation of some of the excited states that do not appear in the absorption spectrum. Among various techniques of nonlinear optical spectroscopy, we describe electroabsorption (EA) and two-photon excitation (TPE) measurements in this chapter, and then show some examples of those measurements on organic semiconductors, including an analysis of the EA spectrum to determine the ratio of the $\pi$ and 5d orbital components [29].

### 4.3.1 Background

To observe nonlinear optical effects, laser pulses with a high peak power are more advantageous than continuous-wave light from a lamp because signal intensity is proportional to the power (square, cube, or so on) of the peak intensity [30]. To investigate excited-state structures of organic semiconductors, it is, however, necessary to cover a spectral range wider than an absorption band. To obtain such a wide spectral data, the measurements have to be repeated many times with changing a wavelength of light. If a fs laser is utilized, an enormous amount of time may be required to change a laser wavelength without changing pulse duration (or peak intensity). In contrast, EA measurements can be carried out with a lamp and a monochromator so that little time has to be spent for tuning the light source [31, 32]. Although TPE measurements need a laser pulse, they can be performed with a tunable ns laser, which can be handled far easier than a fs laser. Thus, EA (and TPE) measurements allow you to focus on physics in organic semiconductors rather than the light source in the spectroscopy.

In EA measurements, small absorption change of a sample due to the applied electric field is measured. When an electric field $\boldsymbol{F}$ is applied to a sample, an energy level, e.g. the ground state, is stabilized by

$$\delta E = -\boldsymbol{\mu} \cdot \boldsymbol{F} - \frac{1}{2}\boldsymbol{F} \cdot \alpha \cdot \boldsymbol{F}, \tag{4.5}$$

where $\boldsymbol{\mu}$ is the dipole moment and $\alpha$ is the polarizability tensor of the molecule. If $\boldsymbol{\mu}$ and $\alpha$ in the ground state are different from those in the excited state, excitation energy is changed due to $\boldsymbol{F}$ by

$$\Delta E(F) = \delta E_{\mathrm{Ex}} - \delta E_{\mathrm{G}} = -\Delta\mu F - \frac{1}{2}\Delta\alpha F^2, \tag{4.6}$$

where $\Delta\mu$ and $\Delta\alpha$ are differences in dipole moment and polarizability, respectively, between the excited and ground states. For simplicity, we assume that a 1D molecule with $\Delta\mu$ and $\Delta\alpha$ parallel to $F$. The first and second terms in the right side of Eq. (4.6) correspond to the linear and quadratic Stark effect, respectively. When you focus on the EA spectral shape, it can be expanded using the first and second derivatives of the absorption spectrum with respect to energy as:

$$\Delta A \approx \frac{\partial A}{\partial E}\Delta E + \frac{1}{2}\frac{\partial^2 A}{\partial E^2}\Delta E^2, \tag{4.7}$$

where the higher order terms are truncated. In the case of randomly oriented molecules, the first term in the right side of Eq. (4.6) vanishes. Thus, substituting $\Delta E = -\Delta\alpha^2 F^2 / 2$ into Eq. (4.7) yields,

$$\Delta A \approx -\frac{1}{2}\Delta\alpha F^2 \frac{\partial A}{\partial E}, \tag{4.8}$$

where we omit the higher term because $\Delta E$ is very small. This equation suggests that an EA spectrum that resembles to the first derivative would be observed.

CT molecules are expected to have large $\Delta\mu$, which is induced by charge transfer from a unit to a spatially separated one. In such a case, isotropic averaging over the randomly oriented molecules yields [33]

$$\langle \Delta E^2 \rangle = \frac{1}{3}\Delta\mu^2 F^2 \tag{4.9}$$

although

$$\langle \Delta E \rangle = 0. \tag{4.10}$$

Thus, by substituting Eqs. (4.9) and (4.10) into Eq. (4.7), the following is obtained

$$\Delta A \approx \frac{1}{6} \Delta \mu^2 F^2 \frac{\partial^2 A}{\partial E^2}. \tag{4.11}$$

This means that, unlike π-π* transitions, CT transitions lead to an EA spectrum similar to the second derivative of the corresponding absorption spectra. Therefore, from a comparison of an EA spectrum and an absorption spectrum, the nature of transitions can be experimentally determined. Note that in either case, the amplitude of the EA signal is proportional to $F^2$.

Organic semiconductors have excited states that cannot be found in their absorption spectra because of their zero transition dipole moments. Let us consider 1D π conjugated polymers, whose wavefunctions can be classified into even- or odd-parity states. In 1D polymers, the HOMO is an even-parity state, and the operator $-er$ in Eq. (4.1) is an odd function. Thus, for even-parity excited states, $\boldsymbol{m}$ will vanish, resulting in dipole-forbidden transitions between the HOMO and those excited states. On the other hand, odd-parity excited states have non-zero $\boldsymbol{m}$ values so that a transition between the HOMO and an odd-parity state becomes dipole-allowed. Even-parity excited states, however, have significant impacts on nonlinear optical processes, which involve a dipole-allowed transition between an odd-parity excited state and a higher even-parity one (not the HOMO).

To explain the importance of even-parity excited states, we will introduce nonlinear optical susceptibility. When the light is sufficiently weak, induced polarization $P(t)$ of a material is proportional to the strength of an optical electric field $F(t)$:

$$P(t) = \varepsilon_0 \chi^{(1)} F(t), \tag{4.12}$$

where $\chi^{(1)}$ is the linear susceptibility, and $\varepsilon_0$ is the permittivity of free space. It is well known that the absorption coefficient and refractive index of a material can be described by the real and imaginary parts of $\chi^{(1)}$. However, when the light intensity becomes strong so that Eq. (4.12) is no longer valid, the corrections using a power series in the field strength $F(t)$ are needed:

$$P(t) = \varepsilon_0 [\chi^{(1)} F(t) + \chi^{(2)} F^2(t) + \chi^{(3)} F^3(t) \cdots], \tag{4.13}$$

where $\chi^{(2)}$ are $\chi^{(3)}$ the second- and third-optical nonlinear susceptibility, respectively. If the material has inversion symmetry, $\chi^{(2)}$ becomes zero, while any materials possess non-zero $\chi^{(3)}$ values [30]. One of nonlinear optical phenomenon that cannot be explained without $\chi^{(3)}$ is the third-harmonic generation (THG), which is a process to convert three photons with a certain photon energy into one photon with the triple photon energy. The EA and two-photon absorption (TPA) processes are also described by $\chi^{(3)}$. In the THG process, the third term of the right side of Eq. (4.13) can be understood as the interaction of a material with three optical electric fields with angular frequency $\omega$ to create nonlinear polarization with angular frequency of $3\omega$; the resultant nonlinear polarization generates a photon with the photon energy of $3\hbar\omega$, where $\hbar$ is a Planck's constant. To describe this process more clearly, the associated third-order nonlinear susceptibility is written as $\chi^{(3)}(-3\omega; \omega, \omega, \omega)$.

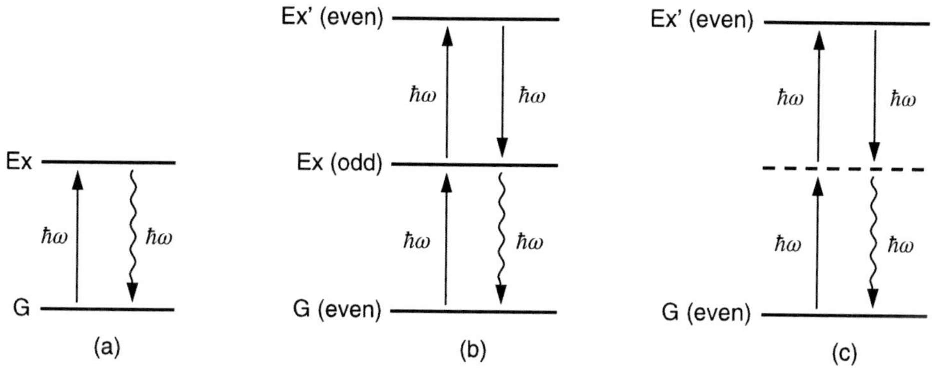


**Figure 4.9** (a) One-photon absorption process between the excited state (Ex) and the ground state (G). (b) TPA process in a system consisting of G and odd-parity (Ex) and even-parity (Ex′) excited states with the energy gaps of $\hbar\omega$. (c) Two-photon absorption process with a system consisting only of G and Ex′ with the energy gap of $2\hbar\omega$. The dashed line represents a virtual state but not a real excited state. In the figures, solid arrows indicate optical electric fields that create (nonlinear) polarizations, whereas wavy arrows indicate optical electric fields that interact with the created (nonlinear) polarizations. Thus, the latter fields do not appear in Eqs. (4.12) and (4.13).

TPA is a process to create an excited state by absorbing two photons simultaneously, and the excitation energy is twice that of the photon energy. The associated nonlinear susceptibility, $\chi^{(3)}(-\omega; \omega, \omega, -\omega)$, is largely enhanced by resonant effects if the energy levels of the odd-parity and even-parity states measured from the ground state are coincident with $\hbar\omega$ and $2\hbar\omega$, respectively (Figure 4.9b) [30]. Since the $\chi^{(3)}(-\omega; \omega, \omega, -\omega)$ value is proportional to the products of four $\boldsymbol{m}$'s represented by the solid and wavy arrows in Figure 4.9b, symmetry of the associated states is important: as mentioned above, $\boldsymbol{m}$ between the even-parity states is zero so that the highest and middle excited states should have the opposite parities. However, in the situation depicted in Figure 4.9b, the TPA process may not be recognized because the one-photon absorption process, which is illustrated in Figure 4.9a, largely surpasses the TPA process. The possible situation where the TPA process is experimentally observed is one illustrated in Figure 4.9c, in which the gap between the even-parity excited state and the ground state is coincident with $2\hbar\omega$, and the odd-parity state lies at higher than $\hbar\omega$ so that the one-photon absorption process does not compete with the TPA process. Such a situation is realized in many $\pi$ conjugated polymers, in which the odd-parity excited state lies at higher in energy than half the gap between the HOMO and the even-parity excited state. Therefore, TPA measurements can be used to find even-parity excited states in $\pi$ conjugated polymers.

The nonlinear susceptibility associated with EA is written as $\chi^{(3)}(-\omega; \omega, 0, 0)$, which indicates that two of the three involved optical electric fields are replaced by the applied electric field with nearly zero angular frequency $\omega_0$ compared to that of a visible light, i.e. $\omega$. The $\chi^{(3)}$ value may be enhanced in several situations, among which two important situations are shown in Figures 4.10a and 4.10b. In Figure 4.10a, the energy gap between the odd-parity excited state and the ground state is coincident with $\hbar\omega$, whereas in Figure 4.10b, the gap between the even-parity excited state and the ground state is coincident with $\hbar\omega + \hbar\omega_0 \approx \hbar\omega$. Therefore, relatively large EA signals would be

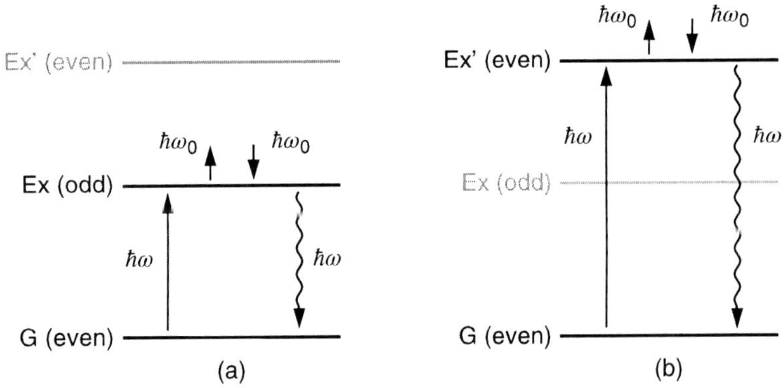


**Figure 4.10** Two possible situations where a large EA signal is observed. In this figure, $\omega_0$ represents angular frequency of the applied electric field so that $\omega_0$ is nearly equal to zero.

observed at the positions of even-parity as well as odd-parity excited states. This is why EA measurements can also be used to study even-parity excited states.

### 4.3.2 Experimental Technique

In the following, we will explain experimental setups for EA and TPE measurements. In the latter measurements, excited states are created in a TPA process but its efficiency (or TPA cross section) is determined from the intensity of the resultant luminescence. Compared to the TPE measurements, it is difficult to determine the TPA cross section in more direct ways [34].

#### 4.3.2.1 EA

For EA measurements, an electric field is applied to a sample with two electrodes. Two types of device have frequently been used for solid-state samples. In one type of device, a thin film of an organic semiconductor is sandwiched with a transparent electrode, such as ITO, and a semitransparent metal electrode, such as a 20-nm-thick aluminum. The latter electrode is usually vapor-deposited on the surface of an organic semiconducting layer. In this case, the gap between the two electrodes is determined by the thickness of the organic semiconducting layer, which is typically around 100 nm. In the other type of device, a pair of metal electrodes with a narrow gap are vapor-deposited on the surface of an organic semiconducting layer or a glass substrate. Typical gap width is a few hundred μm when a shadow mask is used in a deposition process, and for narrower gaps, photolithographic techniques are needed. Each type of device has its merit and demerit: in sandwich devices, high electric field can be realized with a relatively weak voltage, while a signal-to-noise ratio tends to be lowered because of attenuation of light by a semitransparent metal electrode.

An experimental setup for EA measurements is shown in Figure 4.11. A monochromated light from a lamp is irradiated to a device and the transmitted light is detected by an appropriate photodiode. To reduce noise, current output from the photodiode should immediately be converted into voltage output by a metal film resistor or a transimpedance amplifier. An electric field modulated at a frequency $f$ is applied to a device using a function generator and, if necessary, a power amplifier. $f$ is usually in a range

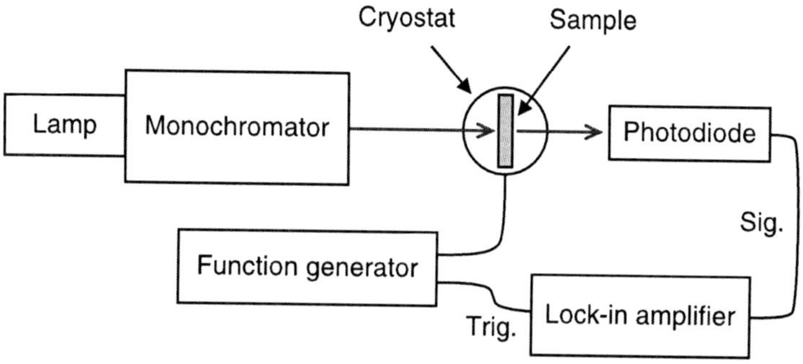


**Figure 4.11** Experimental setup for EA measurements. When the gap between two electrodes is wide, a voltage amplifier may be needed to apply an electric field appropriate for the measurements.

from a few hundred Hz to a few kHz. Since the EA signal is proportional to the square of the field strength, the signal is expected to be modulated at a double frequency, i.e. $2f$. This can be confirmed if $F$ in Eq. (4.8) or (4.11) is replaced with $F\sin(2\pi ft)$. In the case of Eq. (4.8), the following is obtained:

$$\Delta A = \frac{1}{4}\Delta\alpha[\cos(4\pi ft) - 1]F^2\frac{\partial A}{\partial E}. \tag{4.14}$$

The term modulated at a frequency $2f$ is measured with a lock-in amplifier. The measured value with the lock-in amplifier corresponds to the transmittance change $\Delta T$, and then, using the separately measured transmittance $T$, the EA signal $-\Delta T\ /\ T$ is calculated. If needed, $\Delta A$ can be obtained from

$$\Delta A = -\mathrm{Log}\left(1 - \frac{\Delta T}{T}\right). \tag{4.15}$$

#### 4.3.2.2 TPE

In TPE measurements, excited states have to be created at an even-parity state via a TPA process. For this purpose, intense pulses with the photon energy corresponding to half of the excitation energy should be prepared, and then a wide spectral range has to be swept to find even-parity excited states. Thus, a tunable ns laser, e.g. an optical parametric oscillator pumped by a Q-switched Nd:YAG laser, may be suitable. After excitation with the intense pulses, a sample emits very weak luminescence, which is collected and led to a monochromator with a pair of lenses (Figure 4.12). Compared to the luminescence, scattering of the excitation pulse from the sample and windows of a cryostat is even stronger. To prevent the scattering light from entering a detector, in this example, an interference filter (or a bandpass filter) as well as a monochromator are put in front of a detector, e.g. the photomultiplier tube (PMT). In addition, to reduce the scattering as much as possible, a beam damper is used and the lenses are placed so as not to be struck by the excitation pulse. From a Q-switched Nd:YAg laser, light pulses with a pulse duration of 1 ns, and a repetition rate of 10 Hz, are typically outputted. This means that luminescence would be emitted only for 10 ns within 1 s. To eliminate noises coming during the rest of time, a boxcar integrator (or a boxcar average) may be needed. Its operational mechanism is also illustrated in Figure 4.12. In a boxcar integrator, a signal is only integrated over a short period of time that is defined by a boxcar gate, and noises coming from outside the measuring period are ignored. If necessary, an average

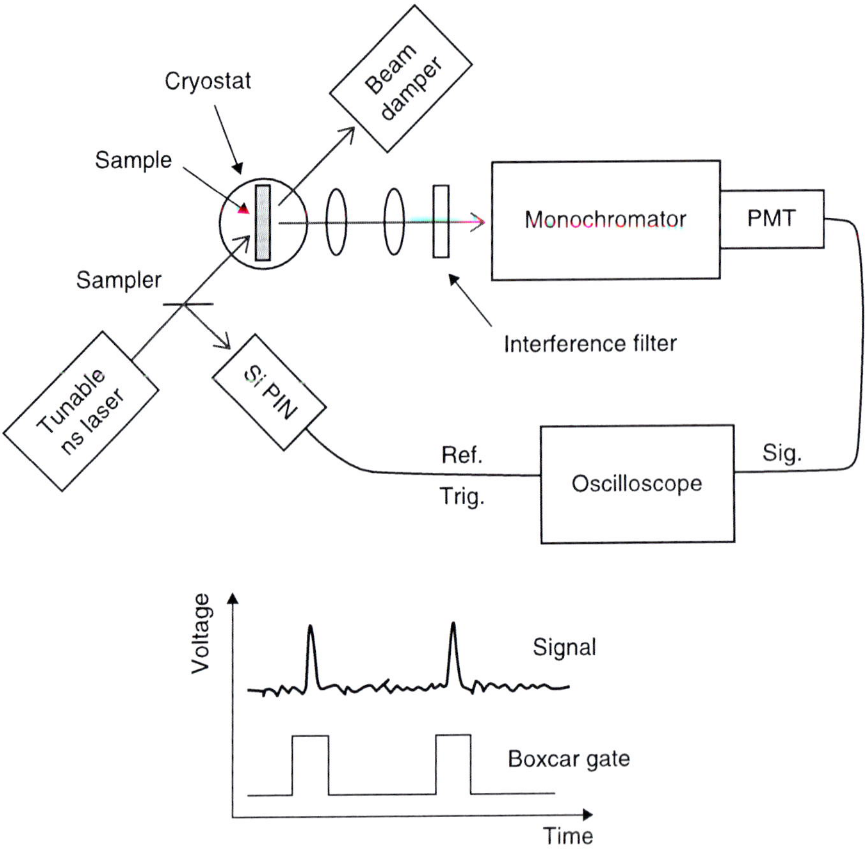


**Figure 4.12** (Upper) Experimental setup for TPE measurements. In this example, a pair of lenses, interferometer filter, and a monochromator are placed in front of a PMT. To simultaneously record intensities of the excitation pulse and luminescence generated by the same pulse, a beam sampler and a Si photodiode are used. An oscilloscope works as a boxcar integrator and is triggered by the signal from the photodiode. (Lower) Timing diagram to explain the function of a boxcar integrator.

over several integrated results is recorded. Recent oscilloscopes, however, have the same function, and thus a boxcar integrate can be replaced by one as shown in Figure 4.12. In this example, in addition to the luminescence intensity, excitation pulse intensity is measured using a beam sampler and a Si photodiode, and is processed similarly to the luminescence intensity in the oscilloscope. Q-switched Nd:YAG lasers usually exhibit relatively large timing jitter in relation to trigger signals (typically a few μs). Thus, a trigger should be generated not from the laser controller but from the outputted light pulses. In this example, the signal from the photodiode also serves as a trigger for the oscilloscope. Although such the configuration may have the disadvantage that a time difference between a signal and a trigger is very short, this does not matter at all if an oscilloscope is used instead of a boxcar integrator; in the latter case, a long coaxial cable may be needed to delay the signal from the PMT. In TPE measurements, the luminescence intensity is proportional to the square of the excitation pulse intensity. Thus, the recorded luminescence intensities are plotted as a function of the excitation pulse intensity in a log-log scale, and the TPA cross section is determined from the data that follow the square law. This process is performed for each photon energy to obtain the TPE spectrum.

### 4.3.3 Experimental Results

In this section, we show some examples of EA measurements for a variety of organic semiconductors and one example of TPE measurements for a π conjugated polymer.

#### 4.3.3.1 DE2

DE2 is a molecule that has tendency to form J-aggregates in solid state [35]. When it is dispersed in an inert polymer, such as poly(methyl methacrylate) (PMMA), J-aggregation can be prevented so that the absorption spectrum similar to that in solution is observed. On the other hand, in vapor-deposited neat thin films, a characteristic feature to J-aggregation, i.e. a redshifted and narrower absorption band, can be observed as shown in Figure 4.13. In the neat thin films, intermolecular coupling not only reduces the excitation energy but also delocalizes the excitation over coupled molecules. As a result, the excitation feels the averaged local inhomogeneities, which leads to a reduction of the width of inhomogeneous broadening [36].

Most of the reported J-aggregates are formed by ionic dyes, and it is known that the EA spectra of those dyes dispersed in a host matrix mainly consist of the second-derivative component [37–39]. On the other hand, DE2 is a neutral dye so that the influence of ionicity can be completely removed. In fact, the EA spectrum of DE2-dispersed thin film closely resembles the first derivative of its absorption spectrum (Figure 4.14) [40]. DE2 has only one excitation energy with non-zero oscillator strength in the observed spectral range, and its Stark shift is the main origin of the observed EA signal. Only the slight discrepancy between the EA and first-derivative spectra can be seen in a range from 2.4 to 3.0 eV. In this spectral range, DE2 has another excitation energy with zero oscillator strength. Interaction between the excitation and the applied electric field may be responsible for the discrepancy.

After J-aggregate formation, ionic dyes also exhibit the EA spectrum resembling the second derivative of their absorption spectrum. The origin of the second-derivative

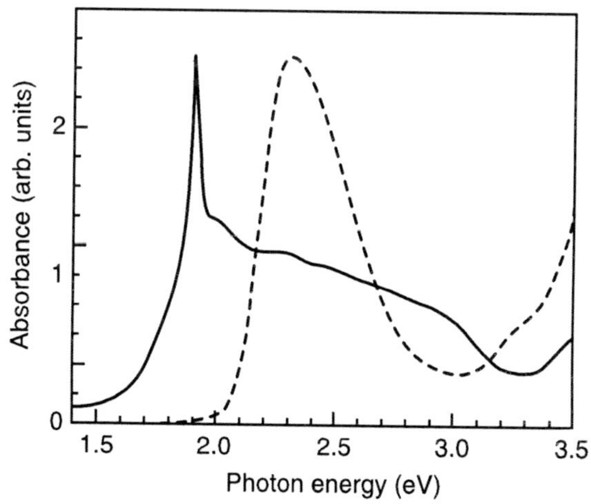


**Figure 4.13** Absorption spectra of (solid line) vapor-deposited neat thin films of DE2 and (dashed line) DE2-dispersed PMMA thin films. Because of its high crystallinity, its absorption spectrum of vapor-deposited thin films is dependent on film thickness[35]. The data is taken using a thin film with a thickness of 150 nm. Source: Modified from Ref. [35].

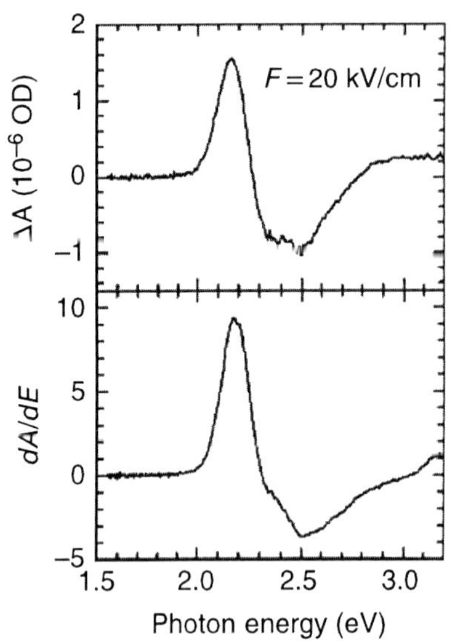


**Figure 4.14** (Upper) EA spectrum of DE2-dispersed PMMA thin films obtained with the applied electric field of 20 kV/cm at room temperature[40]. (Lower) The first derivative of its absorption spectrum with respect to energy. The absorption spectrum is shown in Figure 4.13. Source: Ref. [40].

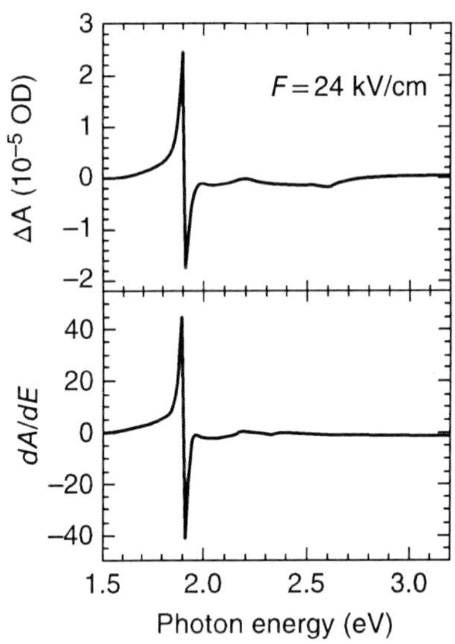


**Figure 4.15** (Upper) EA spectrum of vapor-deposited DE2 neat thin films obtained with the applied electric field of 24 kV/cm at room temperature[40]. (Lower) The first derivative of its absorption spectrum with respect to energy. The original absorption spectrum is shown in Figure 4.13. Source: Ref. [40].

component has been considered as large $\Delta\mu$ induced by the displacement of counter ions in thin films [41]. On the other hand, the EA spectrum of J-aggregates of DE2 is almost the same as the first derivative spectrum as shown in Figure 4.15 [40]. As a result of concentration of the oscillator strength in a very narrow spectral range (J-aggregate formation), the absolute $\Delta A$ value in vapor deposited thin films increases with respect to that in dispersed thin films. A Similar increase in $\Delta A$ due to J-aggregate formation is also observed in ionic dyes.

From the EA spectrum, $\Delta\alpha$ and $\Delta\mu$ can be determined using Eqs. (4.8) and (4.11) (more precise equations can be found in Ref. [31, 32]). It has been proposed that a ratio of $\Delta\alpha$'s of J-aggregate and dispersed samples is proportional to the aggregation number of the dye [37]. In the case of DE2, the enhancement of $\Delta\alpha$ is not significant, and the aggregation number is estimated to be only a few [40]. Such a small number suggests that J-aggregates of DE2 are not well developed in neat thin films although similar small numbers are also reported in some ionic J-aggregates [38, 39]. The small number for DE2 is rationalized as follows: on glass substrates, crystal nuclei with random orientations are formed in the early stages of the vapor-deposition process, and thus there is a limited space for subsequent development of each nucleus. In fact, on oriented polymer films, nuclei with the same orientation are formed and then further development is confirmed by the large absorption band attributable to J-aggregation [42]. Such oriented J-aggregates also exhibit greater EA signals.

From the EA spectrum, it is also possible to obtain the $\chi^{(3)}(-\omega; \omega, 0, 0)$ spectrum. To perform this calculation precisely, the low influence of the refractive index change due to the applied electric field upon the observed EA signal has to be taken into consideration. For that, the refractive index and its change are calculated from the absorption and EA spectra, respectively, using the Kramers-Kronig relations. All the necessary equations can be found in Ref. [38]. Then, the maximum $|\chi^{(3)}(-\omega; \omega, 0, 0)|$ values of DE2-dispersed PMMA and neat thin films are determined to be $1.5\times10^{-13}$ and $7.0\times10^{-10}$ esu, respectively [40]. The enhancement of the $\chi^{(3)}(-\omega; \omega, 0, 0)$ value mainly results from the concentration of the oscillator strength due to the J-aggregation. In the oriented J-aggregates, further enhancement of the $\chi^{(3)}(-\omega; \omega, 0, 0)$ value by a factor of 10 is achieved [42].

#### 4.3.3.2 Ir(ppy)$_3$

In Ir(ppy)$_3$, the absorption bands observed in a visible range are attributed to MLCT transitions, but the HOMO actually contains the $\pi$ orbital as well as the 5d orbital. Using quantum chemical calculations, the ratio of the $\pi$ orbital component in Ir(ppy)$_3$ is estimated to be 48% [16]. Such information is, however, unable to be obtained with any experimental techniques except EA spectroscopy. This means that there was no way to examine the validity of the calculated ratio before the proposal by means of EA spectroscopy by Stamper et al. [29]. They have used the fact that neutral excitations such as $\pi$-$\pi^*$ transitions induce the first-derivative such as EA spectrum whereas CT transitions result in the second-derivative like EA spectrum. As mentioned previously, the LUMO in Ir(ppy)$_3$ is the $\pi^*$ orbital so that, from the detailed analysis of the EA spectrum, the ratio of the two orbital components can be determined.

Figure 4.16 shows the EA spectrum of Ir(ppy)$_3$ neat thin films reported by Stamper et al. [29]. The EA spectrum cannot be reproduced with any of the first-derivative, second-derivative, and a sum of them. This is because the observed absorption band in a range from 2.4 to 4.2 eV consists of several transitions that have different $\Delta\mu$ values. Thus, prior to an analysis of the EA spectrum, Stamper et al. have decomposed the observed absorption bands with eight Gaussian functions. For this procedure, the following fact may be helpful: the second derivative of the absorption spectrum has negative peaks that indicate the peak positions of the constituent absorption bands. The eight Gaussian functions decomposed by Stamper et al. and the original absorption spectrum are shown in Figure 4.17. The EA spectrum can be fitted with a sum of the first and second derivatives of the eight absorption bands. Stamper et al. actually focused on the spectral range from 2.4 to 4.2 eV and thus performed the fitting procedure with

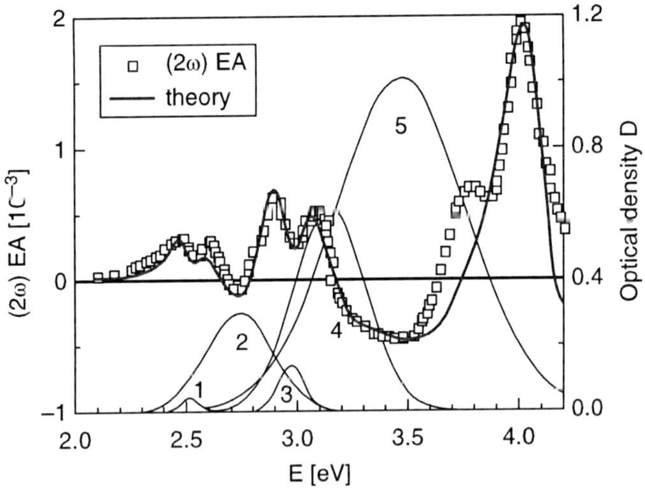


Figure 4.16 EA spectrum of $Ir(ppy)_3$ neat thin films. In this figure, the fitted result with Eq. (4.16) is also plotted with the thick curve. The five Gaussian profiles represent the lowest absorption bands found in the experimental absorption spectrum (see Figure 4.17). Source: Ref. [29].

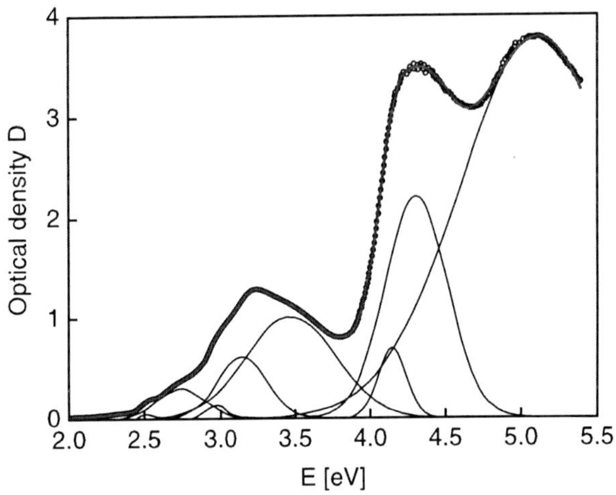


Figure 4.17 Absorption spectrum of $Ir(ppy)_3$ neat thin films, and a fit with a sum of eight Gaussian profiles. In this figure, all of the used profile is depicted. Source: Ref. [29].

only the lowest five Gaussian functions using [29]

$$\Delta A \propto -\frac{1}{2}\Delta\alpha F^2 \frac{\partial A}{\partial E} F^2 + \sum_{n=1}^{5} \frac{1}{6}\Delta\mu_n^2 F^2 \frac{\partial^2 A_n}{\partial E^2}, \tag{4.16}$$

where $n$ indicates the Gaussian number, which is shown in Figure 4.16. In this equation, the five transitions are supposed to have the same $\Delta\alpha$ because $\Delta\alpha$ is virtually independent of the ratio of both orbital components. This is in contrast to $\Delta\mu$ which is strongly dependent on the ratio. As shown in Figure 4.16, the quality of the fitting is good. From the fitting, $\Delta\mu$ for the absorption band 2, which corresponds to the lowest singlet excitation, is found to be 3.8 D [29]. If it is assumed that 50% of *e* is effectively transferred from the metal to the ligands upon excitation, the resultant $\Delta\mu$ would be 3.9 D. This value is in good agreement with the experimental $\Delta\mu$ mentioned above. Thus, the ratio of the 5d orbital component included in the HOMO is determined to be slightly less than 50%, which is very close to the theoretical value (52%) by Hey [16].

#### 4.3.3.3 PFO

In 1D systems, the oscillator strength concentrates on the exciton levels so that it is difficult to find the bottom state of the continuum state in the conventional absorption spectrum. An alternative method utilized to locate the bottom state is one with nonlinear spectroscopy. It is theoretically revealed that in π conjugated polymers, the bottom state is strongly dipole coupled with an even-parity state lying slightly lower in energy than the bottom state [43, 44]. The bottom state and even-parity state are often written as $nB_u$ and $mA_g$, respectively, where $n$ and $m$ are unknown numbers. Based on this theoretical result, the position of the $nB_u$ state (and the $mA_g$ state) can be determined with nonlinear spectroscopy, and in particular, EA measurements have been applied to a various π conjugated polymers [45, 46]. In this chapter, α-phase PFO is dealt with as an example [47]. As the lowest excited state of typical π conjugated polymers is not a CT excited state, it may be expected that the EA spectrum closely resembles the first derivative of the absorption spectrum. The first derivative of the absorption spectrum, however, explains only one part of the experimental EA spectrum. As shown in Figure 4.18, the EA spectrum of α-phase PFO consists of two oscillatory features around 3.0 and 4.0 eV, and only the former is similar to the first derivative. This means that the former feature is due to the Stark shift of the lowest odd-parity excited state (the $1B_u$ state). On the other hand, the feature around 4.0 eV cannot be explained even with the second derivative of the absorption spectrum. Thus, it can be attributed to neither the Stark shift nor the change in the dipole moment (CT transitions).

According to the perturbation theory, the third-order optical susceptibility can be calculated from an assumed excited-state structure with transition dipole moments between the involved states. The equation, consisting of so many terms, can be found in Ref. [30, 45, 48]. For the case of α-phase PFO, a four-level model consisting of the $1B_u$,

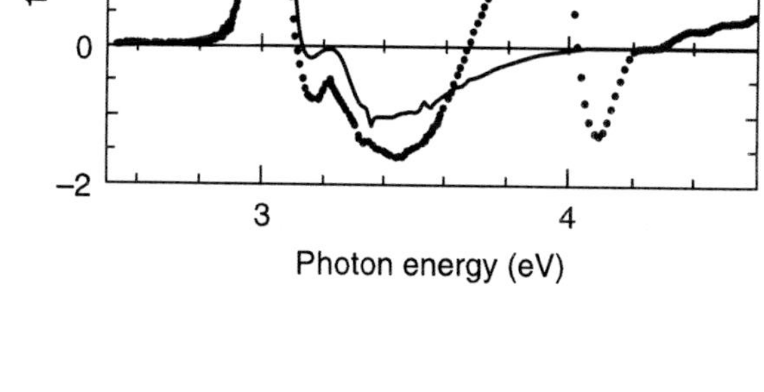


**Figure 4.18** EA spectrum of thin films of α-phase PFO measured at 100 K.. Source: Ref. [47].

$mA_g$, and $nB_u$ states and the ground state ($1A_g$) is needed (see the inset of Figure 4.19). The $\chi^{(3)}(-\omega;\omega,0,0)$ spectrum calculated from the four-level model is finally related to the EA spectrum with the relationship

$$-\frac{\Delta T}{T} = \frac{4\pi\omega}{nc}\mathrm{Im}[\chi^{(3)}(-\omega;\omega,0,0)]F^2 d, \tag{4.17}$$

where $n$ is the refractive index, $c$ is the speed of light, and $d$ is the film thickness. Thus, from the fitting to the resultant EA spectrum, the positions of the three excited states and transition dipole moments between them and the ground state can be determined. To obtain better fitting results, the vibronic structures and an asymmetric shape of absorption bands may also be taken into account [45]. The best result obtained in this manner is shown in Figure 4.19a, and the excited-state structure determined from the fit is depicted in the inset of Figure 4.19b. The fitted result reproduces the experimental EA spectrum much better than the first-derivative spectrum. This means that the $mA_g$ state plays an important role in the nonlinear optical process. The simple four-level model is known to explain the EA spectra of many π conjugated polymers [45–47] well so that the three excited states are sometimes called the *essential states* [43]. However, contributions from other excited states can be recognized in β-phase PFO, where inhomogeneous broadening is significantly reduced [49]. The slight discrepancy between the fitted and experimental results in Figure 4.19a is mainly due to the omission of these additional excited states. In spite of the omission, the determined positions of the $mA_g$ and $nB_u$ states are considered to be reliable because of the following facts: as shown in

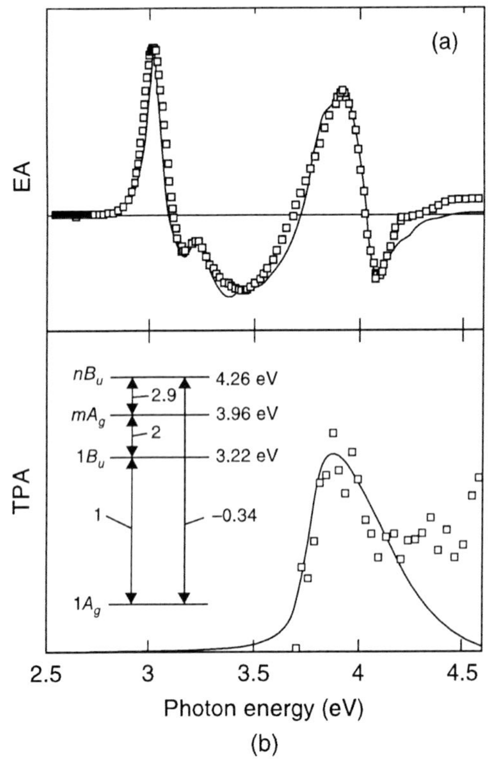


**Figure 4.19** (a) EA and (b) TPE spectra of thin films of α-phase PFO measured at 100 K. Solid lines are the results calculated using Eqs. (4.17) and (4.18) with the same physical constants. The inset in (b) shows the assumed energy diagram with energy levels and transition dipole moments, which are indicated in units of one between the $1B_u$ and $1A_g$ states. Source: Ref. [47].

Figure 4.19b, the peak of the TPE spectrum appears at almost the same position of the determine $mA_g$ state (3.96 eV). Note that similar to the EA spectrum, TPE spectrum can also be simulated from an assumed excited-state structure [30, 48]: From the calculated $\chi^{(3)}(-\omega;\omega,\omega,-\omega)$ spectrum, the TPE spectrum (or the TPA cross section $\sigma^{(2)}$ spectrum) can be calculated using the following relationship

$$\sigma^{(2)}(\omega) = \frac{8\pi^2\hbar\omega^2}{n^2c^2}\mathrm{Im}[\chi^{(3)}(-\omega;\omega,-\omega,\omega)]. \tag{4.18}$$

The simulated TPE spectrum from the excited-state structure determined from the EA spectrum is also shown in Figure 4.19b. Also, the omission of the additional excited states does not influence the position of the $nB_u$ state (4.26 eV). It has been found that the position of the $nB_u$ state is independent of the polymer conformation [49, 50]. The $nB_u$ state is the bottom state of the continuum so that the photoexcitation above it is expected to immediately generate separated charges. In fact, photoconductivity efficiency steeply rises around 4.3 eV in both α- and β-phase PFO [49, 50]. These facts support the validity of the determined energy levels of the $mA_g$ and $nB_u$ states. From the determined excited-state structure, it is also found that the exciton binding energy in PFO is estimated to be around 1 eV.

## 4.4 Decay Mechanism of Excited States

It is important to understand how excited states return to the ground state. In particular, for emitters for light-emitting devices, a ratio of radiative and nonradiative decay rates is directly related to the device performance. However, organic semiconductors have a variety of decay channels, and the decay mechanism is, in general, very complicated. In this section, we will first explain major decay channels in organic semiconductors, and then describe some experimental techniques for time-resolved PL and PLQE measurements. Finally, we will show some examples of experiments and analysis for a π conjugated polymer, phosphorescence emitter, and TADF emitter.

### 4.4.1 Background

Major decay channels in a molecule are summarized in a Jablonski diagram, which is basically an energy diagram with some vibrational levels (Figure 4.20). After a photon is absorbed, a singlet excited state is created. The energy interval between the excited state and the ground state ($S_0$) is exactly the same as the energy of the absorbed photon. The created excited state immediately starts decaying to a lower vibrational level of the same electronic state (vibrational relaxation, VR) and, at the same time, to a lower electronic state (internal conversion, IC). In these processes, the energy that the excited state loses is transferred to vibrational (kinetic) energy. These processes are nonradiative and ultrafast (decay rates are estimated to be $10^{-14} \sim 10^{-13}$ $s^{-1}$) [51]. Thus, the excited state soon reaches the zero vibrational level of the lowest excited state ($S_1$). Since VR and IC are several orders of magnitude faster than any radiative transitions, PL always seems to be emitted from $S_1$ (or $T_1$). This is the reason why PL spectrum does not depend on the excitation photon energy (Kasha's rule) [52].

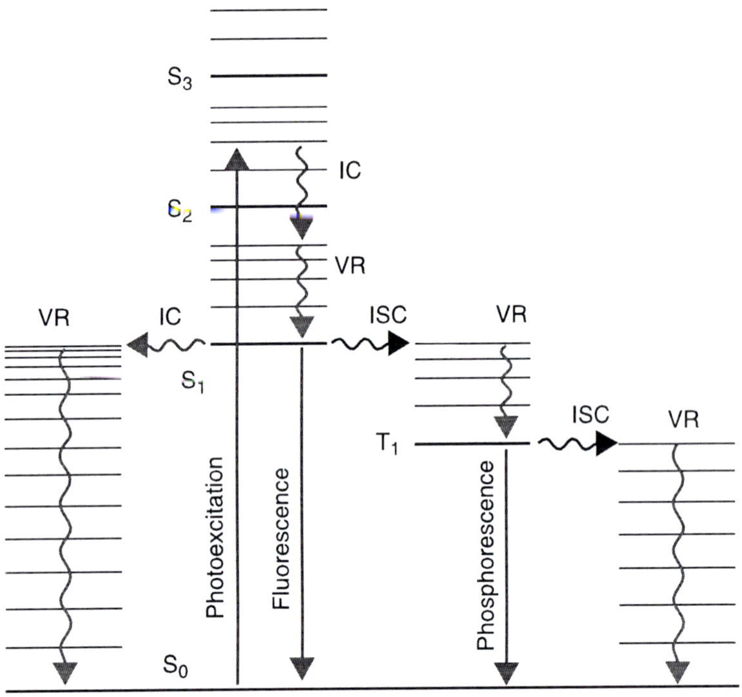


**Figure 4.20** A Jablonski diagram. In this diagram, possible decay channels of the excited state, i.e., internal conversion (IC), vibronic relaxation (VR), intersystem crossing (ISC), and radiative transitions with fluorescence or phosphorescence emission, are illustrated along with the upward one due to photoabsorption. Straight and wavy arrows represent transitions associated with and without light, respectively.

From $S_1$, there are three possible decay channels, i.e. radiative transition into $S_0$ with fluorescence emission, nonradiative transition into $S_0$, and intersystem crossing (ISC) into a triplet excited state. The radiative transition rate, which is proportional to a square of the transition dipole moment between the associated excited state and $S_0$, is in the order of $10^{-9} \sim 10^{-8}$ $s^{-1}$ in fluorescence emitters [51]. Compared to this value, ISC is inefficient in fluorescence emitters, where the spin-orbit coupling is weak, and is thus omitted frequently. As shown in Figure 4.20, the nonradiative transition consists of IC into a higher vibrational level of $S_0$ and subsequent VR, and its rate significantly differs in molecules and surrounding circumstances. As a result, PLQE varies from 0 to 1. On the other hand, phosphorescence emitters have the strong spin-orbit coupling, and their ISC is so efficient that fluorescence from $S_1$ is completely quenched, although their radiative transition rate from $S_1$ to $S_0$ is somewhat smaller than that of fluorescence emitters. After ISC, the excited state reaches the zero vibrational level of $T_1$ via VR and, if necessary, IC. Then, the excited state radiatively and nonradiatively returns to $S_0$. Note that this nonradiative transition includes ISC as shown in Figure 4.20. A radiative transition rate from $T_1$ into $S_0$ ranges from $\sim 10^{-6}$ (phosphorescence emitters) to $\sim 10^{0}$ $s^{-1}$ (fluorescence emitters).

Important decay channels not illustrated in a Jablonski diagram are related to interactions with other molecules. Among them, we briefly describe Förster resonance energy transfer and Dexter electron transfer [53]. In the former process, energy of an excited donor molecule is transferred to an acceptor molecule in its ground state via dipole-dipole coupling. Efficiency of this process depends on the radiative decay rate of a donor molecule and the absorption coefficient of an acceptor molecule as well as the distance between the two molecules. Because of the characteristic inverse sixth power law dependence on the distance, this process is regarded as a long-range process in comparison with the Dexter electron transfer process. In the latter, energy is transferred via electron exchange between donor and acceptor molecules. For this reason, an overlap of their wavefunctions is required. Efficiency of this process decreases exponentially with the distance between them. Thus, this process only takes place between two molecules that are extremely close to each other. Note that in either process, efficiency increases with rising spectral overlap between fluorescence of a donor molecule and absorption of an acceptor molecule.

There are many examples associated with Förster or Dexter processes: in a host-guest system, an excited state is created on a host molecule and its energy is transferred to a guest molecule via the Förster resonance energy transfer mechanism. Phosphorescence from phosphorescence emitters is often quenched by oxygen; this is due to Dexter electron transfer between an excited emitter and oxygen. In OLEDs, at high current injection levels, excited states are quenched by exciton-exciton interactions such as singlet-triplet annihilation (STA) and triplet-triplet annihilation (TTA). These annihilation processes are caused by energy transfer between singlet and triplet excited states via Förster resonance energy transfer (STA) and between excited triplet states via Dexter electron transfer (TTA). STA and TTA will be dealt with again in Section 4.4.3.3. In addition, self-quenching in phosphorescence emitters has been proposed to result from Förster resonance energy transfer between guest molecules [54].

### 4.4.2 Experimental Technique

In this section, the basics of some experimental techniques for time-resolved PL and PLQE measurements will be described. We will not deal with more fundamental techniques, such as PL spectrum measurements. However, as organic semiconductors have very wide PL spectral shapes, attention must be paid to wavelength dependences of the efficiency of a diffraction grating, the focal length of lenses, and the sensitivity of photodetectors. In the case of absorption measurements, conventional spectrophotometers employ a double-beam method, in which absorption coefficient is determined by a ratio of transmittances measured separately for a sample and a reference. Thus, wavelength dependence of used optical components is automatically cancelled. However, in the case of PL measurements, wavelength dependence should be manually corrected using a calibrated light source.

#### 4.4.2.1 Time-resolved PL Measurements

One of common ways for time-resolved PL measurement is to use a streak camera. In this way, PL from a sample, which is photoexcited with an fs laser, is simply led to a slit of a monochromator attached to a streak camera. As shown in Figure 4.21, a photon that arrives at the photocathode of a streak tube is converted into an electron, and then the electron is accelerated and flies toward the micro-channel plate (MCP). Between the photocathode and MCP, there is a pair of sweep electrodes, which generate an electric

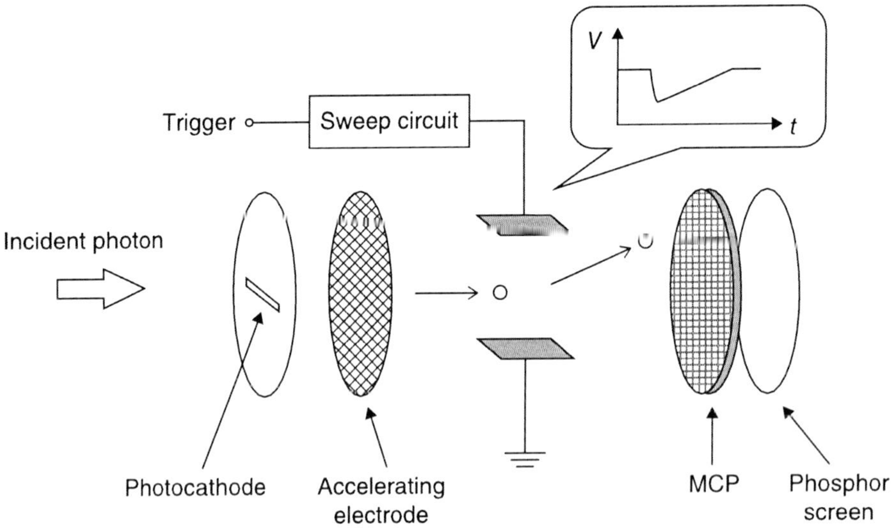


**Figure 4.21** Operational mechanism of a streak tube.

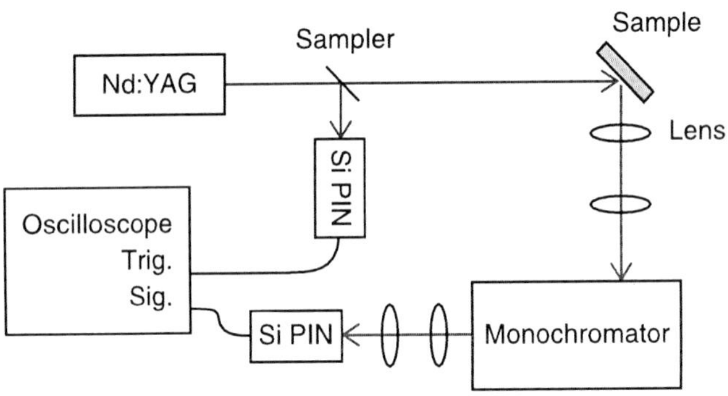


**Figure 4.22** Setup for ns time-resolved PL measurements using a ns laser and an oscilloscope.

field modulated synchronously with a laser pulse so that deflection angle changes from upward to downward in accordance with time after the laser pulse irradiation. Inside the MCP, the electron is multiplied and is finally detected as a signal. Thus, the time variation of PL intensity is transferred into a spatial profile in the vertical direction. Photons incident to a streak tube are already spectrally resolved by the attached monochromator. Thus, the resultant data becomes 2D data (a streak image) where vertical and horizontal axes are time and wavelength, respectively. Time resolution of this type is usually limited to a few ps ~ a few tens of ps by a streak camera.

An easier way for time-resolved PL measurement is to use a high-speed photodiode and an oscilloscope. In this case, the time resolution is limited to a few ns by the electronics. Thus, a Q-switched Nd:YAG laser with a pulse duration of a few ns is suitable as an excitation light source. An experimental setup is illustrated in Figure 4.22. For PL measurements of organic semiconductors, the second harmonic (532 nm) or the third harmonic (355 nm) of a Q-switched Nd:YAG laser is usually selected. In this example, a trigger signal is acquired using an additional photodiode as in the example in Figure 4.12. As shown in Figure 4.22, it is desirable to collect PL from the front side of the sample, from which laser pulses are irradiated. This arrangement is sometimes called reflection geometry. In transmission geometry, PL spectrum is possibly altered by reabsorption effect. Recent oscilloscopes may have a huge memory but the measurable time range is limited to when PL intensity decreases to 1/256 of its maximum value because the conventional oscilloscope is still an 8-bit system.

If a decay rate of PL is much slower, a combination of a PMT and a multichannel scaler (or a photon counting system) is suitable. In this combination, the measured time range is split into a large number of small dwell times, and the number of incident photons within each dwell time is counted. Then, the number of incident photons, as a function of time after laser irradiation, is obtained. In this method, a thermal noise and a signal due to an incident photon are separated by an appropriately tuned threshold and only the number of photons is counted. Thus, noise is significantly suppressed. Note that in a streak tube, mentioned above, the photon number can be counted in the same manner. In the method using a PMT and a multichannel scaler, PL intensity has to be attenuated so that more than two photon do not enter the PMT within the time resolution (much shorter than a dwell time). Thus, the scan must be repeated many times to obtain a smooth PL decay curve. Since the repetition rate of laser pulses may not be able to increase because of the slow PL decay rate, a long measuring time is probably required in this method. An experimental setup is illustrated in Figure 4.23. In this example, a pulse generator is used as a source of triggers, which are sent to a Nd:YAG laser and a multichannel scaler. In this configuration, a multichannel scaler can receive a trigger in advance of receiving any signals due to photons. If a conventional (passive) Q-switched Nd:YAG laser is operated in an external trigger mode, the stability of pulse intensity will be degraded although this does not cause any problem in linear optical spectroscopy such as time-resolved PL measurements.

Time-resolved PL measurements with a far higher time-resolution can be carried out using an optical Kerr gate. In this way, an fs laser pulse is split into pump and gate pulses by a beam splitter. A sample is photoexcited with the pump pulse to generate PL, which passes through a polarizer and is focused on the surface of a Kerr medium. The gate pulse goes through a delay line and is focused on the surface to overlap with the polarized PL. After passing through the Kerr medium, the PL is blocked with another crossed

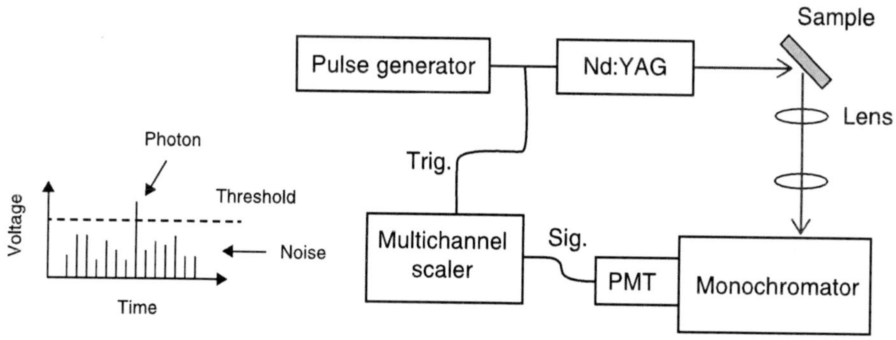


**Figure 4.23** (Left) Mechanism of photon counting. (Right) Setup for time-resolved PL measurements with a photomultiplier tube (PMT) and a multichannel scaler.

polarizer. In the Kerr medium, however, polarization of the PL is rotated during the gate pulse and is irradiated due to a nonlinear optical effect, so that only the rotated PL component can pass through the second polarizer. Finally, by recording the intensity of the transmitted PL component, as a function of delayed time of the gate pulse, a temporal profile of PL can be recorded with a time resolution of a few hundred fs. For a more detailed description of this method, see Refs. [55, 56].

#### 4.4.2.2 PLQE Measurements

Figure 4.24 shows an experimental setup for PLQE measurements using an integrating sphere [57]. As an excitation light source, a continuous-wave laser or a monochromated light from a lamp can be used. In this way, emission spectra of a reference and a sample are first recorded. Here, we assume that a reference and a sample are a quartz substrate and an organic thin film coated on a quartz substrate, respectively. In the spectrum recorded for the reference, only scattered excitation light will appear. On the other hand, in the spectrum recorded for the sample, a reduced excitation light and the emission would be observed. The reduction of the scattered excitation light indicates that an excitation light is partially absorbed by the sample. Thus, in the differential spectrum of the two spectra, areas of the negative and positive bands correspond to the photon numbers being absorbed by the sample and emitted from the sample (Figure 4.24). The ratio of the two areas is, therefore, equal to PLQE of the sample. It should be noted that an integrating sphere is sometimes misunderstood to be a device that collects all photons emitted from a sample inside. The actual function of an integrating sphere is to improve spatial uniformity of light emitted from a sample. In fact, reflectivity of the inner surface of an integrating sphere is not 100%. Therefore, measured PLQE values may be slightly affected by a surface profile of a sample and reflectivity of its surface.

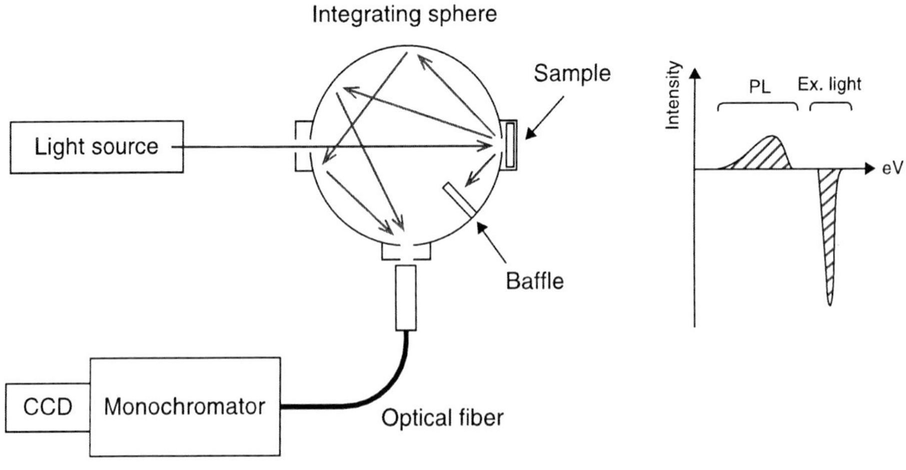


**Figure 4.24** (Left) Experimental setup for PLQE measurements using an integrating sphere. An integrating sphere for this purpose usually has a baffle to prevent first reflections from going out from the sphere. (Right) Expected differential PL spectrum, in which areas of the positive and negative bands are proportional to photon numbers of emission and excitation light absorbed by a sample, respectively.

### 4.4.3 Experimental Results

In this section, we will show some examples of PL measurements on several organic semiconductors.

#### 4.4.3.1 PFO

Figure 4.25 shows the PL spectrum of α-phase and β-phase PFO. PFO consists only of carbon and hydrogen atoms (Figure 4.1d) so that the spin-orbit coupling is very small. Thus, ISC in PFO is ignored for a while. Both spectra in Figure 4.25 can be attributed to fluorescence because their decay rates are in the order of $10^9$ $s^{-1}$, as shown later. Thin films of π conjugated polymers can be regarded as an ensemble of segments with various effective conjugation lengths. Since excited states tend to migrate to segments with lower energy prior to decay into $S_0$ with fluorescence emission, PL originates only from segments with relatively long effective conjugation. This is in striking contrast to the absorption spectrum, to which all segments contained in a sample contribute. Thus, this migration process explains why absorption spectra of π conjugated polymers are always much broader than their PL spectra.

However, excited states do not radiatively decay from only a small number of segments with extremely long effective conjugation. The probability that an excited state finds a segment with longer effective conjugation nearby decreases rapidly as the migration proceeds. Hence, most of the excited states cannot reach segments with extremely long effective conjugation but rather decay radiatively from segments with somewhat shorter segments. A kind of threshold of effective conjugation length may exist; excited states rarely reach segments much longer than the threshold, and PL mainly originates from segments with effective conjugation lengths around the threshold. Therefore, as long as excited states are created in shorter segments than the threshold, the PL spectrum does not depend on the excitation photon energy. The situation is, however, changed if excited states are directly created in longer segments than the threshold using an excitation light with an appropriately lowered photon energy. In such a condition, only segments with similar effective conjugation lengths are selectively excited, and excited states radiatively decay prior to the migration. As a result, a narrower and redshifted PL spectrum is observed and, in addition, the PL spectrum redshifts with decreasing

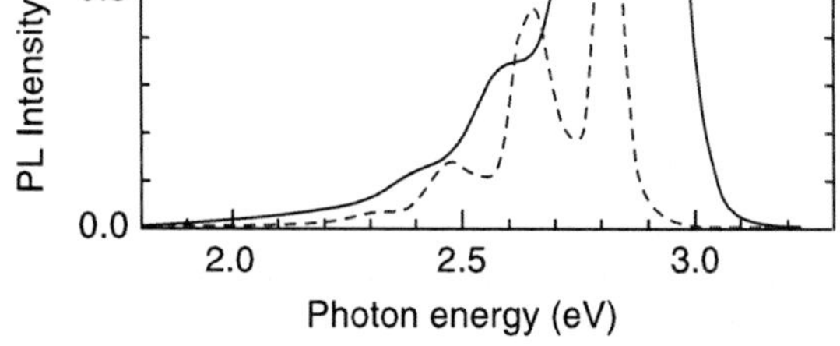


**Figure 4.25** PL spectra of α-phase and β-phase PFO measured using a He-Cd laser (325 nm) at room temperature. The PL spectra are the same as those in Figure 4.8.

excitation photon energy. Such a dependence of the PL spectral shape on the excitation photon energy is confirmed in many $\pi$ conjugated polymers [58]. In the case of PFO, a distribution of effective conjugation length can be largely controlled by changing fabrication conditions [26, 59], and the change in the distribution influences the threshold of effective conjugation length. As shown in Figure 4.25, the PL spectrum of β phase is narrower and redshifted with respect to that of α phase. This spectral difference can be explained by the fact that longer segments are contained in β phase than in α phase.

It is known that β-phase PFO has higher PLQE than α-phase PFO [26]. The PLQE values determined using an integration sphere are $\Phi = 0.54$ for α phase and $\Phi = 0.70$ for β phase. Only from these values, however, the reason of the improvement of PLQE cannot be clarified. The possible reason is an enhancement of radiative decay, a suppression of nonradiative decay, or both. To identify this reason, PL lifetime is determined in addition to PLQE. PL lifetime is the reciprocal of the total PL decay rate $k$, which is the sum of radiative and nonradiative decay rates, $k_r$ and $k_{nr}$. In Figure 4.26, we show PL decay curves of α-phase and β-phase PFO measured using a streak camera. In either case, PL decreases exponentially. Thus, using single exponential fits, PL lifetimes can be determined to be $\tau = 0.46$ ns (α phase) and $\tau = 0.28$ ns (β phase). As well known, $\Phi$ and $\tau$ are related to $k_r$ and $k_{nr}$ as follows:

$$\tau = \frac{1}{k} = \frac{1}{k_r + k_{nr}}, \tag{4.19}$$

$$\Phi = \frac{k_r}{k_r + k_{nr}} = \tau k_r. \tag{4.20}$$

These equations can be rewritten as

$$k_r = \frac{\Phi}{\tau}, \tag{4.21}$$

$$k_{nr} = \frac{1 - \Phi}{\tau}. \tag{4.22}$$

Using these relationships, $k_r$ and $k_{nr}$ can be calculated from the experimentally determined $\tau$ and $\Phi$. The results are summarized in Table 4.1. From these values, it has been found that the improvement of PLQE is caused by the enhancement of $k_r$ but not by the

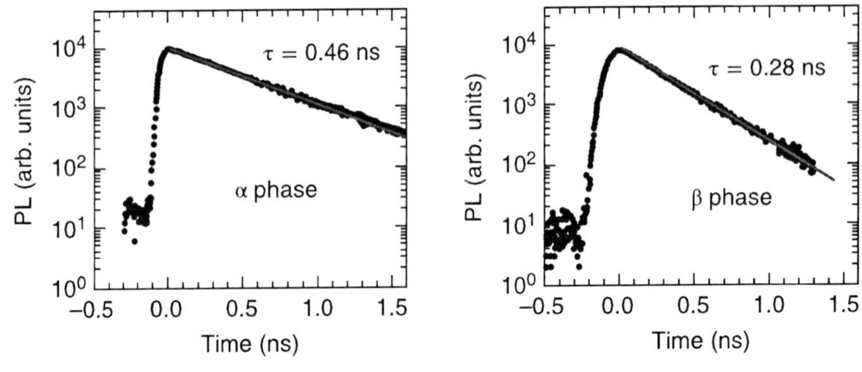


**Figure 4.26** PL decay curves of (left) α-phase and (right) β-phase PFO measured at room temperature using a streak camera and a fs laser. Solid lines are single exponential fits with time constants of 0.46 and 0.28 ns.

**Table 4.1** Measured PLQE (Φ) and PL lifetime (τ) and calculated radiative and nonradiative transition rates ($k_r$ and $k_{nr}$) for α-phase and β-phase PFO.

| | Φ | τ (ns) | $k_r$ ($10^9$ s$^{-1}$) | $k_{nr}$ ($10^9$ s$^{-1}$) |
|---|---|---|---|---|
| α phase | 0.54 | 0.46 | 1.2 | 1.0 |
| β phase | 0.70 | 0.28 | 2.5 | 1.1 |

suppression of $k_{nr}$ [26]. In β phase, excited states decay from longer segments than those in α phase. It is known that transition dipole moment rises with increasing effective conjugation length. Thus, the migration to further longer segments in β phase results in the enhancement of $k_r$.

In π conjugated polymers, ISC is inefficient but its rate is not zero. If nonradiative transitions are well suppressed, very weak phosphorescence may be observed. In fact, phosphorescence has been measured in many π conjugated polymers [60]. In general, phosphorescence lifetime is much longer than fluorescence lifetime. The difference in lifetime allows us to record a very weak phosphorescence spectrum using a PMT with a photon counting system or a CCD with a high sensitivity. In an actual experimental setup, a mechanical shutter is placed in front of the detector to prevent intense fluorescence from entering the detector. The sample is photoexcited by a laser pulse, and after fluorescence becomes weak sufficiently or disappears, the shutter is opened and then a measurement starts. A phosphorescence spectrum of α-phase PFO recorded in this way is shown in Figure 4.27. Although a signal-to-noise ratio is not good compared to its fluorescence spectrum, a clear vibronic structure can be seen. From a comparison between the fluorescence and phosphorescence spectra, it is found that the associated photon energy is almost the same.

#### 4.4.3.2 Ir(ppy)$_3$

In phosphorescence emitters, the spin-orbit coupling is strong so that created singlet excited states efficiently decay into $T_1$ via ISC. Thus, the observed PL can be attributed to phosphorescence. A characteristic feature of phosphorescence is its slow decay rate (or its long PL lifetime), which is several orders of magnitude lower than that of emission

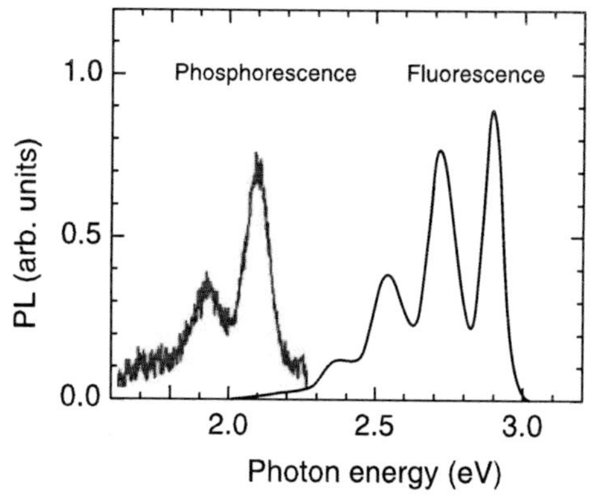


**Figure 4.27** Fluorescence and phosphorescence spectra of α-phase PFO measured at 6.5 K. Note that for phosphorescence measurements, more intense photoexcitation pulses and longer integration time were used than for fluorescence measurements.

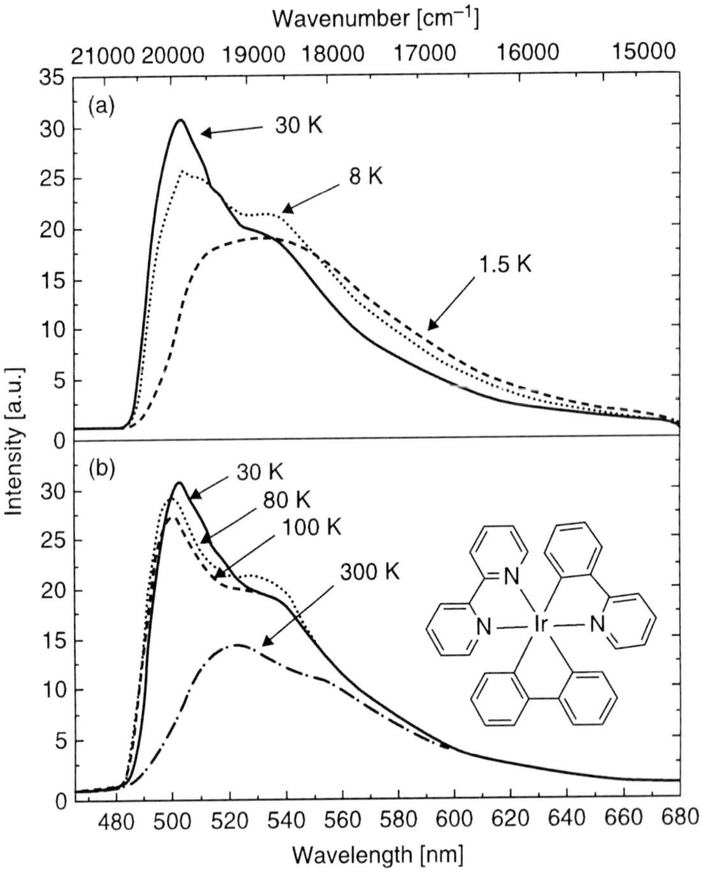


**Figure 4.28** PL spectra of $Ir(ppy)_3$ in THF at various temperatures. Source: Ref. [61].

from $S_1$ in fluorescence emitters; a typical decay rate in phosphorescence emitters is around $10^6$ $s^{-1}$, which corresponds to PL lifetime in the order of μs.

Figure 4.28 shows the temperature dependence of the phosphorescence spectra of $Ir(ppy)_3$ in tetrahydrofuran (THF) reported by Finkenzeller et al. [61]. As temperature increases from 1.5 to 30 K, phosphorescence spectrum shifts and its intensity grows. Above 30 K, however, the spectral shift and intensity increase are no longer observed. The relatively large intensity change between 100 and 300 K is due to the phase transition of THF from the frozen solution to the liquid phase at 165 K. In a liquid solution, $Ir(ppy)_3$ does not exhibit good PLQE [62]. On the other hand, in solid-state matrixes such as 4,4'-N,N'-dicarbazole-biphenyl (CBP) and PMMA, PLQE of $Ir(ppy)_3$ is nearly 1.0 at room temperature, and as such excellent PLQE is maintained across a wide temperature range [63, 64]. As an example of the temperature dependence of PLQE of $Ir(ppy)_3$, we show one reported by Hofbeck et al. [64] in Figure 4.29. The data was taken using $Ir(ppy)_3$ lightly doped in PMMA. In this example, PLQE is almost the constant between 77 and 370 K.

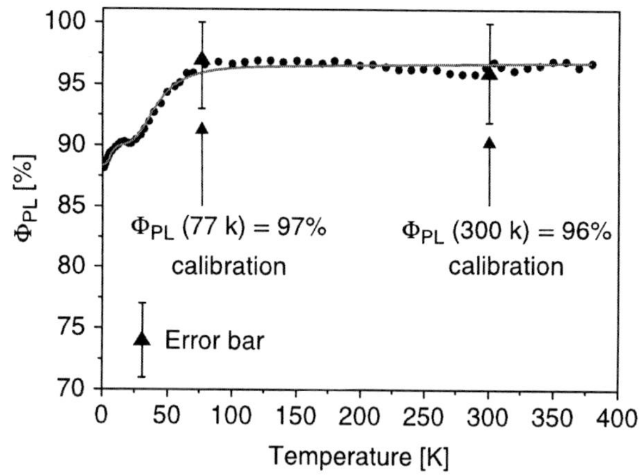


**Figure 4.29** (Circles) Measured PLQE vs temperature of $Ir(ppy)_3$ in PMMA (0.01 wt%). (Solid lines) The fit according to Eq. (4.26). Source: Ref. [64].

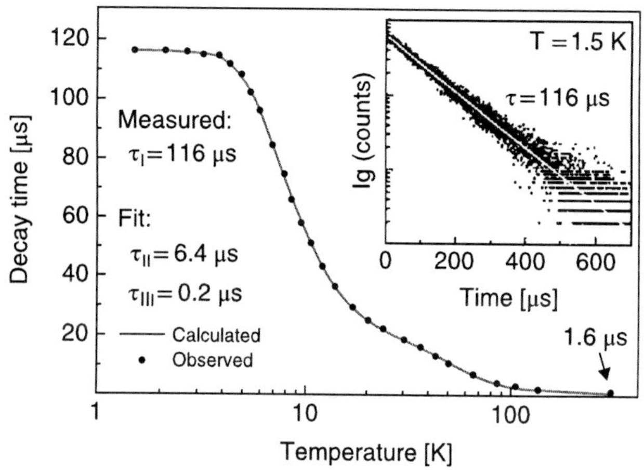


**Figure 4.30** (Circles) Measured PL decay time vs temperature of $Ir(ppy)_3$ in dichloromethane, whose melting point is 175 K. (Solid lines) Fitted results with Eq. (25). The inset shows the PL decay curve measured at 1.5 K and a single exponential fit with a time constant of 116 μs. Source: Ref. [64].

In general, $k_{nr}$ depends on temperature and is decreased at lower temperatures while $k_r$ is virtually independent of temperature. Thus, the resultant improvement of PLQE at lower temperatures is frequently observed [65]. Unlike $Ir(ppy)_3$ in a solid-state host matrix or in a frozen solution, its neat thin films indeed exhibit such temperature dependent $k_{nr}$ and consequent temperature dependent PLQE [66].

From the lack of temperature dependence of PLQE (Figure 4.29), the temperature independence of PL decay rate might be expected. However, the strong temperature dependence of the decay rate has been observed in neat and doped thin films and in frozen solution [61, 63, 64, 66]. In Figure 4.30, we show the reported temperature dependence of the PL decay rate of $Ir(ppy)_3$ in dichloromethane [64]. At 1.5 K the decay time, which is equal to PL lifetime and is the inverse of the PL decay rate, is two orders of magnitude longer than that at 300 K. This strong temperature dependence is explained in terms of three triplet substates, which are split due to the spin-orbit coupling and have different characters (Figure 4.31). Since their energy separations are very small, excited states can populate the higher substates at room temperature and then radiatively decay from those. On the other hand, at very low temperatures, excited states radiatively decay only from the lowest substate. Using the Boltzmann distribution, a population of the substate $i$ at temperature $T$ can be expressed as

$$n_i = N\frac{e^{-\Delta E_{I,i}/(k_B T)}}{\sum_{i=I,II,III} e^{-\Delta E_{I,i}/(k_B T)}}, \tag{4.23}$$

where $N$ is the total population at time $t$, $k_B$ is the Boltzmann constant, and $\Delta E_{I,i}$ is the energy of substate $i$ measured from the lowest substate I; thus $\Delta E_{I,I}$ is zero. Here, we ignore a population at further higher excited states. The total decay rate $k$ (or decay time $\tau$) can be written as

$$\frac{dN}{dt} = -kN\left(= -\frac{1}{\tau}N\right) = -\sum_{i=I}^{III} k_i n_i \tag{4.24}$$

where $k_i$ is the decay rate from substate $i$. By substituting Eq. (E23) into Eq. (E24), the following expression is obtained:

$$k = \frac{k_I + k_{II}e^{-\Delta E_{I,II}/(k_B T)} + k_{III}e^{-\Delta E_{I,III}/(k_B T)}}{1 + e^{-\Delta E_{I,II}/(k_B T)} + e^{-\Delta E_{I,III}/(k_B T)}} \tag{4.25}$$

Note that this expression assumes the Boltzmann distribution is maintained during the decay process. The solid line in Figure 4.30 is the fit with Eq. (4.25), and an excellent agreement can be seen. The determined physical constants in addition to the measured $\tau_I$ are shown in Figures 4.30 and 4.31. It is found that $k_{III}$ is three orders of magnitude slower than the measured $k_I$. This very slow decay rate suggests that the substate III is a pure triplet excited state [64].

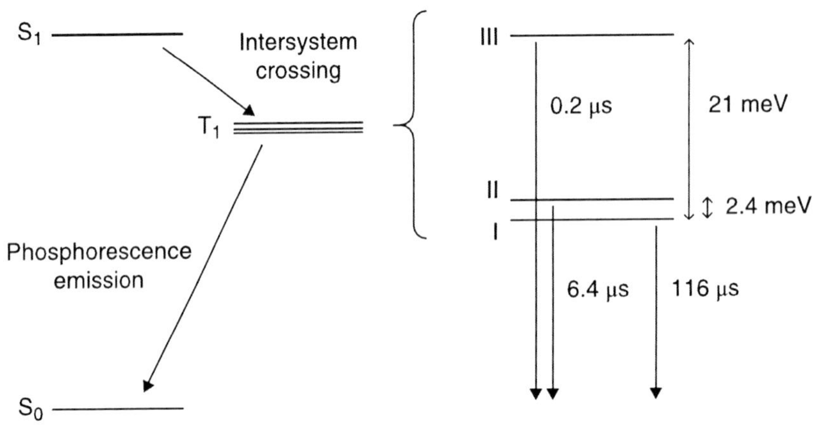


**Figure 4.31** (Left) Energy diagram for Ir(ppy)$_3$. (Right) Magnified diagram for three triplet substates. The PL decay rates and splitting energies are determined from the fit to the experimental data in Figure 4.30. Source: Modified from Ref. [64].

Using the Boltzmann distribution, PLQE at any $T$ can also be simulated. From $\Phi = k_r/k$, the following expression is obtained:

$$\Phi = \frac{k_{r,\mathrm{I}} + k_{r,\mathrm{II}}e^{-\Delta E_{\mathrm{I,II}}/(k_B T)} + k_{r,\mathrm{III}}e^{-\Delta E_{\mathrm{I,III}}/(k_B T)}}{k_{\mathrm{I}} + k_{\mathrm{II}}e^{-\Delta E_{\mathrm{I,II}}/(k_B T)} + k_{\mathrm{III}}e^{-\Delta E_{\mathrm{I,III}}/(k_B T)}} \tag{4.26}$$

where $k_{r,i}$ represents the radiative component of $k_i$. The fitted result with Eq. (4.26) is shown in Figure 4.29, and reproduces the experimental result well. From the fit, $k_{r,i}$ is determined. If PLQE of each substate is defined as $\Phi_i = k_{r,i} / k_i$, $\Phi_{\mathrm{I}} = 0.88$, $\Phi_{\mathrm{II}} = 0.90$, and $\Phi_{\mathrm{III}} = 0.97$ are obtained. Thus, the decrease in PLQE at a very low temperature region in Figure 4.29 can be attributed to the lower $\Phi_{\mathrm{I}}$ and $\Phi_{\mathrm{II}}$ values than the $\Phi_{\mathrm{III}}$ value. Note that these lower $\Phi_{\mathrm{I}}$ and $\Phi_{\mathrm{II}}$ values do not suggest efficient nonradiative decay channels from the substates I and II; these lower $\Phi_{\mathrm{I}}$ and $\Phi_{\mathrm{II}}$ values rather result from the very slow radiative decay rates.

#### 4.4.3.3 4CzIPN

In TADF emitters, the decay process is more complicated than those in fluorescence and phosphorescence emitters because mutual conversion between singlet and triplet excited states must be considered [8, 9]. In TADF emitters, therefore, neither singlet nor triplet excited states can be omitted. In Figure 4.32, a simplified energy diagram for a TADF emitter is illustrated. After being created by photoexcitation, singlet excited states immediately return to $S_0$ through radiative and nonradiative channels and also decay into $T_1$ via ISC. During this initial process, a *prompt* fluorescence with a shorter PL lifetime is observed. Then, triplet excited states gradually decay into $S_0$ through $S_1$ via reverse ISC. In this timescale, a *delayed* fluorescence with a longer PL lifetime is observed [8, 9]. If the reverse ISC is not efficient, triplet excited states may directly return to $S_0$ with or without phosphorescence emission (downward in Figure 4.32). At very low temperatures, reverse ISC is completely suppressed so that those downward transitions become the only decay channels for triplet excited states.

In Figure 4.33, we show two different time-integrated PL spectra of 4CzIPN-doped thin films measured at 6.5 K with a pulse laser. In the measurements, 1,3-bis(9-carbazolyl)benzene (m-CP) was used as a host matrix. One was obtained by integrating the emission from 30 to 980 ms after the pulse irradiation. The resultant spectrum exclusively contains the phosphorescence component. The other one was obtained by integrating the emission from 0 (immediately before the pulse irradiation) to 30 ms.

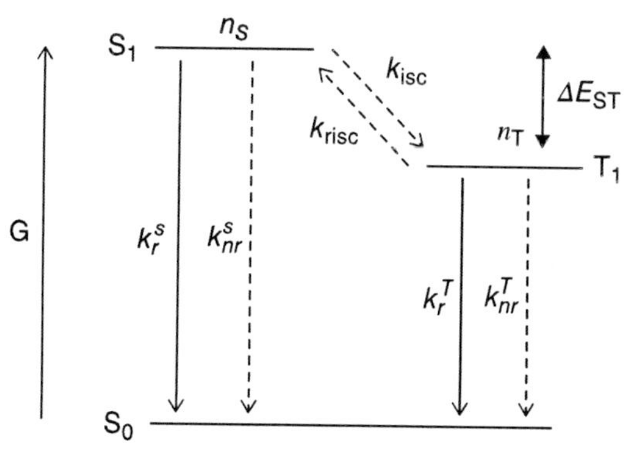


Figure 4.32 A simplified energy diagram of a TADF emitter. After photoexcitation (G), created excited states return to $S_0$ through various channels represented by solid and dashed arrows, which indicate radiative and nonradiative transitions, respectively.

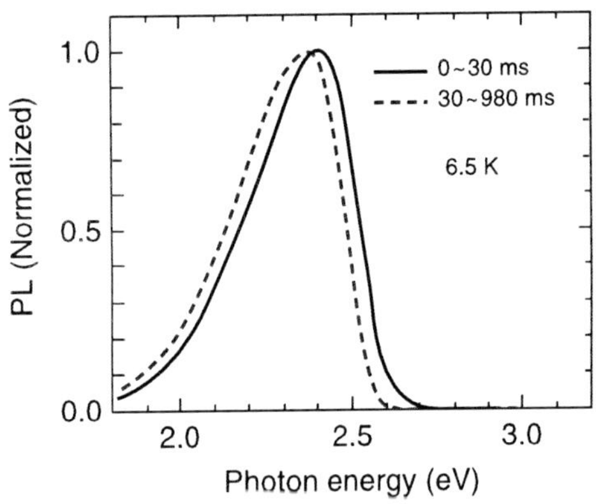


**Figure 4.33** PL spectra of 4CzIPN-doped m-CP thin films measured at 6.5 K upon integrating the emission from 0 to 30 ms and from 30 to 980 ms after the pulse irradiation. The latter spectrum consists only of the phosphorescence component whereas the former mainly consists of the fluorescence component.

Thus, the latter spectrum mainly consists of the fluorescence component but, to a certain extent, the phosphorescence component is still contained. These spectra strongly suggest that the *pure* fluorescence is well overlapped with the phosphorescence spectrum. The small spectral shift reflects its small $\Delta E_{st}$, which is one of the characteristic features of TADF emitters. Because of the small shift, it is difficult to experimentally investigate the temporal evolutions of fluorescence and phosphorescence separately. It should be noted that the prompt and delayed fluorescence has an identical spectral shape so that they can be identified only by their PL lifetimes.

An analysis based on a simple three-level model is helpful to understand the decay process in TADF emitters. We first consider the decay dynamics of excited states uing a three-level model shown in Figure 4.32. Temporal evolution of densities of singlet and triplet excited states, $n_S$ and $n_T$, after photoexcitation are described by the following rate equations [67, 68]:

$$\frac{dn_S}{dt} = -(k_r^S + k_{nr}^S)n_S - k_{isc}n_S + k_{risc}n_T, \tag{4.27}$$

$$\frac{dn_T}{dt} = k_{isc}n_S - (k_r^T + k_{nr}^T)n_T - k_{risc}n_T, \tag{E28}$$

where $k_{isc}$ and $k_{risc}$ are the transition rates via ISC and reverse one, respectively, and $r$ and $nr$ in the subscript mean radiative and nonradiative transitions, respectively; the upper script indicates an associated excited state. These rate equations can be analytically solved. If we assume that $n_S = G$ and $n_T = 0$ at $t = 0$, their solutions become

$$n_S(t) = \frac{k_p - k_{risc} - k^T}{k_p - k_d} Ge^{-k_p t} - \frac{k_d - k_{risc} - k^T}{k_p - k_d} Ge^{-k_d t}, \tag{4.29}$$

$$n_T(t) = -\frac{k_{isc}}{k_p - k_d} G(e^{-k_p t} - e^{-k_d t}), \tag{4.30}$$

where $k^T = k_r^T + k_{nr}^T$ and $k^S = k_r^S + k_{nr}^S$, and

$$k_p, k_d = \frac{1}{2}(k^S + k^T + k_{isc} + k_{risc} \pm \sqrt{(k^S + k^T + k_{isc} + k_{risc})^2 - 4(k_{risc}k^S + (k^S + k_{isc})k^T)}). \tag{4.31}$$

Both $n_S(t)$ and $n_T(t)$ are in the form of a sum of two exponential functions with decay rates of $k_p$ and $k_d$ [67, 68]. Because of this feature, Eq. (4.31) can be applied, whether the observed PL is a sum of prompt fluorescence and delayed fluorescence or a sum of prompt fluorescence and phosphorescence.

In Figure 4.34, we show PL decay curves of 5wt% 4CzIPN-doped m-CP thin films. Since the curves have long tails, the data were taken in two different ways: one with an oscilloscope for (a), and another using a multichannel scaler and a PMT for (b) and (c). The data are presented in a log-log scale, in which PL decay components with well-separated time constants can be found more easily. Indeed, the PL decay curves at both temperatures consist of two decay components. The prompt components at 300 and 6.5 K are observed in the same time range, and their PL lifetimes are estimated to be

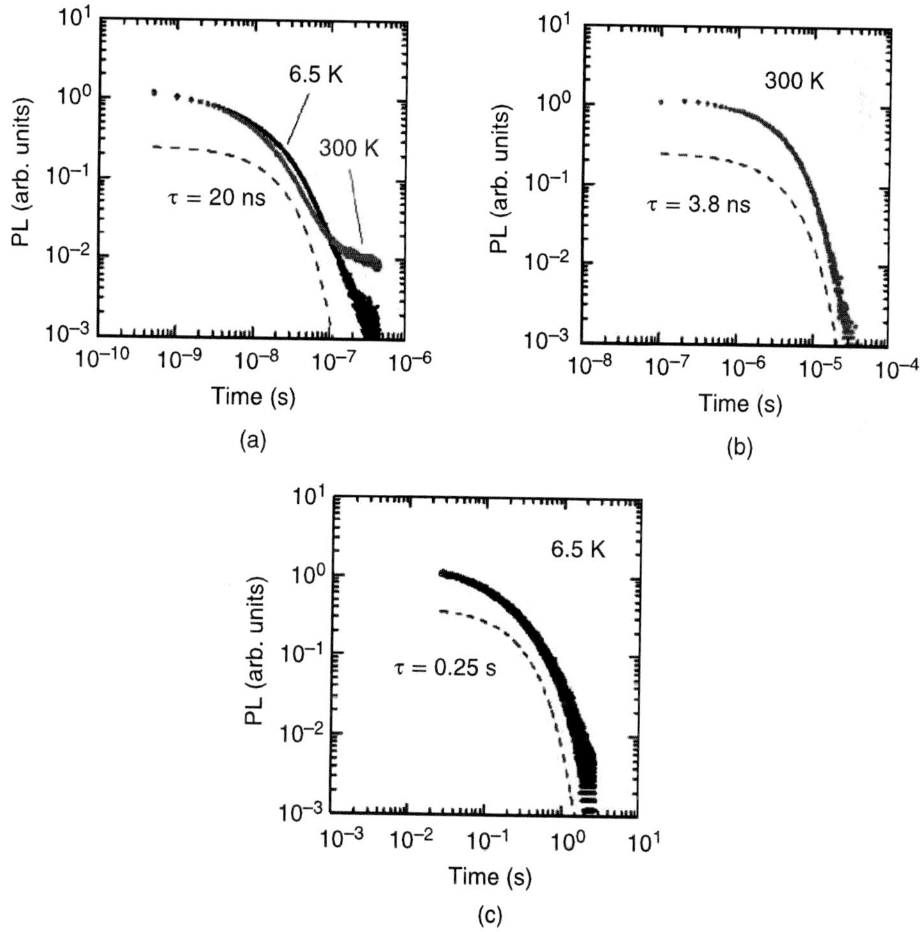


**Figure 4.34** PL decay curves of 4CzIPN in three different time ranges. The prompt components at 6.5 and 300 K appear in the same shortest time range, while the delayed fluoresce at 300 K and the phosphorescence at 6.5 K are observed in different time ranges. The dashed lines represent exponential decays with the described lifetimes; only the delayed fluorescence in panel (b) is well fitted with a single exponential function.

a few tens of ns. On the other hand, the delayed components appear in a different time range. PL lifetime of delayed fluorescence at 300 K can be determined to be 3.8 μs from a single exponential fit, whereas PL lifetime of phosphorescence at 6.5 K is five orders of magnitude slower than that. Note that, unfortunately, it is difficult to determine all the decay constants, such as $k_{isc}$ and $k^S$, even with an analysis using Eqs. (4.29) ~ (4.31).

PLQE of 4CzIPN thin films is almost constant (~ 0.7) in a temperature region from 6.5 to 300K as long as the used photoexcitation light is sufficiently weak [69]. Since phosphorescence is not observed in 4CzIPN at 300 K, the ratio of $k_r^S$ / $k^S$ is estimated to be around 0.7 from the PLQE value. Similarly, from the fact that PLQE is independent of temperature, it is suggested that the ratio of $k_r^T$ / $k^T$ is also around 0.7. In Figure 4.35, we show temperature dependence of PLQE measured with slightly stronger excitation intensities [70]. In this figure, it is found that below 100 K, PLQE decreases with increase of the excitation intensity [70]. At lower temperatures, the reverse ISC is suppressed so that the density of triplet excited states significantly increases under a continuous-wave photoexcitation. As a result, STA, TTA, or both may reduce PLQE.

In a STA process, the singlet excited state ($S$) disappears and its energy is transferred to the triplet excited state ($T$). This process can be described as [12]

$$T + S \rightarrow T + S_0. \tag{4.32}$$

On the other hand, TTA can take place in two manners,

$$T + T \rightarrow S + S_0, \tag{4.33}$$

$$T + T \rightarrow T + S_0. \tag{4.34}$$

Quintets may be generated by TTA, but this process can be negligible at room temperature [71]. Although in π conjugated polymers (and fluorescence emitters) TTA is regarded as a technique to improve PLQE [72], TTA simply reduces PLQE in phosphorescence and TADF emitters. If TTA, STA, or both are included in the rate equations, the analytic solutions are obtained only under the steady-state condition. Let us consider

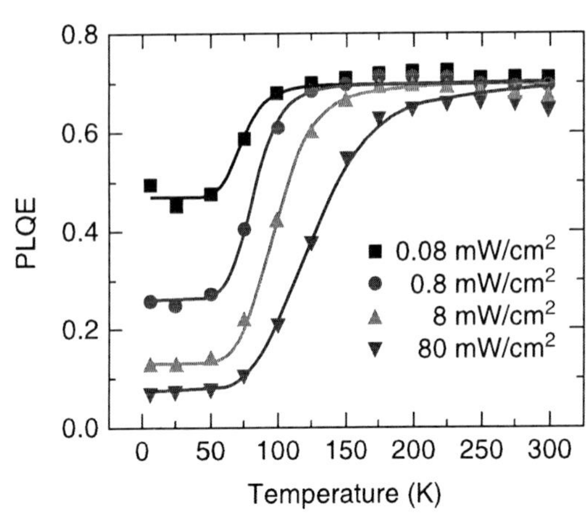


**Figure 4.35** Temperature dependence of PL quantum efficiencies of 5wt% 4CzIPN-doped m-CP thin films measured at various excitation intensities. The solid lines are fitted results obtained using Eq. (39). Source: Ref. [70].

the TTA case. Then, the following rate equations become

$$\frac{dn_S}{dt} = G - k^S n_S - k_{isc} n_S + k_{risc} n_T + \alpha \gamma_{TT} n_T^2, \tag{4.35}$$

$$\frac{dn_T}{dt} = k_{isc} n_S - k^T n_T - k_{risc} n_T - (1+\alpha)\gamma_{TT} n_T^2, \tag{4.36}$$

where $\gamma_{TT}$ is the TTA rate and $\alpha$ represents the number of generated $S$ per a TTA event; we assume $\alpha = 0.5$. The solutions under the steady-state condition, $n_{T0}$ and $n_{S0}$, are

$$n_{T0} = \frac{-k^T(k^S + k_{isc}) - k^S k_{risc} + \sqrt{4Gk_{isc}\{k^S(1+\alpha) + k_{isc}\}\gamma_{TT} + \{k^T(k^S + k_{isc}) + k^S k_{risc}\}^2}}{2\{k^S(1+\alpha) + k_{isc}\}\gamma_{TT}} \tag{4.37}$$

$$n_{S0} = \frac{G + k_{risc} n_{T0} + \alpha \gamma_{TT} n_{T0}^2}{k^S + k_{isc}}. \tag{4.38}$$

Thus, PLQE can be expressed using $n_{S0}$ and $n_{T0}$ as

$$\Phi = \frac{k_r^S}{k^S + k_{isc}} + \frac{k_r^S(k_{risc} n_{T0} + \alpha \gamma_{TT} n_{T0}^2)}{G(k^S + k_{isc})} + \frac{k_r^T n_{T0}}{G}. \tag{4.39}$$

In this equation, the first, second, and third terms correspond to contributions of prompt fluorescence, delayed fluorescence, and phosphorescence, respectively. If we assume the following relationship between $k_{risc}$ and $T$,

$$k_{risc} = k_{isc} \exp\left(-\frac{\Delta E_{st}}{k_B T}\right) \tag{4.40}$$

temperature and excitation intensity dependences of PLQE can be calculated.

Fitted results of Eq. (4.39) to the experimental results are also shown in Figure 4.35. From the figure, it is found that the experimental results are almost completely explained with the model expressed by Eqs. (4.35) ~ (4.39). This means that at very low temperatures, the density of the triplet excited states becomes high enough that the annihilation processes take place efficiently. To achieve such high density, in addition to the suppression of the reverse ISC, 4CzIPN is needed to have a high ISC rate in spite of the absence of the heavy atom effect as well as well-suppressed nonradiative decay from $T_1$. These features are necessary conditions for efficient TADF emitters, and thus, similar temperature and excitation intensity dependences of PLQE should be commonly observed in efficient TADF emitters. In fact, we have obtained similar results from several kinds of TADF emitters. Note that at very low temperatures, reduction of PLQE in 4CzIPN can be observed even with an excitation density as weak as a few tens of $\mu W/cm^2$. Therefore, to know the intrinsic PLQE value, further weak excitation density is required [69].

All of the fitted results shown in Figure 4.35 are obtained from the single set of physical constants except $G$. However, similar fitting results can also be obtained with different sets of constants; in particular, there is huge arbitrariness in decay rates. In addition, if STA is included in the rate equations instead of TTA, or if both are included, the experimental results are also thoroughly reproduced. Therefore, Eq. (4.39) cannot be used to determine physical constants and also to identify the mechanism of the annihilation process. For these purposes, the model described by the rate equations, i.e. Eqs. (4.35) and (4.36), has to be further improved. For instance, since the importance of another excited states other than $S_1$ and $T_1$ has been recently pointed out [19, 73, 74], the contribution of the excited state may have to be included [75].

As demonstrated above, the rate equations of Eqs. (4.35) and (4.36) can be analytically solved under the steady-state condition (although in the case that both of TTA and STA are included, a cubic equation has to be solved). However, if you want to know the temporal evolutions of the singlet and triplet densities, the rate equations have to be solved numerically. In an actual calculation, in particular, for a case of a very low temperature, the time evolution has to be simulated with a very short time pitch width (maybe $< 1$ ns) over a few s after the photoexcitation (see Figure 4.34c), and thus an enormous amount of time will be needed. Therefore, an analysis of experimental temporal evolutions with such numerical simulation would be very inefficient, or an additional approximation (simplification) is required despite the fact that improvement (complication) of the model is needed to express more realistic dynamics of excited states in a TADF emitter as mentioned above. Hence, we conclude that the analysis of the temporal evolution of PL decay is not practical for TADF emitters. Note that this example of the numerical simulation reminds us that the omissions of triplet excited states in fluorescence emitters ($\pi$ conjugated polymers) and of singlet excited states in phosphorescence emitters are reasonable and very efficient for quantitative understanding of the dynamics of the excited states. Alternatively, we expect that the detailed investigation of steady-state PL measurements will give us a further deeper understanding of excited-state dynamics in a TADF emitters, and have carried out our research on this line.

## 4.5 Summary

This chapter considered several topics on organic semiconductors related to photoabsorption and PL in a visible range. Although various organic compounds are classified into the same category, i.e. organic semiconductors, their optical properties are rich in variety. The diversity of the properties may make it difficult to understand the underlying physics, but it demonstrates that organic semiconductors have the strong potential to meet any demands for applications by appropriate material design.

## Acknowledgement

The authors acknowledge Sumitomo Chemical Co. Ltd., for supplying PFO and Prof. S. Matsumoto for supplying DE2. We also acknowledge collaboration with Prof. C. Adachi and Prof. K. Goushi. This work was supported by JSPS KAKENHI Grant Number 18H03902, 19H02599, 20K21007, and 20H02716.

## References

1 Tsujimura, T. (2012). *OLED Display Fundamentals and Applications*. New Jersey: John Wiley & Sons Inc.
2 Scharber, M.C. and Sariciftci, N.S. (2013). Efficiency of bulk-heterojunction organic solar cells. *Prog. Polym. Sci.* 38: 1929.

3 Liang, Y., Xu, Z., Xia, J. et al. (2010). For the Bright Future–Bulk Heterojunction Polymer Solar Cells with Power Conversion Efficiency of 7.4%. *Adv. Mater.* 22: E135.

4 Brown, P.J., Sirringhaus, H., Harrison, M. et al. (2001). Optical spectroscopy of field-induced charge in self-organized high mobility poly(3-hexylthiophene). *Phys. Rev. B* 63: 125204.

5 Manaka, T., Lim, E., Tamura, R. and Iwamoto, M. (2007). Direct imaging of carrier motion in organic transistors by optical second harmonic Generation. *Nat. Photonics* 1: 581.

6 Baldo, M.A., Lamansky, S., Burrows, P.E. (1999). Very high-efficiency green organic light-emitting devices based on electrophosphorescence. *Appl. Phys. Lett.* 75: 4.

7 Yersin, H. (ed.) (2006). *Highly Efficient OLEDs with Phosphorescent Materials.* Weinheim: Wiley-VCH.

8 Uoyama, H., Goushi, K., Shizu K. et al. (2012). Highly efficient organic light-emitting diodes from delayed fluorescence. *Nature* 492: 235.

9 Adachi, C. (2014). Third-generation organic electroluminescence materials. *Jpn. J. Appl. Phys.* 53: 060101.

10 Atkins, P. and de Paula, J. (2010). *Physical Chemistry, 10e.* Oxford: Oxford University Press.

11 Gierschner, J., Ehni, M., Egelhaaf, H. et al. (2005). Solid-state optical properties of linear polyconjugated molecules: π-stack contra herringbone. *J. Chem. Phys.* 123: 144914.

12 Pope, M. and Swenberg, C.E. (1999). *Electronic Processes in Organic Crystals and Polymers, 2e.* Oxford: Oxford University Press.

13 Kobayashi, T. (ed.) (1996). *J-aggregates.* Singapore: World Scientific.

14 Spano, F.C. (2006). Excitons in Conjugated Oligomer Aggregates, Films, and Crystals. *Annu. Rev. Phys. Chem.* 57: 217.

15 Adachi, C., Baldo, M.A., Thompson, M.E. and Forrest, S.R. (2001). Nearly 100% internal phosphorescence efficiency in an organic light-emitting device. *J. Appl. Phys.* 90: 5048.

16 Hay, P.J. (2002). Theoretical Studies of the Ground and Excite Electronic States in Cyclometalated Phenylpyridine Ir(III) Complexes Using Density Functional Theory. *J. Phys. Chem. A* 106: 1634.

17 Endo, A., Suzuki, K., Yoshihara, T. et al. (2008). Measurement of photoluminescence efficiency of Ir(III) phenylpyridine derivatives in solution and solid-state films. *Chem. Phys. Lett.* 460: 155.

18 Lakowicz, J.R. (2006). *Principles of Fluorescence Spectroscopy, 3e.* New York: Springer.

19 Dias, F.B., Bourdakos, K.N., Jankus, V. et al. (2013). Triplet Harvesting with 100% Efficiency by Way of Thermally Activated Delayed Fluorescence in Charge Transfer OLED Emitters. *Adv. Mater.* 25: 3707.

20 Ishimatsu, R., Matsunami, S., Shizu, K. et al. (2013). Solvent Effect on Thermally Activated Delayed Fluorescence by 1,2,3,5-Tetrakis(carbazol-9-yl)-4,6-dicyanobenzene. *J. Phys. Chem. A* 117: 5607.

21 Kastner, J., Kuzmany, H., Paloheimo, J. and Dyreklev, P. (1993). Resonance Raman Scattering from Spincoated and Langmuir-Blodgett Poly(3-alkylthiophene) Films. *Synth. Met.* 55–57: 558.

22 Kishino, S., Ueno, Y., Ochiai, K. et al. (1998). Estimate of the effective conjugation length of polythiophene from its $|\chi^{(3)}(\omega; \omega, \omega, -\omega)|$ spectrum at excitonic resonance. *Phys. Rev. B* 58: R13430.

23 Gierschner, J., Cornil, J. and Egelhaaf, H. (2007). Optical Bandgaps of π-Conjugated Organic Materials at the Polymer Limit: Experiment and Theory. *Adv. Mater.* 19: 173.

24 Thémans, B., Salaneck, W.R. and Brédas, J.L. (1989). Theoretical study of the influence of thermochromic effects on the electronic structure of poly(3-hexylthiophene). *Synth. Met.* 28: C359.

25 Klaerner, G. and Miller, R.D. (1998). Polyfluorene Derivatives: Effective Conjugation Lengths from Well-Defined Oligomers. *Macromolecules* 31: 2007.

26 Asada, K., Kobayashi, T. and Naito, H. (2006). Control of Effective Conjugation Length in Polyfluorene Thin Films. *Jpn. J. Appl. Phys.* 45: L247.

27 Grell, M., Bradley, D.D.C., Ungar, G. et al. (1999). Interplay of Physical Structure and Photophysics for a Liquid Crystalline Polyfluorene. *Macromolecules* 32: 5810.

28 Ariu, , M., Sims, M., Rahn, M.D. et al. (2003). Exciton migration in β-phase poly(9,9-dioctylfluorene). *Phys. Rev. B* 67: 195333.

29 Stampor, W., Mężyk, J., and Kalinowski, J. (2004). Electroabsorption study of metal-to-ligand charge transfer in an organic complex of iridium (III). *Chem. Phys.* 300: 189.

30 Boyd, R.W. (1992). *Nonlinear Optics*. San Diego: Academic Press.

31 Liptay, W. (1974). Dipole Moments and Polarizabilities of Molecules in Excited Electronic States. In: *Excited States* (ed. E. C. Lim). New York: Academic Press.

32 Bublitz, G.U. and Boxer, S.G. (1997). Stark Spectroscopy: Applications in Chemistry, Biology, and Materials Science. *Annu. Rev. Phys. Chem.* 48: 213.

33 Sebastian, L., Weiser, G. and Bässler, H. (1981). Charge Transfer Transitions in Solid Tetracene and Pentacene Studied by Electroabsorption. *Chem. Phys.* 61: 125.

34 Sheik-Bahae, M., Said, A.A., Wei, T.-H. et al. (1990). Sensitive measurement of optical nonlinearities using a single beam. *IEEE J. Quantum Electron.* 26: 760.

35 Matsumoto, S., Kobayashi, T., Aoyama, T. and Wada, T. (2003). J-Aggregates in vapor deposited films of a bisazomethine dye. *Chem. Commun.* 2003: 1910.

36 Knapp, E.W. (1984). Lineshapes of Molecular Aggregates, Exchange Narrowing and Intersite Correlation. *Chem. Phys.* 85: 73.

37 Misawa, K. and Kobayashi, T. (1995). Giant Static Dipole Moment and Polarizability in Highly-Oriented J-Aggregates Prepared by the Vertical Spin-Coating. *Nonlin. Opt.* 14: 103.

38 Murakami, H., Morita1, R., Watanabe, T. et al. (2000). Determination of Third-Order Optical Nonlinearity Dispersion of 1-Methyl-1'-Octadecyl-2, 2'-Cyanine Perchlorate Langmuir-Blodgett Films Using Electroabsorption Spectroscopy. *Jpn. J. Appl. Phys.* 39: 5838.

39 Chowdhury, A., Wachsmann-Hogiu, S., Bangal, P.R. et al. (2001). Characterization of Chiral H and J Aggregates of Cyanine Dyes Formed by DNA Templating Using Stark and Fluorescence Spectroscopies. *J. Phys. Chem. B* 105: 12196.

40 Kobayashi, T., Matsumoto, S., Aoyama, T. and Wada, T. (2006). Optical nonlinearity of J-aggregates in vapor-deposited films of a bisazomethine dye. *Thin Solid Films* 509: 145.

41 Misawa, K., Minoshima, K., Ono, H. and Kobayashi, T. (1994). Giant static dipole moment change on electronic excitation in highly oriented J-aggregates. *Chem. Phys. Lett.* 220: 251.

42 Tanaka, T., Matsumoto, S., Kobayashi, T. et al. (2011). Highly Oriented J-Aggregates of Bisazomethine Dye on Aligned Poly(tetrafluoroethylene) Surfaces. *J. Phys. Chem. C* 115: 19598.

43 Kawabe, Y., Jarka, F., Peygambarian, N., et al. (1991). Roles of band states and two-photon transitions in the electroabsorption of a polydiacetylene. *Phys. Rev. B* 44: 6530.

44 Guo, D., Mazumdar, S., Dixit, S.N. et al. (1993). Role of the conduction band in electroabsorption, two-photon absorption, and third-harmonic generation in polydiacetylenes. *Phys. Rev. B* 48: 1433.

45 Liess, M., Jeglinski, S., Vardeny, Z.V. et al. (1997). Electroabsorption spectroscopy of luminescent and nonluminescent π-conjugated polymers. *Phys. Rev. B* 56: 15712.

46 Martin, S.J., Bradley, D.D.C., Lane, P.A. et al. (1999). Linear and nonlinear optical properties of the conjugated polymers PPV and MEH-PPV. *Phys. Rev. B* 59: 15133.

47 Ikame, S., Kobayashi, T., Murakami, S. and Naito, H. (2007). Electronic structure of a glassy poly(9,9-dioctylfluorene) thin film determined using linear and nonlinear spectroscopies. *Phys. Rev. B* 75: 035209.

48 Orr, B.J. and Ward, J.F. (1971). Perturbation theory of the non-linear optical polarization of an isolated system. *Mol. Phys.* 20: 53.

49 Endo, T., Ikame, S., Kobayashi, T. et al. Electroabsorption study of ordered polyfluorene thin films: Origin of oscillatory structure near the bottom of the continuum state. *Phys. Rev. B* 81: 075203.

50 Kobayashi, T., Nagase, T. and Naito, H. (2015). Electronic Structures of Planar and Nonplanar Polyfluorene. In: *Excitonic and Photonic Processes in Materials* (eds. J. Singh and R.T. Williams). Singapore: Springer.

51 Jaffe, H.H. and Miller, A.L. (1966). The Fates of Electronic Excitation Energy. *J. Chem. Educ.* 43: 469.

52 EcRae, E.G. and Kasha, M. (1958). Enhancement of Phosphorescence Ability upon Aggregation of Dye Molecules. *J. Chem. Phys.* 28: 721.

53 Baleur, B. and Berberan-Santos, M.N. (2013). *Molecular Fluorescence: Principles and Applications, 2e*. Weinheim: Wiley-VCH.

54 Kawamura, Y., Brooks, J., Brown, J.J. et al. (2006). Intermolecular Interaction and a Concentration-Quenching Mechanism of Phosphorescent Ir(III) Complexes in a Solid Film. *Phys. Rev. Lett.* 96: 017404.

55 Wang, L., Ho, P.P., Liu, C. et al. (1991). Ballistic 2-D imaging through scattering walls using an ultrafast optical Kerr gate. *Science* 253: 769.

56 Arzhantsev, S. and Maroncelli, M. (2005). Design and Characterization of a Femtosecond Fluorescence Spectrometer Based on Optical Kerr Gating. *Appl. Spectrosc.* 59: 206.

57 Kawamura, Y., Sasabe, H. and Adachi, C. (2004). Simple Accurate System for Measuring Absolute Photoluminescence Quantum Efficiency in Organic Solid-State Thin Films. *Jpn. J. Appl. Phys.* 43: 7729.

58 Bässler, H. and Schweitzer, B. (1999). Site-Selective Fluorescence Spectroscopy of Conjugated Polymers and Oligomers. *Acc. Chem. Res.* 32: 173.

59 Ariu, M., Sims, M., Rahn, M.D. et al. (2003). Exciton migration in β-phase poly(9,9-dioctylfluorene). *Phys. Rev. B* 67: 195333.

60 Rothe, C., King, S.M., Dias, F. and Monkman, A.P. (2004). Triplet exciton state and related phenomena in the β-phase of poly(9,9-dioctyl)fluorene. *Phys. Rev. B* 70: 195213.

61 Finkenzeller, W.J. and Yersin, H. (2003). Emission of $Ir(ppy)_3$. Temperature dependence, decay dynamics, and magnetic field properties. Chem. Phys. Lett. 377: 299.

62 Tamayo, A.B., Alleyne, B.D., Djurovich, P.I. et al. (2003). Synthesis and Characterization of Facial and Meridional Tris-cyclometalated Iridium(III) Complexes. *J. Am. Chem. Soc.* 125: 7377.

63 Goushi, K., Kawamura, Y., Sasabe, H. and Adachi, C. (2004). Unusual Phosphorescence Characteristics of $Ir(ppy)_3$ in a Solid Matrix at Low Temperatures. *Jpn. J. Appl. Phys.* 43: L 937.

64 Hofbeck, T. and Yersin, H. (2010). The Triplet State of fac-$Ir(ppy)_3$. *Inorg. Chem.* 49: 9290.

65 Curie, D. (1963). *Luminescence in Crystals*. London: Mehuen.

66 Kobayashi, T., Ide, N., Matsusue, N. and Naito, H. (2005). Temperature Dependence of Photoluminescence Lifetime and Quantum Efficiency in Neat fac-$Ir(ppy)_3$ Thin Films. *Jpn. J. Appl. Phys.* 44: 1966.

67 Baldo, M.A. and Forrest, S.R. (2000). Transient analysis of organic electrophosphorescence: I. Transient analysis of triplet energy transfer. Phys. Rev. B 62: 10958.

68 Goushi, K., Yoshida, K., Sato, K. and Adachi, C. (2012). Organic light-emitting diodes employing efficient reverse intersystem crossing for triplet-to-singlet state conversion. *Nat. Photon.* 6: 253.

69 Niwa, A., Takaki, K., Kobayashi, T. et al. (2016). Photoluminescence in a thermally activated delayed fluorescence emitter for organic light-emitting diodes. *J. Imaging Soc. Jpn.* 55: 143.

70 Niwa, A., Kobayashi, T., Nagase, T. et al. (2014). Temperature dependence of photoluminescence properties in a thermally activated delayed fluorescence emitter. *Appl. Phys. Lett.* 104: 213303.

71 Köhler, A. and Bässler, H. (2009). Triplet states in organic semiconductors. *Mat. Sci. Eng. R* 66: 71.

72 Wallikewitz, B.H., Kabra, D., Gélinas, S. and Friend, R.H. (2012). Triplet dynamics in fluorescent polymer light-emitting diodes. *Phys. Rev. B* 85: 045209.

73 Sato, T., Uejima, M., Tanaka, K. et al. (2015). A light-emitting mechanism for organic light emitting diodes: molecular design for inverted singlet-triplet structure and symmetry-controlled thermally activated delayed fluorescence. *J. Mater. Chem. C* 3: 870.

74 Yao, L., Zhang, S., Wang, R. et al. (2014). Highly Efficient Near-Infrared Organic Light-Emitting Diode Based on a Butterfly-Shaped Donor–Acceptor Chromophore with Strong Solid-State Fluorescence and a Large Proportion of Radiative Excitons. *Angew. Chem.* 126: 2151.

75 Kobayashi, T., Niwa, A., Takaki, K. et al. (2017). Contributions of a Higher Triplet Excited State to the Emission Properties of a Thermally Activated Delayed-Fluorescence Emitter. *Phys. Rev. Applied* 7: 034002.

# 5

# Characterization of Transport Properties of Organic Semiconductors Using Impedance Spectroscopy

*Kenichiro Takagi and Hiroyoshi Naito*

*Department of Physics and Electronics, Osaka Prefecture University, Sakai, Japan*

**CHAPTER MENU**



## 5.1 Introduction

Electronic transport properties in organic semiconductors are characterized as drift mobilities, localized-state distributions and deep trapping lifetimes, and the determination of the transport properties is essential to the design of organic devices such as organic light-emitting diodes (OLEDs), organic field-effect transistors (OFETs) and organic solar cells (OSCs) [1]. Recent performance of such organic devices can now compete with that of some traditional inorganic materials, for example, hydrogenated amorphous silicon [2]. In addition, information concerning the transport properties is also essential to understand the device physics of such organic devices [3]. The transport properties of organic devices can be studied using impedance spectroscopy (IS).

The transport measurements using IS have several advantages over conventional methods:

1) Drift mobility can be determined from the analysis of capacitance-frequency or conductance-frequency characteristics of organic space-charge-limited (SCL) diodes [4–9] (The term "diode" in this chapter means metal/organic semiconductor/metal structures with parallel plates.). Space charges in single-charge carrier SCL diodes are formed by injected electrons or holes, which are often called electron-only devices (EODs) or hole-only devices (HODs). Hence, the drift-mobility measurements are fully automatic if we use a computer-controlled frequency response analyzer such as a Solartron 1260 and 1296, and, in principle, a drift mobility ranging from $10^{-14}$ to $10^{-2}$ $cm^2V^{-1}s^{-1}$ can be measured for an organic device with the semiconducting layer thickness of ~100 nm and applied

electric field of $10^4$ V/cm. We note that IS can detect the mobility of a 100-nm thick semiconducting layer, comparable to the active-layer thickness of OLEDs and OSCs, and that the time-of-flight (TOF) drift-mobility measurement (the most popular method for measuring drift mobility in disordered semiconductors [10–16]) cannot be applied to the measurements of such thin samples.

2) Electron and hole mobilities can be determined simultaneously in double-injection devices such as working OLEDs and OSCs [17–22].
3) Drift mobilities [4–7], localized-state distributions [23, 24] and deep trapping lifetimes [25] can be determined simultaneously.

Such unique features of IS will be demonstrated using recent experimental data and we will stress the importance of IS for electrical characterization of transport properties in organic devices.

In this chapter, we will describe the theoretical basis and experimental results for the measurement of drift mobilities, localized-state distributions and deep trapping lifetimes of organic semiconducting materials. In addition, we will review conventional methods for measuring these transport properties before describing and highlighting the advantages of IS.

## 5.2 Charge-Carrier Mobility

### 5.2.1 Methods for Mobility Measurements

TOF transient photocurrent method is one of the most widely used techniques for the determination of carrier mobility of high-resistivity materials [10–16]. In this technique, the carriers, generated by the irradiation of a short pulse of strongly-absorbed light beneath the illuminated electrode, are drifted toward the counter electrode along the applied electric field. The transit time (i.e. time needed for carriers to drift from the illuminated electrode to the counter electrode) can be obtained in the photocurrent transients. The carrier mobility is determined by using the following relation:

$$\mu_d = \frac{d^2}{t_t V}, \tag{5.1}$$

where $\mu_d$ is drift mobility of charge carriers, $d$ is the thickness of the semiconductor layer, $t_t$ is transit time, and $V$ is applied voltage. However, the spatial extent of photocarriers generated by the light pulse is at least 100 nm in semiconductors, and, therefore, this technique generally requires thick samples (>1 μm) to obtain a well-defined flight distance. It has been known that the electronic and optical properties of organic semiconducting materials are drastically changed depending on the film thickness [26, 27], and physical quantities should therefore be characterized in organic devices whose film thickness is comparable to the active layer thickness of OLEDs and OSCs.

The determination of mobility using a steady-state space-charge-limited current (SCL current or SCLC) method has also been widely used. The mobility is extracted by fitting current density–voltage ($J$-$V$) characteristics of HODs or EODs with the Mott-Gurney law [28, 29]

$$J = \frac{9}{8}\varepsilon\mu\frac{V^2}{d^3}, \tag{5.2}$$

where $\varepsilon$ is the permittivity of a semiconductor. The law assumes ohmic contact and a trap-free semiconductor with an electric field independent mobility. However, the $J$-$V$ characteristics do not obey Eq. (5.2) ($J \propto V^n$ where $n > 2$) for most organic semiconductors because of the existence of traps and imperfect ohmic contacts. In the case where Eq. (5.2) cannot be fitted to experimentally obtained $J$-$V$ characteristics, the following expression with Poole-Frenkel type of field dependence of mobility has been conventionally used to fit the $J$-$V$ characteristics for mobility determination [30].

$$J = \frac{9}{8}\varepsilon\mu_0 \exp\left(0.89\beta\sqrt{\frac{V}{d}}\right)\frac{V^2}{d^3}, \tag{5.3}$$

where $\mu_0$ denotes the mobility at zero field and $\beta$ is the Poole-Frenkel coefficient. A number of reports have been made on the determination of carrier mobility of organic semiconductors by using Eq. (5.3) [31–35]. However, the slope of double logarithmic $J$-$V$ characteristics can be above two even when injection-limited current (ILC) from non-ohmic contact is dominant [36] and Eq. (5.3) can be fitted to such $J$-$V$ characteristics. The magnitude of ILC is generally much less than that of SCLC, and the mobility is significantly underestimated [37, 38]. It should be noted that the good agreement between the theory and the experiment does not guarantee the SCL condition, and it is quite difficult to determine whether the current is due either to SCLC or ILC only from $J$-$V$ characteristics. The SCL condition can be confirmed when the $J$-$V$ characteristics of EODs or HODs with different semiconducting layer thicknesses obey the scaling law. The scaling law can be examined by a plot of $J/d$ versus $V/d^2$; we observe a straight line on the plot when the conduction is due to SCLC as obvious from Eq. (5.2) [39–41] (the examination of the scaling behavior is somewhat different in case of Eq. (5.3) [30]). Unfortunately, it is rather difficult to confirm the scaling law if electronic properties of organic semiconducting layers are strongly dependent on the layer thickness [26, 27].

One method to determine carrier mobility using IS proposed in literature is based on equivalent circuit analysis of the complex impedance of organic devices [42]. The determination of equivalent circuits is, however, not always straightforward because different equivalent circuits can be fitted to a complex impedance spectrum [43]. Another method using IS is based on the determination of transit time of charge carriers [4, 17]. This method is valid even in the presence of localized states of semiconductors as described below. We stress that this method is reliable and applicable to a wide range of organic devices. In fact, this method has been successfully used to study carrier mobility of organic semiconductors [5, 6, 8, 9, 18–21]. In the following, we will describe the IS method in detail and discuss the influence of barrier height for carrier injection [44], contact resistance [45], and localized-state distributions [7] on the determination of carrier mobility.

### 5.2.2 Theoretical Basis for Determination of Charge-Carrier Mobility

The determination of mobility using the IS method is based on a single-carrier injection SCLC model [46–48], which assumes that either electrons or holes are injected into a semiconductor. The device structure studied here is a thin film of planar geometry of a semiconductor sandwiched by two electrodes. The formalism described below has been originally developed for insulating semiconductors in the 1960s. Insulating semiconductors are insulators without optical or electrical excitation while once charge carriers are injected optically or electrically, injected carriers move freely with finite mobilities in the semiconductors. Organic semiconductors are indeed insulating semiconductors. The basic equations describing the single-injection model are the Maxwell's continuity equation and the Poisson's equation [46–48].

$$J(t) = q\rho(x,t)\mu F(x,t) + \varepsilon \frac{\partial F(x,t)}{\partial t}, \tag{5.4}$$

$$\frac{\partial F(x,t)}{\partial x} = \frac{q\rho(x,t)}{\varepsilon}, \tag{5.5}$$

where $q$ is the elementary charge and $\rho(x, t)$ is the time-dependent local carrier density. The diffusion of charge carriers is neglected for simplicity. This assumption is reasonable when the applied voltage is greater than several $kT/q$, where $k$ is the Boltzmann constant and $T$ is temperature. The localized states in semiconductors are also neglected, and an ohmic contact is assumed, where the boundary condition of $F(0, t) = 0$ is satisfied. Another contact is blocking for the countercharge carriers. The mobility in the semiconductor is field and time independent under small ac perturbation. As the intrinsic charge-carrier density of an undoped organic semiconductor is usually negligible compared with the charges injected from the electrode, the electric field in the semiconductor is determined solely by the density of space charges present in the semiconducting layer. The steady-state solution of Eqs. (5.4) and (5.5) under the boundary condition of $F = 0$ at the injecting contact gives the Mott-Gurney law [Eq. (5.2)].

For the analysis of complex admittance, which is the reciprocal of complex impedance, we consider small-signal conditions,

$$F(x,t) = F_{dc}(x) + F_1(x,t), \tag{5.6}$$

$$V(x,t) = V_{dc}(x) + V_1(x,t), \tag{5.7}$$

$$\rho(x,t) = \rho_{dc}(x) + \rho_1(x,t), \tag{5.8}$$

$$J(t) = J_{dc} + J_1(t). \tag{5.9}$$

We show, to the first order, the ac current density $J_1(t)$ is written as

$$J_1(t) = q\mu\rho_{dc}F_1(x,t) + \varepsilon\mu F_{dc}\frac{\partial F_1(x,t)}{\partial x} + \varepsilon\frac{\partial F_1(x,t)}{\partial t}. \tag{5.10}$$

The first, second, and third terms in Eq. (5.10) give, respectively, the response of the dc charge-carrier density in the device, the current which stems from the additional time-dependent injected charge-carrier density, and the dielectric displacement current. An analytical solution of the Fourier transform of Eq. (5.10) gives complex admittance $Y_1 = G_1 + jB_1 = G_1 + j\omega C_1 = J_1/V_1$ written as [46–48]

$$G_1 = \frac{g\Omega^3}{6}\frac{\Omega - \sin\Omega}{(\Omega - \sin\Omega)^2 + \left(\frac{\Omega^2}{2} + \cos\Omega - 1\right)^2}, \tag{5.11}$$

$$B_1 = \omega C_1 = \frac{g\Omega^3}{6}\frac{\frac{\Omega^2}{2} + \cos\Omega - 1}{(\Omega - \sin\Omega)^2 + \left(\frac{\Omega^2}{2} + \cos\Omega - 1\right)^2}, \tag{5.12}$$

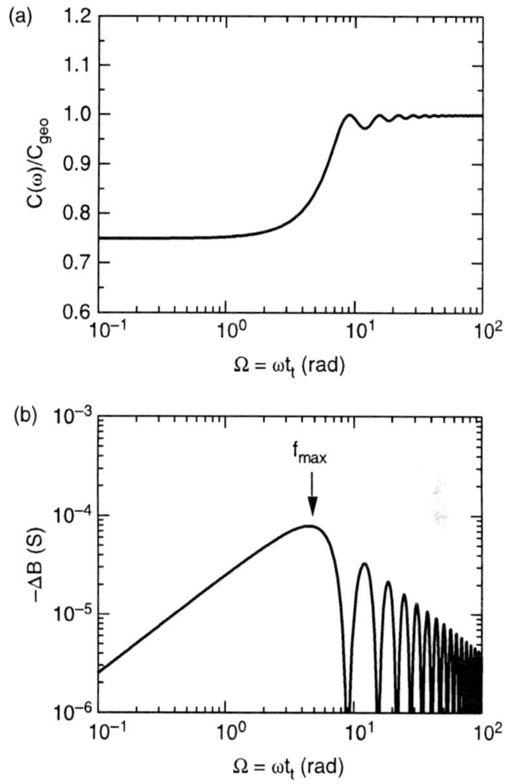


**Figure 5.1** Frequency dependence of normalized capacitance in a trap-free single injection SCL diode calculated from Eq. (5.12) (a). Frequency dependence of $-\Delta B$ obtained from (a) (b).

where $g$ and $\Omega$ are, respectively, the steady current incremental conductance obtained from Eq. (5.2), and the transit angle given by

$$g = \frac{\partial J_{dc}}{\partial V_{dc}} = \frac{9\varepsilon\mu V_{dc}}{4d^3}, \tag{5.13}$$

$$\Omega = \omega t_t. \tag{5.14}$$

The frequency dependence of normalized capacitance obtained from Eq. (5.12) is shown in Figure 5.1a. At high frequency, where the oscillation period of small ac voltage is shorter than transit time, the carriers injected by the ac signal cannot reach the equilibrium space-charge distribution, which results in the observation of geometrical capacitance. At lower frequency, where the oscillation period is longer than transit time, the injected carriers relax into the equilibrium and lag behind the ac signal, which leads to an inductive contribution to the capacitance. Thus, capacitance shows stepwise feature at $\Omega \sim 1$. Hereafter, this stepwise change in capacitance is referred to as the transit time effect. Note that the observation of the transit time effect is itself the evidence for SCL condition (the transit time effect is not observed in ILC condition [48]).

### 5.2.3 Determination of Charge-Carrier Mobility

The transit time effect is more clearly resolved by plotting the negative differential susceptance $-\Delta B = -\omega[C_1(\omega) - C_{geo}]$ as a function of frequency [4, 17]. The frequency dependence of $-\Delta B$ derived from Figure 5.1a is shown in Figure 5.1b. Figure 5.1b yields a peak in $-\Delta B$ at a frequency $f_{max}$ as highlighted by an arrow. Transit time is related to $f_{max}$ by

$$t_t = 0.72 f_{max}^{-1}. \tag{5.15}$$

Equation (5.15) is derived from Eq. (5.12). Transit time in SCL condition is expressed by the following relation [49]:

$$\mu = \frac{4}{3}\frac{d^2}{t_t|V_{dc} - V_{bi}|}, \tag{5.16}$$

where $V_{bi}$ is the built-in voltage of an SCL diode. The carrier mobility is thus determined by Eqs. (5.15) and (5.16), which we call $-\Delta B$ method.

The frequency dependence of conductance obtained from Eq. (5.11) also shows the transit time effect with stepwise change in conductance at $\Omega{\sim}1$. We have found that mobility is obtained by the following relation [7]

$$t_t = 0.48 f_{max}^{-1}, \tag{5.17}$$

where $f_{max}$ here is a peak observed in the frequency dependence of $\omega\Delta G$ $\{=\omega[G_1(\omega) - G_1(\infty)]\}$ [7]. The mobility determination using Eqs. (5.16) and (5.17) is referred to as $\omega\Delta G$ method.

### 5.2.4 Influence of Barrier Height for Carrier Injection on Determination of Charge-Carrier Mobility

In general, the formation of perfect ohmic contact between an electrode and an organic semiconductor is difficult mainly because of its large forbidden gap. It is, therefore, necessary to clarify the influence of barrier for carrier injection on the determination of carrier mobility using IS. We assumed a Schottky barrier and carried out the numerical calculation of IS measurements using a device simulator, ATLAS [44]. The simulation revealed that the carrier mobility can be correctly determined even in the presence of injection barrier. The upper limit of the barrier height which yields observations of transit time effects and gives correct mobility determination is ~0.2 eV for the $-\Delta B$ method and ~0.4 eV for the $\omega\Delta G$ method [44]. The results show that the mobility determination using IS method is valid even in the case where the injecting electrode is not perfectly ohmic.

The simulation implies that when the barrier height becomes higher, the mobility is underestimated especially at low dc bias (i.e. low electric field) [44]. Figure 5.2 shows the hole mobility as a function of electric field extracted from the frequency spectra of the negative differential susceptance of α-NPD HODs with indium tin oxide (ITO), ITO/$MoO_3$, and chlorinated indium tin oxide (Cl-ITO) anodes [50] ($\alpha$-NPD is N,N′-bis(naphthalen-1-yl)-N,N′-bis(phenyl)-2,2′-dimethylbenzidine and its chemical structure is shown in Figure 5.2). The mobility measured by TOF is also shown as reference [51]. As predicted, the hole mobility extracted from HOD with ITO anode using IS is significantly lower than the value measured using TOF. In addition, the electric field dependence of the hole mobility is significantly different than that obtained from the TOF measurements. This is consistent with the numerical calculation [44] in the case of the HODs with non-negligible injection barrier at the ITO/α-NPD interface [50]. The mobility extracted from α-NPD HODs with ITO/$MoO_3$ anode using IS is

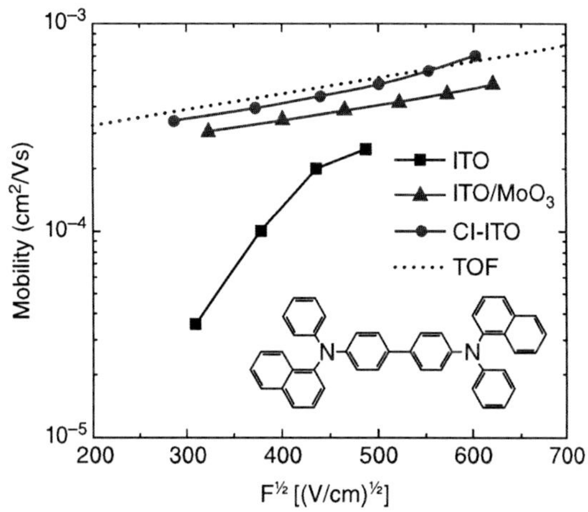


**Figure 5.2** Electric field dependence of hole mobility as a function of the square root of electric field extracted from α-NPD HOD with ITO, ITO/$MoO_3$, and Cl-ITO (chlorinated indium tin oxide) anodes using IS. The electric field ($F$) is taken as $F = (V - V_{bi})/d$, where $V$ is the applied voltage, $V_{bi}$ is the built-in potential determined from photovoltaic measurements and $d$ is the device thickness (536 nm). The dashed line is the mobility measured by TOF shown as Ref. [51]. Source: Ref. [50].

much closer to the value measured using TOF because $MoO_3$ has been reported to form an ohmic contact with α-NPD [52]; the ohmic contact is one of the important premises for the mobility determination by means of IS. In addition, the hole mobility extracted from the HODs with Cl-ITO anode using IS is also closer to the TOF mobility. Figure 5.2 clearly demonstrates that IS measurements be carried out in HODs or EODs with low or no injection barrier for accurate determination of drift mobility.

### 5.2.5 Influence of Contact Resistance on Determination of Charge-Carrier Mobility

Contact resistance at an electrode/semiconductor interface can also affect mobility measurements by IS. Figure 5.3 shows the frequency dependence of capacitance of the ITO/2.0 wt.% $Ir(ppy)_3$-doped CBP/Au HOD, where $Ir(ppy)_3$ is tris(2-phenylpyridine) iridium and CBP is 4,4′-*N*,*N*′-dicarbazole-biphenyl. Capacitance decreases with increasing frequency at $\Omega > 1$ in contrast to the theoretical frequency dependent capacitance shown in Figure 5.1a. This behavior has often been found in literature and is attributed to the contact resistance [6, 8, 53]. Thus, it is difficult to strictly extract geometrical capacitance of a device when calculating $-\Delta B$ for mobility determination in the presence of contact resistance.

We numerically investigated the influence of contact resistance on the determination of carrier mobility [45]. We used a peak value of capacitance at $\Omega > 1$ for geometrical capacitance (in Figure 5.3, the peaks are located at 5–200 Hz depending on the applied voltages). The numerical calculation showed that mobility can be accurately determined when the mobility is $<10^{-5}$ $cm^2V^{-1}s^{-1}$ in the case of the thickness of the sample being 100 nm and the contact resistance 100 Ω. This is due to the fact that the contact resistance is negligible in comparison to the bulk resistance. On the other hand, when the mobility is $>10^{-4}$ $cm^2V^{-1}s^{-1}$ (the thickness and the contact resistance are the same as those mentioned above), mobility is not accurately determined because of non-negligible contact resistance. In IS measurements, the contact and bulk resistances can be readily obtained in Nyquist plot of complex impedance, and one can examine the influence of contact resistance on the mobility determination before the measurement of mobility.

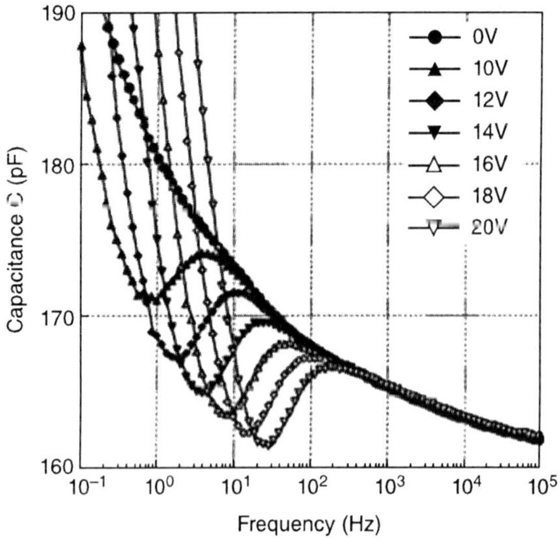


**Figure 5.3** Frequency dependence of capacitance (C) in 2.0 wt.% $Ir(ppy)_3$-doped CBP film at dc bias voltages of 0, 10, 12, 14, 16, 18, and 20 V, where $Ir(ppy)_3$ is tris(2-phenylpyridine) iridium and CBP is 4,4′-*N*,*N*′-dicarbazole-biphenyl. Source: Ref. [8].

### 5.2.6 Influence of Localized States on Determination of Charge-Carrier Mobility

Since organic semiconductors inherently contain localized states in their forbidden gaps, we examine the influence of localized states in semiconductors on the determination of mobility [7]. We assume hole transport here (the same description is obtained for electron transport). The expression for complex impedance of single-injection SCL diodes (HODs) in the presence of localized states has been derived on the basis of trap-controlled band transport and is given by the following infinite series [54–56]:

$$Z_1 = 6\psi R_i \sum_{k=0}^{\infty} \frac{1}{k+3} \frac{\Gamma(\psi+1)}{\Gamma(\psi+k+2)} \left(\frac{\psi}{\delta}\right)^k (-j\Omega)^k, \tag{5.18}$$

where $\Gamma$ is the Euler gamma function, and $R_i$ is the differential resistance of the diodes, which is expressed by

$$R_i = \frac{4}{9} \frac{d^3}{\varepsilon \mu \delta V_{dc} S}, \tag{5.19}$$

where $\mu$ is microscopic mobility of charge carriers and $S$ is the active area. $\delta$ and $\psi$ are trapping parameters respectively given by

$$\delta = \left[1 + \int_{E_v}^{E_c} \frac{\gamma_c(E)}{\gamma_t(E)} dE\right]^{-1}, \tag{5.20}$$

$$\psi(\omega) = \left[1 + \int_{E_v}^{E_c} \frac{\gamma_c(E)}{\gamma_t(E) + j\omega} dE\right] \delta, \tag{5.21}$$

where $E_c$ is the conduction-band mobility edge, $E_v$ is the valence-band mobility edge, $\gamma_t(E)$ {$=\nu \exp[-(E-E_v)/(kT)]$} is the release rate from the localized state located at the energy $E$, $\gamma_c(E)dE$ [$\cong c_p N_t(E)dE$] is the capture rate of the localized state at the energy $E$, $\nu$ ($=c_p N_v$) is the attempt-to-escape frequency, $c_p$ is the hole capture coefficient, $N_t(E)$

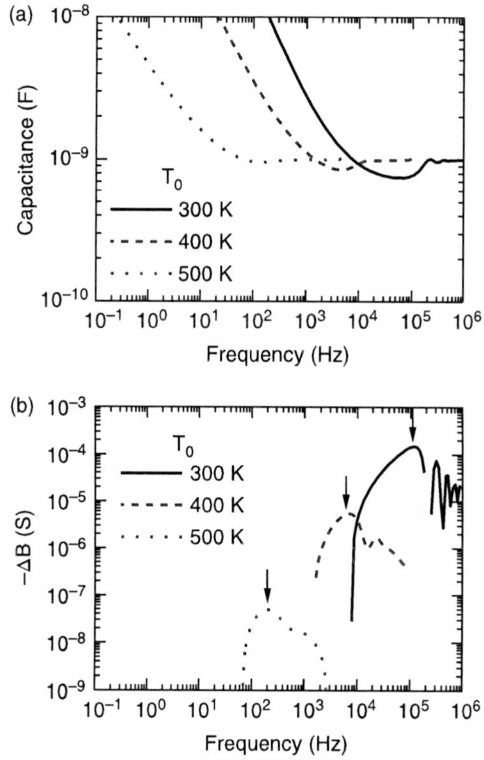


**Figure 5.4** Simulated frequency-dependent capacitance in single injection SCL diodes in the presence of exponentially distributed localized states (a). Frequency dependence of −ΔB calculated from (a). The thickness of semiconducting layer and the electrode area used in this numerical simulation were 100 nm and 4 mm$^2$, respectively (b). Source: Ref [57] Reproduce with the permission of Springer Nature.

is the energetic distribution of localized-state density, and $N_v$ is the effective density of states of the valence band.

We assume an exponential distribution of localized states, which has often been used to describe the electronic states of organic semiconductors in the forbidden gap:

$$N_t(E) = N_0 \exp\left(-\frac{E - E_v}{kT_0}\right), \tag{5.22}$$

where $N_0$ is the density of localized states at the valence band edge, and $T_0$ is the characteristic temperature. Figures 5.4a and 5.4b show the frequency dependences of capacitance and of $-\Delta B$ calculated from Eq. (5.18), respectively [57]. The transit time effects are seen at different characteristic temperatures. In Figure 5.4a, capacitance increases with decreasing frequency ($\Omega < 1$), which reflects the process of carrier trapping in deep localized states [7]. Such behaviors have been experimentally observed in organic semiconductors [5, 8, 9, 18, 21, 53].

We carried out the numerical calculation of TOF transient photocurrent for the same physical quantities used in the calculation of IS using a fast inverse Laplace transform method [58, 59]. The mobility obtained by IS is consistent with that obtained by TOF with different characteristic temperatures as shown in Figure 5.5 [57]. It is quite natural that these mobilities are in agreement with each other because both the mobilities are drift mobilities.

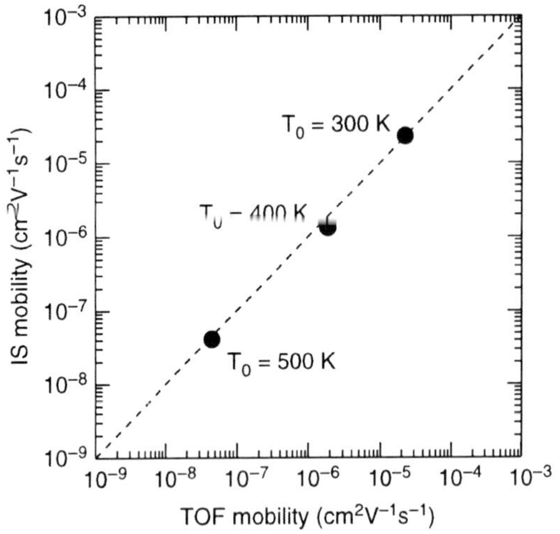


**Figure 5.5** Plot of carrier mobility in the presence of exponentially distributed localized states extracted from IS method vs that from TOF method obtained by the numerical calculation. Source: Ref [57] Reproduce with the permission of Springer Nature.

### 5.2.7 Demonstration of Determination of Charge-Carrier Mobility

In this section, we demonstrate the determination of carrier mobility in an MDP using IS measurements. The reason for this is that MDPs have been well characterized by TOF technique and hence the materials are suitable for the present demonstrative purpose in the sense that drift mobility measurements are possible using the IS technique. In addition, HODs can easily be fabricated using MDPs. The fabrication of some HODs with polymeric materials [6, 60] is difficult in the sense that electrons were unintentionally injected in the HODs, manifested from negative capacitance in the frequency dependence of capacitance, which is an indication of double injection [61]. A hole transport material of organic photoreceptor 7-(2,2-diphenylvinyl)-1,2,3,3a,4,8b-hexahydro-4-phenylcyclopenta[b]indole was used. The layer of the hole transport material dispersed in an electrically inactive binder polymer was prepared by a bar coater. The thickness of the hole transport layer (HTL) was 8.6 μm. The structure of the HOD was ITO/HTL/Al. The IS measurements were carried out by a Solartron 1260 impedance analyzer with a 1296 dielectric interface.

Figures 5.6a and 5.6b show the frequency dependences of capacitance and $-\Delta B$ of the ITO/HTL/Al HOD, respectively [57]. In Figure 5.6a, the transit time effect can clearly be observed. Here, we calculate hole mobility using the $-\Delta B$ method (we confirmed that the similar results were also obtained from the $\omega\Delta G$ method). We also confirmed that the influences of contact resistance and injection barrier are negligible.

Applied electric field dependence of hole drift mobilities of the MDP calculated from Figure 5.6b is shown in Figure 5.7 [25]. The hole drift mobilities are consistent with those measured in terms of TOF measurements and we stress that, unlike TOF experiments, it is easy to determine the drift mobilities of organic semiconductors with a variety of film thickness ranging from 50 nm to several hundred micron.

An important advantage of IS for drift mobility determination is that the measurement is fully automatic; if we use a computer-controlled frequency response analyzer,

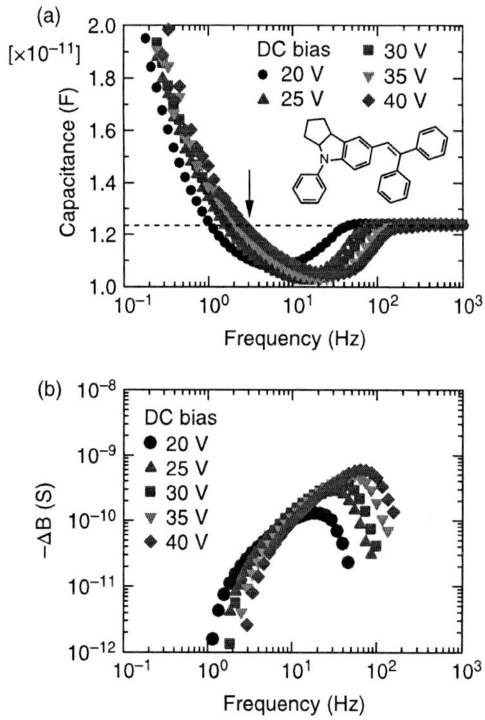


**Figure 5.6** Frequency dependences of a capacitance measured at 290 K (a) and $-\Delta B$ of the ITO/HTL/Al HOD (b). In (a), the inset shows the chemical structure of the hole transport material and the dashed line denotes $C_{geo}$. The sample thickness and the electrode area were 8.6 μm and 4 $mm^2$, respectively. Source: Ref [57] Reproduce with the permission of Springer Nature.

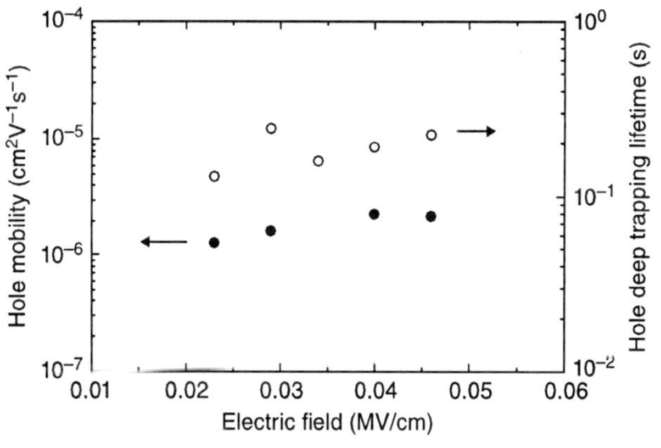


**Figure 5.7** Hole drift mobilities (solid circles) and hole deep trapping lifetimes (open circles) extracted in the MDP HODs as a function of electric field. Source: Ref [25] Reproduced with the permission of AIP Publishing LLC.

the data acquisition of impedance spectra is automatic at different applied voltages and temperatures. On the other hand, automatic measurements are difficult in the TOF method because search processes for transit signal in photocurrent transient are required by changing the time base of a digital synchroscope.

## 5.3 Localized-State Distributions

### 5.3.1 Methods for Localized-State Measurements

Although the importance of studying distributions of localized states in organic semiconductors is widely recognized at present, the study on this topic has still been limited. The thermally stimulated current (TSC) method has been employed for the studies of localized states of semiconductors. Using this measurement, the carriers injected by voltage bias or optical excitation are trapped in localized states at low temperature. The sample is then heated in the dark at a constant heating rate and the resulting thermostimulated current is measured to obtain the densities and energy depths of localized states. Although only discrete localized states were assumed in the early studies of the TSC method, it has been developed for the determination of continuously distributed localized states of disordered semiconductors [62]. In the studies of organic semiconductors, TSC signal has often been analyzed in terms of a TSC theory for discrete localized states [63–66], while there are a few reports on the study of energetically distributed localized states in organic semiconductors using the TSC method [67–70]. The TSC technique, however, inherently involves the scanning of temperature with a constant heating rate. It has been pointed out that this technique is valid only at high temperature in which carrier hopping between localized states in the forbidden gap can be neglected [71]. In addition, thermal scanning sometimes causes mechanical damage such as peel-off of the organic semiconducting layer.

Deep level transient spectroscopy (DLTS) is also one of the most common methods to study the density of localized states in crystalline and amorphous semiconductors [72]. This method is based on the measurement of the transient capacitance of a Schottky contact at an electrode/semiconductor interface (or a p-n junction). Initially, a constant reverse bias is applied to a sample to deplete carriers, and then a forward pulsed bias is applied to fill localized states. The transient capacitances after the cease of the injection pulse are measured at different temperatures, and the analysis of the differences in capacitance between two fixed times at different temperatures gives an emission time constant. For the case of continuous distributions of localized states, more complicated analyses are required [73, 74]. There have been a few studies on the characterization of localized states in organic devices using the conventional DLTS [75, 76] and DLTS-based techniques [77–79]. One of the disadvantages of the DLTS technique is that the measurement requires a wide range of temperature scanning, although a constant heating rate for temperature scanning is not imposed. The DLTS technique can therefore suffer from the disadvantages similar to those of the TSC technique.

The information concerning density of localized states can be obtained from transient photocurrents. We developed methods to determine localized-state distributions using transient photocurrent (TPC) [80–82] or TOF [58, 59] techniques, which are based on the analysis of transient photocurrents using Laplace transforms. These methods are valid for all-time domains of transient photocurrents encompassing a recombination lifetime for TPC or a transit time for TOF, and therefore localized state distributions can be characterized in a wider energy range as compared with the conventional techniques [83–86]. However, these methods cannot be applied to working organic devices such as OLEDs and OSCs.

### 5.3.2 Theoretical Basis for Determination of Localized-State Distribution

The theory for the determination of localized-state distributions by the IS method is based on a single-injection SCL current theory under small ac voltage perturbation in the presence of localized states as described in Section 5.2.6. Here, we assume HODs and derive an analytical solution for the determination of localized-state distributions using Eq. (5.18). The solution is also valid in the case of EODs.

At low frequency of $\Omega \ll 1$, where the carrier trapping in deep localized states is concerned, Eq. (5.18) can be approximated only by the first term, and the impedance is thus expressed by

$$Z_1 = \frac{2R_i\psi}{1+\psi}. \tag{5.23}$$

The conductance and capacitance $[1/Z_1 = G + j\omega C]$ in the low frequency region are then respectively given by

$$G(\omega) = \frac{1}{2R_i\delta}\left[\frac{1+A}{(1+A)^2+B^2}+\delta\right], \tag{5.24}$$

$$C(\omega) = \frac{1}{2R_i\delta\omega}\frac{B}{(1+A)^2+B^2}, \tag{5.25}$$

where

$$A = \int_{E_v}^{E_c} \frac{c_pN_t(E)\gamma_t(E)}{\gamma_t(E)^2+\omega^2}dE, \tag{5.26}$$

$$B = \omega\int_{E_v}^{E_c} \frac{c_pN_t(E)}{\gamma_t(E)^2+\omega^2}dE, \tag{5.27}$$

which are obtained when $\psi$ is defined as $\psi = (1+A-jB)\delta$. Differentiating $\omega B$ with respect to $\omega$, we have

$$\frac{\partial \omega B}{\partial \omega} = 2\omega\int_{E_v}^{E_c} c_pN_t(E)h(\omega,E)dE, \tag{5.28}$$

where

$$h(\omega,E) = \left[\frac{\gamma_t(E)}{\gamma_t(E)^2+\omega^2}\right]^2. \tag{5.29}$$

This function can be approximated using a delta function,

$$h(\omega,E) = \frac{kT}{2\omega^2}\delta(E-E_0), \tag{5.30}$$

where

$$E_0 - E_v = kT\ln\left(\frac{\nu}{\omega}\right). \tag{5.31}$$

Substituting Eq. (5.30) into Eq. (5.28), we have

$$\frac{\partial \omega B}{\partial \omega} = c_pN_t(E_0)\frac{kT}{\omega}. \tag{5.32}$$

The localized-state distribution is then obtained from Eqs. (5.24), (5.25), and (5.32) as

$$N_t(E_0) = \frac{2R_i\omega}{c_p\delta kT}\frac{\partial}{\partial\omega}\left\{\frac{\omega^2 C(\omega)}{[2R_i\omega C(\omega)]^2 + [2R_i G(\omega) - 1]^2}\right\}. \quad (5.33)$$

The localized-state distribution is also obtained from Eq. (5.30) as

$$N_t(E_0) = \frac{2\omega}{c_p\delta kT\pi}\left\{\frac{2R_i G(\omega) - 1}{[2R_i\omega C(\omega)]^2 + [2R_i G(\omega) - 1]^2} - \delta\right\}, \quad (5.34)$$

using the delta function approximation, $\gamma_t(E)/\{\gamma_t^2(E) + \omega^2\} \cong (kT\pi/2\omega)\delta(E - E_0)$, in Eq. (5.26). The localized-state distributions in EODs or HODs can be determined from the frequency dependences of conductance and capacitance using Eqs. (5.31) and (5.33) or Eqs. (5.31) and (5.34).

We stress that the method described above gives higher energy resolution as compared with the TPC and TOF methods. For instance, in the case of an exponential distribution of tail states with $T_0$, the energetically resolvable $T_0$ is $T/2$ in case of Eq. (5.33), while $T$ in case of TPC [80–82], TOF [58, 59] and Eq. (5.34).

### 5.3.3 Demonstration of Determination of Localized-State Distribution

We examined HODs with the structure of ITO/PEDOT:PSS/TFB/Au, where PEDOT:PSS is poly(3,4-ethylenedioxythiophene)/polystyrenesulphonic acid, and TFB is poly(9,9-dioctyl-fluorene-co-N-(4-butylphenyl)-diphenylamine). The area and the thickness of the HODs were $2 \times 2\,\text{mm}^2$ and 450 nm, respectively. Figures 5.8a and 5.8b show the frequency dependences of capacitance and conductance of the HOD at different measurement temperatures obtained by means of IS measurements, respectively [24]. The increase in capacitance and decrease in conductance with decreasing frequency below 10 kHz are observed in Figures 5.8a and 5.8b, respectively. The frequency dependences below 10 kHz are due to the presence of distributed localized states as shown in Figure 5.4 and in Ref. [7] and were used for the determination of localized-state distributions (the transit time effects were observed in the frequency region of 10 kHz to 200 kHz, depending on measurement temperatures).

The localized-state distributions above $E_v$ calculated from the frequency dependences of capacitance and conductance of HOD shown in Figure 5.9 at different measurement temperatures from 293 K to 338 K using Eqs. (5.31) and (5.33) or Eqs. (5.31) and (5.34) are shown in Figure 5.9. Here, we assumed that $\nu$ is independent of the energy, and the value of $\nu$ is determined by calculating localized-state distributions at different temperatures and then by analyzing the Arrhenius plot of $\gamma_t$(E) versus 1/T in Figure 5.10. The value of $\gamma_t$(E) is equal to $\omega$ at the bump in the localized-state distribution, and $\omega$, hence, $\gamma_t$(E) at the bump is dependent on temperature as shown in Figure 5.4. The value of $\nu$ is determined to be $10^{12}$ Hz according to $\gamma_t(E) = \nu \exp\{-(E - E_v)/kT\}$.

We observe an exponential distribution of localized states, Eq. (5.22), from the analysis using Eq. (5.34) in the whole energy range. The value of $T_0$ is 378 K. On the other hand, we observe a structured distribution of localized states, a Gaussian distribution superimposed on a continuously decaying distribution of localized states, from the analysis using Eq. (5.33). The Gaussian bump in the localized-state distribution in TFB appears at 0.7 eV above $E_v$. The structured distribution can be approximated to be an exponential distribution in the energy range between 0.47 eV and 0.55 eV and its characteristic

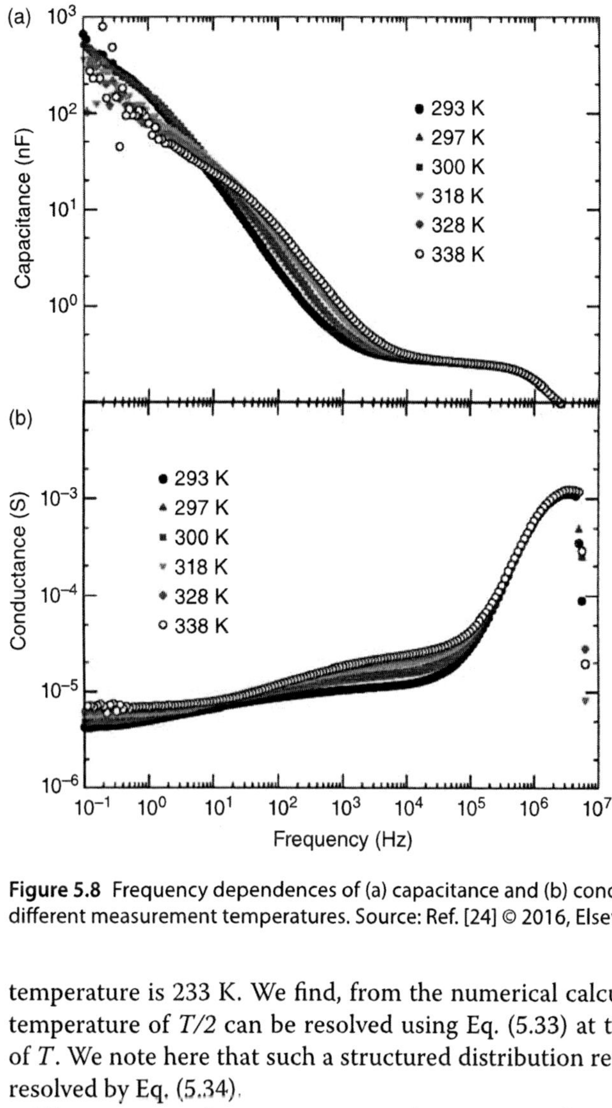


**Figure 5.8** Frequency dependences of (a) capacitance and (b) conductance of TFB HOD by IS with different measurement temperatures. Source: Ref. [24] © 2016, Elsevier.

temperature is 233 K. We find, from the numerical calculation, that the characteristic temperature of $T/2$ can be resolved using Eq. (5.33) at the measurement temperature of $T$. We note here that such a structured distribution revealed by Eq. (5.33) cannot be resolved by Eq. (5.34).

We carry out the computer simulation to confirm high energetic resolution of localized-state distributions determined by Eq. (5.33). We assumed a Gaussian distribution superimposed on an exponential distribution as a structured distribution similar to the observed distribution in Figure 5.9 and as shown in the solid line in Figure 5.11.

The frequency dependences of conductance and capacitance were calculated from Eq. (5.18) in the case of the distribution (the solid line in Figure 5.11) as the input distribution, and the localized-state distributions determined by Eqs. (5.33), (5.34) are compared to the input distribution. The following physical quantities appropriate for disordered organic semiconductors were used in the calculation: $N_v = 10^{20}$ cm$^{-3}$ and $c_p = 10^{-8}$ cm$^3$/s [1]. All calculations were carried out at $T$ = 300 K, $d$ = 450 nm, $S = 4$ mm$^2$, $V_{dc} = 10$ V, and $\mu_0 = 10^{-3}$ cm$^2$/Vs.

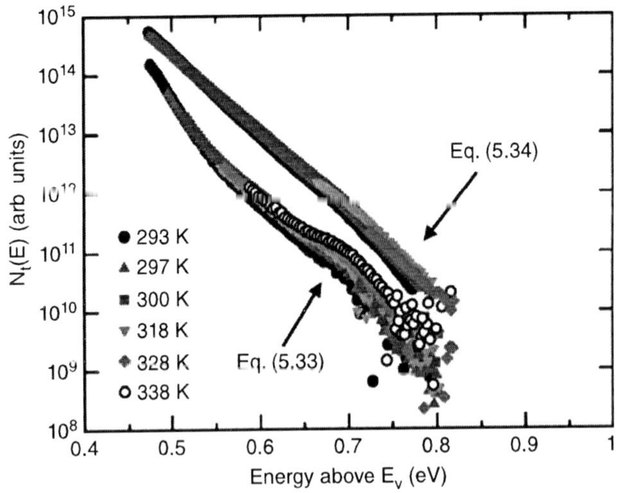


**Figure 5.9** Localized-state distributions from the valence-band mobility edge in TFB determined from the capacitance and conductance data in Fig. 5.8 using Eqs. (5.31), (5.33) and Eqs. (5.31), (5.34). Source: Ref. [24] © 2016, Elsevier.

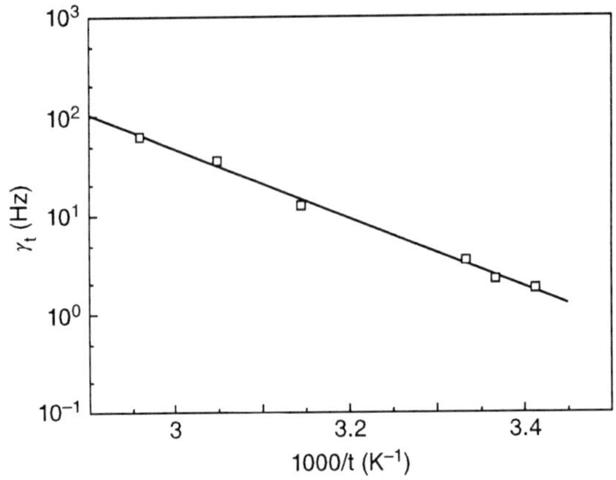


**Figure 5.10** Arrhenius plot of $\gamma_t(E)$ vs 1/T. The value of $\gamma_t(E)$ is equal to $\omega$ at the bump of the localized-state distribution in Figure 5.9, and the value of $\nu$ is determined to be $10^{12}$ Hz. Source: Ref. [24].

Figure 5.11 shows the localized-state distributions determined from the calculated capacitance and conductance data using Eqs. (5.31), (5.33) and Eqs. (5.31), (5.34). It can be seen that the input localized-state distribution is not reconstructed correctly when we analyze the data using Eq. (5.34) (the localized state-distribution determined using Eq. (5.34) looks like a single exponential distribution). On the contrary, Eq. (5.33) can accurately reconstruct the input distribution. We note, from these results, that the structured distribution obtained using Eq. (5.33) in Figure 5.9 reflects the true distribution of localized states in TFB. We stress that higher energetic resolution of Eq. (5.33) can be shown both experimentally and numerically in comparison with that of Eq. (5.34).

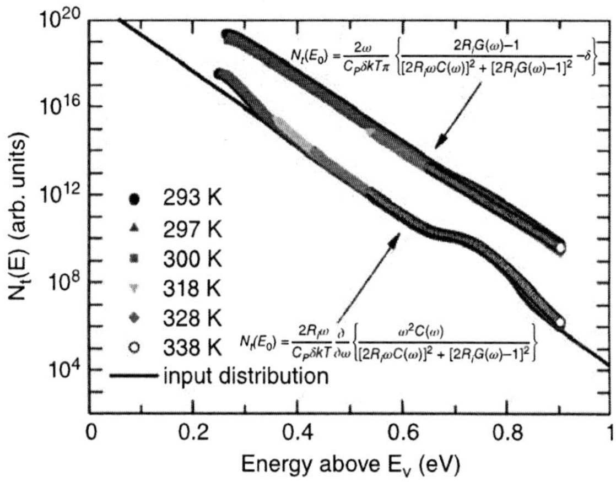


**Figure 5.11** Localized-state distributions from the valence-band mobility edge calculated from computer-generated capacitance and conductance data using Eqs. (5.31), (5.33) and Eqs. (5.31), (5.34). The capacitance and the conductance data are calculated from Eq. (5.18) using the solid line as the input distribution of localized states. Source: Ref. [24].

## 5.4 Lifetime

### 5.4.1 Methods for Deep-Trapping-Lifetime Measurements

Deep trapping lifetime is determined by TOF [87], TOF with Hecht-type analysis [88], and interrupted TOF [89–91] methods, where the layer of a semiconductor should be thicker than ~1 μm. In the transient grating method [92], the diffusion coefficient and the lifetime of photoexcited carriers are determined by observing the time decay of diffracted probing light caused by photocarrier grating. The deep trapping lifetime is also obtained by means of TPC measurements of samples with coplanar ohmic electrodes as mentioned below. These methods are not applicable to the lifetime measurement of working organic devices. One exception is deep trapping lifetime measurements in electrophotoreceptors; deep trapping lifetime is obtained from the analysis of photoinduced discharge curves [93].

In this section, we describe a method to determine deep trapping lifetime using IS. The method is based on the analytical solution of Eq. (5.18). We show the validity of the method by numerical calculations and demonstrate the applicability of the method to experimental results.

### 5.4.2 Determination of Deep-Trapping-Lifetime using the Proposed Method

In single-injection SCL diodes (HODs or EODs), with decreasing frequency of the small ac signal, capacitance first decreases from geometrical capacitance at $\Omega$~1 owing to the inductive contribution from the transit time effect, and then increases at $\Omega < 1$ owing to the presence of deep localized states, as shown in Figure 5.6a. We focus on the impedance spectrum at $\Omega < 1$, which contains the information on deep localized states [25]. At $\Omega < 1$, the impedance can be expressed only by the first term of the infinite series of Eq. (5.18) as described in Section 5.3.2 (or sum of the first several terms to ensure greater accuracy). Taking the first three terms and assuming a single level of deep trapping states, we obtain the following expression for the real and imaginary parts of complex impedance $[Z_1 = \mathrm{Re}(Z_1) + j\,\mathrm{Im}(Z_1)]$:

$$\mathrm{Re}(Z_1) = R_i \left\{ 2 - \frac{2P}{P^2+Q^2} + \frac{3\Omega}{2}\left(1+\frac{\gamma_c}{\gamma_t}\right)\left[\frac{Q}{P^2+Q^2} - \frac{4Q}{(P+1)^2+Q^2}\right] \right.$$
$$\left. -\frac{3\Omega^2}{5}\left(1+\frac{\gamma_c}{\gamma_t}\right)^2\left[2 - \frac{P}{P^2+Q^2} + \frac{16(P+1)}{(P+1)^2+Q^2} - \frac{27(P+2)}{(P+2)^2+Q^2}\right]\right\}, \tag{5.35}$$

$$\mathrm{Im}(Z_1) = R_i \left\{ -\frac{2Q}{P^2+Q^2} - \frac{3\Omega}{2}\left(1+\frac{\gamma_c}{\gamma_t}\right)\left[1 + \frac{P}{P^2+Q^2} - \frac{4(P+1)}{(P+1)^2+Q^2}\right] \right.$$
$$\left. +\frac{3\Omega^2}{5}\left(1+\frac{\gamma_c}{\gamma_t}\right)^2\left[\frac{Q}{P^2+Q^2} - \frac{16Q}{(P+1)^2+Q^2} + \frac{27Q}{(P+2)^2+Q^2}\right]\right\}, \tag{5.36}$$

where

$$P = 1 + \frac{\gamma_t}{\gamma_t+\gamma_c} + \frac{1}{\gamma_t^2+\omega^2}\frac{\gamma_t^2\gamma_c}{\gamma_t+\gamma_c}, \tag{5.37}$$

$$Q = \frac{\omega}{\gamma_t^2+\omega^2}\frac{\gamma_t\gamma_c}{\gamma_t+\gamma_c}. \tag{5.38}$$

Here, $\gamma_t$ and $\gamma_c$ are, respectively, release rate from the deep trapping states and capture rate of the deep trapping states with the density of $N_t(E_0)$ located at $E_0$ from $E_v$, expressed as $\gamma_t(E_0) = c_p N_v \exp[-(E_0 - E_v)/kT]$ and $\gamma_c(E) \cong c_p N_t(E_0)$, respectively.

We have found that deep trapping lifetime can be determined by the following procedure:

1) Read frequency $f_\alpha$ $(=\omega_\alpha/2\pi)$ at which $C = C_{geo}$ in the frequency region of $\Omega < 1$ [highlighted by an arrow in Figure 5.6a in case of applied voltage of 40 V]
2) Read the values of $\mathrm{Re}(Z_1)$ and $\mathrm{Im}(Z_1)$ at $f = f_\alpha$
3) Extract transit time $t_t$ and obtain $\Omega_\alpha = \omega_\alpha t_t$
4) Read the value of differential resistance $R_i$ obtained from $\mathrm{Re}(Z_1)$ at $\Omega \ll 1$
5) Substitute the values obtained from procedure 1–4 into Eqs. (5.35)–(5.38) and obtain a simultaneous equation for $\gamma_t$ and $\gamma_c$
6) Solve the simultaneous equation and determine deep trapping lifetime $\tau_c$ $(=1/\gamma_c)$

### 5.4.3 Validity of the Proposed Method

To show the validity of the present method, we apply the method to the impedance spectra obtained by numerical calculations using Eq. (5.18). In the numerical calculations, a single discrete deep trapping level superimposed on an exponential distribution [Eq. (5.22)] is assumed $[N_t(E) = N_0 \exp(-(E - E_v)/kT_0) + N_{t0}\delta(E_0 - E_v)]$. We also carried out the numerical calculation of transient photocurrent obtained by means of the TPC technique, known as an established method for measuring deep trapping lifetime. After a thin-film sample with ohmic coplanar electrodes to which dc bias is applied is excited by a short light pulse, time dependence of the photocurrent is measured. We assume here that the photocurrent is due to unipolar conduction, and numerically solve the basic equations using the fast inverse Laplace transform technique to obtain the transient photocurrent waveforms [80–82].

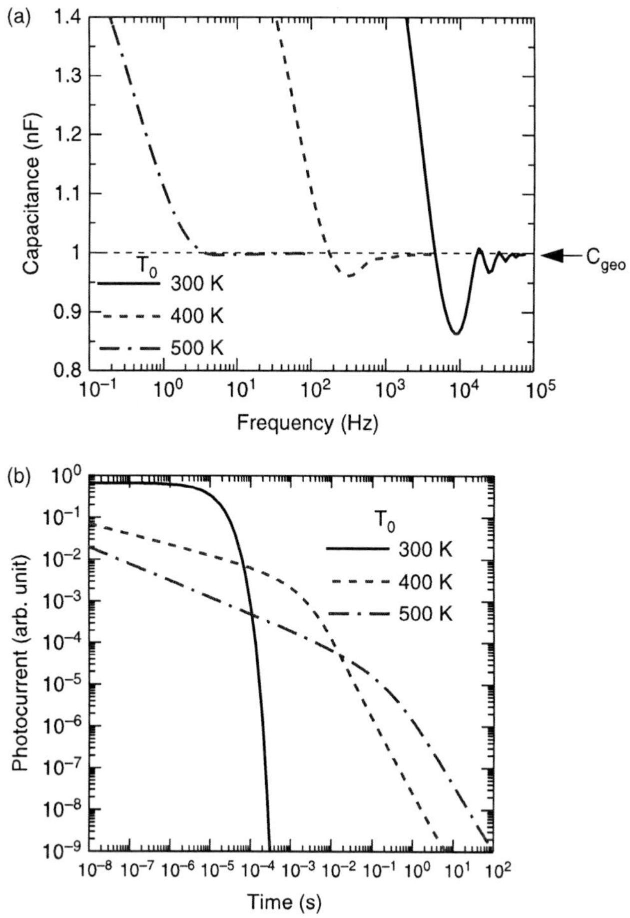


**Figure 5.12** Simulated frequency-dependent capacitance in single injection SCL (a) and simulated transient photoconductivity in a sample with coplanar electrode configuration (b) in the presence of exponentially distributed localized states. Source: Ref [25] Reproduce with permission of AIP Publishing LLC.

Figures 5.12a and 5.12b show the calculated frequency dependence of capacitance spectra and photocurrent transients, respectively. The relation between deep trapping lifetimes obtained from the impedance spectra using the present method and those obtained from the TPC method is examined. The lifetimes obtained by IS are in good agreement with those obtained by the TPC method at different characteristic temperatures, as shown in Figure 5.13. We conclude from Figure 5.13 that the present method is valid even in the presence of exponential distributions of localized states. We have also found that this conclusion is valid in the presence of Gaussian distribution of localized states.

### 5.4.4 Demonstration of Determination of Deep-Trapping-Lifetime

We experimentally demonstrate the applicability of the present method. We analyzed the data of the ITO/HTL/Al HOD (Figure 5.6a). Deep trapping lifetime and mobility

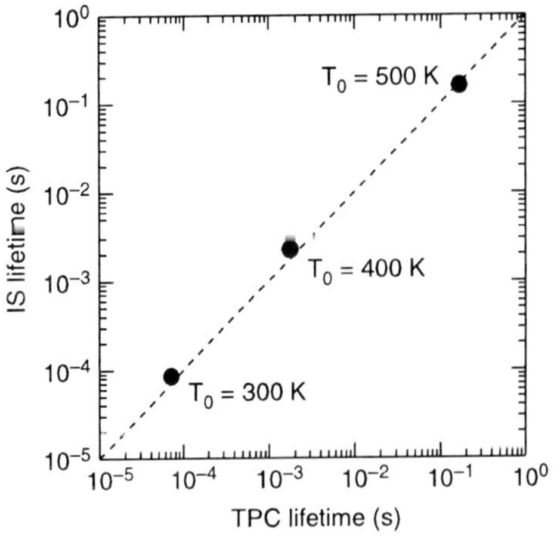


**Figure 5.13** Plot of deep trapping lifetime extracted from IS method vs that from TPC method obtained by the numerical calculation. A single discrete level superimposed on exponential distributions with different characteristic temperatures is assumed for deep trapping states. Source: Ref [25] Reproduce with permission of AIP Publishing LLC.

are simultaneously determined, as shown in Figure 5.7, and the value of deep trapping lifetime is 0.23 s at $4.6 \times 10^4$ V/cm. The relatively long lifetime has often been found in MDPs. The average distance that injected holes can travel in the HTL, known as mean range or schubweg expressed as $\lambda = \mu\tau_c E$ [94], can be obtained. The schubweg of holes in the HTL is calculated to be 230 μm, which significantly exceeds the thickness of conventional organic photoreceptors. This indicates that the hole transport material shown in Figures 5.6 and 5.7 exhibits good transport properties for electrophotographic application.

## 5.5 IS in OLEDs and OPVs

IS mentioned above was carried out in HODs or EODs. We note that IS can be applied to double-injection devices such as OLEDs. We have shown that the transport properties in double-injection devices (electron and hole drift mobilities, tail-state distributions from the valence band edge and from the conduction band edge, and bimolecular recombination constant) can be determined [22]. Unfortunately, IS cannot be always successfully applied to the characterization of the transport properties of OPVs because OSCs are not double-injection devices but charge-carrier extraction devices [95]. The transport properties of OSCs can easily be determined using modulated photocurrent techniques [96].

## 5.6 Conclusions

We reviewed the basic theories for measuring drift mobilities, localized-state distributions, and deep trapping lifetimes in organic semiconductor HODs or EODs using IS. As we discussed, IS measurements have a number of advantages over conventional methods reviewed in this chapter. The important advantages are fully automatic and simultaneous measurements of the transport parameters. We do not show the results but electron and hole drift mobilities can be measured in OLEDs under operation. IS measurements for the characterization of transport properties can be carried out in a variety of insulating semiconductors such as hydrogenated amorphous silicon and oxide semiconductors, not restricted to organic semiconductors because the theories for IS are based on trap-controlled band transport.

## Acknowledgments

This work is partly supported by a Grant-in-Aid for Scientific Research on Innovative Areas "New Polymeric Materials Based on Element-Blocks (No. 2401)" (No. 24102011) and by a Grant-in-Aid for Scientific Research (A) (JP17H01265), (B) (JP19H02599) and (B) (JP20H02716).

## References

1 Brutting, W. and Adachi, C. (2012). *Physics of Organic Semiconductors*. Weinheim: WileyVCH.
2 Street, R.A. (1991). *Hydrogenated Amorphous Silicon*. Cambridge: Cambridge University Press.
3 Baranovski, S. (ed.) (2006). *Charge Transport in Disordered Solids with Applications in Electronics*. Chichester: Wiley.
4 Martens, H.C.F., Pasveer, W.F., Brom, H.B. et al. (2001). *Phys. Rev. B* 63: 125328.
5 Tsang, S.W., So, S.K. and Xu, J.B. (2006). *J. Appl. Phys.* 99: 013706.
6 Nguyen, N.D., Schmeits, M. and Loebl, H.P. (2007). *Phys. Rev. B* 75: 075307.
7 Okachi, T., Nagase, T., Kobayashi, T. and Naito, H. (2008). *Jpn. J. Appl. Phys.* 47: 8965.
8 Ishihara, S., Okachi, T. and Naito, H. (2009). *Thin Solid Films* 518: 452.
9 Ishihara, S., Hase, H., Okachi, T. and Naito, H. (2011). *Org. Electron.* 12: 1364.
10 Lebedev, E., Dittrich, T., Petrova-Koch, V. et al. (1997). *Appl. Phys. Lett.* 71: 2686.
11 Bettenhausen, J., Strohriegl, P., Brutting, W. et al. (1997). *J. Appl. Phys.* 82: 4957.
12 Campbell, I.H., Smith, D.L., Neef, C.J., and Ferraris, J.P. (1999). *Appl. Phys. Lett.* 74: 2809.
13 Redecker, M., Bradley, D.D.C., Inbasekaran, M., and Woo, E.P. (1999). *Appl. Phys. Lett.* 74: 1400.
14 Juska, G., Genevicius, K., Arlauskas, K. et al. (2002). *Phys. Rev. B* 65: 233208.
15 Markham, J.P.J., Anthopoulos, T.D., Samuel, I.D.W. et al. (2002). *Appl. Phys. Lett.* 81: 3266.
16 Laquai, F., Wegner, G., Im, C. et al. (2006). *J. Appl. Phys.* 99: 023712.
17 Martens, H.C.F., Huiberts, J.N. and Blom, P.W.M. (2000). *Appl. Phys. Lett.* 77: 1852.
18 Poplavskyy, D. and So, F. (2006). *J. Appl. Phys.* 99: 033707.
19 Schmeits, M. (2007). *J. Appl. Phys.* 101: 084508.
20 Ishihara, S., Hase, H., Okachi, T. and Naito, H. (2011). *J. Appl. Phys.* 110: 036104.
21 Ishihara, S., Okachi, T. and Naito, H. (2014). *Thin Solid Films* 554: 213.
22 Takada, M., Nagase, T., Kobayashi, T. and Naito, H. (2019). *J. Appl. Phys.* 125: 115501.
23 Okachi, T., Nagase, T., Kobayashi, T. and Naito, H. (2009). *Appl. Phys. Lett.* 94: 043301.

24 Hase, H., Okachi, T., Ishihara, S. et al. (2014). *Thin Solid Films* 554: 218.
25 Takagi, K., Nagase, T., Kobayashi, T. and Naito, H. (2016). *Appl. Phys. Lett.* 108: 053305.
26 Vissenberg, M.C.J.M. and Blom, P.W.M. (1999). *Synth. Met.* 102: 1053.
27 Azuma, H., Asada, K., Kobayashi, T. and Naito, H. (2006). *Thin Solid Films* 509: 182.
28 Mott, N.F. and Gurney, R.W. (1940). *Electronic Processes in Ionic Crystals*, 2e. London: Oxford University Press
29 Lampert, M.A. and Mark, P. (1970). *Current Injection in Solids*, p. 16. New York: Academic.
30 Murgatroyd, P.N. (1970). *J. Phys. D* 3: 151.
31 Blom, P.W.M., de Jong, M.J.M. and van Munster, M.G. (1997). *Phys. Rev. B* 55: R656.
32 Yasuda, T., Yamaguchi, Y., Zou, D.C. and Tsutsui, T. (2002). *Jpn. J. Appl. Phys.* 41: 5626.
33 Chu, T.Y. and Song, O.K. (2007). *Appl. Phys. Lett.* 90: 203512.
34 So, S.K., Tse, S.C. and Tong, K.L. (2007). *J. Disp. Technol.* 3: 225.
35 Cheung, C.H., Kwok, K.C., Tse, S.C. and So, S.K. (2008). *J. Appl. Phys.* 103: 093705.
36 Shen, Y., Hosseini, A.R., Wong, M.H. and Malliaras, G.G. (2004). *Chem. Phys. Chem.* 5: 16.
37 Wang, Z.B., Helander, M.G., Greiner, M.T. et al. (2009). *Phys. Rev. B* 80: 235325.
38 Wang, Z.B., Helander, M.G., Greiner, M.T. et al. (2010). *J. Appl. Phys.* 107: 034506.
39 Brutting, W., Berleb, S. and Muckl, A.G. (2001). *Synth. Met.* 122: 99.
40 Brutting, W., Berleb, S. and Muckl, A.G. (2001). *Org. Electron.* 2: 1.
41 Nataliand, D. and Sampietro, M. (2002). *J. Appl. Phys.* 92: 5310.
42 Garcia-Belmonte, G., Munar, A., Barea, E.M. et al. (2008). *Org. Electron.* 9: 847.
43 Barsoukov, E. and Macdonald, J.R. (2005). *Impedance Spectroscopy: Theory, Experiment, and Applications*, 2e. New York: Wiley-Interscience.
44 Okachi, T., Nagase, T., Kobayashi, T., and Naito, H. (2008). *Thin Solid Films* 517: 1331.
45 Takata, M., Kouda, N., Ishihara, S. et al. (2014). *Jpn. J. Appl. Phys.* 53: 02BE02.
46 Shao, J. and Wright, G.T. (1961). *Solid State Electron.* 3: 291.
47 Wright, G.T. (1966). *Solid State Electron.* 9: 1.
48 Kao, K.C. and Hwang, W. (1981). *Electrical Transport in Solids.* Oxford: Pergamnon Press.
49 Lampert, M.A. and Mark, P. (1970). *Current Injection in Solids*, p. 46. New York: Academic.
50 Helander, H.G., Wang, Z.B. and Lu, Z.H. (2011). *Org. Electron.* 12: 1576.
51 Tsang, S.W., Denhoff, M.W., Tao, Y., and Lu, Z.H. (2008). *Phys. Rev. B* 78: 081301.
52 Matsushima, T., Kinoshita, Y., and Murata, H. (2007). *Appl. Phys. Lett.* 91: 253504.
53 Martens, H.C.F., Brom, H.B., and Blom, P.W.M. (1999). *Phys. Rev. B* 60: R8489.
54 Dascalu, D. (1966). *Int. J. Electron.* 21: 183.
55 Dascalu, D. (1968). *Solid State Electron.* 11: 491.
56 Kassing, R. (1975). *Phys. Status Solidi A* 28: 107.
57 Takagi, K., Abe, S., Nagase, T. et al. (2015). *J. Mater. Sci. Mater. Electron.* 26: 4463.
58 Nagase, T., Kishimoto, K. and Naito, H. (1999). *J. Appl. Phys.* 86: 5026.
59 Nagase, T. and Naito, H. (2000). *J. Appl. Phys.* 88: 252.
60 Chua, L.L., Zaumseil, J., Chang, J.F. et al. (2005). *Nature* 434: 194.
61 Gommans, H.H.P., Kemerink, M., and Janssen, R.A.J. (2005). *Phys. Rev. B* 72: 235204.

62 Simmons, J.G., Taylor, G.W. and Tam, M.C. (1973). *Phys. Rev. B* 7: 3714.
63 Stallinga, P., Gomes, H.L. and Rost, H. (1999). *Phys. B* 273: 923.
64 Nakahara, M., Minagawa, M., Oyamada, T. et al. (2007). *Jpn. J Appl. Phys.* 46: L636.
65 Kawano, K. and Adachi, C. (2009). *Adv. Funct. Mater.* 19: 3934.
66 Yu, P., Migan-Dubois, A., Alvarez, J. et al. (2012). *J. Non-Cryst. Solids* 358: 2537.
67 Schmechel, R. and von Seggern, H. (2004). *Phys. Status Solidi A* 201: 1215.
68 Schafferhans, J., Baumann, A., Deibel, C. and Dyakonov, V. (2008). *Appl. Phys. Lett.* 93: 093303.
69 Schafferhans, J., Baumann, A., Wagenpfahl, A. et al. (2010). *Org. Electron.* 11: 1693.
70 Schafferhans, J., Deibel, C. and Dyakonov, V. (2011). *Adv. Energy Mater.* 1: 655.
71 Baranovski, S. (ed.) (2006). *Charge Transport in Disordered Solids with Applications in Electronics,* p. 83. Chichester: Wiley.
72 Lang, D.V. (1974). *J. Appl. Phys.* 45: 3023.
73 Lang, D.V., Cohen, J.D. and Harbison, J.P. (1982). *Phys. Rev. B* 25: 5285.
74 Cohen, J.D. and Lang, D.V. (1982). *Phys. Rev. B* 25: 5321.
75 Campbell, A.J., Bradley, D.D.C., Werner, E. and Brutting, W. (2000). *Synth. Met.* 111: 273.
76 Yang, Y.S., Kim, S.H., Lee, J.I. et al. (2002). *Appl. Phys. Lett.* 80: 1595.
77 Stallinga, P., Gomes, H.L., Rost, H. et al. (2000). *Synth. Met.* 111: 535.
78 Nguyen, T.P. (2008). *Phys. Status Solidi A* 205: 162.
79 Neugebauer, S., Rauh, J., Deibel, C. and Dyakonov, V. (2012). *Appl. Phys. Lett.* 100: 263304.
80 Naito, H., Ding, J. and Okuda, M. (1994). *Appl. Phys. Lett.* 64: 1830.
81 Naito, H. and Okuda, M. (1995). *J. Appl. Phys.* 77: 3541.
82 Naito, H., Nagase, T., Ishii, T. et al. (1996). *J. Non-Cryst. Solids* 198: 363.
83 Simmons, J.G. and Tam, M.C. (1973). *Phys. Rev. B* 7: 3706.
84 Orenstein, J. and Kastner, M. (1981). *Phys. Rev. Lett.* 46: 1421.
85 Seynhaeve, G.F., Barclay, R.P., Adriaenssens, G.J. and Marshall, J.M. (1989). *Phys. Rev. B* 39: 10196.
86 Song, H.Z., Adriaenssens, G.J., Emelianova, E.V. and Arkhipov, V.I. (1999). *Phys. Rev. B* 59: 10607.
87 Shein, L.B. (1977). *Phys. Rev. B* 15: 1024.
88 Street, R.A., Zesch, J. and Thompson, M.J. (1983). *Appl. Phys. Lett.* 43: 672.
89 Kasap, S.O., Polischuk, B. and Dodds, D. (1990). *Rev. Sci. Instrum.* 61: 2080.
90 Kasap, S.O., Aiyah, V., Polischuk, B. et al. (1991). *Phys. Rev. B* 43: 6691.
91 Nemeth-Buhin, A. and Juhasz, C. (1997). *J. Phys. Condens. Matter* 9: 4831.
92 Komuro, S., Aoyagi, Y., Segawa, Y. et al. (1983). *Appl. Phys. Lett.* 43: 968.
93 Kanazawa, K.K. and Batra, I.P. (1972). *J. Appl. Phys.* 43: 1845.
94 Lampert, M.A. and Mark, P. (1970). *Current Injection in Solids,* p. 150. New York: Academic.
95 Kumoda, Y., Nakatsuka, E., Mori, K. et al. (2020). *Jpn. J. Appl. Phys.* 59 (064002).
96 Nojima, H., Kobayashi, T., Nagase, T., and Naito, H. (2019). *Sci. Rep.* 9: 20346.

# 6

# Time-of-Flight Method for Determining the Drift Mobility in Organic Semiconductors

*Masahiro Funahashi*

*Program in Advanced Materials Science, Faculty of Engineering and Design, Kagawa University, 2217-20 Hayashi-cho, Takamatsu, Kagawa 761-0396, Japan*

**CHAPTER MENU**



## 6.1 Introduction

Electronic properties of organic semiconductors are determined by the carrier density and carrier mobility. Unlike the carrier density, which depends upon extrinsic parameters, the carrier mobility value is specific to a material and is significant for the characterization of organic semiconductors and the design of devices and molecules of organic electronic materials. The time-of-flight (TOF) method is one of the techniques used for the measurement of the drift mobility of charge carriers, unlike the microwave conductivity method and Hall measurement, which provide the band mobility. The mobility measured by the TOF method is that of the bulk under a low charge carrier density, whereas the field-effect mobility is measured in a two-dimensional (2D) interface under a high carrier concentration. From the integration of transient photocurrent curves, the amount of photogenerated charge carriers can also be determined [1].

In organic semiconductors and inorganic amorphous semiconductors, charge carriers have a tendency to localize on molecules or in the tail of the density of states, resulting in hopping transport. The Hall measurement is not effective for determining the carrier mobility in the incoherent charge hopping process [2]. The TOF method can be applied to insulators or semiconductors with low electrical conductivities. Spear and co-workers first measured the carrier mobility of chalcogenide glass using this method [3].

In 1960, the carrier mobility of single crystals of anthracene was determined by Kepler [4] and LeBlanc [5] using this method. The carrier transport characteristics in single crystals of aromatic compounds with high purity were studied by Karl and co-workers [6]. In addition to molecular crystals, the carrier mobilities in organic amorphous solids including polyvinylcarbazole and triphenylamine derivatives were investigated extensively considering the application to xerographic photoreceptors and organic electroluminescence devices [7]. The bulk carrier transport characteristics of conjugated polymers, such as poly(3-alkylthiophene), poly(9,9-dialkylfluorenone), and related materials, were studied by implementing the TOF method [8, 9]. Other than solid-state materials, the TOF method was applied to carrier mobility measurements in dielectric liquids [10] and liquid crystal phases [11–13]. Liquid crystal molecules comprised of extended π-conjugated systems work as 1D or 2D semiconductors. Electronic charge carrier transport was studied in columnar phases, which have 1D π-stacking structures and smectic phases with layered structures.

## 6.2 Principle of the TOF Method

### 6.2.1 Carrier Mobility and Transient Photocurrent

The carrier mobility $\mu$ is expressed by Eq. 6.1, where $j$ is current density, $E$ electric field, $n$ carrier density, $e$ the elemental charge.

$$j = \sigma E = ne\mu E \tag{6.1}$$

The carrier mobility $\mu$ has the same dimensions as the quotient of the average velocity of charge carriers divided by the electric field and can be regarded as a proportional constant of the acceleration of a charge carrier under the application of a DC bias. In the TOF experiment, the average velocity of charge carriers is measured considering the dimensions of the charge carrier mobility. For a sample whose thickness is known, the measurement of the transit time, $t_T$, in which carriers drift across a sample, reveals the average velocity of the carriers, $v$. As shown in Eq. 6.2, the carrier mobility can be calculated using the electric field, $E$, sample thickness, $d$, applied voltage, $V$, and transit time, $t_T$.

$$\mu = \frac{v}{E} = \frac{d}{Et_T} = \frac{d^2}{Vt_T} \tag{6.2}$$

The transit time is determined as follows. Figure 6.1a displays the schematic illustration of a TOF sample and the electric circuit of the measurement [2, 3]. A sandwich-type sample, in which the material is placed between two electrodes, is prepared. Then a DC bias is applied to the sample. Pulse light is illuminated on one side of the sample, inducing a displacement current through an external serial resistor. The voltage drop across the external resistor is recorded in an oscilloscope. For precise determination of the transit time, charge carriers have to be generated on one electrode simultaneously. In order to satisfy this condition, pulse light illumination produces a sheet of charge carriers on one side of the sample. In the old days, an Xe flash lamp was used for this purpose. Recently, pulse lasers, such as the $N_2$ and Nd:YAG lasers, have been used for the generation of the charge carrier sheet. The pulse width should be sufficiently shorter than the transit time, $t_T$. The amount of drifting carriers can be monitored by measurement of the displacement current, $i_d$, flowing through the external serial resistor as a function of the time. This displacement current is called the transient photocurrent. Figures 6.1b and 6.1c show the schematic illustration of the distribution of charge carriers in the

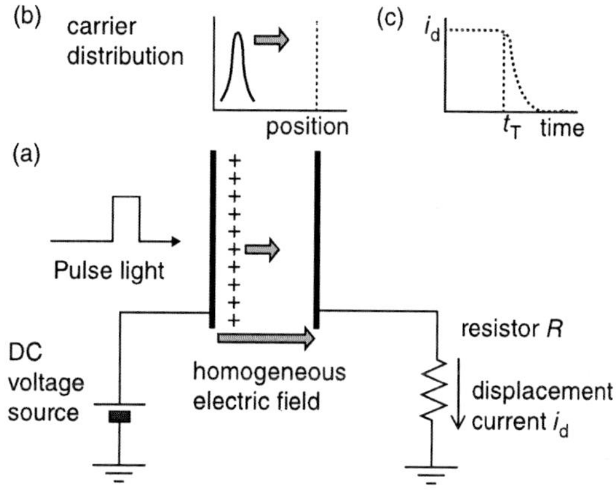


**Figure 6.1** Illustrations for the principle of TOF measurement. (a) A schematic of a sample and an electrical circuit for the measurement, (b) distribution of photogenerated carriers in the sample, and (c) observed transient photocurrent.

sample and the displacement current, $i_d$, flowing through the external serial resistor, respectively. If the carriers drift at a constant velocity, the transient photocurrent is constant. When the carriers arrive at the counter electrode and stop, the transient current drops to zero. The time, in which this constant current retains, corresponds to the transit time $t_T$. The carrier sign can be determined by considering the polarity of the DC bias. The mobility of positive charge carriers is obtained when the illuminated electrode is biased positively. When a negative DC bias is applied to the illuminated electrode, the mobility of negative charge carriers is determined.

### 6.2.2 Standard Setup of the TOF Measurement

Figure 6.2 shows the standard setup of the TOF measurement. The system consists of a hot stage to maintain the sample temperature, pulse light as the excitation source, DC voltage source, which applies a DC electric field to the sample, serial resistor, trigger unit, and oscilloscope to record the photocurrent. For the low-temperature measurements, a cryostat is used as a temperature-controlling unit. Recently, the whole system has become commercially available, but it is costly. The author recommends constructing this system by combining the parts.

The pulse light excites the absorption band of the sample. The sample has to exhibit a sufficiently high absorption coefficient for the excitation light. For organic semiconductors, which usually absorb near-UV light, an Xe flash lamp, $N_2$ laser, or third harmonic generation of a Nd:YAG laser is used. They are relatively inexpensive. The author used a Nd:YAG laser, combined with a nonlinear optical crystal, for the third harmonic generation (the wavelength is 356 nm). For samples, which have the absorption bands in the visible area, a dye laser unit combined with a $N_2$ laser is convenient.

The author also used a handmade hot stage whose temperature was retained with an accuracy of ±0.1° C through a PID controller. In the hot state, cables were connected to the sample via BNC connectors for the reduction of current noise. For the low-temperature measurements, a cryostat with a quartz window for the introduction of the excitation light and connectors for the application of a DC voltage and the measurement of a photocurrent was used. The DC voltage supplied a stable voltage to the sample. The author adopted an electrometer with a DC voltage source, which could monitor

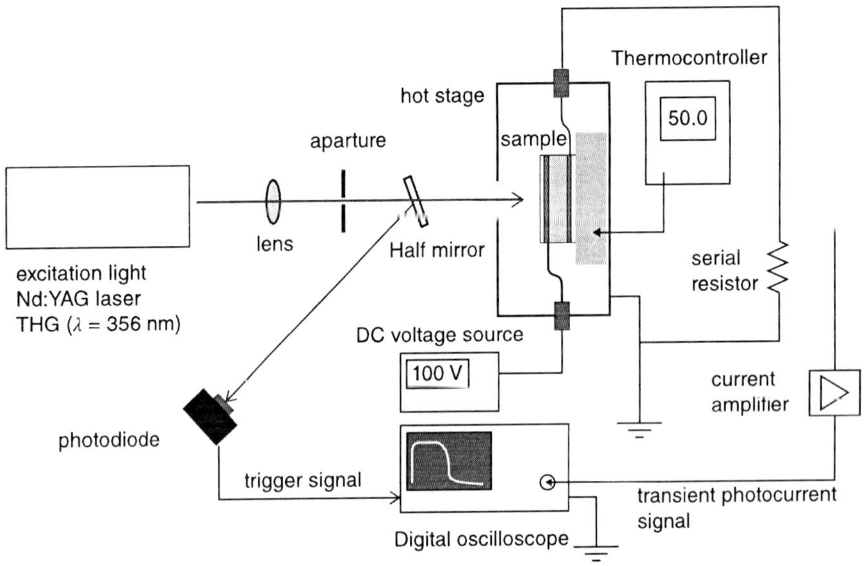


**Figure 6.2** Standard setup of TOF measurement system.

the applied voltage and dark current of the sample. The oscilloscope should be selected by considering the voltage resolution and frequency range. For the measurement of high-mobility samples, the frequency should be in the GHz range. For an apparatus working in a higher-frequency range, the voltage resolution becomes lower.

The external serial resistor is selected considering the carrier mobility of the sample. In the case of weak current signals, we need to use a high-resistance resistor, which causes a delayed current response to induce a distortion of the transient photocurrent. We should use an optimized resistor considering the time range of the measurement and the strength of the current signal. High-speed current amplifiers are sometimes used when the photocurrent is weak. The amplifier should be selected considering the upper limit of the input current and the response speed.

### 6.2.3 Sample Preparation

In principle, the sample for TOF measurement is of the sandwich type, in which the thickness is around several micrometers. For organic amorphous solids and liquid crystals with low molecular weights, a sample is prepared using capillary filling of a molten material into a cell consisting of two transparent electrodes, which is cooled down to room temperature. The cell is fabricated by sticking the two transparent electrodes with glue containing silica particles. ITO-coated glass plates are often used as the transparent electrodes [15, 16]. A vacuum-deposited semi-transparent Al electrode on a glass plate can be used for this purpose. The cell thickness can be determined by the periodicity of the interference fringes in its transmission spectrum. The thickness can also be estimated from the cell capacitance.

For organic amorphous solids with low molecular weights, vacuum-deposited thin films, with a thickness of several micrometers, are also used for TOF measurement. On an Al-deposited glass plate, an organic semiconductor layer is vacuum-deposited, and semi-transparent Al film is vacuum-deposited on the organic layer [17].

For polymers and high-molecular weight oligomers, the organic semiconductor layer is deposited on an ITO- or Al-coated substrate by means of the spin-coating method [18]. After the organic thin film is sufficiently dried, the counter electrode is vacuum-deposited on the organic layer. DEKTAK stylus surface profiler is convenient to determine the thickness of the organic layer.

The cells and vacuum-deposited thin films can also be used for the measurement of crystalline samples. Polycrystalline samples are usually obtained using these methods. In polycrystalline samples, transient photocurrent curves often become featureless because localized states are formed on the grain boundaries and the carrier mobilities cannot be determined clearly. Therefore, single-crystal samples are sometimes used for TOF experiments. First, crude crystals are purified using recrystallization, vacuum-sublimation, and zone-melt methods. Single crystals with high purities are grown by implementing the Bridgeman method or the Czochralski process. Single crystals are cut or polished to provide crystalline plates. On both sides of the crystalline plates, semi-transparent metal films are vacuum-deposited [19]. The vacuum crystal growth technique flowing inert carrier gas is also used to produce single-crystal samples [20].

To obtain sufficiently strong photocurrent signals, the efficiency of the photocarrier generation in the samples should be high. For samples with low photocarrier generation efficiencies, a carrier generation layer is sometimes inserted between the illuminated electrode and the organic semiconductor layer. As carrier generation layers, materials, which can produce charge carriers under light illumination with a high efficiency, are used. Vacuum-deposited chalcogenide glass films, such as amorphous Se and $As_2Se_3$ [21, 22], vacuum-deposited organic dyes including phthalocyanine, perylene bisimide, and diazo pigments, are often used as carrier generation layers [23, 24]. As discussed in Section 6.5, the carrier generation layer must be sufficiently thinner than the organic semiconductor layer for a sheet-like distribution of photogenerated charge carriers. Conventional organic semiconductors have the absorption bands in the near-UV area, while materials used as carrier generation layers absorb visible light efficiently. Excitation pulse light is usually illuminated on the carrier generation layer through the carrier transport layer.

### 6.2.4 Current Mode and Charge Mode

In TOF experiments, the movement of photogenerated or injected charge carriers is monitored by observing the voltage drop across the external serial resistor. The current induced through the serial resistor is the displacement current, which is produced by the change in the distribution of charge carriers in the sample [25]. Figure 6.3a is the schematic illustration of the distribution of the electric field in a TOF sample. Before the photogeneration or injection of charge carriers, a homogeneous electric field, $E = V/d$, is formed when a DC Voltage $V$ is applied to a sample with a thickness of $d$. When $N$ charges are photogenerated and move at position $x$ from the illuminated electrode at a velocity of $v$, the electric field increases in front of moving charges and that behind the charges decreases. If the electric field in front of the charges (at the side of the counter electrode) is $E_1$ and that behind the charges at the side of the illuminated

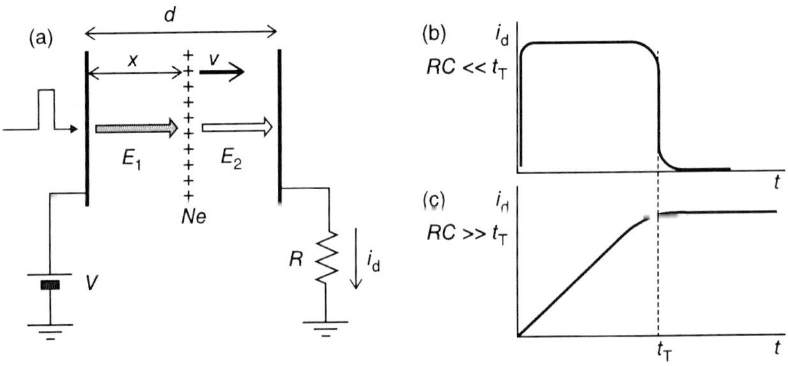


**Figure 6.3** (a) Schematic of distribution of an electric field in a TOF sample. Transient photocurrent signals obtained under a condition of (b) the current mode and (c) the charge mode.

electrode) is $E_2$, they are expressed by Eqs. (6.3) and (6.4), respectively, where $\varepsilon$ is the dielectric constant of the sample and $S$ is the electrode area of the sample [1].

$$E_1 = E - \frac{Ne}{\varepsilon S}\left(1 - \frac{x}{d}\right) \tag{6.3}$$

$$E_2 = E + \frac{Ne}{\varepsilon S}\frac{x}{d} \tag{6.4}$$

The characteristic response time of the TOF system including the sample (capacitance $C$) and serial resistor $R$ is $RC$. When $RC$ is sufficiently shorter than the transit time $t_T$, charge $\pm Q$ ($= CV$) is induced on the electrodes when the DC voltage is applied to the sample, and the charges induced on the electrodes retain during the measurement. As Eqs. (6.3) and (6.4) indicate, the movement of photogenerated charges attempt to change the distribution of the electric field within the sample. However, the displacement current flows through the serial resistor so that charge $Q$ on the electrodes is constant, eliminating the change in the distribution of the electric field. As shown in Figure 6.3b, a rectangular-shaped transient photocurrent signal is observed as the voltage drop across the serial resistor (the current mode).

In contrast, as characteristic response time $RC$ is sufficiently longer than $t_T$, charges accumulated on the counter electrode are not retained during the measurement. The change in the charge on the counter electrode, $\Delta Q$, is determined by the change in $E_2$. Therefore, $\Delta Q$ is described by $N$, $e$, $x$, and $d$ as Eq. (6.5).

$$\Delta Q = \frac{Nex}{d} \tag{6.5}$$

The voltage drop through the serial resistor, $\Delta V$ is expressed by the carrier average velocity $v$, as Eq. (6.6).

$$\Delta V = \frac{\Delta Q}{C} = \frac{Ne}{Cd}vt \tag{6.6}$$

The transient photocurrent monotonically increases with time and becomes constant after $t_T$, as shown in Figure 6.3c. This transient photocurrent curve indicates the charges arrived at the counter electrode. In addition, it corresponds to the integral of the curve displayed in Figure 6.3b. This measurement is called the charge mode.

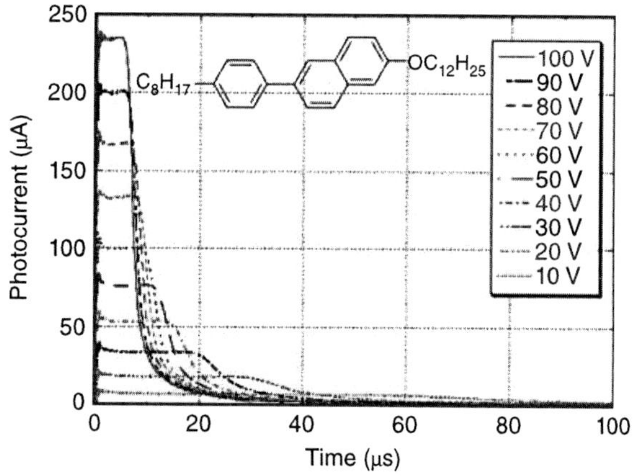


**Figure 6.4** Typical current-mode transient photocurrent curves for holes in the smectic B phase of phenylnaphthalene derivatives (inset) at 90° C using an Al/Al liquid crystal cell with thickness of 12 μm.

Figure 6.4 exhibits typical transient photocurrent curves in the current mode. They are those for holes in the smectic phase of a liquid crystalline phenylnaphthalene derivative at 90° C [15]. The resistance of the serial resistor was 100 Ω, and the sample capacitance was 0.2 nF, affording a response time, $RC$, of 0.02 μs, which is sufficiently shorter than the transit time. Therefore, the transient photocurrent curves are displayed in the current mode.

### 6.2.5 Instructions in the TOF Measurements

1) Sample thickness
   For the generation of ideal transient photocurrent curves, a sheet of charge carriers has to be produced at the illuminated electrode. The area of the photocarrier generation is determined by the absorption area of an excitation light pulse. When the absorption coefficient of the sample for the excitation light is sufficiently high, the generation area of photogenerated charge carriers is regarded as a sheet because the excitation light is absorbed at the interface between the illuminated electrode and the organic layer. The penetration depth $\delta$ of the excitation light is an indicator for the generation area of photocarriers. According to the Lambert–Beer law, the penetration depth $\delta$ is described by the absorption coefficient $\alpha$, transmittance $T$, and length of the optical path, $l$, as shown in Eq. (6.7).

$$\alpha l = -\ln T, \quad \delta = \frac{1}{\alpha} \tag{6.7}$$

   In the TOF measurement, $\delta$ must be much shorter than the sample thickness $d$. Usually, the sample should be prepared so that $\delta$ should be less than 10 % of $d$.
2) Amount of photogenerated charge carriers and space charge limited current
   In principle, the samples used for TOF measurements are insulators or intrinsic semiconductors with a high resistance. Therefore, a sandwich-type sample works as a capacitor. When a voltage, $V$, is applied to the sample, charges $Q$ (= $CV$) are accumulated on the electrodes, forming a homogeneous electric field, $E$ (= $V/d = Q/(\varepsilon S)$), across the sample. For a constant velocity of charge carriers under the homogeneous electric field, the amount of photogenerated charge carriers, $q$, has to be sufficiently smaller than $Q$. Under this condition, the transient photocurrent is constant before the transit time.

However, homogeneity of the electric field is not always assumed. In dark conductivity measurements of the material with low mobility under the high electric field, injected charge carriers from the electrode are accumulated near the electrode in the sample to weaken the electric field near the electrode, compared to that in the bulk. The weakened electric field at the electrode, promotes the accumulation of the injected charge carriers. In the limit of the zero-field at the interface, space charge limited current is observed. Under the condition without traps, the current is proportional to the square of the applied voltage and inversely proportional to the thickness to the power of three, called as Child's law, where $j$ is the current density, $\varepsilon$ is the relative dielectric constant, $\varepsilon_0$ is the vacuum dielectric constant, and $\mu$ is the carrier mobility (Eq. [6.8]).

$$j = \frac{9}{8}\mu\varepsilon_0\varepsilon\frac{V^2}{d^3} \tag{6.8}$$

In TOF measurements, transient photocurrent is also affected by photogenerated space charge under a strong illumination condition. Figure 6.5a exhibits the distributions of the electric field and carrier density under the space charge limited current condition. As the intensity of the excitation light is increased, the amount of photogenerated charge carriers $q$ usually increases. Under the condition of $q \gg Q$, the carrier generation rate becomes higher than the transfer rate of photogenerated charge carriers, resulting in the accumulation near the illuminated electrode. As a result, the electric field near the electrode is weakened to promote the accumulation of photogenerated charge carriers [26]. The transient photocurrent increases with the time before the transit time, resulting in the formation of a peak in the transient photocurrent curve. The peak time $t_\mathrm{p}$ depends on the conventional transit time, $t_\mathrm{T}$, which is observed under the condition of $q \ll Q$ as shown in Eq. (6.9) [27].

$$t_\mathrm{p} = 0.787t_\mathrm{T} \tag{6.9}$$

The peak current $I_\mathrm{p}$ is proportional to the square of the applied voltage $V$ under this space charge limited condition if the carrier generation efficiency is independent of the electric field.

Figure 6.5b is the transition from a conventional rectangular transient photocurrent curve to a space charge-limited transient photocurrent curve when the excitation light intensity is increased. The measurement was carried out in the liquid crystal phase of a phenylquaterthiophene derivative (the inset Figure 6.5b). When the intensity of the excitation pulse was low, a constant conventional transient photocurrent before the transit time was observed. Under the condition of high-intensity excitation, transient photocurrent curves with a peak were obtained. The peak time is shorter than the transit time.

Figures 6.5c and 6.5d display the transient photocurrent curves measured under high- and low-excitation condition, respectively. Under low-excitation condition, ideal rectangular transient photocurrent curves are obtained, while those with a cusp are observed in the high-excitation mode. The peak current is proportional to the square of the applied voltage, as shown in Figure 6.5c.

3) Electrode

The carrier density of the TOF samples is considerably low and they are usually insulators with a high resistance. In TOF measurements, charge carriers are generated under excitation light. Therefore, the desired dark current of the samples is

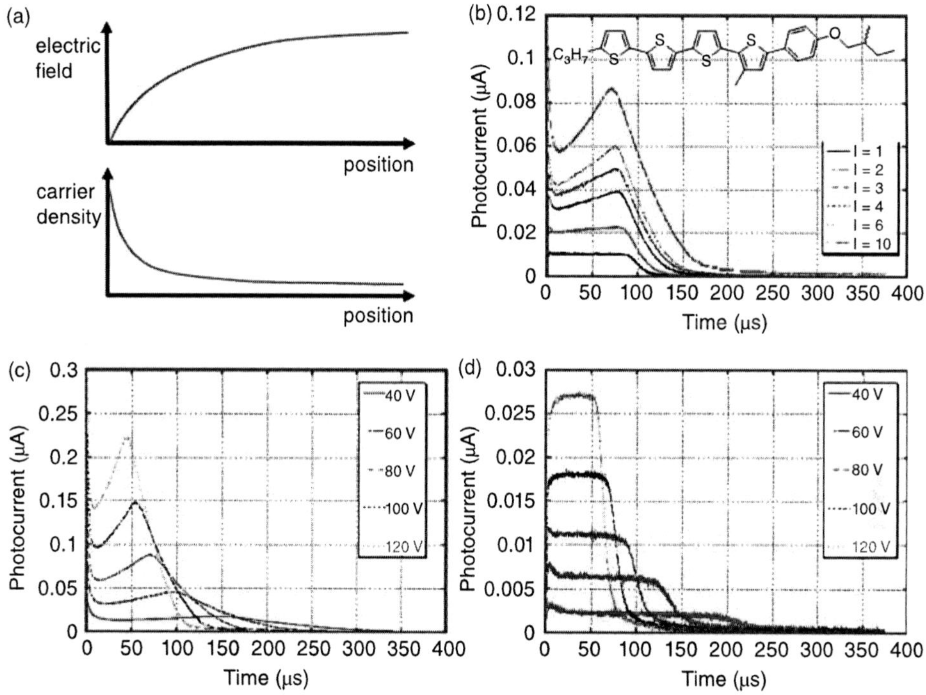


**Figure 6.5** (a) Distribution of the electric field and carrier density under the space charge limited current condition. (b) Transient photocurrent curves for holes in the nematic phase of phenylquaterthiophene derivative (inset) at 120° C and 80 V. 'I' in the legend indicates relative intensity of the excitation pulse. (c) Transient photocurrent curves for holes in the same sample under the space charge limited current condition. (d) Transient photocurrent curves under weak illumination intensity. A liquid crystal cell consisting of two ITO-coated glass plates with a thickness of 9 μm is used for the measurement. Serial resistor is 1 kΩ.

low. If charge carriers are injected from the electrode into the sample, the contrast between the dark and illuminated states is degraded and it is difficult to provide good photocurrent signals. Photo-induced carrier injection (secondary photocurrent) distorts the transient photocurrent curves making it difficult to analyze the transient photocurrent. Therefore, blocking electrodes, on which charge carrier injection is inhibited, should be used for TOF measurements.

For TOF measurements of organic amorphous semiconductors, semitransparent aluminum electrodes are used as the blocking electrodes. On the surface of an aluminum film, an insulative aluminum oxide layer is formed and it inhibits charge carrier injection. This electrode is easily produced using the vacuum deposition method. The carrier injection process is affected by the injection barrier at the interface between the electrode and the organic layer. The injection barrier is mainly determined by the difference between the Fermi level of the electrode and the HOMO or LUMO levels of the organic layer [28]. An ITO electrode is also used for the TOF measurement when the work function of the ITO electrode differs sufficiently from the HOMO or LUMO levels of the sample. Unlike metal electrodes, it is difficult to exfoliate ITO electrodes from substrates and surface treatment, such as rubbing, can be performed. However, ITO is polycrystalline oxide and contains defects on the surface, which often promote local carrier injection and electrochemical reaction.

4) Nondispersive and dispersive transport

In semiconductors with high purities and low defect densities, charge carriers can drift through the samples without being trapped by impurities and defects. In such cases, the photogenerated carrier sheet diffuses as it moves forward at a constant velocity, as shown in Figure 6.6a. As a result, a constant current flows through the serial resistor, and the current drops to zero when the carrier sheet arrives at the counter electrode, as shown in Figure 6.6b.

Organic semiconductor samples usually contain impurities and defects. When contaminated impurities and defects form localized states distributed between the HOMO and LUMO levels of the samples, charge carriers repeat trapping and detrapping as they move forward at a constant velocity. During this charge carrier movement, the carrier sheet diffuses to induce a constant displacement current through the serial resistor. When the carrier sheet arrives at the counter electrode, the current signal drops to zero, as shown in Figure 6.6b.

In these cases, the carrier sheets spread with Gaussian diffusion during the constant velocity movement through the samples. The transient photocurrents first indicate constant values and drop to zero immediately when the carrier sheets arrive at the counter electrode. This type of transient photocurrent curves is called nondispersive. The transit time is defined as the midpoint of the decaying part of the transient photocurrent, $t_T$, or the point, from which the photocurrent decays, $t_T'$, as shown in Figure 6.6b. The value $t_T$ means the average transit time of all the charge carriers, while $t_T'$ indicates that of the fastest charge carriers. The transient photocurrent curves displayed in Figures 6.4 and 6.5 (d) are typical nondispersive curves.

The aspect of the charge carrier transport is different from the nondispersive process, which is characterized by the Gaussian diffusion of the carrier sheets, when the

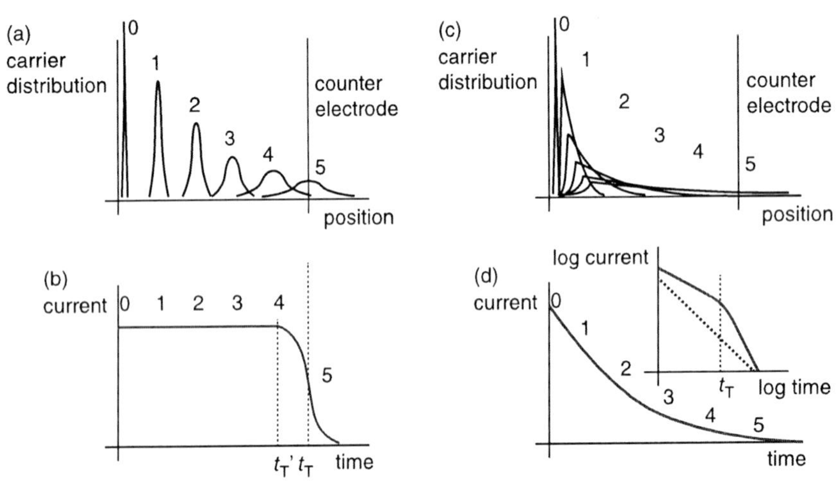


**Figure 6.6** Schematics of current signals obtained by the TOF measurement. (a) Distribution of the photogenerated carriers in the sample and (b) a current signal for non-dispersive transport. (c) Distribution of the photogenerated carriers in the sample and (d) a current signal for dispersive transport.

samples contain traps with various depths and some of them form relatively deep levels. In this case, a few charge carriers move forward without being trapped while most of the charge carriers are captured by deep traps and not released during the measurement. Therefore, the spatial distribution of charge carriers is asymmetric unlike the Gaussian distribution, as shown in Figure 6.6c. Because most of the charge carriers are trapped near the starting electrode and only a small number of charge carriers can arrive at the counter electrode, the transient photocurrent decays monotonically with the time, as shown in Figure 6.6d. Such a transient photocurrent pattern is called dispersive. In dispersive transient photocurrent curves, the transit time cannot be defined clearly. For the dispersive transport, Scher and Montroll proposed a theory based on the random walk of charge carriers in a system where the distribution of the trap depth is described by the power law [29].

For dispersive transport, kink points sometimes appear in the double logarithmic plots of transient photocurrent curves, as shown in the inset of Figure 6.6d. This kink point can be defined expediently as the transit time in the dispersive transport. In the double logarithmic plot of the dispersive transient photocurrent curves, the transit time is proportional to the power of the time and the exponent changes at the expediential transit time, as described by Eqs. (6.10) and (6.11).

$$I \propto t^{-(1-\alpha)} \quad t < t_{\mathrm{T}} \tag{6.10}$$

$$I \propto t^{-(1+\alpha)} \quad t > t_{\mathrm{T}} \qquad 1 \geq \alpha \geq 0 \tag{6.11}$$

The parameter $\alpha$ indicates the dispersivity of the transport. The dispersivity increases as $\alpha$ decreases. When $\alpha$ is 1, the transport process is nondispersive. The transport process is completely dispersive when $\alpha$ is 0, and the kink point indicating the transit time cannot be obtained [26].

Figure 6.7 displays typical transient photocurrent curves obtained in the columnar phase of a perylene tetracarboxylic bisimide derivative at 30° C. In the linear plots, the photocurrent signals decay monotonically and clear kink points are not observed. However, the kink points are clearly defined in the double logarithmic plots of the transient photocurrent curves, as shown in Figure 6.7b, and the transit times are determined [30].

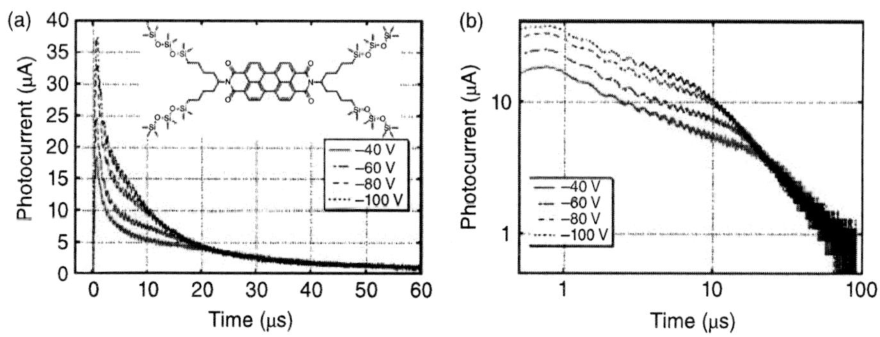


**Figure 6.7** Dispersive transient photocurrent curves for electrons in a disordered columnar phase of liquid crystalline perylene tetracarboxylic bisimide derivative (inset) at room temperature using an ITO cell with a thickness of 9 μm. (a) linear plot (b) double logarithmic plot.

## 6.3 Information Obtained From the TOF Experiments

In TOF experiments, charge carriers move from one edge of the samples to another. During the charge carrier movement, they interact with localized states formed by impurities and defects. Therefore, the TOF experiment affords the drift mobility of charge carriers in the bulk of the samples. This is different from other techniques used to determine the carrier mobility, e.g., the microwave absorption method and Hall measurement provide the band mobility, which is not affected by localized states [31]. The TOF measurement provides the charge carrier mobility in the bulk of the sample in contrast to that obtained from a field-effect transistor, in which the charge carrier mobility is in a 2D space on the dielectric layer. Therefore, TOF mobility is not influenced by localized states formed on the electrode surfaces [32].

The integrals of the transient photocurrent $i_{\mathrm{ph}}$, collected in the current mode experiments as shown in Figure 6.3b, indicate the number of the photogenerated charge carriers $Q$ (Eq. [6.12]).

$$Q = \int_0^{\infty} i_{\mathrm{ph}} dt \tag{6.12}$$

In charge mode experiments, as shown in Figure 6.3c, the saturated current value after the transit time, multiplied by the resistance of the serial resistor and the capacitance of the sample, produces the number of photogenerated charge carriers. The photocarrier generation process in amorphous organic semiconductors was studied through TOF measurements [33].

The decaying part of the transient photocurrent curve is associated with the spatial distribution of photogenerated charge carriers in the sample, as shown in Figures 6.8a

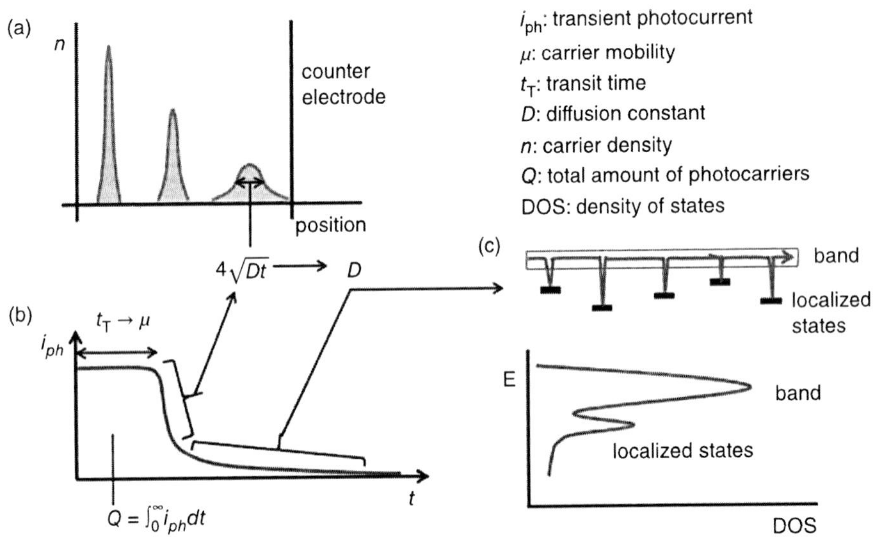


**Figure 6.8** Information abstracted from transient photocurrent curves obtained in the TOF experiments. Schematic illustrations for (a) distribution of charge carriers, (b) transient photocurrent curve, and (c) density of states of the band and localized states.

and 6.8b. The theoretical method to obtain the diffusion constant of photogenerated charge carriers from the transient photocurrent curves was proposed by Hirao and Nishizawa [34].

Photogenerated charge carriers repeat the trapping and detrapping processes during movement in the sample. In particular, the tail parts of the transient photocurrent curves are determined by the trapping events of localized states, as shown in Figures 6.8b and 6.8c. The theoretical method to determine the energetic distribution of localized states in the bulk based on the multiple trapping model was proposed by Naito and co-workers [35].

## 6.4 Techniques Related to the TOF Measurement

### 6.4.1 Xerographic TOF Method

Until the 1980s, the main application of organic semiconductors was xerographic photoreceptors. The carrier transport characteristics of molecularly doped polymers were extensively studied. The xerographic TOF method can reveal the charge carrier mobilities of organic semiconductors deposited on conductive substrates under the same condition as the operation of a xerographic photocopier [36].

In the conventional TOF method, two electrodes are required to produce a sandwich-type sample. In contrast, an organic semiconductor thin film, deposited on a conductive substrate, is used for the xerographic TOF method. The second electrode is not required. As shown in Figure 6.9, the apparatus consists of a corona charger, excitation pulse light source, thin film sample with a conductive substrate, Kelvin probe to measure the surface potential of the sample, and an oscilloscope. The corona discharger needs a high-voltage source producing several thousand volts. By corona discharge, the surface of the organic semiconductor film is charged and the electric field is produced by the surface charges. In the case of a double-layer sample consisting of the carrier generation and transport layers, which is the standard structure of xerographic photoreceptors, pulse light is illuminated on the surface of the carrier generation layer through the carrier transport layer. Photogenerated charges are injected into the carrier

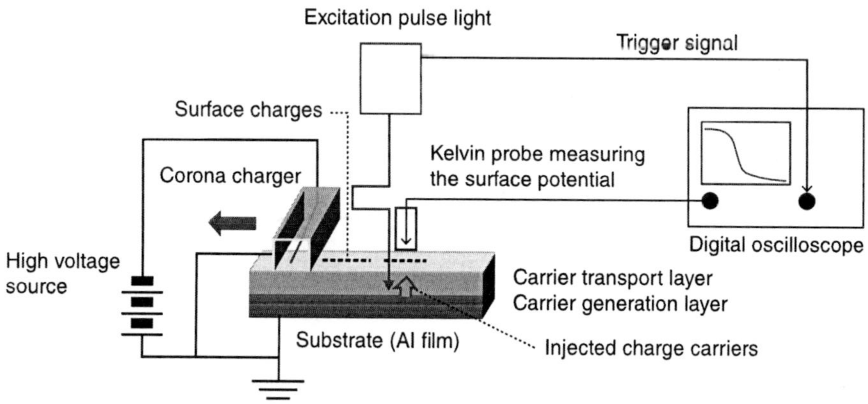


**Figure 6.9** Schematic illustration of the xerographic TOF measurement using a double-layer sample.

transport layer and are transported to the sample surface under the homogeneous electric field produced by surface charges generated by the corona discharge. The transient photocurrent curves are monitored as the change in the surface potential of the organic semiconductor films. The induced transient photocurrent $i_{\mathrm{ph}}$ is expressed as Eq. (6.13), using surface charge $Q$, surface potential $V$, and sample capacitance $C$.

$$i_{\mathrm{ph}} = -\frac{dQ}{dt} = -C\frac{dV}{dt} \tag{6.13}$$

In this method, an external DC voltage is not applied to the sample stationary and the charge carrier mobility under a high voltage can be determined avoiding breakdowns of the sample and apparatus. Conversely, this method is not appropriate for a sample with a high dark conductivity because such a sample cannot retain surface charges [37].

### 6.4.2 Lateral TOF Method

For the TOF measurement of samples with high charge carrier mobilities, such as molecular crystals of aromatic compounds, the transit times are sometimes comparable to the time constant of the measurement circuit. In such a case, the rising time of the transient photocurrent curve overlaps with the transit time, resulting in the distortion of the transient photocurrent curves. To avoid this problem, samples with a thickness of around 1 mm can be used. However, such thick samples are usually difficult to fabricate using the vacuum deposition and spin-coating methods.

Figure 6.10 exhibits the schematic illustration of a lateral TOF measurement. In the lateral TOF method, photogenerated charge carriers move perpendicularly to the normal of thin films, using semiconductor thin films, on which planar electrodes are deposited, instead of a sandwich-type sample. This method is also effective for studying the carrier transport mechanism in organic field-effect transistors, in which charge carriers move in the direction perpendicular to the normal of the organic semiconductor layer.

As shown in Figure 6.10a, illumination of an excitation pulse induces a transient photocurrent curve indicating the kink point corresponding to the transit time in the ideal

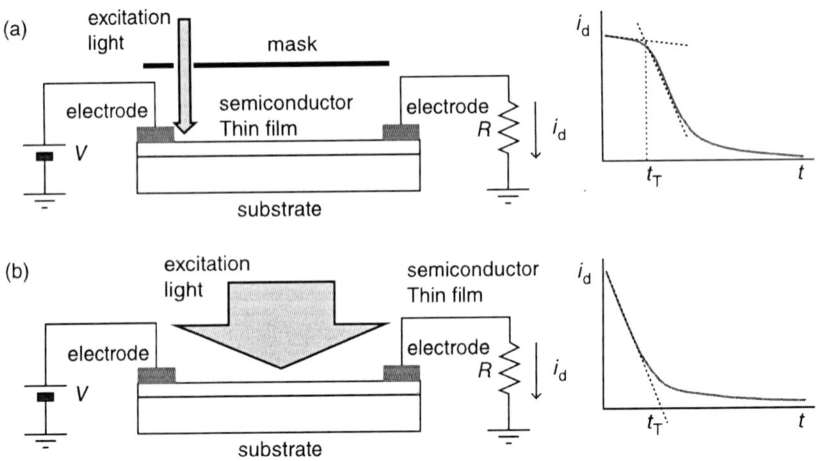


**Figure 6.10** Schematic of the setup of lateral TOF measurement system and induced transient photocurrent curves illuminating (a) near one electrode using a mask and (b) whole the channel area.

sample. Kitamura and co-workers vacuum-deposited planar electrodes with gaps of 6, 8, and 25 μm on polycrystalline thin films of Cu phthalocyanine; excitation light was focused near one electrode [38]. Naka *et al.* used a mask to excite near one electrode. In these cases, the kink points corresponding to the transit times were observed in the transient photocurrent curves [39]. Bratina proposed a method to analyze transient photocurrent curves, considering the distribution of charges and field dependence of the charge carrier mobility [40]. Iwamoto and co-workers measured the carrier mobility in field-effect transistors based on a polycrystalline thin film of pentacene, using this lateral TOF technique [41]. Unlike these measurements, Naito *et al.* determined charge carriers by illumination of the whole channel area in the lateral alignment of electrodes as shown in Figure 6.10b. In this case, photocurrent decay was observed. They also proposed a theoretical method to extract the charge carrier mobilities from photocurrent decay curves, considering the presence of localized states [42].

### 6.4.3 TOF Measurements Under Pulse Voltage Application

In the TOF measurement using amorphous silicon, chalcogenide glasses, and organic polycrystals, space charges produced by charge carrier trapping in impurities and defects affect the carrier transport. Space charges form local electric fields that often distort transient photocurrent curves and cause their instability. To avoid these space charge effects, pulse DC voltages are applied to the sample. The DC bias is applied to the sample only in the measurement of transient photocurrent curves. Between the measurements, the two electrodes are short-circuited and space charges are discharged [1, 2].

For liquids and liquid crystals containing ionic impurities, ions are accumulated on the electrodes when the DC bias is applied to the samples. When the concentration of ionic species is high, the electrical double layers are formed on the electrodes to cause electrochemical reaction at the interface between the electrodes and the organic layer. In such cases, the alternating voltage is applied to the samples to escape the formation of the electrical double layers.

Ellman *et al.* characterized the hole mobility in the columnar phase of hexapentyloxytripheneylene containing ionic impurities with the TOF measurement using alternating voltage. By considering the distribution of ions, the effective electric field in the sample is calculated and the transient photocurrent curves are simulated. Figure 6.11 shows the schematic illustration of the transient photocurrent measurements in columnar liquid crystals containing ionic impurities. The arrows numbered from 1 to 3 in Figure 6.11a indicate the change in the pulse light illumination time. As the pulse excitation time changes, the transit times depend upon the period between the polarity inversion of the electric field and excitation pulse illumination because of the screening effect due to ionic species as shown in Figure 6.11b [43]. This method should be effective for the characterization of the carrier transport of conjugated electrolytes [44] and $\pi$-conjugated liquid crystals bearing ionic moieties [45, 46].

### 6.4.4 Dark Injection Space Charge-Limited Transient Current Method

Figure 6.12 displays the schematic illustration of the dark injection space charge-limited transient current (DI-SCLC) method. Unlike conventional TOF measurements using

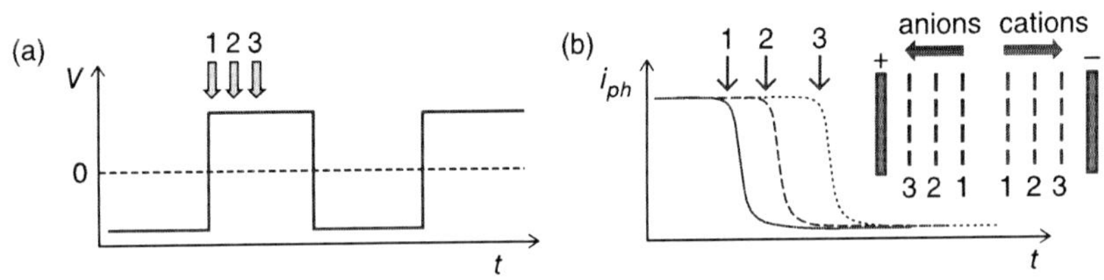


**Figure 6.11** Schematic illustration for the screening effect by ionic impurities and the change of the transient photocurrent by the delay of the excitation pulse light illumination. Source: Based on Ref. [43].

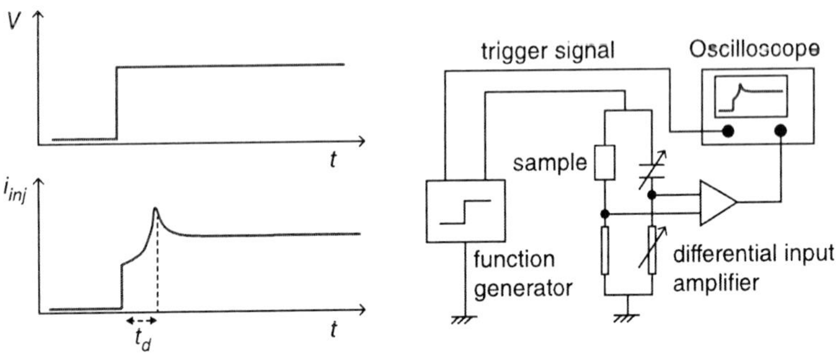


**Figure 6.12** Schematic illustrations for the measurement system of the DI-SCLC method.

the pulse light excitation, a DC voltage step is applied to the sample using ohmic contact electrodes in the DI-SCLC method [47, 48]. Because of the ohmic contact between the injection electrode and the organic semiconductor layer, charge carriers are injected from the electrode and form space charges near the electrode. After the arrival of the first component of injected charge carriers, the current becomes constant under the space charge-limited condition. When the first component of injected charge carriers reaches the counter electrode, the current indicates a peak. The peak time $t_\mathrm{d}$ is related to the transit time $t_\mathrm{T}$ as shown in Eq. (6.14).

$$t_d = 0.787\ t_\mathrm{T} \tag{6.14}$$

The advantage of this method is the applicability to thin films with a thickness of less than 1 μm. The thicknesses of organic semiconductor layers in electroluminescence devices and solar cells are typically from several tens to a few hundred nanometers, which are considerably thinner than the typical thickness of the samples for the conventional photoexcitation TOF method. Using the DI-SCLC method, the charge carrier mobility of super thin film states can be determined, while the conventional TOF method is applied to the measurement in the bulk states.

However, the charging current to the sample often interferes with the observation of the injection current based on the charge carrier movement. In order to avoid this effect, a variable capacitor and resistors are connected to the sample, and the charging current can be eliminated using a differential input amplifier.

## 6.5 Conclusion

In this chapter, the principle, experimental setup, sample preparation, and characteristic features of TOF measurement are described. TOF measurements reveal the bulk drift mobility of photogenerated charge carriers, which is a significant parameter for understanding electronic processes in organic semiconductors. This TOF measurement can reveal the number of photogenerated charge carriers, diffusion constant of charge carriers, and distribution of localized states other than the carrier mobility. In addition to the conventional TOF method, the xerographic TOF, lateral TOF, pulse-voltage TOF, and DI-SCLC techniques are used for the measurements of the charge carrier mobilities of organic semiconductors.

## References


1 A. Köhler, A. and Bässler, H. (2015). *Electronic Processes in Organic Semiconductors*. Weinheim: Wiley-VCH.
2 Spear, W.E. (1968). *J. Non-Cryst. Solid.* 1: 197.
3 Spear, W.E. (1957). *Proc. Phys. Soc.* B70: 669.
4 Kepler, R.G. (1960). *Phys. Rev.* 119: 1226.
5 LeBlanc, O.H. (1960). *J. Chem. Phys.* 33: 626.
6 Warta, W., Stehle, R. and Karl, N. (1985). *Appl. Phys.* A36: 163.
7 Shirota, Y. and Kageyama, H. (2007). *Chem. Rev.* 107: 953.
8 Pandey, S.S., Takashima, W., Nagamatsu, S. et al. (2000). *Jpn. J. Appl. Phys.* 39: L94.
9 Campbell, A.J., Bradley, D.D.C. and Antoniadis, H. (2001). *Appl. Phys. Lett.* 79: 2133.
10 Schmidt, W.F. and Allen, A.O. (1969). *J. Chem. Phys.* 50: 5037.
11 Pisula, W., Zorn, M., Chang, J.Y. et al. (2009). *Rapid Commun.* 30: 1179.
12 O'Neill, M., Kelly, S.M. (2011). *Adv. Mater.* 23: 566.
13 Funahashi, M., Shimura, H., Yoshio, M. and Kato, T. (2008). *Struct. Bond.* 128: 151.
14 Schmidt, W.F. and Allen, A.O. (1969). *J. Chem. Phys.* 50: 5037.
15 Funahashi, M. and Hanna, J. (1997). *Appl. Phys. Lett.* 71: 602.
16 Plint, T.G., Kamino, B.A. and Bender, T.P. (2015). *J. Phys. Chem. C* 119: 1676.
17 Borsenberger, P.M., Pautmeier, L., Richert, R. and Bässler, H. (1991). *J. Chem. Phys.* 94: 8276.
18 Redecker, M., Bradley, D.D.C., Inbasekaran, M. and Woo, E.P. (1998). *Appl. Phys. Lett.* 73: 1565.
19 Tripanthi, A., Heinrich, M., Siegrist, T. and Pflaum, J. (2007). *Adv. Mater.* 19: 2097.
20 Laudise, R.A., Kloc, Ch., Simpkins, P.G. and Siegrist, T. (1998). *J. Cryst. Growth.* 187: 449.
21 Murakami, S., Naito, H., Okuda, M. and Sugimura, A. (1995). *J. Appl. Phys.* 78: 4533.
22 Borsenberger, P.M., Pautmeier, L. and Bässler, H. (1991). *J. Chem. Phys.* 94: 5447.
23 Laquai, F., Wegner, G., Im, C. et al. (2006). *J. Appl. Phys.* 99: 023712.
24 Nomura, S., Nishimura, K. and Shirota, Y. (1996). *Thin Solid Film.* 273: 27.
25 Pope, M. and Swenberg, C.E. (1999). *Electronic processes in organic crystals and polymers*. Oxford: Oxford University Press.

26 Mark, P. and Helfrich, W. (1962). *J. Appl. Phys.* 33: 205.
27 Gibbons, D.J. and Papadakis, A.C. (1968). *J. Phys. Chem. Solids.* 29: 115.
28 Arkhipov, V.I., Wolf, U. and Bässler, H. (1999). *Phys. Rev. B* 59: 7514.
29 Scher, H. and Montroll, E.W. (1975). *Phys. Rev. B* 12: 2455.
30 Funahashi, M. and Sonoda, A. (2012). *Org. Electron.* 13: 1633.
31 Podzorov, V., Menard, E., Rogers, J.A. and Gershenson, M.E. (2005). *Phys. Rev. Lett.* 95: 226601.
32 Sirringhaus, H. (2005). *Adv. Mater.* 17: 2411.
33 Pai, D.M. (1970). *J. Chem. Phys.* 52: 2285.
34 Hirao, A. and Nishizawa, H. (1996). *Phys. Rev. B* 54: 4755.
35 Nagase, T. and Naito, H. (2000). *J. Appl. Phys.* 88: 252.
36 Mikla, V.I. and Mikla, V.V. (2011). *J. Non-Cryst. Solids* 357: 3675.
37 Reghu, R.R., Grazulevicius, J.V., Simokaitiene, J. et al. (2012). *J. Phys. Chem. C* 116: 15878.
38 Kitamura, M., Imada, T., Kako, S. and Arakawa, Y. (2004). *Jpn. J. Appl. Phys.* 43: 2326.
39 Kuwahara, A., Naka, S., Okada, H. and Onnagawa, H. (2006). *Appl. Phys. Lett.* 89: 132106.
40 Pavlica, E. and Bratina, G. (2012). *Appl. Phys. Lett.* 101: 093304.
41 Weis, M., Lin, J., Taguchi, D. et al. (2009). *J. Phys. Chem. C* 113: 18459.
42 Yoshikawa, T., Nagase, T., Kobayashi, T. et al. (2008). *Thin Solid Films* 516: 2595.
43 Pokhrel, C., Shakya, N., Purtee, S. et al. (2007). *J. Appl. Phys.* 101: 103706.
44 Yang, R., Xu, Y., Dang, X.-D. et al. (2008). *J. Am. Chem. Soc.* 130: 3282.
45 Yazaki, S., Funahashi, M. and Kato, T. (2008). *J. Am. Chem. Soc.* 130: 13206.
46 Yazaki, S., Funahashi, M., Kagimoto, J. et al. (2010). *J. Am. Chem. Soc.* 132: 7702.
47 Lambert, M.A. and Mark, P. (1970). *Current Injection in Solids.* New York: Academic Press.
48 Tsue, S.C., Tsang, S.W. and So, S.K. (2006). *J. Appl. Phys.* 100: 063708.

# 7

# Microwave and Terahertz Spectroscopy

*Akinori Saeki*

*Department of Applied Chemistry, Graduate School of Engineering, Osaka University, 2-1 Yamadaoka, Suita, Osaka 565-0871, Japan*

**CHAPTER MENU**



## 7.1 Introduction

The electromagnetic wave including radio wave, light, and high-energy radiation is a form of energy that exhibits both wave-like and particle-like properties [1]. The wave-like properties are identified by the wavelength (m) or frequency ($Hz = s^{-1}$), where their ranges vary by orders of magnitudes (Figure 7.1). The ultraviolet, visible, and infrared lights are the most familiar electromagnetic waves that have not only a direct and indirect influence on life on Earth, but also an access to the electronic and structural properties of chemical species. The latter is the steady-state and transient optical spectroscopies reviewed in this book, where the particle-like behavior with 0.1–10 eV photon energy is essential for absorption (electronic excitation in molecular/atomic orbitals) and subsequent photophysical and photochemical reactions. Infrared spectroscopy as well as Raman spectroscopy [2] allows for detailing the chemical form of substituent and weak hydrogen bond interaction, since the energy of infrared photon corresponds to the vibrational transition specific to the individual chemical bonds.

In contrast, the photon energies of terahertz ($10^{12}$ Hz = THz) and gigahertz radiations ($10^9$ Hz = GHz) are in the order of $10^{-4}$ to $10^{-6}$ eV, even lower than thermal energy at room temperature (0.025 eV). However, these energies match the rotational energy of polar molecules (e.g. $H_2O$), so that the microwave oven (e.g. 2.45 GHz microwave from a magnetron) [3] was invented to heat water/foods and advanced to an industrial or laboratory-scale microwave reactor [4]. Microwave is also of particular importance

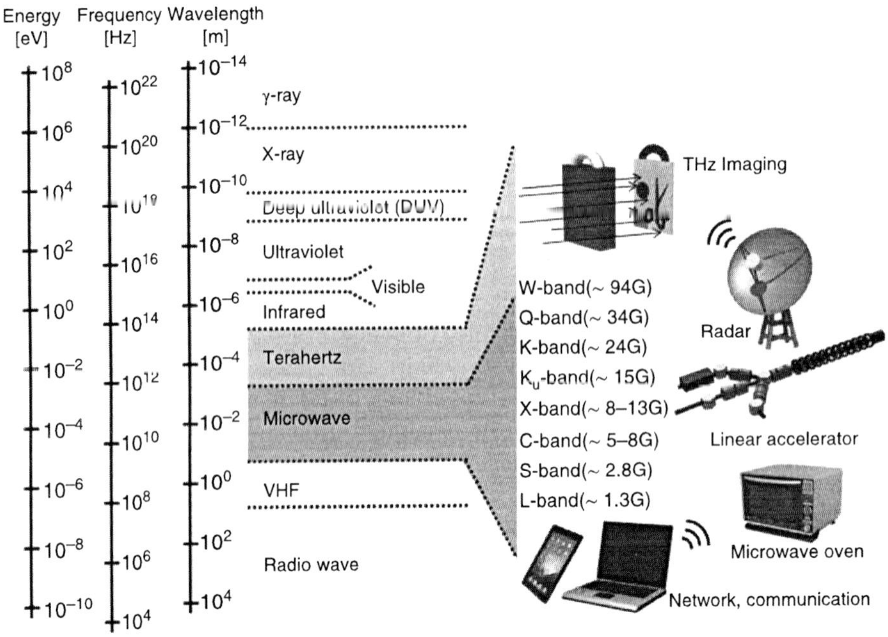


**Figure 7.1** Electromagnetic waves sorted by energy, frequency, and wavelength. The terahertz and microwave (gigahertz) regions are magnified along with the examples of application.

in wireless communication, radar, particle linear accelerator, and measurement (e.g. electron spin resonance, ESR). The use of microwave band in a free space is regulated by a government or international organization, and is often named as an abbreviation (e.g. L (long)-band and S (short)-band), according to the IEEE standard [5].

Owing to recent advances in generation and detection of THz radiation, it has gained much attention as an imaging tool, which is leveraged on balance with the high transmittance, short wavelength relating to the spatial resolution, and absorption coefficient of polar molecule and metal. As understood from the spark from metal overheated in a microwave oven, the gigahertz and terahertz radiations are highly susceptible to metals and electric conducting materials. More specifically, charge carriers (hole and electron) in these conductors are oscillated in the alternating current (AC) electric field at the velocity dependent on their mobilities, which leads to the absorption (decrease in amplitude) and/or phase shift of electromagnetic wave. They are linked to the imaginary part ($\varepsilon''$) and real part ($\varepsilon'$) of dielectric constant, respectively. In a microwave oven, the absorbed power of microwave is converted to thermal energy for heating. In an analogy to the refractive index (real part) and attenuation coefficient (imaginary part) in the optical region, the dielectric constant is expressed as a complex form ($\varepsilon = \varepsilon' + i\varepsilon''$) as a function of frequency, namely frequency dispersion. The imaginary part of $\varepsilon$ is referred to as dielectric loss and associated with the "real" part of conductivity, while the real part of $\varepsilon$ is associated with the "imaginary" part of conductivity. Therefore, gigahertz and terahertz spectroscopies reveal the local complex electric conductivity ($\sigma$) equal to the product of elementary of charge ($e$), charge carrier mobility ($\mu$), and charge carrier concentration ($n$). In this chapter, recent advances in microwave conductivity are reviewed, in particular on the bulk heterojunction organic photovoltaic (OPV) and organic-inorganic hybrid perovskite, along with the terahertz and gigahertz spectroscopy of titanium dioxide nanoparticles. The theory and analysis underlying the observed complex photoconductivity enable the elucidation of the charge transport, dynamics (recombination, transfer, and relaxation), and energetic state (trap), which opens up perspectives on the inherent nature of organic electronic materials.

## 7.2 Instrumental Setup of Time-Resolved Gigahertz and Terahertz Spectroscopies

A generation source and detector are indispensable for spectroscopy using electromagnetic wave. This is the reason that optical (ultraviolet to infrared) spectroscopy has been advanced at the early stage of modern science. The advent of commercially-available femtosecond Ti:Sapphire laser (< 100 fs, repetition rate: 1 KHz to 100 MHz) in the 1980s to 1990s created a fruitful research field of ultrafast transient optical spectroscopy based on pump-and-probe method [6], and is applied for generating ultrashort terahertz radiation [7]. Figure 7.2 shows the schematic of THz time-domain spectroscopy system composed of an fs laser system and terahertz generation/detection antennas [8]. Terahertz radiation is generated by focusing an fs laser pulse (e.g. 800 nm of fundamental light as the pump) into a photoconductive switching antenna made of inorganic semiconductor (e.g. GaAs). The photogenerated charge carriers are accelerated at the narrow gap of electrode (a few μm) of the antenna biased by a few tens of volts, generating a coherent and short-pulsed terahertz radiation (e.g. 0.1–10 THz). The terahertz pulse is guided using the pairs of paraboloidal mirrors to a sample and to another terahertz antenna (detector). The terahertz detector is driven in the same manner with the generator, where the simultaneous irradiation of terahertz and probe laser pulse in turn generates a current signal. An optical delay located in the probe light optics is used to

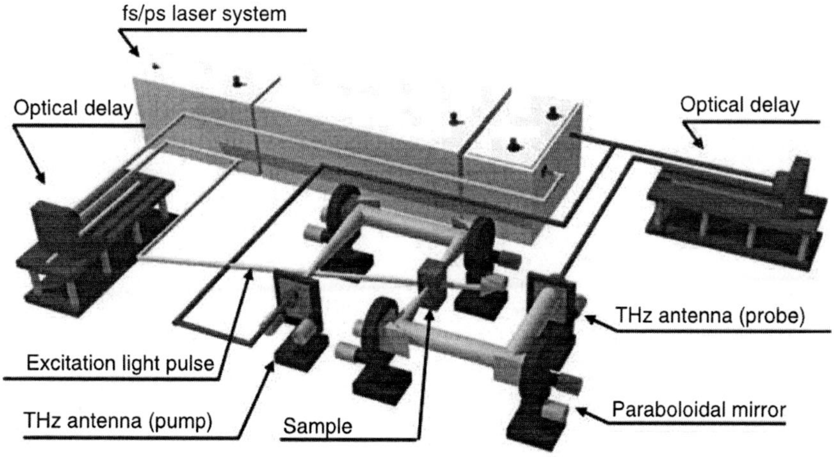


**Figure 7.2** Schematic of THz time-domain spectroscopy system using fs/ps light pulse as the excitation and a pair of THz antenna trigger by fs/ps light pulse as the probe.

change the time difference ($\Delta t$) between the terahertz and probe pulse. Fourier transform of the plot of intensity vs $\Delta t$ gives a terahertz spectrum with real and imaginary components. This is the time-domain spectroscopy (TDS) technique which allows for measuring terahertz spectrum without Kramers-Kronig analysis and a bolometer operated at liquid-He temperature. To perform a time-resolved spectroscopy, an excitation laser pulse (e.g. 400 nm, second harmonic generation of a fundamental light) is exposed to the sample by changing the time difference from the terahertz pulse. Consequently, the plot of terahertz spectrum as a function of optical delay time of the excitation laser produces the kinetics of photogenerated charge carriers.

The instrumental setup of gigahertz spectroscopy, which can also evaluate the local charge carrier motion without electrodes, is considerably different and rather simple compared to terahertz spectroscopy [9, 10]. Figure 7.3 shows the schematic of flash-photolysis (FP) time-resolved microwave conductivity (TRMC) setup composed of a light source and a microwave circuit including source, waveguide, a resonant cavity, amplifier, and detector [11]. Owing to the moderate scale of wavelength (ca. 3 cm for 10 GHz X-band microwave), microwave circuit is able to utilize a resonant cavity to improve a signal-to-noise (S/N) ratio. The microwave frequency is tuned to a resonant frequency which is defined by the cavity geometry and sample properties (size, position, and dielectric constant). The quality of the cavity (sensitivity and response time) is quantified by a $Q$ value relating to the number of reflection. This is in contrast to the THz-TDS technique where the interaction is only passing through the sample once. Furthermore, the microwave source is simple and compact, for example, a Gunn oscillator and a signal generator. The microwave power from these sources is small and attenuated to a few to one hundred milliwatt, so that the electric field of the microwave neither disturbs the motion of charge carriers nor heats a sample. For the time-resolved measurement triggered by excitation of the light pulse, a wide range of light sources is available such as a Nd:YAG laser (ca. 5–10 ns, repetition rate: 10 Hz) and visible

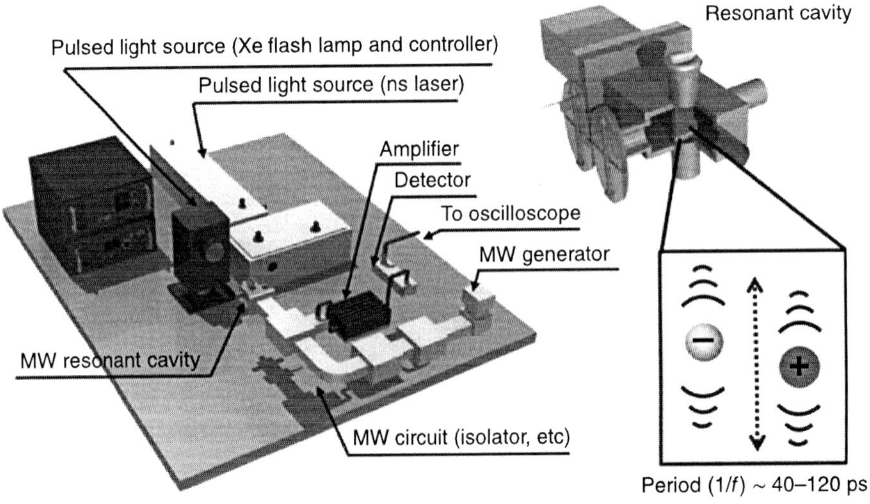


**Figure 7.3** Schematic of flash-photolysis time-resolved microwave conductivity (TRMC) using a light pulse as the excitation and continuous microwave as the probe. It also utilizes a resonant cavity to increase the sensitivity.

**Figure 7.4** Illustration of flash-photolysis TRMC systems with the options of frequency modulation, pressure control, excitation light source, and temperature control.

light from an optical parametric oscillator seeded by the third harmonic generation of a Nd:YAG laser. The author and coworkers have also utilized an in-house-built Xe-flash lamp (10 μs pulse duration, 10 Hz) for solar cell evaluation [12]. In addition, a high-energy electron beam pulse from an accelerator is another option, so-called pulse radiolysis, which allows "homogeneous" and "quantitative" generation of charges in non-polar –solution [13–15], gas phase [16], and block (powder) materials [17, 18].

As the trade-off of the high sensitivity of resonant cavity, the measurement frequency is limited to the specific frequency for each microwave circuit. Therefore, frequency dispersion is not obtained at one time like THz-TDS. The author and coworkers have developed frequency-modulated TRMC systems using $K_u$-band (ca. 15 GHz), K-band (ca. 23 GHz), and Q-band (or $K_a$-band, 33 GHz) in addition to the common X-band (ca. 9 GHz) [19], as shown in Figure 7.4. The increase of resonant frequency ($f_0$) also profits the instrumental response time, since it is proportional to $Q/f_0$. Furthermore, other FP-TRMC systems have been developed such as pressure control in a nonpolar hydrostatic media (<0.15 GPa) [20] and temperature control (ca. 100 K-400 K) in a chamber [21].

## 7.3 Theory of Complex Microwave Conductivity in a Resonant Cavity

Slater developed the cavity perturbation theory [22] which relates the resonant angular frequency, $\omega_0$ ($=2\pi f_0$) and the change of cavity quality factor ($Q = f_0/\Delta f_{1/2}$, $\Delta f_{1/2}$ is the full bandwidth at half maximum of the reflected microwave power near the $f_0$) to the complex conductivity, as given by

$$\Delta\left(\frac{1}{Q}\right) - i\frac{2\Delta\omega_0}{\omega_0} = \frac{F}{\varepsilon_0\omega_0}(\Delta\sigma' - i\Delta\sigma''), \tag{7.1}$$

where $\varepsilon_0$, $\Delta\omega_0$, and $F$ are the permittivity in vacuum, the shift of resonant angular frequency, and the geometry factor considering the spatial overlap of electric field and sample in a resonant cavity. Grabtchak and Cocivera have pointed out that the Slater expression is the first perturbation where a sample is placed in an empty cavity [23]. They extended the second perturbation theory which causes a change in the complex dielectric constant of a sample. The changes of real and imaginary parts of dielectric constant, $\Delta\varepsilon'$ and $\Delta\varepsilon''$ are cast in the following form.

$$\Delta\varepsilon' \equiv -\frac{\Delta f_0}{f_0 F} \tag{7.2}$$

$$\Delta\varepsilon'' \equiv \frac{\Delta(\Delta f_{1/2})}{2f_0 F} \tag{7.3}$$

The relationship between the conductivity and dielectric constant is described as follows:

$$\Delta\sigma' = \varepsilon_0 \omega \Delta\varepsilon'' \tag{7.4}$$

$$\Delta\sigma'' = -\varepsilon_0 \omega \Delta\varepsilon'' \tag{7.5}$$

The combination of Eqs. (7.2) through (7.5), together with the approximation, $\Delta(\Delta f_{1/2}) \approx f_0 \Delta(1/Q)$ [23] yields the same expression as Eq. (7.1) (the factor of 2 is convoluted into $F$). In the experiments, the sample geometry factor $F$ was calculated by the hand-made software incorporating the results of electrostatic simulation in the cavity performed by MW-studio [11].

The real photoconductivity transient $\Delta\sigma'$ is converted to the product of the quantum efficiency: $\phi$ and the sum of charge carrier mobilities: $\Sigma\mu$ $(= \mu_+ + \mu_-)$ by

$$\phi \sum \mu = \frac{1}{e \cdot I_0 \cdot F_{\text{Light}}} \cdot \Delta\sigma', \tag{7.6}$$

where $e$, $I_0$, and $F_{\text{Light}}$ are the unit charge of a single electron, the excitation photon density of laser (photons $\text{cm}^{-2}$) and a correction factor ($\text{m}^{-1}$). The $F_{\text{Light}}$ was calculated in the same matter with $F$, but the depth profile of exciting laser is convoluted [11]. When the $\Delta\sigma'$ is small (this is usually true for photo-induced transient conductivity of organic electronics) and the $\Delta\omega_0$ (i.e. $\Delta\sigma''$) is negligible, $\Delta\sigma'$ can be approximated by

$$\Delta\sigma' = \frac{1}{A}\frac{\Delta P_r}{P_r}, \tag{7.7}$$

$$A = \frac{\mp Q(1/\sqrt{R_0} \pm 1)}{\pi f_0 \varepsilon_0 \varepsilon_r}, \tag{7.8}$$

where $A$, and $\Delta P_r/P_r$ are the sensitivity factor and the change in the fractional microwave power observed in the reflection from the cavity at the resonant frequency, respectively [10]. The $R_0$ is the ratio of reflected ($P_r$) and incident ($P_i$) microwave power ($R_0 = P_r/P_i$), and the $\varepsilon_r$ is the real part of the relative dielectric constant inside the cavity. Since the right term of Eq. (7) is experimentally determined, transient decays of $\Delta\sigma'$ and $\phi\Sigma\mu$ are readily evaluated for each sample.

Using the FP-TRMC, the author and collaborators have investigated the charge dynamics in various materials [24–26] as shown in Figure 7.5, such as powder [27], liquid crystal [28, 29], self-assembled nanotubes [30, 31], gels [32, 33]. molecular crystals [34, 35], $\pi$-conjugated polymers [36, 37], and donor-acceptor blends [38, 39]. The obtained 9 GHz-AC mobilities of charge carriers are in the orders of $10^{-3}$ to $10^{0}$ $\text{cm}^2\text{V}^{-1}\text{s}^{-1}$, which is, in most of the cases, a few orders of magnitude higher than those estimated by direct-current (DC) technique, demonstrating the characteristics of AC mobility and the potential feasibility for the use as efficient organic semiconducting materials. Decay kinetics were found to be sensitive to the film morphology, grain

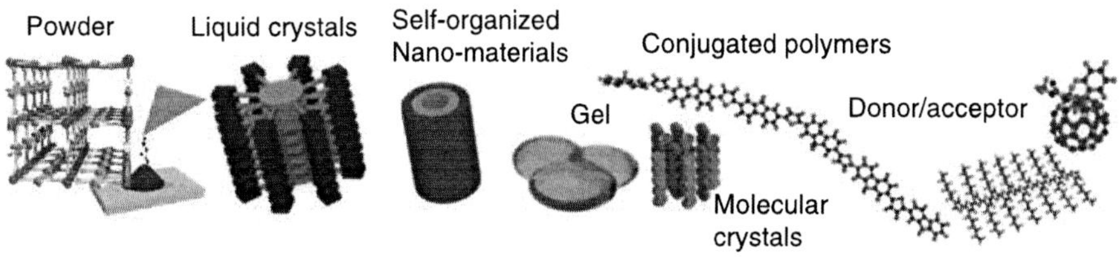


**Figure 7.5** Illustrations of organic/inorganic electronic materials evaluated by TRMC.

size of thin films, bicontinuous ordering of liquid crystal, and $\pi$-stacking distance (e.g. side-chain length of conjugated polymer). The photoconductivity amplitudes were also significantly dependent on the nanometer-scale ordering such as amorphous vs crystal and random stacking vs ordered stacking. These differences are sometime hidden in DC measurements by unexpected factors like impurities and film/device conditions.

## 7.4 Microwave Spectroscopy for Organic Solar Cells

Organic photovoltaics (OPV) have attracted special attention in applications to convert sunlight energy to electricity [40, 41]. Several innovations during the last decade have resulted in raising the power conversion efficiency (PCE) up to 10–11% [42, 43]. The introduction of bulk heterojunction (BHJ) architecture [44] followed by the development of low-band gap polymers [45] and soluble fullerene derivatives [46] have revolutionized this area. Several permutations such as the p/n blend ratio, solvents, solvent additives, and processing conditions could be optimized to realize the best efficiency from new materials; however, this can be a tedious procedure that has suppressed the rapid development of OPVs.

There have only been limited attempts toward achieving rapid development of OPVs using techniques that include transient absorption spectroscopy (TAS) [47], time-resolved terahertz spectroscopy (TRTS) [48], and time-resolved microwave conductivity (TRMC) [12, 39, 49]. TAS only gives information about the time-evolutional concentration of transient species, while TRTS is limited to the femtosecond to nanosecond time domain and requires expensive laser systems and terahertz optics.

The author and coworkers have reported the use of flash-photolysis TRMC to evaluate the representative BHJ of regioregular poly(3-hexylthiophene) (P3HT) and [6, 6]-phenyl-$C_{61}$-butyric acid methyl ester (PCBM) blend films [39]. The optimization of donor and acceptor blend ratio is one of the important parameters to maximize the charge separation and collection efficiencies [50]. Figure 7.6a shows the product of the photoconductivity maximum ($\Delta\sigma_{max}$) and half-lifetime of TRMC decay ($\tau_{1/2}$) found for P3HT:PCBM films upon excitation at 515 nm. One peak at around P3HT:PCBM = 1:1 is observed, which has been widely recognized as the best blend ratio of P3HT: PCBM system.

Figure 7.6b displays the correlation between TRMC signal ($\phi\Sigma\mu_{max}$) of P3HT:PCBM = 1:1 blend films excited at 355 nm and PCE of corresponding OPV devices under AM1.5G presudo-sunlight (100 mW cm$^{-2}$), both of which were simultaneously prepared from *o*-dichlorobenzene (*o*DCB), chlorobenzene (CB), or chloroform (CF) solution in a nitrogen glove box. The evaporation rates of solvents were controlled by

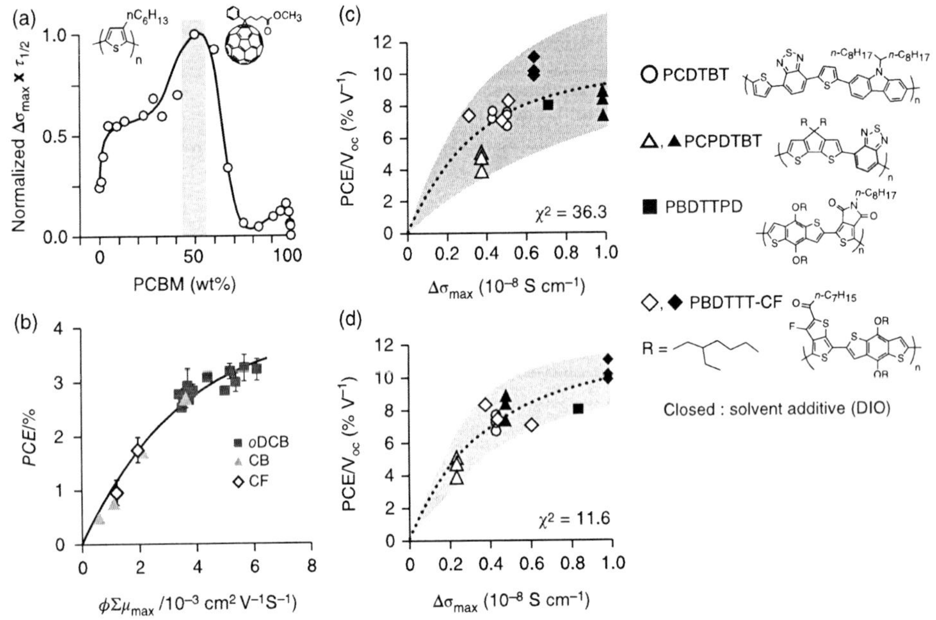


**Figure 7.6** (a) $\Delta\sigma_{max} \times \tau_{1/2}$ of P3HT:PCBM blend films evaluated using TRMC excited at 515 nm. The gray area represents the device-optimal blend ratio (50 wt% PCBM, P3HT:PCBM = 1:1). (b) Correlation of PCE of P3HT:PCBM = 1:1 devices with $\phi\Sigma\mu_{max}$ evaluated using TRMC excited at 355 nm. (c) PCE/$V_{oc}$ of LBP OPV cells plotted by $\Delta\sigma_{max}$ (355 nm laser excitation). The polymer structures and symbols are shown in the right panel. (d) PCE/$V_{oc}$ of LBP OPV cells plotted by $\Delta\sigma_{max}$ (a pulsed white light from a Xe-flash lamp). The solid and dotted lines are a guide for the eyes. Sources: (a,b) Ref .[39] Reproduce with permission of John Wiley & Sons. (c,d) Ref [12] Reproduce with permission of American Chemical Society.

complete or partial covering of substrate in a petri dish, referred to as solvent annealing [51]. The substrates were thermally annealed on a hot plate at different times and temperatures (no anneal, 80 – 180 °C). Notably, the transient photoconductivity was directly correlated with the PCEs of the OPV devices. Therefore, this technique allows for the facile and stable optimization of BHJ films in terms of processing and blend ratio without the need for device fabrication, and is thus not subject to influences such as BHJ/metal interfacial issues, impurities, and degradation of the buffer and active layers.

Motivated by these results, the correlation between TRMC and PCE were examined for high performing low bandgap polymers. However, the correlation is considerably degraded as shown in Figure 7.6c (355 nm excitation, the vertical axis is PCE divided by open circuit voltage, $V_{oc}$). The results were similar for other excitation wavelength (e.g. 500 and 600 nm), regardless of multiplying the half lifetime of decay. The degradation in correlation is possibly due to the variation in the photoabsorption properties of polymers and uncorrelated bulk charge recombination rates at the high excitation intensity. Accordingly, a Xe-flash TRMC system was developed as a rapid and reliable evaluation scheme specific to the OPV characteristics of low-band gap polymers [12]. A BHJ film on a quartz substrate is exposed to a microsecond white light pulse (WLP) from a Xe flash-lamp, which gives rise to short-lived charge carriers. The response time of Xe-flash TRMC is governed by the pulse width of WLP (10 μs), while that of nanosecond laser-flash TRMC is determined by the $Q$ value and resonant frequency (ca. 5–100 ns).

Therefore, the conductivity maxima of Xe-flash TRMC convolute the charge separation yield, the sum of charge carrier mobilities, and the lifetime of charge carries at the low excitation density. It should be noted that the excitation density of WLP is three orders of magnitude smaller than that of nanosecond laser. As shown in Figure 7.6d, $\Delta\sigma_{max}$ of Xe-flash TRMC demonstrates an improved correlation with PCE/$V_{oc}$ of low-bandgap polymer:PCBM blend films. Investigation of the p/n blend ratio of these polymers also highlighted the advantage of Xe-flash TRMC over laser-flash TRMC, where the former gave a single peak close to the optimized blend ratio found in actual devices. Thus, Xe-flash TRMC can predict photovoltaic performance without the need for device fabrication, which should facilitate the rapid development of OPV materials [52–54].

In order to access the reason that the TRMC signal correlates with OPV efficiencies, $\phi$ and $\Sigma\mu$ at 355 nm excitation were experimentally separated and examined for low bandgap polymers [55]. Figure 7.7a shows PCE/$V_{oc}$ as a function of $\Sigma\mu$ for seven pure polymers, where the $\Sigma\mu$ is scattered over three orders of magnitude ($10^{-2}$–$10^{0}$ $cm^2\ V^{-1}\ s^{-1}$). In particular, PBTTT-$C_{14}$ has an exceptionally high TRMC mobility in the absence of PCBM. A similarly poor correlation has been reported between PCE and long-range mobilities estimated using FETs and the space-charge limited current (SCLC) [56].

In contrast, the plots of $\Sigma\mu$ for polymer:PCBM blends demonstrate an improved correlation, where the local mobility of the *blend* is more important for the OPV output than those of the *pure* polymers (Figure 7.7b). Moreover, it was noted that the correlation includes two trends: that for crystalline polymers (P3HT, PBTTT-$C_{14}$, PQT-12, PTzBT-BOHD, PTzBT-14HD, and PTzBT-14OD) and that for semi-crystalline polymers (PCPDTBT, PCDTBT, PBDTTT-CF, and PSBTBT). This trend is not simply linked to whether the polymer is a push-pull type, because the crystalline PTzBT polymer is composed of electron-donating (bithiophene) and electron-withdrawing (thiazolothiazole) units.

To quantitatively examine these categories, X-ray diffraction (XRD) measurements of the blend films were performed. The mean size of polymer crystallite (correlation length) in the inter-lamellar direction, $L_{(100)}$ were evaluated using the Scherrer equation [57]. The crystalline polymers have larger $L_{(100)}$ of 9–36 nm than those of semi-crystalline (or amorphous) polymers (< 8 nm). In addition, the peak intensity, which corresponds to an inter-lamellar distance of 1.4–2.8 nm, can also be used to approximately distinguish the crystalline and semi-crystalline polymers. Thus, it is speculated that the observed positive correlation between the TRMC local charge carrier mobility and PCE/$V_{oc}$ can be linked to the correlation length of crystallite and resultant charge delocalization.

## 7.5 Frequency-Modulation: Interplay of Free and Shallowly-Trapped Electrons

Despite the intriguing optoelectronic properties of $TiO_2$, the immobilization of electron by trapping on the surface of mp-$TiO_2$ and $TiO_2$ nanoparticle (NP) has often been one of the most critical issues for photovoltaic applications. The trap density, typically in the order of $10^{18}$ $cm^{-3}$, has been evaluated by current extraction [58] and infrared spectroscopy [59], and found to correlate with the size of $TiO_2$-NP. Accordingly, the traps are assumed to associate with surface defects and can be filled by a continuous and/or

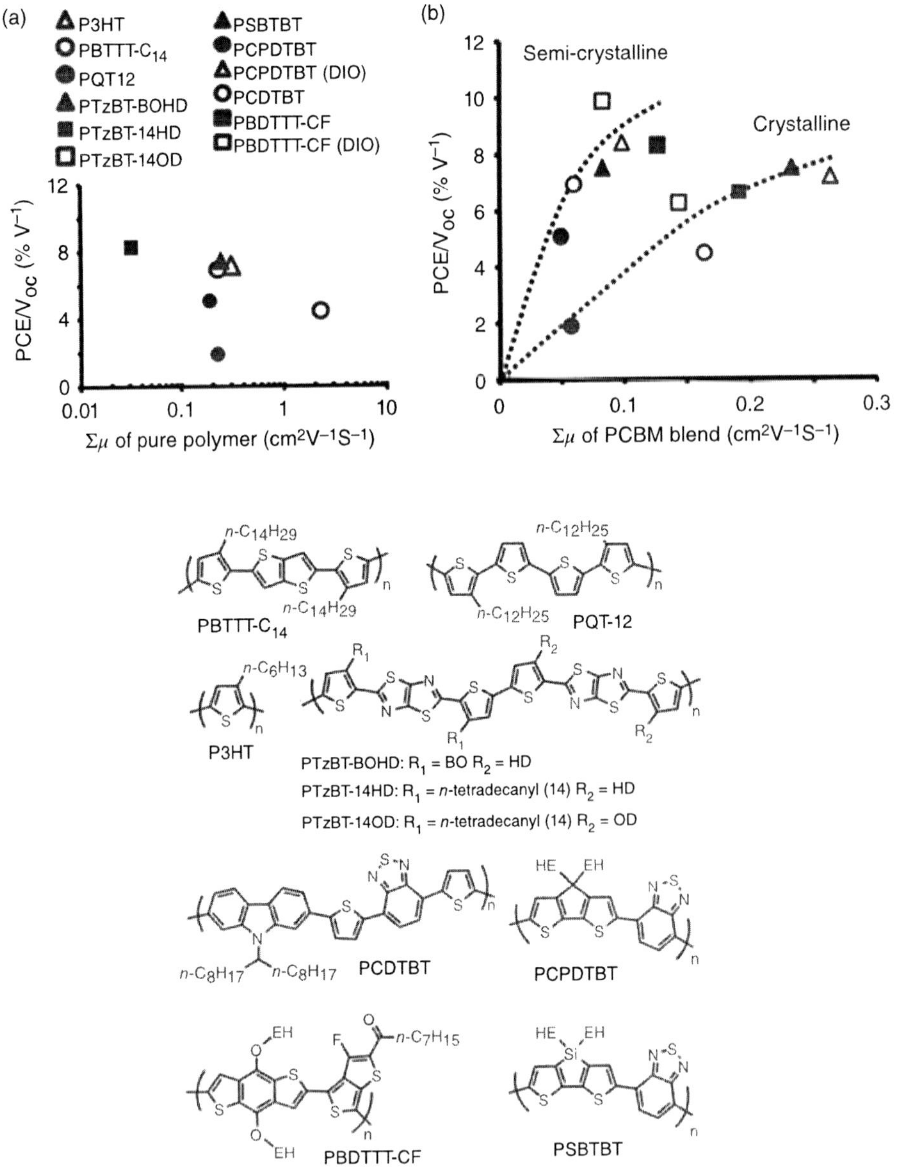


**Figure 7.7** (a) PCE/$V_{oc}$ *vs.* $\Sigma\mu$ for the pure polymers. (b) Correlation of PCE/$V_{oc}$ with $\Sigma\mu$ for the polymer:PCBM blends. The polymer:PCBM blend ratio was similar for both OPV and $\Sigma\mu$ evaluations. Source: Ref [55] Reproduce with permission of Royal Chemical Society.

high power illumination of UV light. Consequently, understanding the energetic landscape of traps and the dynamics of electrons is the focus of great attention with the aim of facilitating the progress of $TiO_2$-based OPV research.

Electrode-less time-resolved spectroscopies of $TiO_2$-NP such as terahertz conductivity [60] and TRMC [61, 62] have been extensively explored to provide a detailed picture on the local electron mobility and trap filling intimately related to the size, surface

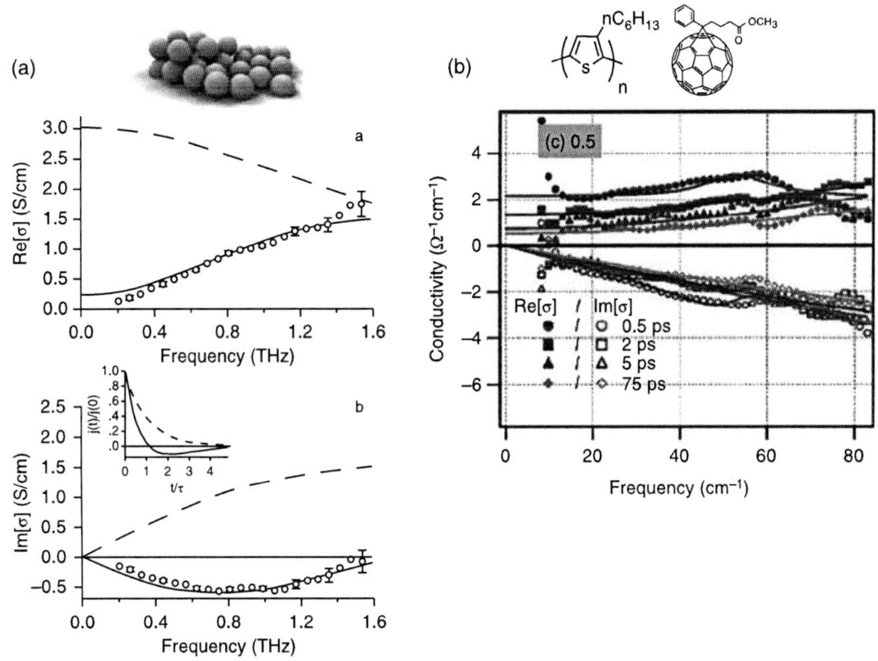


**Figure 7.8** (a) THz spectrum of $TiO_2$ nanoparticles. The upper and lower panels are real and imaginary parts of THz spectrum. The solid and dashed lines are the fits of Drude-Smith and Drude models. Source: Ref [60] Reproduce with permission of American Chemical Society. (b) THz spectrum of P3HT:PCBM films. The solid lines are theoretical curves produced from Drude-Smith model. Source: Ref [64] Reproduce with the permission of American Chemical Society.

area, and electronic quality of $TiO_2$-NP. Moreover, the THz spectrum of complex conductivity allows for gaining information on the transport model in semiconductors. In most cases, the frequency dispersion of THz transient photoconductivity ($\Delta\sigma$) of $TiO_2$-NP is analyzed by the Drude-Smith model [63] which successfully rationalized the observed spectrum and polarity of positive real part $\Delta\sigma'$ and negative imaginary part $\Delta\sigma''$ (Figure 7.8a) [60]. The same analysis is applicable to the organic photovoltaics (P3HT:PCBM, Figure 7.8b) [64]. In the microwave region, characteristic dependence of $\Delta\sigma'$ maxima on excitation laser intensity has been reported, which is attributed to the interplay between increase of local electron mobility due to trap filling and decrease of charge-carrier generation efficiency arising from a high-order charge recombination and/or exciton deactivation processes [61]. Nonetheless, a study on their frequency dependences has not been explored, and thus less is known about the distribution of complex $\Delta\sigma$ in the GHz area, which might hinder the nature of electron traps and their kinetic processes.

As a link in the continuing chain of this issue, the author and coworkers investigated frequency-modulated complex $\Delta\sigma$ of $TiO_2$-NP at four kinds of frequencies [19]. In the presence of imaginary conductivity, the change of the reflected microwave power, $\Delta P_r$, at the resonant frequency involves contributions from both $\Delta\sigma'$ and $\Delta\sigma''$. In this case, they appear as the change of the reflected microwave power level between "light" and "dark" at the resonant frequencies and the shift of the resonant frequency, respectively

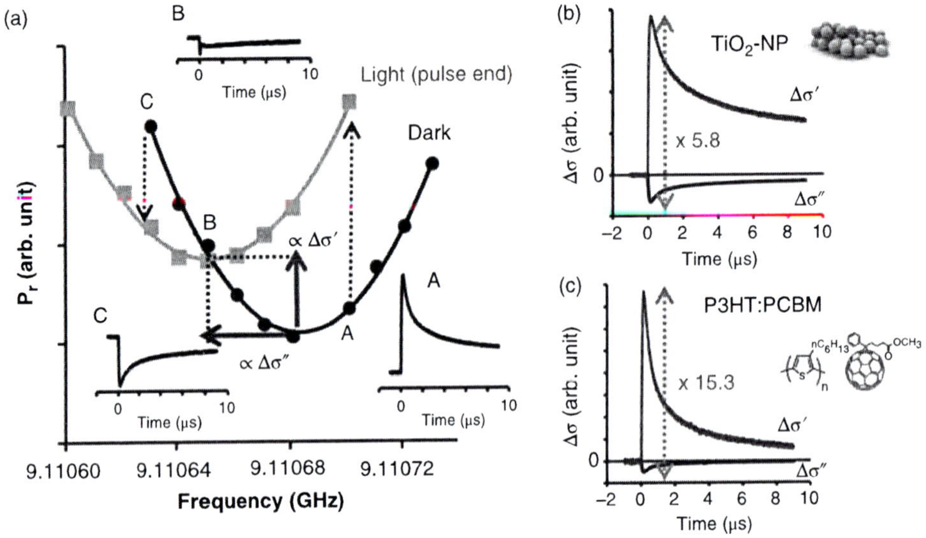


**Figure 7.9** (a) Transient Q curve of $TiO_2$ nanoparticles under dark (black circles) and light (peak-top, gray squares) exposure to the laser pulse. The solid lines are the least mean square curves of the second order function. The vertical and horizontal shifts of the minimum point are proportional to the real ($\Delta\sigma'$) and imaginary ($\Delta\sigma''$) conductivities, respectively. The kinetics at the frequencies of A, B, and C are shown in the inset. Positive real ($\Delta\sigma'$) and negative imaginary ($\Delta\sigma''$) transient photoconductivities of (b) $TiO_2$ nanoparticles and (b) P3HT:PCBM probed at 9.1 GHz. The ratio of $|\Delta\sigma'/\Delta\sigma''|$ is indicated (× 5.8 or 15.3). Source: Ref [19] Reproduce with permission of American Chemical Society.

(see Eq. [7.1]). Therefore $\Delta\sigma'$ and $\Delta\sigma''$ can be separated by analyzing the kinetics at multiple frequencies near the resonant frequency.

Figure 7.9a shows the $Q$ curves in the dark and just after the light pulse irradiation, which are evaluated from the frequency dependence of the reflected microwave power near the dark resonant frequency (ca. 0.1 MHz). Least mean square fit by the second or third order function indicated a good approximation for the dip of transient $Q$ curve. This analysis was performed for all the time delays, and consequently $\Delta\sigma'(t)$ was evaluated by Eq. (7.7) and $\Delta P_r(t)$, enabling us to remove the contribution from the shift of the resonant frequency [65]. Then $\Delta\sigma''(t)$ was obtained from the extracted $\Delta\omega_0(t)$ and Eq. (7.1). The frequency dependence of $\Delta P_r$ for a wide frequency area (ca. 10 MHz) can be fitted by a Lorentzian function, which also allows for the separation between the frequency shift and change of the reflected power [62]. Since the $Q$ curve dip is well-reproduced by the abovementioned simple low-order function, the off-resonant frequency by only ca. 0.1 MHz allows to separate the power change and frequency shift. Such a small scanning area of frequency has a merit for almost unchanged time resolution, a fast measurement, and a high sensitivity.

Figure 7.9b shows the kinetic traces of $\Delta\sigma'(t)$ and $\Delta\sigma''(t)$ of $TiO_2$-NP at 9.1 GHz. The positive $\Delta\sigma'$ corresponds to the increase of dielectric loss $\varepsilon''$ (Eq. [7.4]) and decrease of $Q$ value (Eq. [7.1]), while the negative $\Delta\sigma''$ represents the increase in the real part of dielectric constant $\varepsilon'$ (Eq. [7.5]) and negative shift of the resonant frequency (Eq. [7.1]). P3HT:PCBM blend film (1:1 in weight% fraction) exhibits a positive $\Delta\sigma'$ and negative $\Delta\sigma''$ as well; however, $|\Delta\sigma'_{max}/\Delta\sigma''_{max}|$ was further increased to 15.3 (Figure 7.9c).

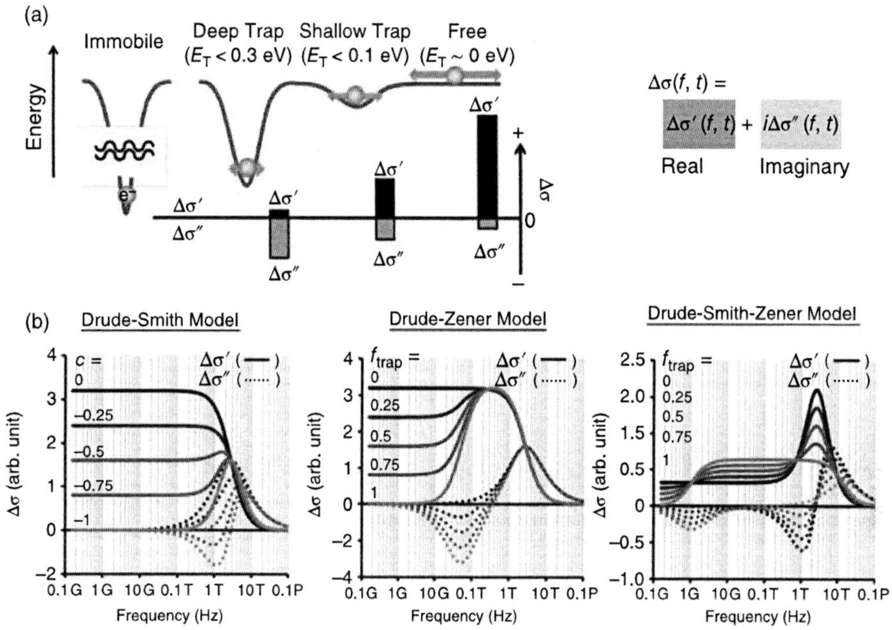


**Figure 7.10** (a) Schematic drawing of the interplay between real ($\Delta\sigma'$) and imaginary ($\Delta\sigma''$) conductivities for free, shallowly trapped, deeply trapped, and immobile electrons. (b) Frequency dispersion of real (solid lines) and imaginary (dotted line) conductivities of (left) Drude–Smith model, (center) Drude-Zener model, and (right) Drude-Smith-Zener model. From dark to bright gray lines represent the changes in $c$ (Drude-Smith) and trap density $f_{trap}$ (Drude-Smith and Drude-Smith-Zener). Source: Ref [19] Reproduce with Permission of American Chemical Society.

The real and imaginary conductivities of free and trapped electrons are schematized in Figure 7.10a, which is based on the Drude-Smith [63] and Drude-Zener [66] models for free and trapped electrons, respectively (vide infra). The complex conductivity of free electrons is mainly composed of its real part, while trapped electrons have substantial fraction of imaginary conductivity. Transition from deeply- to shallowly-trapped state decreases the imaginary part and, in turn, increases the real part, corresponding to the observed decrease of $\Delta\sigma''(t)/\Delta\sigma'(t)$. The model rests on the notion that immobile trapped electrons, roughly deeper than 0.3~1 eV, cannot contribute to either real or imaginary conductivities [67], and thus become silent in GHz conductivity.

Smith introduced a coefficient $c_n$ (hereafter n = 1) into the standard Drude model, where the $c$ represents the fraction of the retained carrier velocity to its original one after the first collision [63]. In other words, c = −1 means the backscattering of charge carrier under the AC electric field. The formula is given by

$$\Delta\sigma(\omega) = \frac{\varepsilon_0 \omega_p^2 \tau}{1 - i\omega\tau}\left[1 + \frac{c}{1\text{-}i\omega\tau}\right]. \tag{7.9}$$

$$\omega_p^2 = \frac{Ne^2}{\varepsilon_0 m^*}, \tag{7.10}$$

$$\mu = \frac{e\tau}{m^*}, \tag{7.11}$$

in which the $\omega_p$ is the plasma frequency, $N$ is the carrier density, $m^*$ is the effective carrier effective mass, and $\tau$ is the carrier relaxation (collision) time. Rearrangement yields the real and imaginary transient conductivities, $\Delta\sigma_{DS}'$ and $\Delta\sigma_{DS}''$, respectively, as follows

$$\Delta\sigma_{DS}'(\omega) = \varepsilon_0\omega_p^2\frac{\tau}{1+(\omega\tau)^2}\left[1+c\frac{1-(\omega\tau)^2}{1+(\omega\tau)^2}\right]. \tag{7.12}$$

$$\Delta\sigma_{DS}''(\omega) = \varepsilon_0\omega_p^2\frac{\omega\tau^2}{1+(\omega\tau)^2}\left[1+\frac{2c}{1+(\omega\tau)^2}\right]. \tag{7.13}$$

Grabtchak and Cocivera use the Drude-Zener expression of complex permittivity for free and trapped electrons [66]. The complex conductivity of trapped electrons (the real part: $\Delta\sigma_{DZ,\,trap}'$ and the imaginary part: $\Delta\sigma_{DZ,\,trap}''$) is derived as follows

$$\Delta\sigma_{DZ,trap}'(\omega) = \frac{\varepsilon_0\omega_p^2\omega^2}{\tau}\frac{1}{(\omega_0^2-\omega^2)^2+\left(\frac{\omega}{\tau}\right)^2}. \tag{7.14}$$

$$\Delta\sigma_{DZ,trap}''(\omega) = -\varepsilon_0\omega_p^2\omega\frac{\omega_0^2-\omega^2}{(\omega_0^2-\omega^2)^2+\left(\frac{\omega}{\tau}\right)^2}. \tag{7.15}$$

$$\omega_0^2 = \frac{2}{m^*}\frac{(4\pi\varepsilon_0)^2}{e^4}E_T^3. \tag{7.16}$$

The $\omega_0$ is a characteristic frequency associated with the trap depth $E_T$. Herein, we introduce the fraction of trapped carrier, $f_{trap}$, to modulate the contributions from both free and trapped charge carriers.

$$\Delta\sigma'(\omega) = (1-f_{trap})\cdot\Delta\sigma_{DS}'(\omega) + f_{trap}\cdot\Delta\sigma_{DZ,trap}'(\omega), \tag{7.17}$$

$$\Delta\sigma''(\omega) = (1-f_{trap})\cdot\Delta\sigma_{DS}''(\omega) + f_{trap}\cdot\Delta\sigma_{DZ,trap}''(\omega), \tag{7.18}$$

The combined model of Drude-Smith for free carrier and Drude-Zener for trapped carrier expressed by Eqs. (7.17) and (7.18) was used to analyze the ratio of observed frequency-modulated transient conductivity, $\Delta\sigma''/\Delta\sigma'$, where the main fitting parameters are $E_T$ and $f_{trap}$.

Figures 7.10b shows the frequency dispersions based on (left) Drude-Smith, (center) Drude-Zener, and (right) Drude-Smith-Zener models with stepwise increment of $c$ from 0 to -1 and $f_{trap}$ from 0 to 1. An emphasis is given to that for $f_{trap}$ < ca. 0.9, the imaginary conductivity of Drude-Smith-Zener model is always negative from sub-GHz to THz regions, in line with the experimental observation. This threshold is raised by increasing free electron mobility and/or decreasing trapped electron mobility. It is striking to note that the distinct dispersion appears in the GHz region, relevant to the effect of trapped electrons.

Figure 7.11a is the $|\Delta\sigma''/\Delta\sigma'|$ of $TiO_2$-NP at 0, 1, and 5 μs delay upon exposure to 355 nm laser pulse ($I_0$ = 9.1 × $10^{14}$ photons $cm^{-2}$), accompanied by the fitting curves using Drude-Smith-Zener models. In the analysis, free and trapped electron mobilities must be assumed as well as their effective masses ($m^*$). Accordingly, the remaining parameters are $E_T$, $c$, and $f_{trap}$. The mobility of 20 $cm^2V^{-1}s^{-1}$ and $m^*$ = 10 $m_e$ ($m_e$: electron mass) [68] were applied for the free electron. A similar high intrinsic

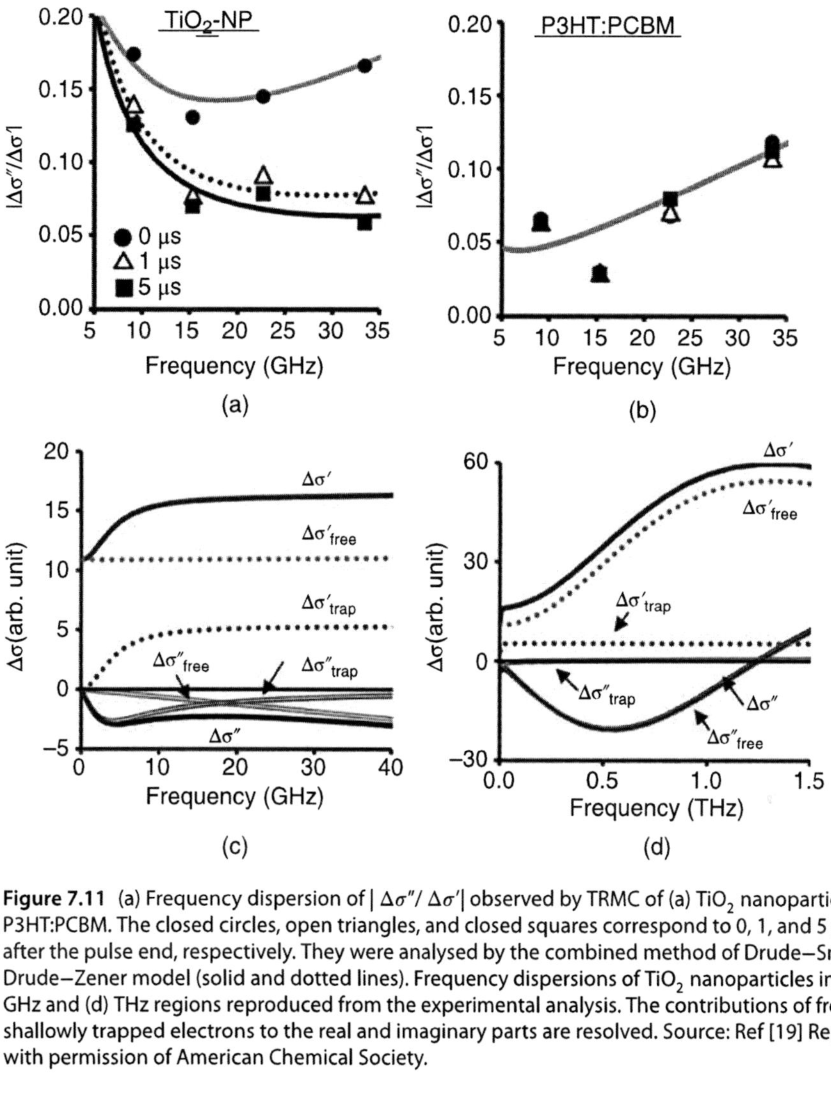


**Figure 7.11** (a) Frequency dispersion of $|\Delta\sigma''/\Delta\sigma'|$ observed by TRMC of (a) $TiO_2$ nanoparticle and (b) P3HT:PCBM. The closed circles, open triangles, and closed squares correspond to 0, 1, and 5 μs delay after the pulse end, respectively. They were analysed by the combined method of Drude–Smith and Drude–Zener model (solid and dotted lines). Frequency dispersions of $TiO_2$ nanoparticles in the (c) GHz and (d) THz regions reproduced from the experimental analysis. The contributions of free and shallowly trapped electrons to the real and imaginary parts are resolved. Source: Ref [19] Reproduce with permission of American Chemical Society.

mobility has been adopted in the Drude-Smith model of $TiO_2$-NP to analyze the THz [60] and GHz [62] complex conductivities. This, in turn, yields the relaxation time, $\tau$ of 114 fs, consistent with the reported analysis and much smaller than the period of GHz microwave (30~110 ps). The Smith coefficient, $c$, was fixed at −0.9 used in the THz analysis [60]. With respect to the shallowly-trapped electrons, the mobility of $0.2\,cm^2V^{-1}s^{-1}$ was assumed, because this value is the experimentally observed mobility at the higher GHz and lower $I_0$ region. From the analysis of complex conductivity, the GHz mobility of $TiO_2$-NP is most likely influenced by these "*shallowly*" trapped electrons. The mobility seems high for "trapped" electrons, but here this term is used to distinguish from completely-free electrons which simply obey the Drude-Smith framework. The $m^*$ of the trapped electron is taken as the same (10 $m_e$), which is actually insensitive to the analysis of $|\Delta\sigma''/\Delta\sigma'|$.

The GHz dispersion of $|\Delta\sigma''/\Delta\sigma'|$ of $TiO_2$-NP at the end-of-pulse (0 μs) was successfully reproduced by using $E_T = 110$ meV and $f_{trap} = 0.83$. This trap depth is equivalent to the reported activation energy of 0.1 ~ 0.15 eV in nanostructured $TiO_2$-based DSSC [69]. The calculated trap density by assuming $\phi = 0.2$ at $I_0 = 9.1 \times 10^{14}$ photons $cm^{-2}$ (Figure 7.6) is $\sim 10^{18}$ $cm^{-3}$, in the same order with the literatures [58, 59]. The photon density required to fill this trap density is $10^{13}$ ~ $10^{14}$ $cm^{-2}$, which coincides with the $I_0$ at the observed maxima of $\phi\Sigma\mu_{max}$. The presence of such a large contribution of shallowly-trapped electrons might be due to the use of bare nanoparticles without intricate networking by sintering with $TiO_2$ precursor solution.

At 1 μs delay time, the frequency dispersion considerably changed, in particular at the higher frequency. The $E_T$ was decreased to 78 meV, while the $f_{trap}$ was slightly increased to 0.97. These parameters did not significantly change at 5 μs delay ($E_T$ = 74 meV and $f_{trap}$ = 0.98). Such a minor fraction of free electron after the equilibrium (2%) is consistent with the previous evaluation [60].

The frequency dispersion was drastically changed for P3HT:PCBM (Figure 7.11b). No temporal change of $|\Delta\sigma''/\Delta\sigma'|$ was observed, in line with the identical decays of $\Delta\sigma''$ and $\Delta\sigma'$. We applied the combined Drude-Smith-Zener model in the same fashion with $TiO_2$-NP. The $f_{trap}$ was found to be 0.25, about three times smaller than $TiO_2$-NP. The $E_T$ was also reduced by about half, down to 60 meV. Notably, the activation energy of 65~75 meV has been reported in the Arrhenius plot of transient photoconductivity maxima of P3HT:PCBM film studied by TRMC at 8.4 GHz [70]. This highlights the intimate link between the complex conductivity and activation energy reminiscent of charge transport property.

The frequency dispersion of $TiO_2$-NP reconstructed from the Drude-Smith and Drude-Zener model analysis is shown in Figure 7.11c (< 40 GHz) and Figure 7.11d (< 1.5 THz). In spite of the small fraction of free electron (17%), its real conductivity ($\Delta\sigma'_{free}$) in GHz region occupies the majority of about 70% of the total real component, $\Delta\sigma'$ (= $\Delta\sigma'_{free} + \Delta\sigma'_{trap}$). On the other hand, the total imaginary conductivity is roughly composed of equal contributions from free and trapped electrons in GHz domains, and besides, a crossover of their amplitudes appears at ca. 18 GHz. With an increase of frequency, free electrons become dominant in both real and imaginary parts. As a consequence in the THz region, the frequency dispersion is converged into the Drude-Smith curve, where only the free electrons play the decisive role. Indeed, this curve is in good agreement with the observed complex conductivity of THz spectroscopy [60], demonstrating that our Drude-Smith-Zener model is vital over the wide frequency domain.

The minor contribution of trapped electron regardless of its population is underlined by the aspect that GHz and THz spectroscopies are not so susceptible to the traps, in particular deep one. Even though, we demonstrated how to extract the depth and population of shallowly-trapped charge carriers by detailed analysis of complex conductivity. It should be reiterated that the frequency dispersion analysis on $|\Delta\sigma''/\Delta\sigma'|$ is more convenient for determining $E_T$ and $f_{trap}$ than the respective analysis of real and imaginary component, because $|\Delta\sigma''/\Delta\sigma'|$ can compensate the time resolution among the respective resonant cavities. The presence of a trapped charge carrier is roughly judged from the mismatch of decay profiles of $\Delta\sigma'$ and $\Delta\sigma''$.

## 7.6 Organic-Inorganic Perovskite

Organometallic lead halide perovskites (PVK), first introduced as sensitizers in dye-sensitized solar cells in 2009 [71], have shown great promise since the advent of fully solid-state mesoporous superstructure solar cells [72, 73]. These solar cells mainly consist of an electron transport layer (ETL) of mesoporous titanium dioxide or alumina (mp-$TiO_2$ or mp-$Al_2O_3$), a methylammonium lead halide ($CH_3NH_3PbX_3$, Figure 7.12a) perovskite layer, and a spiro-OMeTAD hole transport layer (HTL). The perovskite layer plays a central role in photoabsorption and charge separation. In addition to its unprecedented ambipolar charge transport [73–75], its low energy loss between the optical bandgap and open circuit voltage has enabled notable improvements in the photovoltaic output, leading to an impressive boost in power conversion efficiencies, of solar cells contained these layers. However, the fundamental optoelectronic aspects underlying the operation of this type of solar cell remain elusive, rendering this theme prevalent in the study of the dynamics of excitons and charge carriers.

As shown in Figure 7.12b, atomic force microscopy (AFM) of the perovskite indicates that the crystal size of two-step PVK/$TiO_2$ (420 ± 50 nm) is obviously larger than those of one-step PVK/$TiO_2$ (300 ± 80 nm) and both PVK/$Al_2O_3$ (70–100 nm). Importantly, Figure 7.12c presents a general correlation with the minimum mobility ($\Sigma\mu$) at

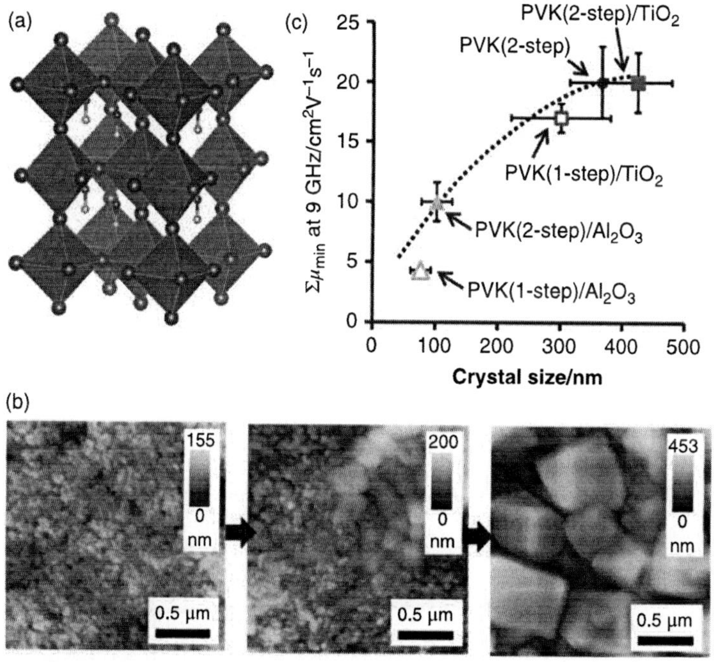


**Figure 7.12** (a) Crystal structure of $CH_3NH_3PbI_3$ perovskite. (b) AFM topographic images of (left) PVK(1-step)/$Al_2O_3$, (center) PVK(1-step)/$TiO_2$, and (right) PVK(2-step)/$TiO_2$. (c) Plot of $\Sigma\mu_{min}$ measured by TRMC at 9 GHz versus averaged crystal size. The dotted line is a visual guide. Source: Modified from Ref. [21]. 2014 American Chemical Society.

9 GHz and the crystal size, irrespective of one- and two-step methods and mesoporous scaffolds [21]. This finding exemplifies the relationship between inherent charge carrier mobility (mainly hole) and crystal size of perovskite affected by the preparation process and under layer. It is of interest to note that one-step PVK/$TiO_2$ contains both large and small crystals. Electronic property of nano-sized semiconductor such as $TiO_2$ nanoparticles is often susceptible to the trapping at surface defects and inter-structural barrier for charge transport, which are reduced with increasing the crystal size [61]. In this work, the two-step PVK/$TiO_2$ having the largest size showed the highest TRMC mobility and PCE [21]. The analogy infers that the crystal size relating to the polarization of perovskite [76], rather than incomplete conversion of $PbI_2$, limits the TRMC mobility and overall device output.

The dynamics and local transport properties of charge carriers were investigated using FM-TRMC. Remarkably, high mobilities of 60–75 $cm^2V^{-1}s^{-1}$ were revealed at 23 GHz in these composites and explained by band-like transport with a $T^{-1.4}$ dependence due to phonon scattering [21]. Charge recombination in PVK was suppressed by 4–5 orders of magnitude from the Langevin rate [77], and was further reduced by another order of magnitude in the presence of a mesoporous $TiO_2$ layer. Analysis of the GHz complex conductivity revealed a very low $E_T$ of 10 meV and $f_{trap}$ of 0.03–0.1 in these PVK frameworks. In addition to the prominent optoelectronic properties of PVK itself, these values were improved by the use of a mesoporous layer, which provided a scaffold on which electronically high-quality crystalline PVK, with a high mobility, a reduced recombination rate, and a low trap density, could be formed. More specifically, whether the scaffold is $Al_2O_3$ or $TiO_2$ determines the mobility and dynamics of charge carriers and the energetic landscape of the trap state.

Since the hole mobility in PVK is more than two orders of magnitude higher than those of organic electronics, hole transfer from PVK to molecular or polymeric HTL is directly evaluated as the decrease of TRMC signal. The hole transfer process is important to improve the short circuit current density and fill factor, in addition to the hole transport in HTL. Currently, spiro-OMeTAD has been used as the predominant hole transport layer (HTL) material, which is a milestone molecule introduced in solid-state dye sensitized solar cells since 1998 [78]. This is because spherically-shaped spiro-OMeTAD exhibits excellent compatibility with mesoporous super-structured solar cells, most likely owing to its amorphous features with high pore-filling ability [79] and moderate hole mobility [80]. HTL materials doped with LiTFSI and CoTFSI together with 4-*tert*-butylpyridine (TBP) significantly outperform non-doped materials [81]; however, the control and stabilization of doped HTLs are remained issues [82].

Therefore, development of novel efficient HTL material is a vital issue, and stable and fast evaluation using TRMC could be appropriate for this purpose. Wakamiya, the author, Scott, and Murata et al. have recently reported bridged-tripenylamine(TPA)-appended azulene molecules for efficient HTL of perovskite solar cell [83]. As shown in Figure 7.13a, four-TPA-appended azulene exhibits a 58% decrease in the $\phi\Sigma\mu_{max}$ of $CH_3NH_3PbI_3$ perovskite upon coating of HTL with dopants. The decay is also accelerated and recovered to the ground level at ca. 0.72 μs, indicating that all of the holes are transferred to HTL within this time scale. The decrease in $\phi\Sigma\mu_{max}$ and decay speed is improved in comparison with spiro-OMeTAD (41% and 0.97 μs, Figure 7.13b) and two-TPA-appended azulene (19% and 2.2 μs, Figure 7.13c). Indeed, the perovskite solar cell using four-TPA azulene as HTL showed 15.7% PCE, outperforming the reference

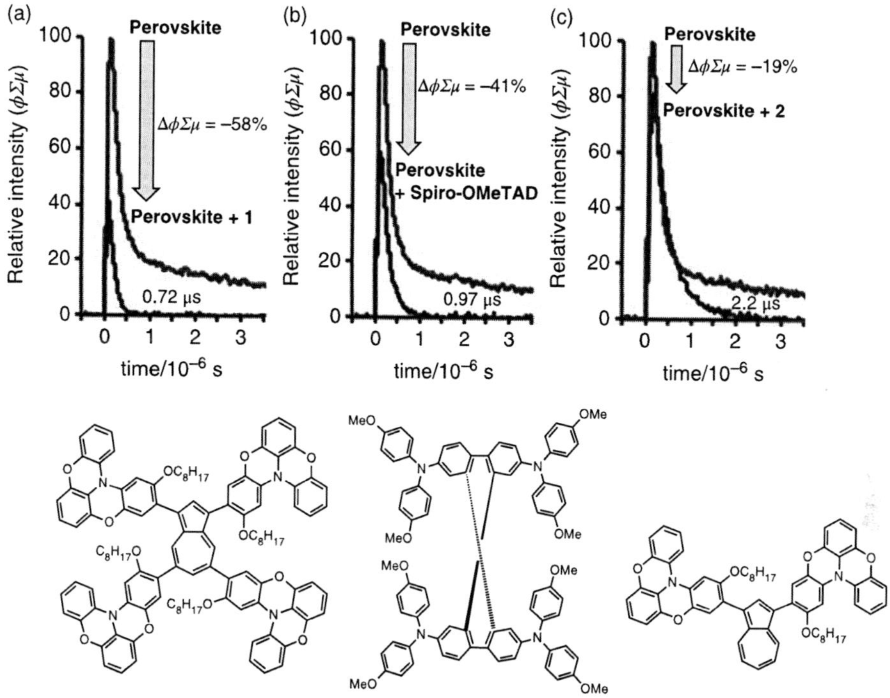


**Figure 7.13** TRMC decays of $CH_3NH_3PbI_3$ perovskite with and without doped HTL layer of (a) four-TPA appended azulene, (b) Spiro-OMeTAD, and (c) two-TPA appended azulene. The arrows represent the decrease of $\phi\Sigma\mu$ maxima upon HTL coating (normalized by $CH_3NH_3PbI_3$ perovskite without HTL). The time indicated in μs unit is the ground-recovery time. Source: Ref [83] Reproduced with permission of American Chemical Society

Spiro-OMeTAD (13.6%). It is clearly illustrated how the device performance is influenced by hole transfer process directly evaluated through TRMC. This highlights the versatility of this approach applicable to a wide range of polymeric and molecular HTLs.

## 7.7 Conclusions

This chapter reviews the basis of time-resolved gigahertz (microwave) and terahertz spectroscopies for organic and organic-inorganic electronic materials. The electrode-less evaluation of these techniques provide an access to the inherent nature of charge carrier dynamics, mobility, and energetics. The unique aspect lies at the frequency dispersion of complex conductivity composed of real and imaginary parts, which are associated with oscillational motion of charge carriers in a confined space and in a shallow trap potential. The microwave region is in particular sensitive to the shallowly-trapping charges and their population. Laser-flash or Xe-flash TRMC offer a versatile approach to the development and fundamental of organic solar cell, as corroborated by the correlation between TRMC signals and device efficiencies. TRMC is also expanded to the investigation on cutting-edge perovskite solar cell, elucidating their excellent optoelectronic properties including high charge carrier mobility

(75 cm$^2$V$^{-1}$s$^{-1}$), reduced charge recombination (five order smaller than Langevin rate), and low trap density (~3%). From the decease of TRMC signals, hole transfer process from perovskite to HTL is directly observed, which is useful for exploring new HTL materials. Further evolution of gigahertz and terahertz spectroscopies is a challenge, which could facilitate deeper understanding of (opto-)electronic properties of organic and hybrid materials.

## Acknowledgement

This work was supported by Kakenhi (Grant-in-aid for Scientific Research A: 16H02285 and JP20H00398, Scientific Research B: 252880840, Innovative Areas "Element block" 15H00747, "π-figuration" 26620104, Transformative Research Areas (A) "Dynamic Exciton" JP20H05836) from the Ministry of Education, Culture, Sports, Science, and Technology (MEXT) of Japan, and the PRESTO (JPMJPR15N6) and ALCA (JPMJAL1603) programs from the Japan Science and Technology Agency (JST).

## References

1 Einstein, A. (1921). *The Nobel Prize in Physics.*
2 Raman, C.V. (1930). *The Nobel Prize in Physics.*
3 Invented by P.L. Spencer at Raytheon Company (1947).
4 Gedye, R., Smith, F., Westaway, K. et al. (1986). *Tetrahedron Lett.* 27: 279.
5 IEEE. Standard 521-1984. *IEEE Standards Association.* https://standards.ieee.org/.
6 Zewail, A.H. (1999). *The Nobel Prize in Chemistry.*
7 Fattinger, C. and Grischkowsky, D. (1989). *Appl. Phys. Lett.* 54: 490.
8 Schmuttenmaer, C.A. (2004). *Chem. Rev.* 104: 1759.
9 Biondi, M.A. and Brown, S.C. (1949). *Phys. Rev.* 75: 1700.
10 Infelta, P.P., de Haas, M.P. and Warman, J.M. (1977). *Radiat. Phys. Chem.* 10: 353.
11 Saeki, A., Seki, S., Sunagawa, T. et al. (2006). *Phil. Mag.* 86: 1261.
12 Saeki, A., Yoshikawa, S., Tsuji, M. et al. (2012). *J. Am. Chem. Soc.* 134: 19035.
13 Warman, J.M., Gelinck, G.H. and de Haas, M.P. (2002). *J. Phys. Condens. Matter* 14: 9935.
14 Grozema, F.C. and Siebbeles, L.D.A. (2011). *J. Phys. Chem. Lett.* 2: 2951.
15 Bird, M.J., Reid, O.G., Cook, A.R. et al. (2014). *J. Phys. Chem. C* 118: 6100.
16 Shimamori, H. and Hatano, Y. (1977). *Chem. Phys.* 21: 187.
17 Warman, J.M., de Hass, M.P., Grätzel, M. and Infelta, P.P. (1984). *Nature* 310: 306.
18 Patwardhan, S., Sengupta, S. and Siebbeles, L.D.A. (2012). *J. Am. Chem. Soc.* 134: 16147.
19 Saeki, A., Yasutani, Y., Oga, H. and Seki, S. (2014). *J. Phys. Chem. C* 118: 22561.
20 Noguchi, Y., Saeki, A., Fujiwara, T. et al. (2015). *J. Phys. Chem. B* 119: 7219.
21 Oga, H., Saeki, A., Ogomi, Y. et al. (2014). *J. Am. Chem. Soc.* 136: 13818.
22 Slater, J.C. (1964). *Rev. Mod. Phys.* 18: 441.
23 Grabtchak, S. and Cocivera, M. (1998). *Phys. Rev. B* 58: 4701.
24 Saeki, A. and Tagawa, S. (2009). *Pure Appl. Chem.* 81: 45.
25 Saeki, A., Koizumi, Y., Aida, T. and Seki, S. (2012). *Acc. Chem. Res.* 45: 1193.

26 Seki, S., Saeki, A., Sakurai, T. and Sakamaki, D. (2014). *Phys. Chem. Chem. Phys.* 16: 11093.
27 Wan, S., Gándara, F., Asano, A. et al. (2011). *Chem. Mater.* 23: 4094.
28 Motoyanagi, J., Yamamoto, Y., Saeki, A. et al. (2009). *Chem. Asian J.* 4: 876.
29 Yeh, M.-C., Su, Y.-L., Tzeng, M.-C. et al. (2013). *Angew. Chem. Int. Ed.* 52: 1031.
30 Yamamoto, Y., Fukushima, T., Suna, Y. et al. (2006). *Science* 314: 1761.
31 Saeki, A., Yamamoto, Y., Koizumi, Y. et al. (2011). *J. Phys. Chem. Lett.* 2: 2549.
32 Prasanthkumar, S., Saeki, A., Seki, S. and Ajayaghosh, A. (2010). *J. Am. Chem. Soc.* 132: 8866.
33 Wu, L., Ohtani, M., Takata, M. et al. (2014). *ACS Nano* 8: 4640.
34 Saeki, A., Seki, S., Takenobu, T. et al. (2008). *Adv. Mater.* 20: 920.
35 Saeki, A., Seki, S., Shimizu, Y. et al. (2011). *J. Chem. Phys.* 132: 134509.
36 Saeki, A., Fukumatsu, T. and Seki, S. (2011). *Macromolecules* 44: 3416.
37 Fukumatsu, T., Saeki, A. and Seki, S. (2015). *Polym. Chem.* 6: 5860.
38 Saeki, A., Ohsaki, S., Seki, S. and Tagawa, S. (2008). *J. Phys. Chem. C* 112: 16643.
39 Saeki, A., Tsuji, M. and Seki, S. (2011). *Adv. Energy Mater.* 1: 661.
40 Coughlin, J.E., Henson, Z.B., Welch, G.C. and Bazan, G.C. (2014). *Acc. Chem. Res.* 47: 257.
41 Hardin, B.E., Snaith, H.J. and McGehee, M.D. (2012). *Nat. Photo.* 6: 162.
42 Liu, Y., Zhao, J., Li, Z. et al. (2014). *Nat. Commun.* 5: 5293.
43 Zhao, J., Li, Y., Yang, G. et al. (2016). *Nat. Energy* 1: 15027.
44 Yu, G., Gao, J., Hummelen, J.C. et al. (1995). *Science* 270: 1789.
45 Peet, J., Kim, J.Y., Coates, N.E. et al. (2007). *Nat. Mater.* 6: 497.
46 Zhao, G., He, Y. and Li, Y. (2010). *Adv. Mater.* 22: 4355.
47 Clarke, T.M. and Durrant, J.R. (2010). *Chem. Rev.* 110: 6736.
48 Cooke, D.G., Krebs, F.C. and Jepsen, P.U. (2012). *Phys. Rev. Lett.* 108 056603.
49 Rance, W.L., Ferguson, A.J., McCarthy-Ward, T. et al. (2011). *ACS Nano* 5: 5635.
50 Xin, H., Ren, G., Kim, F.S. and Jenekhe, S.A. (2008). *Chem. Mater.* 20: 6199.
51 Li, G., Yao, Y., Yang, H. et al. (2007). *Adv. Funct. Mater.* 17: 1636.
52 Tsuji, M., Saeki, A., Koizumi, Y. et al. (2014). *Adv. Funct. Mater.* 24: 28.
53 Gopal, A., Saeki, A., Ide, M. and Seki, S. (2014). *ACS Sustainable Chem. Eng.* 2: 2613.
54 Ide, M., Koizumi, Y., Saeki, A. et al. (2013). *J. Phys. Chem. C* 117: 26859.
55 Yoshikawa, S., Saeki, A., Saito, M. et al. (2015). *Phys. Chem. Chem. Phys.* 17: 17778.
56 Lecover, R., Williams, N., Markovic, N. et al. (2012). *ACS Nano* 6: 2865.
57 Erb, T., Zhokhavets, U., Gobsch, G. et al. (2005). *Adv. Funct. Mater.* 15: 1193.
58 Bailes, M., Cameron, P.J., Lobato, K. and Peter, L.M. (2005). *J. Phys. Chem. B* 109: 15429.
59 Nguyen, T.T.O., Peter, L.M. and Wang, H. (2009). *J. Phys. Chem. C* 113: 8532.
60 Turner, G.M., Beard, M.C. and Schmuttenmaer, C.A. (2002). *J. Phys. Chem. B* 106: 11716.
61 Kroeze, J.E., Savenije, T.J. and Warman, J.M. (2004). *J. Am. Chem. Soc.* 126: 7608.
62 Fravventura, M.C., Deligiannis, D., Schins, J.M. et al. (2013). *J. Phys. Chem. C* 117: 8032.
63 Smith, N.V. (2001). *Phys. Rev. B* 64: 155106.
64 Ai, X., Beard, M.C., Knutsen, K.P. et al. (2006). *J. Phys. Chem. B* 110: 25462.
65 Saeki, A., Seki, S. and Tagawa, S. (2006). *J. Appl. Phys.* 100 023703.

66 Grabtchak, S.Y. and Cocivera, M. (1998). *Phys. Rev. B* 58: 4701.
67 Grabtchak, S.Y. and Cocivera, M. (1996). *J. Appl. Phys.* 79: 786.
68 Enright, B. and Fitzmaurice, D. (1996). *J. Phys. Chem.* 100: 1027.
69 Boschloo, G. and Hagfeldt, A. (2005). *J. Phys. Chem. B* 109: 12093.
70 Grzegorczyk, W.J., Savenije, T.J., Dykstra, T.E. et al. (2010). *J. Phys. Chem. C* 114: 5182.
71 Kojima, A., Teshima, K., Shirai, Y. and Miyasaka, T. (2009). *J. Am. Chem. Soc.* 131: 6050.
72 Kim, H.-S., Lee, C.-R., Im, J.-H. et al. (2012). *Sci. Rep.* 2: 591.
73 Lee, M.M., Teuscher, J., Miyasaka, T. et al. (2012). *Science* 338: 643.
74 Stranks, S.D., Eperon, G.E., Grancini, G. et al. (2013). *Science* 342:341.
75 Xing, G., Mathews, N., Sun, S. et al. (2013). *Science* 342: 344.
76 Kim, H.-S. and Park, N.-G. (2014). *J. Phys. Chem. Lett.* 5: 2927.
77 Wehrenfennig, C., Eperon, G.E., Johnston, M.B. et al. (2014). *Adv. Mater.* 26:1584.
78 Bach, U., Lupo, D., Comte, P. et al. (1998). *Nature* 395: 583.
79 Docampo, P., Hey, A., Guldin, S. et al. (2012). *Adv. Funct. Mater.* 22: 5010.
80 Poplavskyy, D. and Nelson, J. (2003). *J. Appl. Phys.* 93: 341.
81 Noh, J.H., Jeon, N.J., Choi, Y.C. et al. (2013). *J. Mater. Chem. A* 1: 11842.
82 Abate, A., Hollman, D.J., Teuscher, J. et al. (2013). *J. Am. Chem. Soc.* 135: 13538.
83 Nishimura, H., Ishida, N., Shimazaki, A. et al. (2015). *J. Am. Chem. Soc.* 137: 15656.

# 8

# Intrinsic and Extrinsic Transport in Crystalline Organic Semiconductors: Electron-Spin-Resonance Study for Characterization of Localized States

*Andrey S. Mishchenko*

*RIKEN Center for Emergent Matter Science (CEMS), 2-1 Hirosawa, Wako 351-0198, Japan*

**CHAPTER MENU**



Field effect transistor (FET) is the cornerstone structure of technology using organic materials for development of electronic devices. Unlike the traditional inorganic transistors, the organic FETs use non-doped intrinsic semiconductors because carriers can be introduced by gate voltage applied to the FET structure. However, modern technology of fabrication of organic materials does not produce crystals and films which are absolutely free of imperfections. As a result, the mobility of carriers in organic FETs is strongly modified by crystal imperfections which create traps for carriers. The mobility in real organic crystal with imperfections is called an *"extrinsic"* one while the word *"intrinsic"* stands for a mobility in a hypothetical perfect material. Both mobilities can be measured by modern experimental technique and its ratio gives a rough estimate of the concentration of traps if the energy of trap centers is known.

A novel technique for the study of traps, based on the fine analysis of the electronic spin resonance spectra, is reviewed. It is capable of obtaining the distributions of trapped carriers over both their degrees of localization and their binding energies in semiconductor crystals or films composed of regularly aligned organic molecules. The first stage of the method is a mathematical analysis of the shape of the ESR spectra giving the distribution of the trap density of states over degrees of localization. The second stage is the transformation of the distribution over the degrees of localization into a distribution over the binding energies. The transformation uses the relationship between the binding energies and the localization parameters of the trapped carriers obtained by diagrammatic Monte Carlo method. We illustrate the practical aspects of the method applying it to the study of pentacene organic thin-film FET.

Small molecule organic semiconductors, such as rubrene or pentacene, for example, are in the focus of intensive research activity because of the importance of these materials for organic electronic devices and, in particular, for a rapidly developing field of flexible electronics [1–5]. FET is the main structure of the technology using small molecule organic semiconductors for development of electronic devices. The FET is a transistor that uses an electric field to control the electrical conductivity of a channel of one type of charge carrier in a semiconductor material. The conductivity of the FET transistor is regulated by a voltage applied to a terminal (the gate) which is insulated from the device. The applied gate voltage imposes an electric field into the device which, in turn, attracts or repels charge carriers to or from the region between a source and a drain terminal. In turn, the density of charge carriers influences the conductivity between the source and drain.

Traditional inorganic transistors use extrinsic doping to introduce carriers of definite signs, holes or electrons into the electronic devices. To the contrary, the organic FETs use non-doped intrinsic semiconductors. In this case, carriers are introduced by gate voltage applied to the FET structure. Hence, the presence of impurities is not an inevitable feature of organic FETs. However, modern technology of fabrication of organic materials does not produce crystals and films which are absolutely free of imperfections. Therefore, as confirmed by experimental and theoretical studies, the mobility of carriers in organic FETs is strongly modified by crystal imperfections which create traps for carriers.

Section 8.1 is a brief introduction into the transport properties of small molecule organic semiconductors showing the crucial role of impurities in the suppression of the observed mobility. The presence of the defects has a negative influence on the functionality of electronic devices because the traps *decrease* the mobility of carriers. Hence, further improvement of the materials for their applications in electronics requires thorough study of the states which can trap carriers. According to the adopted terminology, mobility in real organic crystal with imperfections is called an *"extrinsic"* one. On the other hand, the word *"intrinsic"* stands for a mobility in a hypothetical perfect material. Both mobilities can be measured by modern experimental technique and its ratio gives important information about properties of imperfections. It is shown that one can roughly determine the concentration of traps if the energy of trap centers is known. Hence, additional information on the energy of traps is required.

In section 8.2, we review a novel mathematical technique for study of traps. It is based on the fine analysis of the electronic spin resonance (ESR) spectra. The method can obtain the distributions of trapped carriers over both their degrees of localization and their binding energies in semiconductor crystals or films composed of regularly aligned organic molecules. The first stage of the method is a mathematical analysis of the shape of the ESR spectra giving the distribution of the trap density of states over the degrees of localization. We relate the distribution of the degree of localization to the shape of the ESR spectra and solve the integral equation using the stochastic optimization numeric technique. The second stage is the mapping of the distribution over the degrees of localization into a distribution over the binding energies. We derive a mathematical form of the transformation and show which kind of relationship between the binding energies and the localization parameters is required for mapping. We obtain this relationship using unbiased diagrammatic Monte Carlo method and illustrate the method studying trap states in pentacene organic thin-film transistor.

## 8.1 Intrinsic and Extrinsic Transport in Crystalline Organic Semiconductors

The state of electrons of an isolated atom can be described as atomic orbitals localized in space around the atom. These orbitals are no longer stable quantum states when an atom is a part of a molecule and one has to consider linear combinations of atomic orbitals forming the molecular orbitals. However, the molecular orbitals are no longer good quantum states when the molecule is a part of a molecular crystal which is a pattern of molecules forming periodic structure in space. The stable states of electrons in a crystal are collective states which can be described as linear combinations of molecular orbitals. These states are plane waves with given momentum whose energies span over an energy band of some width $W_B$. The bandwidth $W_B$ is proportional to the transfer intermolecular integral $t$ characterizing the intermolecular coupling. For example, the bandwidth of hypercubic crystal is $W_B = 4dt$ where $d$ is dimensionality.

Because of the weak intermolecular bonding in small molecule organic semiconductors, the electronic bands are comparatively narrow. The transfer integral $t$ in these compounds is around 0.1 eV [6, 7] which is one order of magnitude smaller than the typical value in inorganic materials. Due to narrow bands, the electron-phonon interaction (EPI) plays a crucial role in organic materials [8–12]. To understand the above point, one can rely to the Holstein model

$$H = -t \sum_{\langle \mathrm{i,j} \rangle} c_\mathrm{i}^\dagger c_\mathrm{j} + \omega_{\mathrm{ph}} \sum_\mathrm{i} b_\mathrm{i}^\dagger b_\mathrm{i} - \gamma \sum_i (b_\mathrm{i}^\dagger + b_\mathrm{i}) c_\mathrm{i}^\dagger c_\mathrm{i} \,. \tag{8.1}$$

Here, $c_i^\dagger$ ($b_i^\dagger$) is the creation operator for the carrier (phonon) in the $i$-th molecule, $\omega_{\mathrm{ph}}$ is the frequency of the dispersionless phonon, and the local Holstein coupling to the phonons is $\gamma$.

The Holstein model was introduced more than half of century ago [13] to study the properties of molecular crystals. It is the simplest model for EPI having only local coupling to lattice displacement. The manifestation of the EPI can be characterized by the dimensionless electron-phonon coupling constant $\lambda = 2\gamma^2/(W_B\omega \mathrm{ph})$.

The dimensionless coupling constant $\lambda$ is the ratio of the elastic energy and kinetic energy of electrons. The elastic energy $\gamma^2/\omega\mathrm{ph}$ is the relaxation energy gained by an isolated molecule coupled to the lattice displacement field due to change of the equilibrium position of the displacement. Strictly speaking, an isolated molecule corresponds to a system with zero bandwidth. The kinetic energy of electrons is characterized by half of the electronic bandwidth $W_B/2$. Qualitatively, the nature of dimensionless coupling constant $\lambda$ is similar for any type of EPI and, hence, narrower bandwidth always means stronger EPI.

Temperature dependence of the mobility of organic compounds is nonmonotonic showing increase at low temperatures and decrease at high ones (see open circles in the upper panel of Figure 8.1). Experimentally, low temperature transport (up to 160 K) in pentacene or rubrene looks like thermally activated [14–16] whereas for higher temperatures the mobility is bandlike, i.e., it decreases with temperature as $\mu \sim T^{-\gamma}$, where $\gamma \approx 2$ [15, 17–19].

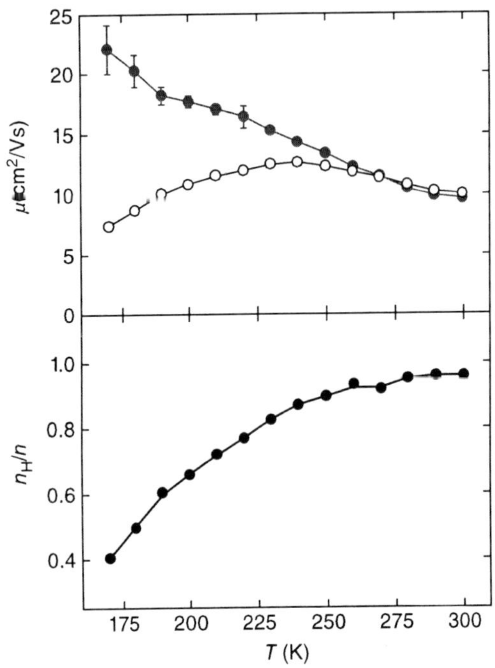


**Figure 8.1** Upper panel: The temperature dependence of the trasverse Hall "intrinsic" mobility $\mu_{tH}(T)$ (solid circles) and the "extrinsic" mobility extracted from the conductivity using the density $n_{gcc}$ calculated from the gate-channel capacitance (open circles). Lower panel: Experimental relative density of free carriers (8.4). Source: Ref [15] Reproduced with permission of American Physical Society.

There are two considerably different approaches explaining this nonmonotonic behavior. The first one explains it in terms of the trap and release model where decrease of mobility with cooling is associated with trapping of carriers by imperfections [15]. The second one suggests that the nonmonotonic behavior is associated with strong-coupling polaron mobility [19]. Below, we show that the trap-mediated mechanism is correct whereas the stong-coupling polaron scenario is in contradiction with experimental data.

The mobility in the strong coupling polaron limit $\mu_{SCL}(T)$ was calculated analytically [13, 20–24] and these results were confirmed by approximation free numeric method [25]. It has the following form

$$\mu_{SCL}(T) \sim T^{-\kappa} \exp(-\Delta/T) , \tag{8.2}$$

where $1 < \kappa < 3/2$. Indeed, nonmonotonic behavior can be easily modeled in this limit Figure 8.2c.

However, there are two arguments against interpretation of the mobility in terms of strong coupling polaron. First, the measurements of transport and infrared properties of rubrene single crystal find small masses of carriers [26] which is in profound contradiction with the large masses expected in case of the strong coupling polaron regime. The second argument, which is even more convincing one, is that the intrinsic transport of carriers does not show any nonmonotonic behavior. Measurements of the transverse Hall conductivity [15, 27] allowed to extract the intrinsic, trap-free mobility that always increases with cooling (traps do not contribute to the Hall voltage since the Lorentz force is zero for these charge carriers). These measurements show that intrinsic mobility $\mu_{int}(T)$, which is equal to the transverse Hall mobility $\mu_{int}(T) = \mu_{tH}(T)$, is monotonic for all temperatures (solid circles in the upper panel of Figure 8.1). This temperature

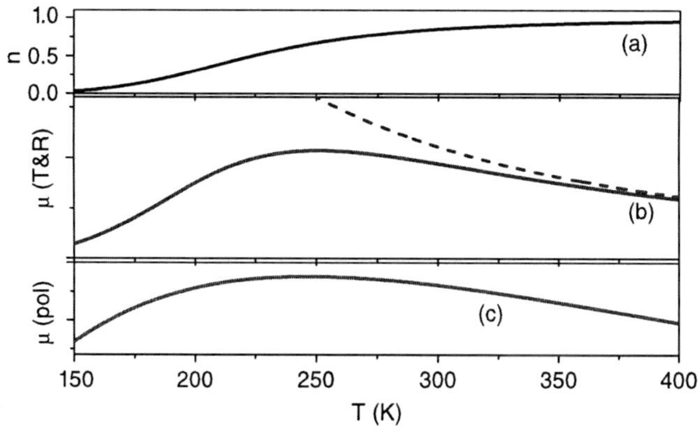


**Figure 8.2** (b) Release factor of the trap and release model from Eq. (8.4) for $c_t = 10^{-3}$ and $E_t = 1550$K, (b) intrinsic mobility (dashed line) and resulting extrinsic mobility of trap and release model (solid line), (c) mobility of a polaron in a strong coupling regime from Eq. (8.2) for $\Delta = 370K$ and $\kappa = 3/2$.

dependence can be roughly approximated as (dashed blue line in Figure 8.2b)

$$\mu_{\text{int}}(T) \sim T^{-2} \,. \tag{8.3}$$

Then, one can ask where nonmonotonic behavior comes from? The answer follows from the description of the procedure how the mobility is determined when one measures it applying the gate voltage to FET. The observed mobility $\mu_{\text{ob}}$ (which is identical to extrinsic mobility $\mu_{\text{ob}} \equiv \mu_{\text{ext}}$) is estimated from conductivity $\sigma$ as $\mu_{\text{ob}} = \sigma/(n_{\text{gcc}}e)$ where electron density $n_{\text{gcc}}$ is calculated from the gate voltage and gate channel capacitance of the FET [15]. Hence, one includes into free carriers all electrons accumulated by gate voltage neglecting the fact that some of them are trapped in crystal imperfections. Such procedure decreases the observed mobility when carriers freeze in trap levels at low temperatures. Indeed, only delocalized carriers with temperature-dependent relative density $n(T)$ (lower panel in Figure 8.1) participate in the transport whereas the total number $n_{\text{gcc}}$ of accumulated in FET channel carriers is almost temperature independent.

However, the presence of traps with relative concentration $c_t$ and depth $E_t$ excludes trapped carriers from the conduction. To reconcile intrinsic and extrinsic mobilities one needs to take into account the relative density of free carriers. Assuming that all traps have the same trap depth $E_t$ one can express the relative density as [28, 29] (compare Figure 8.2a and the lower panel in Figure 8.1)

$$n(T) = \frac{1}{1 + c_t \exp\left[\frac{E_t}{T}\right]} \,. \tag{8.4}$$

Then, the observed extrinsic mobility $\mu_{\text{ob}}$

$$\mu_{\text{ob}} = n(T)\mu_{\text{int}}(T) \tag{8.5}$$

gains nonmonotonic behavior (solid red line in Figure 8.2b). We note, that to reproduce the experimental maximum for rubrene single crystal at $\approx$ 240K one can use, for example, $c_t = 10^{-3}$ and $E_t = 1550\text{K} \approx 140\text{meV}$. Indeed, these values are ambiguous because the maximum at the same temperature can be obtained for various pairs $c_t$ and $E_t$, increase of $E_t$ can be compensated by decrease of $c_t$ and vice versa. Hence, at least one of two quantities has to be determined from another source and we took the trap depth ≈ 140meV equal to the energy of deep-trap states found in pentacene during study by ESR technique (see Section 8.2). Then, the value $c_t = 10^{-3}$ becomes unambiguous telling us the density of traps with such energy. This value is quite reasonable meaning that there is one trap per one thousand rubrene molecules.

An additional difficulty to describe the whole temperature dependence of mobility by Eqs. (8.3–8.5) is the complex dependence of the intrinsic mobility which deviates from the simple relation (8.3). The intrinsic mobility of small molecule organic materials was studied in ref. [30] by unbiased Monte Carlo approaches. This study considers the Su-Schrieffer-Heeger (SSH) model [31, 32] which is the standard model to describe the EPI in soft organic materials. The SSH model is the tight-binding model with hopping integral $t_{i,i+1}$ between sites $i$ and $i+1$ which involves Einstein phonons with the lattice displacements $u_i$ and $u_{i+1}$ modifying the rigid lattice electronic hopping integrals $t^{(0)}_{i,i+1}$

$$t_{i,i+1} = t^{(0)}_{i,i+1}[1 + \alpha(u_{i+1} - u_i)] \ . \tag{8.6}$$

It was in shown in [30] that the intrinsic mobility switches between different regimes at temperature $T_c \approx 200$K. The results of calculations are in perfect agreement with the experimental temperature dependence of the intrinsic mobility. The mean free path of carrier decreases when temperature increases and becomes smaller than intermolecular distance at $T_c \approx 200$K. This temperature also signifies change of carrier motion from superdiffusive at low temperature to subdiffusive at $T > T_c$. Mean square displacement of carrier in superdiffusive (subdiffusive) regime increases as a function of time $\sim t^{\kappa}$ faster $\kappa > 1$ (slower $\kappa < 1$) than in the standard random walker diffusion regime with $\kappa = 1$.

It is, therefore, clear that the trapping on localized states play a crucial role in the transport properties of organic semiconductors and the study of these states is vital to achieve the understanding of the compounds and to improve devices based on organic materials. It is also clear that dramatic changes of intrinsic mobility with temperature prevent reasonable factorization of impurity effects and only a rough estimate of the trap concentration $c_t \approx 10^{-3}$ can be given provided one knows the value of trap depth $E_t$ from some other source. In the next section, which is the main part of this chapter, we introduce a novel method of trap studies and present results of the characterization of the traps in organic material by fine analysis of the electronic spin resonance spectra.

## 8.2 Electron Spin Resonance Study for Characterization of Localized States

### 8.2.1 Introduction into ESR Study

The electron spin resonance (ESR) technique offers a microscopic probe of carriers in semiconductors with unpaired spins. It measures transitions between the quantum levels $m_s = \pm 1/2$ in the presence of a magnetic field [33–36]. The spectrum of the transition should be a $\delta$-function for an isolated system, though, interactions with the environment broaden the spectrum. The broadening is a superposition of two fundamentally different contributions. Decay of the quantum spin levels, due to interaction with the environment, leads to the Lorentzian shape of the spectral line. The second contribution is due to the inhomogeneities of the medium. For example, hyperfine interaction of electrons with spins broadens the signal. The energy of a signal from a particular site is shifted by the local magnetic field depending on the configuration of neighboring nuclear spins. Indeed, the signal summed from different sites is broadened.

For example, isolated in a solution pentacene a molecule exhibits a broad ESR signal arising due to hyperfine coupling with 14 proton nuclear spins. Each individual line of the signal corresponds to a particular configuration of nuclear spins and the envelope function of the summary signal is roughly reproduced by a Gaussian [37–39]. Such shape of the envelope is the consequence of the central limit theorem (CLT). The local magnetic field for each electronic spin is the sum of independent contributions from many particular nuclear spins whose orientations are random. Thus, according to CLT, the sum of identical independent random variables has Gaussian distribution. The individual lines become unresolved when the width of individual line is broader than the interline energy separation. The ESR spectrum in this case is just the Gaussian.

Recent major developments in the field of organic thin-film transistors (TFTs) allow high-precision field-induced ESR measurement for the carriers in semiconductor crystals or films composed of regularly aligned organic molecules [40, 41]. The ESR signal observed in pentacene TFT appeared to be narrower than that observed in solution. The narrowing was supposed to be evidence of a spread of the electron wavefunction on $N$ neighboring pentacene molecules leading, according to CLT, to a decrease of signal width by the factor of $1/\sqrt{N}$. Assuming the Gaussian signal shape it was estimated that $N \approx 10$ molecules.

Subsequent thorough studies showed that the ESR signal shape is temperature dependent [42–44] but never Gaussian. Pentacene TFTs and rubrene single-crystal transistors exhibit Lorentzian shape ESR signal showing motional narrowing [45] with increase of T above $T = 50$K. In contrast, the narrowing effect is not observed below 50 K which implies carriers localization. Besides, continuous wave saturation experiments confirm that all carriers in the pentacene TFTs are localized at $T < 50$K [46]. Hence, the origin of the broadening at $T < 50$K is purely inhomogeneous. However, in contrast with naive expectations, the ESR spectrum is not Gaussian. Obviously, this anomaly is associated with the properties of localized carrier states in TFTs.

Below, we analyze the case when the ESR signal of a semiconducting organic molecular system is a smooth curve that deviates from the Gaussian at very low temperatures where all carriers are localized. We assume that peculiar lineshape is associated with the inhomogeneous distribution of localized carrier states and develop a method to extract the distribution of trapped carriers over its binding energy from the ESR signal [46, 47]. The energy resolution of the method is much higher than that by other methods [48–54].

In Section 8.2.2, we establish the reason of the deviation of the ESR signal from the Gaussian shape. We show that the signal from a single pentacene molecule is very similar to a Gaussian showing almost no individual lines from hyperfine splittings. It is proved that the sum of signals from many identical independent traps is a Gaussian too (see 8.2.2.1–8.2.2.2). The width of this signal is uniquely determined by an effective number of sites $N_{eff}$ characterizing the spatial extend of carrier in the trap. Hence, the non-Gaussian lineshape can be ascribed only to a superposition of signals from different kinds of traps, each is characterized by different localization parameters $N_{eff}$. In Section 8.2.2.3, we derive an explicit relation between the shape of the ESR signal and the distribution of the traps over the localization parameters $N_{eff}$. This relationship is a Fredholm integral equation of the first kind. Section 8.2.3 outlines the stochastic optimization method (SOM) [55–59] capable of solving this equation. We describe the SOM and test its stability with respect to the experimental noise in Sections 8.2.3.1 and 8.2.3.2, respectively. Section 8.2.3.3 introduces practical recipes on how to handle realistic noisy experimental data and presents results for traps distributions in pentacene TFTs. Section 8.2.3.4 analyzes the limits of the reliability of the obtained distributions.

In Section 8.2.4, we introduce a general method to map distribution of the traps over the localization parameter $N_{eff}$ to a distribution over the binding energies $E_B$. Particular implementation of the mapping requires knowledge of the explicit $E_B - N_{eff}$ relation between the binding energy $E_B$ and the localization parameter $N_{eff}$. This relation is calculated in Section 8.2.4.1 for two dimensional the 2D Holstein model by exact numeric diagrammatic Monte Carlo method [60], though, similar results can be obtained by many other accurate methods [61–66]. The distribution of the trapped states over the binding energies $E_B$ in pentacene TFTs are shown in Section 8.2.4.2. Sections 8.2.5 presents discussion and Section 8.2.6 provides conclusive remarks.

### 8.2.2 ESR Spectra of Trapped Carriers

In this section, we consider a molecular crystal in which a single molecule contains many nuclear spins. The typical width of the individual spectral lines of the split with hyperfine interaction quantum levels is greater than the typical energy distance between these levels [35, 38] and the lineshape of the ESR signal is Gaussian. The temperature is assumed to be sufficiently low that we can neglect the "homogeneous" relaxation leading to the Lorentzian shape of the ESR signal. It is also low enough to avoid self-averaging of the inhomogeneities by the "motional narrowing" mechanism.

We introduce the Gaussian ESR spectra of a single large molecule and a cluster containing several molecules (8.2.2.1). We then consider a noticeably different case when carrier is localized on a single impurity in a crystal (8.2.2.2) but the ESR lineshape is still Gaussian. Finally, we relate the non-Gaussian lineshape of the experimental ESR signal to the presence of several types of impurities having different localization parameters $N_{eff}$ (8.2.2.3).

#### 8.2.2.1 ESR Spectra for Single Molecule and a Cluster Containing Several Molecules

The standard expression for the ESR signal of one molecule is [38]

$$R(B) = \sum_{m_1=-n_1I_1}^{n_1I_1} \cdots \sum_{m_k=-n_kI_k}^{n_kI_k} \frac{P(m_1, \ldots, m_k)\times}{\pi} \frac{\Gamma}{\left(B - \sum_{i=1}^{k} A_i m_i\right)^2 + \Gamma^2}. \tag{8.7}$$

Here, $k$ is the number of groups of equivalent nuclei, $n_i$ is the number of equivalent nuclei in the $i$th group, $I_i$ is the nuclear spin in the $i$th group, $\Gamma$ is the linewidth of each peak, P is the intensity of each peak and $B$ is the magnetic field. If protons ($I = 1/2$) are the only paramagnetic nuclei, as is the case for pentacene molecules, $P$ is given as

$$P(m_1, \ldots, m_k) = \Pi_{i=1}^{k} \frac{C_{2n_iI_i}^{m_i+n_iI_i}}{(2I_i + 1)^{-n_i}}, \tag{8.8}$$

where $C_{2n_iI_i}^{m_i+n_iI_i}$ are binomial coefficients.

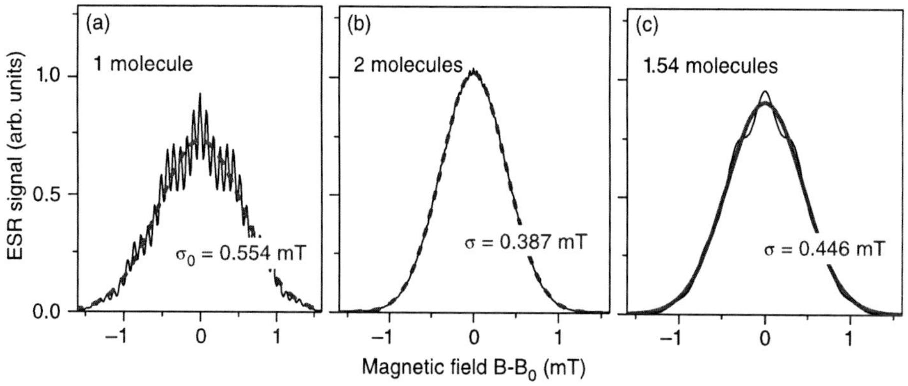


**Figure 8.3** Simulated ESR spectrum (black solid lines) of 1, 2, and 1.54 pentacene molecules based on experimental hyperfine splitting of one pentacene molecule with fits by Gauss distribution (red dashed lines).

For the particular case of the pentacene molecule, we set $\Gamma = 0.02\text{mT}$ and use reported in [37] coupling constants $\{A_i; i = 1, \ldots, 4\}$ and numbers of equivalent nuclei $\{n_i; i = 1, \ldots, 4\}$. The ESR signal obtained from Eqs. (8.7) and (8.8) can be represented (see Figure 8.3a) as a curve fluctuating around the Gaussian envelope

$$G_0(B) = \sqrt{\frac{1}{2\pi\sigma_0^2}} \exp\left[-\frac{(B - B_0)^2}{2\sigma_0^2}\right] \tag{8.9}$$

with standard deviation $\sigma_0 = 0.554$ mT.

The standard situation, known from the physics of gases and solutions, is the case where the carrier is localized in a cluster containing $N$ molecules and its density is uniformly spread over $N$ molecules. In this case, the signal retains its Gaussian shape with the width of the distribution reduced by the factor $N^{1/2}$. The hyperfine structure of the $N$ molecules can be simulated by Eqs. (8.7) and (8.8) by replacing $n_i \to Nn_i$ and $A_i \to A_i/N$. This transformation is related to the increase in the number of nuclear spins and consequent decrease of electron density interacting with one nuclear spin. Figure 8.3b shows an example of the spectrum for $N = 2$ with standard deviation $\sigma = \sigma_0/\sqrt{2}$. It is clear that the oscillations around the Gaussian envelope are quickly suppressed as $N$ increases.

The shape and the narrowing of the ESR signal in a cluster of $N$ molecules follow from the CLT. The shifts of the signal, $y_i = (B_i - B_0)$ for the $i$-th molecule, is an independent random variable with Gaussian distribution $R(y) = G_0(y)$ having standard deviation $\sigma_0$. According to CLT, the distribution $R(\overline{y})$ of the random variable $\overline{y} = N^{-1}\sum_{i=1}^{N} y_i$ is also Gaussian with $\sigma_N = \sigma_0/\sqrt{N}$.

#### 8.2.2.2 ESR Spectra for a Trap in Crystal

The situation for a carrier localized in a trap in a crystal is different from the cluster case. In the cluster, one assumes the uniform charge distribution whereas the distribution over molecules $i$ in a trap $\{p_i\}$ is nonuniform with the normalized $\sum_i p_i \equiv 1$ probabilities $p_i$. However, it can be proved [67] and explicitly demonstrated [47] that the lineshape in this case is also Gaussian.

The probability distribution $\{p_i\}$ unambiguously determines the linewidth of the ESR signal. The width is the standard deviation $\sigma$ which is the root square of the second moment of the line. If we consider the standard deviation of a signal in a trap $\sigma$ and compare it with that from a single molecule $\sigma_0$ we can introduce the effective number of molecules $N_{eff}(\{p_i\})$ to describe the linewidth of the ESR signal from a carrier in a trap. The distribution of the ESR shift $B$ for each molecule $i$ has the mean value $\langle B \rangle = B_0$ and variance $\sigma_0^2 = \langle (B - B_0)^2 \rangle$. Since the hyperfine configuration of molecules are independent of each other, the variables $y_i = (B_i - B_0)$ are independent for different molecules $i$. Hence, the standard deviation $\sigma(\{p_i\})$ of the sum of random variables $\overline{y} = \sum_i p_i(B_i - B_0)$ is related to the single-molecule standard deviation $\sigma_0$ by the expression

$$o(\{p_i\})/\sigma_0 = \sqrt{\sum_i p_i^2} \,. \tag{8.10}$$

It is natural to define the effective $N_{eff}(\{p_i\})$, corresponding to $\{p_i\}$, as

$$\sigma(\{p_i\}) = \sigma_0 / \sqrt{N_{eff}(\{p_i\})} \,. \tag{8.11}$$

Then, the effective number of molecules $N_{eff}(\{p_i\})$ is unambiguously determined by the charge distribution in a trap $\{p_i\}$

$$N_{eff}(\{p_i\}) = \left[ \sum_i p_i^2 \right]^{-1} \,. \tag{8.12}$$

We conclude that the ESR signal from carriers localized in a set of identical *independent* traps is Gaussian whose width $\sigma$ is given by Eqs. (8.11) and (8.12) which relate the width to the density distribution in trap $\{p_i\}$.

#### 8.2.2.3 ESR Spectra for Several Kinds of Traps

The last possible source of non-Gaussian shape is to assume that it originates from a superposition of the signals from different kinds of traps $\xi$, each having different probability distribution from the trapped carriers. We describe the experimental spectrum $\mathcal{E}_{exp}(B)$ by the superposition of the ESR spectra $\int d\xi \mathcal{G}(B, \xi)$ from traps of different kinds $\xi$. We showed in Section 8.2.2.1 that the lineshape of spectrum from single trap is parametrized by single parameter $N_{eff}$ and, hence, it is natural to chose $\xi = N_{eff}$. Then, $\mathcal{G}(B, N_{eff})$ is a Gaussian with width $\sigma = \sigma_0 / \sqrt{N_{eff}(\{p_i\})}$. Hence, the ESR signal from the trap of type $\xi = N_{eff}$ is

$$G(B, N_{eff}) = \sqrt{\frac{N_{eff}}{2\pi\sigma_0^2}} \exp\left[ -\frac{(B - B_0)^2}{2(\sigma_0^2 / N_{eff})} \right] \,. \tag{8.13}$$

Introducing the distribution function $D(N_{eff})$ of the traps over $N_{eff}$, we can express the experimental signal $\mathcal{E}_{exp}(B)$ in terms of the superposition

$$\mathcal{E}_{exp}(B) = \int_1^{\infty} G(B, N_{eff}) D(N_{eff}) dN_{eff} \tag{8.14}$$

where signals from different traps are weighted by $D(N_{eff})$.

In practice, ESR measure the derivative $\mathcal{X}_{exp}(B) = d\mathcal{E}_{exp}(B)/dB$ and $\mathcal{X}_{exp}(B)$ is related to the distribution function of traps $D(N_{eff})$ via

$$\mathcal{X}_{exp}(B) = \int_1^\infty \frac{dG(B, N_{eff})}{dB} D(N_{eff}) dN_{eff} \,. \tag{8.15}$$

To obtain $D(N_{eff})$ we must solve the integral equations (8.14) and (8.15). The quantities $\mathcal{X}_{exp}(B)$ ($\mathcal{E}_{exp}(B)$) and the kernel $dG(B, N_{eff})/dB$ ($G(B, N_{eff})$) are known functions while the distribution $D(N_{eff})$ is a function to determine.

### 8.2.3 From ESR Spectrum to Trap Distribution Over Degree of Localization

Equation (8.14) [(8.15)], where $\mathcal{E}_{exp}(B)$ [$\mathcal{X}_{exp}(B)$] and $G(B, N_{eff})$ [$dG(B, N_{eff})/dB$] are known and $D(N_{eff})$ is to be determined, is the Fredholm equation of the first kind. Naively, true solution $\widetilde{D}(N_{eff})$ has to minimize the objective function

$$Q = \sum_{i=1}^{M} |\Delta(i)|^2 \tag{8.16}$$

where residuals

$$\Delta(i) = \widetilde{\mathcal{X}}(B_i) - \mathcal{X}_{exp}(B_i), \quad i = 1, M \tag{8.17}$$

represent the difference between the experimental data $\mathcal{X}_{exp}(B_i)$ and quantities $\widetilde{\mathcal{X}}(B_i)$ which are obtained from the solution $\widetilde{D}(N_{eff})$

$$\widetilde{\mathcal{X}}(B) = \int_0^\infty \frac{dG(B, N_{eff})}{dB} \widetilde{D}(N_{eff}) dN_{eff} \,. \tag{8.18}$$

However, this naive approach fails since one faces complicated "ill-posed problem" which can be handled only by sophisticated numeric techniques. Section 8.2.3.1 gives a description of the stochastic optimization method (SOM) [55–59] to solve Eqs. (8.14) and (8.15). In Section 8.2.3.2, we verify the applicability of the method to the analysis of the ESR data and check the influence of the experimental noise on the reliability of the results. Section 8.2.3.3 introduces practical aspects of the implementation of the the SOM to ESR analysis and presents the results for the trap distribution in pentacene TFTs. Finally, Section 8.2.3.4 demonstrates the limits of the reliability of the distribution obtained by SOM.

#### 8.2.3.1 Method to Solve Inverse Problem

The Fredholm equations of the first kind (8.14) and (8.15) belong to "ill posed" problems. Naively, the true solution of the Eq. (8.15) $\widetilde{D}(N_{eff})$, being convoluted with the kernel, produces the set of values $\{\widetilde{\mathcal{X}}(B_i)\}$ which coincide with the measured $\{\mathcal{X}_{exp}(B_i)\}$. However, the knowledge about the function $\mathcal{X}_{exp}(B)$ is not exact and values $\{\mathcal{X}_{exp}(B_i)\}$ contain experimental noise. Naturally, to extract maximal information about the "solution" one targets minimization of the objective function (8.16). For noisy data, the residuals $\Delta(i)$ are never equal to zero at all points $i$, $Q \neq 0$, and, hence, there is no solution in the traditional sense. Moreover, a straightforward attempt to minimize (8.16) leads to an unreasonable "solution". Typically, it has huge fluctuations, called "saw tooth" noise, which exceed the true values of $D(N_{eff})$ by several orders of magnitude [56]. Then, suppression of this "saw tooth" noise is essential and it is achieved by two different groups of approaches. Historically, the first group is called "regularization approach". The first such method was developed by Tikhonov in early 1940s [56, 69, 70]. Nowadays, the most popular method of this group is the maximal entropy method [68]. In the "regularization approach"one minimizes the measure (8.16) with additional regularization terms which smooths the "saw tooth" noise. The main drawback here is that the solution is corrupted by the smoothing regularization term. The second strategy applies the "stochastic approach" to obtain many statistically independent solutions [55, 56, 59] whose linear combination smooths the "saw tooth" noise without corrupting any of the individual solutions. Here, we apply the SOM [55–58] as one of most effective methods of the "stochastic approach".

#### 8.2.3.2 Tests of SOM Stability Against the Noise in Experimental Data

The SOM has been successfully applied integral equations with many different kernels. The exponential kernel $K(y,x) = \exp[-yx]$ was examined in [55, 60, 71–88], and various kernels ranging from the Fermi distribution to the Matsubara frequency representation are considered in [25, 89–92]. In this section, we test the stability of the SOM to experimental noise for the kernels (8.14) and (8.15).

We introduce a normalized to unity function $D(N_{eff})$ (see the dotted line in Figure 8.4) and generate a set of "experimental" data $\{\mathcal{X}_{exp}(B_i), i = 1, 200\}$ using relations (8.14)

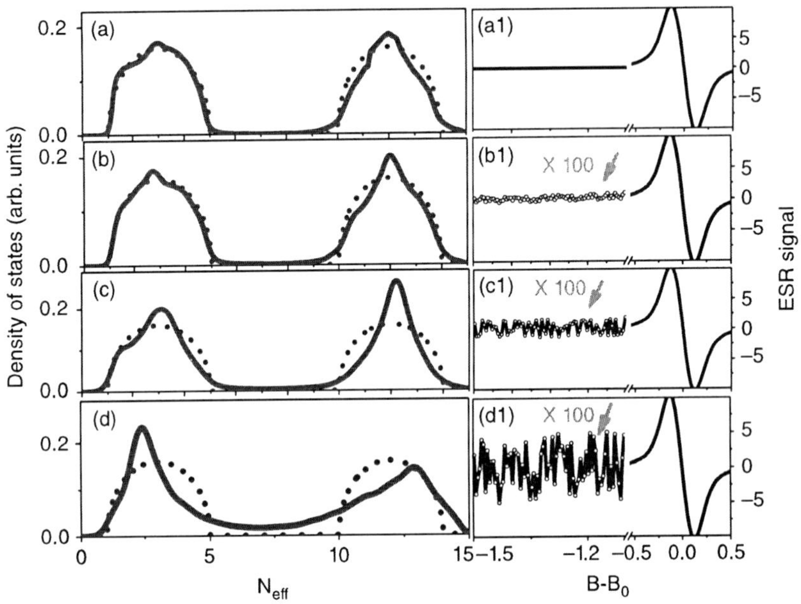


**Figure 8.4** Test of SOM procedure to restore distribution of spatial extent of traps $D(N_{eff})$ from "experimental" ESR data $\mathcal{X}_{exp}(B)$ (8.15) under different levels of noise. "Experimental" ESR data with noise are given in panels (a1)-(d1) and actual (restored) spectrum $D(x)$ is indicated by dashed (solid) line in corresponding panels (a)-(d). Noise level is (a,a1) $f = 0.0$, (b,b1) $f = 0.01$ [$s_n = 10^{-3}$], (c,c1) $f = 0.03$ [$s_n = 3 * 10^{-3}$], and (d1,d) $f = 0.1$ [$s_n = 10^{-2}$]. Restored spectrum is obtained by solving Eq. (8.15) using SOM.

and (8.15). Then, we try to find $\tilde{D}(N_{eff})$ by solving Eqs. (8.14) and (8.15). For ideal data without any noise in the set $\{\mathcal{X}_{exp}(B_i), i = 1{,}200\}$ one restores $D(N_{eff})$ quite successfully (see Figure 8.4a). There was no significant difference between the results obtained by solving Eqs. (8.14) or (8.15).

We introduce the noise by a sequence of random numbers $\mathcal{R}(i)$ uniformly distributed in the range $[-1{,}1]$ and generate data sets $\{\mathcal{X}_{exp}(B_i) + (f/2)\mathcal{R}(i), i = 1{,}200\}$, to create "experimental" ESR data with different amplitudes of noise $f$. The noise-to-signal ratio is defined as a ratio of the amplitude $f$ and maximal absolute value $MAX\{| \mathcal{X}_{exp}(B_i) |\}$ of the given signal $\mathcal{X}_{exp}(B_i)$: $s_n = f/MAX\{| \mathcal{X}_{exp}(B_i) |$. The results are shown in Figure 8.4. Increase of the noise-to-signal ratio $s_n$ corrupts the solution for large values of $N_{eff}$ first: the shape of the high-energy peak is not reproduced but its position is still correct. Also, a higher values of $s_n$, the shape of the low-energy peak is not reproduced. Note, although the shapes of the high- and low-energy peaks are not reproduced, their positions are still approximately correct even for large values of the noise-to-signal ratio $s_n$.

#### 8.2.3.3 Practical Implementation of Method: Distribution of Traps in Pentacene TFT

Equation (8.15) is preferable for the practical implication of the SOM because it resolves the problem of unknown background. There is always some uncertainty in the normalization and background of the experimental data. Implying the normalization of the distribution $D(N_{eff})$ to unity and the condition $\int_{-\infty}^{\infty} dN_{eff} \int_{-\infty}^{B} dz[dG(z, N_{eff})/dz] = 1$, we must normalize the experimental data as

$$\int_{-\infty}^{\infty} dB \int_{-\infty}^{B} dz\mathcal{X}_{exp}(z) = I\,, \tag{8.19}$$

where I=1. However, there is uncertainty in normalization of the experimental data because of noise. An additional problem comes from the unknown background if the data are in the form $\mathcal{E}_{exp}(B)$. However, this problem disappears for Eq. (8.15) because $\lim_{B\to\pm\infty}\mathcal{X}_{exp}(B) = 0$ for any constant background. Therefore, the equation (8.15) is preferable because one has only to tune normalization.

To handle the normalization uncertainty we normalized to unity $I = 1$ the experimental signal as shown in (8.19) and varied the normalization $I$ (Figure 8.5) of the

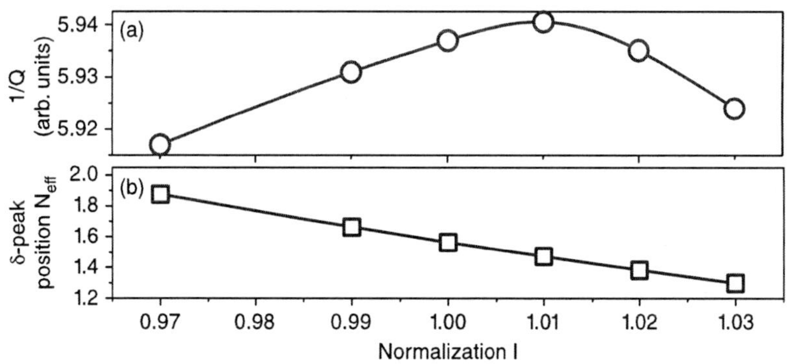


**Figure 8.5** (a) Dependence of the (a) fit quality $Q^{-1}$ and (b) position of low-energy sharp peak $\delta(N - N_{eff})$ on normalization $I$ of the spectral function.

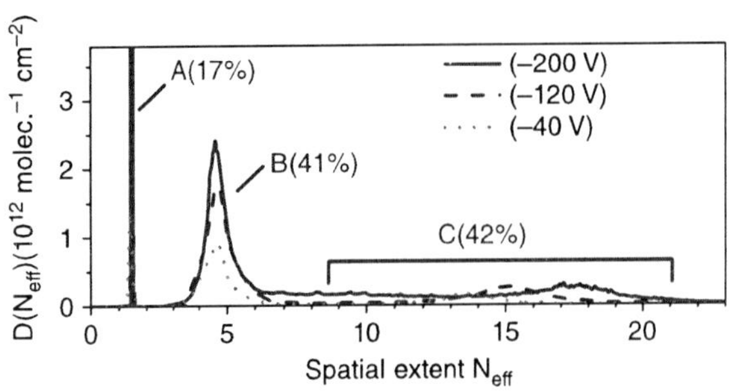


**Figure 8.6** Distribution of trap states in pentacene TFT vs spatial extent $N_{eff}$ of charge distribution in traps obtained from ESR spectrum of pentacene TFT at 20 K with gate voltage −200 V (solid line), −120 V (dashed line), and −40 V (dotted line).

distribution density $\widetilde{D}(N_{eff})$. For example, the procedure to determine normalization for the result shown in Figure 8.6 for gate voltage −200V is as follows. The equation (8.15) is solved for different normalizations $I$ and the functional dependence $Q^{-1}(I)$ of inverse of the objective function (8.16) is plotted (Figure 8.5a) It can be seen that the position of the sharp peak (Figure 8.5b) is very sensitive to the value of the normalization $I$. Finally, the result corresponding to the maximal $Q^{-1}(I)$ is chosen and the probability density $\widetilde{D}(N_{eff})$ is renormalized back to physical value equal to unity.

We can show that the above procedure is robust and stable. Let us compare results at different gate voltages (Figure 8.6). The best normalizations are different at different gate voltages: $I(V = -200) = 1.01$, $I(V = -120) = 1.038$, and $I(V = -40) = 1.03$, respectively. However, the position of the sharp peak at low values of $N_{eff}$ does not depend on the voltage $V$. Physically, the sharp peak at low values of $N_{eff}$ corresponds to deep impurity levels that depend only slightly on the gate voltage. Therefore, its independence on the gate voltage in the fine analysis of the ESR spectra indicates the high stability of the procedure based on the suggested approach.

#### 8.2.3.4 Reliability of Trap Distribution Result

It is important to to check how many details of the resulting distribution of impurities are reliable. The reliability can be analyzed by plotting the residual function (8.17). The spectrum $\widetilde{D}(N_{eff})$ (Figure 8.6) obtained by solving Eq. (8.15) has three peaks. Figure 8.7a shows the fit of the ESR signal using the distribution $\widetilde{D}(N_{eff})$ in Figure 8.6. It also shows the separate contributions of the A, B, and C components of the distribution. To clarify which features of the distribution $\widetilde{D}(N_{eff})$ are reliable for the given level of noise in the experimental data we studied the residuals (8.17) $\mathcal{X}_{exp}(B) - \widetilde{\mathcal{X}}(B)$ (Figures 8.7b-e). The quality of the fit by the SOM (Figure 8.7(b)) is much better than that obtained by e.g., the Lorentzian (Figure 8.7c) and two $\delta$-functions (Figure 8.7d). The fit from three $\delta$-functions gives a residual function (Figure 8.7e) as good as that obtained from $\widetilde{D}(N_{eff})$ in Figure 8.6. Therefore, we conclude that, within the limits of the noise of the experimental data, the existence of at least three kinds of traps is a reliable result. However, a data analysis with less noise could, in principle, reveal more fine structure in the distribution function $D(N_{eff})$.

### 8.2.4 Transformation From Spatial Distribution to Energy Distribution

Here, we find the distribution of the traps $\mathcal{Z}(E_B)$ over the binding energies $E_B$ using the distribution $D(N_{eff})$ and known functional dependence

$$N_{eff} = N_{eff}(E_B)\,. \tag{8.20}$$

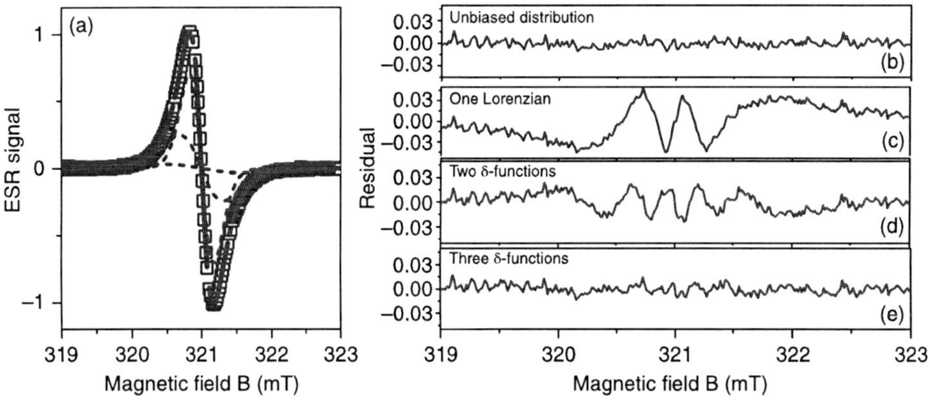


**Figure 8.7** (a) Experimental signal (squares), fit by spectrum in Figure 8.6 at the gate voltage −200 *V* (solid lines) and contributions from (see Figure 8.6) A, B, and C components (dashed lines). Residuals (8.17) $\mathcal{X}_{exp}(B) - \widetilde{\mathcal{X}}(B)$ from (b) unbiased distribution (Figure 8.6), (c) best fit by one Lorentzian, (d) best fit by two $\delta$-functions $0.68\delta(N-4.5)$ and $0.32\delta(N-20)$, and (e) best fit by three $\delta$-functions $0.31\delta(N-1.4)$, $0.51\delta(N-7.5)$, and $0.18\delta(N-25.0)$.

Indeed, there is the relation

$$\mathcal{Z}(\overline{E})[E_B^{(i+1)} - E_B^{(i)}] = D(\overline{N}_{eff})[N_{eff}(E_B^{(i+1)}) - N_{eff}(E_B^{(i)})] \tag{8.21}$$

between $D$ and $\mathcal{Z}$ for $E_B^{(i+1)} \to E_B^{(i)}$. Here, $\overline{E} = (E_B^{(i+1)} + E_B^{(i)})/2$ and $\overline{N}_{eff} = (N_{eff}(E_B^{(i+1)}) + N_{eff}(E_B^{(i)}))/2$. Hence

$$\mathcal{Z}(E_B) = D(N_{eff})\frac{dN_{eff}(E_B)}{dE_B} . \tag{8.22}$$

One can obtain (8.20) as a parametric function calculating $N_{eff} = N_{eff}(\{\mathcal{P}\})$ and $E_B = E_B(\{\mathcal{P}\})$ for a range $\{\mathcal{P}\}$ of parameters $\mathcal{P}$ of some relevant model.

Transfromation (8.22) uses relation (8.20) dependent on the model chosen to describe the localization of a carrier in a trap. We consider a 2D Holstein polaron with an attractive center in Section 8.2.4.1 and show results of the transformation for pentacene TFTs in Section 8.2.4.2.

#### 8.2.4.1 Trap Model: 2D Holstein Polaron and On-Site Attractive Center

Charge carrier in pentacene TFTs can be described as a coupled to phonons particle [93] in a system with attractive impurities [42]. It can be modeled by a 2D Holstein polaron close to an on-site attractive center

$$H = -t\sum_{\langle \mathrm{i,j}\rangle} c_\mathrm{i}^\dagger c_\mathrm{j} + \omega_\mathrm{ph}\sum_\mathrm{i} b_\mathrm{i}^\dagger b_\mathrm{i} - \gamma\sum_i (b_\mathrm{i}^\dagger + b_\mathrm{i})c_\mathrm{i}^\dagger c_\mathrm{i} - Uc_0^\dagger c_0. \tag{8.23}$$

Here, $c_i^\dagger$ ($b_i^\dagger$) is the creation operator for the carrier (phonon) in the $i$-th molecule, $U$ is the attractive impurity potential for the carrier $c_0^\dagger$ at site 0 and $\omega_\mathrm{ph}$ is the frequency of the dispersionless phonon. The amplitude $t$ is the electron transfer between nearest neighbor sites and the local Holstein coupling to the phonons is $\propto \gamma$. The dimensionless electron-phonon coupling constant $\lambda$ is defined as $\lambda = \gamma^2/(4t\omega\mathrm{ph})$. Parameter domain to determine relation (8.20) is $\mathcal{P} = \{U, \lambda\}$.

To calculate $E_B = E_B(U, \lambda)$ and $N_{eff} = N_{eff}(U, \lambda)$, we used the direct space diagrammatic Monte Carlo (DSDMC) [60]. The data for $E_B = E_B(U, \lambda)$ and $N_{eff} = N_{eff}(U, \lambda)$ are presented in Figures 8.8a and 8.8b. The values of $N_{eff}$ were determined using relation (8.12) from the charge distribution in the trap which was calculated by the DSDMC technique (see Figure 8.9). To determine the function $N_{eff} = N_{eff}(E_B)$ from relations $E_B = E_B(U, \lambda)$ and $N_{eff} = N_{eff}(U, \lambda)$ (Figures 8.8a and 8.8b) one has to decide which parameter is responsible for variation of $E_B$ and $N_{eff}$. It is natural to assume that $\lambda$ is fixed and the distributions of $E_B$ and $N_{eff}$ are due to different values of trapping potential $U$.

#### 8.2.4.2 Energy Distribution of Traps in Pentacene TFTs

We chose $\lambda = 0.15$, as determined in optical absorption experiments [93], and $t = 0.1$eV, as obtained in band-structure calculations [6, 94]. Figure 8.10 shows the distribution

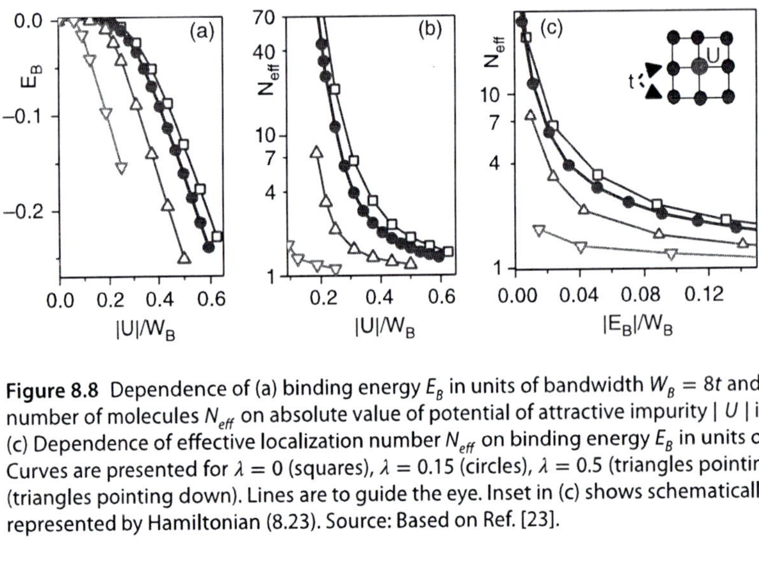


**Figure 8.8** Dependence of (a) binding energy $E_B$ in units of bandwidth $W_B = 8t$ and (b) effective number of molecules $N_{eff}$ on absolute value of potential of attractive impurity | $U$ | in units of $W_B$. (c) Dependence of effective localization number $N_{eff}$ on binding energy $E_B$ in units of bandwidth $W_B$. Curves are presented for $\lambda = 0$ (squares), $\lambda = 0.15$ (circles), $\lambda = 0.5$ (triangles pointing up), and $\lambda = 1$ (triangles pointing down). Lines are to guide the eye. Inset in (c) shows schematically model represented by Hamiltonian (8.23). Source: Based on Ref. [23].

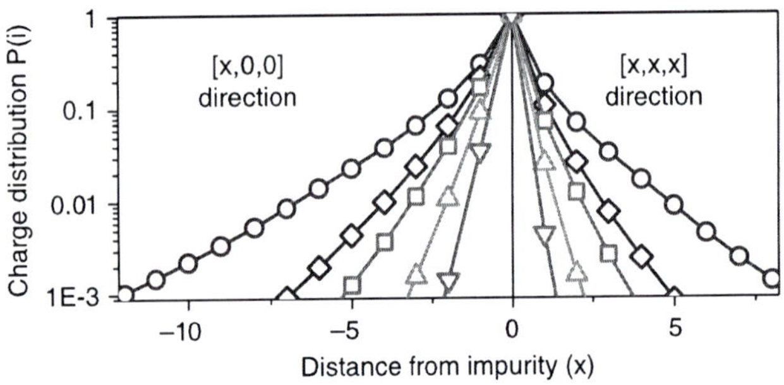


**Figure 8.9** Charge distributions around attractive impurity with potential $U$ at $\lambda = 0.15$ along [100] and [111] directions: $U/W_B = -0.206$ [$E_B/W_b = 0.0032$, $N_{eff} = 44.3$] (circles); $U/W_B = -0.25$ [$E_B/W_b = 0.0113$, $N_{eff} = 11.2$] (rhombus); $U/W_B = -0.281$ [$E_B/W_b = 0.0213$, $N_{eff} = 6.05$] (squares); $U/W_B = -0.375$ [$E_B/W_b = 0.0703$, $N_{eff} = 2.38$] (triangles pointing up); $U/W_B = -0.59$ [$E_B/W_b = 0.2397$, $N_{eff} = 1.35$] (triangles pointing down).

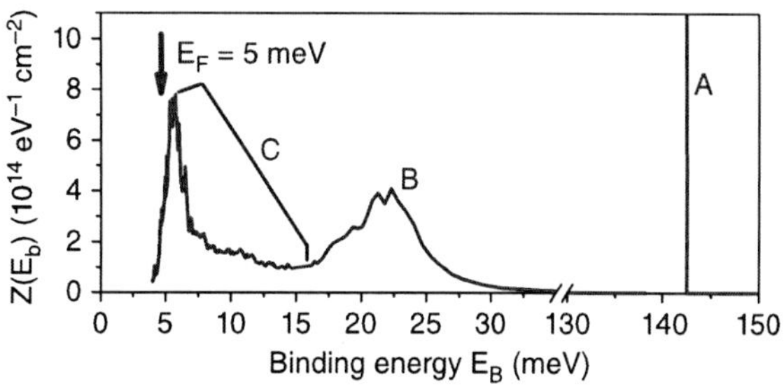


**Figure 8.10** Distribution of trap states in pentacene TFTs as a function of trap energy $E_B$ for gate voltage −200 V.

$\mathcal{Z}(E_B)$ of the trapped carriers over the binding energies in TFT at the gate voltage −200 V. To obtain $N_{eff}(E_B)$ (bold line in Figure 8.8c), we fixed $\lambda = 0.15$ and used the dependence of $E_B$ (bold line in Figure 8.8a) and $N_{eff}$ (bold line in Figure 8.8b) on the attractive potential $U$.

We used transformation (8.22) to obtain data shown in Figure 8.10. There are two discrete trap levels (A and B) with peaks at 140 ± 50 meV and 22 ± 3 meV, respectively, and a broad feature (C) distributed between 5 and 15 meV. The low-energy profile points to the existence of tail states located below the band edge which are partially occupied up to the Fermi level at $E_B = 5$ meV.

### 8.2.5 Discussion

Weakly-localized in-gap states play a crucial role in the intrinsic charge transport along semiconductor/gate dielectric interfaces in organic transistors. Weak temperature dependence of mobility is often observed in devices with high mobility and highly-ordered molecular interfaces, which indicates that the Fermi energy is just below the band edge [96].

To date, interfacial trap density has been studied by deep-level transient spectroscopy (DLTS) [97], photocurrent yield [98], gate-bias stress [99], and thermally-stimulated current [100] experiments. However, these measurements are based on the charge transport, and it is strongly affected by the "extrinsic" potential barriers at grain boundaries and/or channel/electrode interfaces. In striking contrast, our method has a crucial advantage due to its ability to disclose the spatial and energy distribution of shallow traps down to a few meV, based on a unique microscopic probe using electronic spins. In addition, the g tensor can be used to identify the molecular species around the trap sites. For the three types (A, B, and C) of trap states, the g tensor should be common, considering the highly symmetric nature of the ESR spectra. This indicates that the trap states are extended over pentacene molecules of regular orientation [40, 42].

The deep discrete trap level (A) might be attributed to structural defects such as grain boundaries [101]. The shallow discrete level (B) and the broad feature (C) might be ascribed to small defects such as molecular sliding along the long axes of the molecules [102], disorder induced by random dipoles in the amorphous gate dielectrics [103, 104], thermal off-diagonal electronic disorder [105, 106], and the fluctuations of the band edge [107].

Strongly localized states with small $N_{eff}$ are not sensitive to the gate voltage (Figure 8.11), which is obvious from the physical point of view. An increase of the gate voltage adds low-energy states with large $N_{eff}$ and small $E_B$ which participate in the

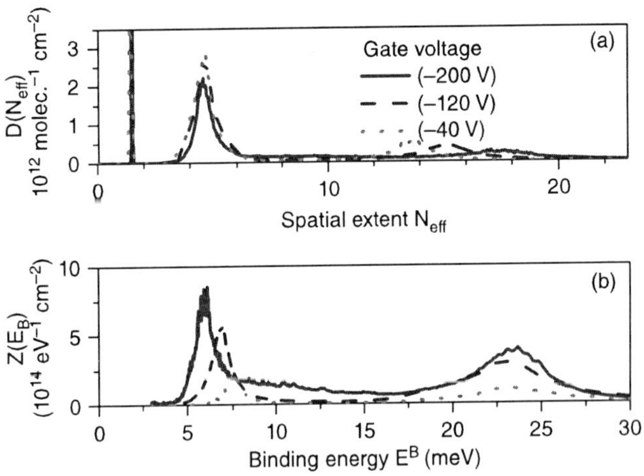


**Figure 8.11** Distributions of trap states in pentacene TFTs ((a) is spatial distribution and (b) is binding energy distribution) as a function of trap energy $E_B$ for gate voltage −200 V (solid line), −120 V (dashed line), and −40 V (dotted line).

creation of the ESR signal (Figure 8.11). This shift of the border where states are visible by the ESR probe indicates that these states are filled when the gate voltage increases. Hence, it can be interpreted as a movement of the Fermi level which shifts to lower binding energies when the bias voltage increases.

It is important to mention that the sharp peak of $D(N_{eff})$ at $N_{eff} = 1.54$ does not contradict the assumptions used to derive the integral Eqs. (8.14) and (8.15). The very essence of these equations implies that the contribution from each state with a given $N_{eff}$ is a Gaussian ESR signal. On the other hand, the signal at small $N_{eff}$ is a more complicated function with fine features (see e.g. Figure 8.3a for $N_{eff} = 1$). The results for distribution $D(N_{eff})$ at small $N_{eff}$ can be unreliable. However, as can be seen in Figure 8.3c, the ESR signal for reasonable parameters is close to Gaussian even at $N_{eff} = 1.54$. Therefore, the results for the distribution of the traps $D(N_{eff})$ are valid even for small values of the localization parameter $N_{eff}$.

### 8.2.6 Summary of Trap Study

We presented an unbiased method for the analysis of high-precision ESR spectra which is capable of obtaining the distribution of trapped carriers over the degree of localization and the binding energy. The first step is the spectral analysis which splits the spectrum into a distribution of Gaussian components each of which corresponds to a different spatial extension of the trapped carriers. The second step is the transformation of the distribution over the degree of localization into a distribution over the binding energies using the relation between the binding energies and characteristics of the trapped carriers. We presented the basics of the spectral analysis, details required for practical applications, and showed the stability of the analysis with respect to experimental noise.

The method is appropriate for ESR spectra of organic TFTs for the following reasons.

First, the channel materials are composed of regularly aligned organic molecules that involve multiple degrees of freedom of nuclear spin moments. This feature clearly justifies our basic assumption that a single type of trap gives the Gaussian lineshape of the ESR spectrum. Second, it is possible to measure the high-precision ESR spectrum because of the fairly small spin-orbit interactions of organic materials. Third, the field-effect device structure also enables control of the carrier density without introducing any randomness in the channel semiconductors.

Such a direct probe is quite unique in investigating the microscopic carrier dynamics in the organic TFTs that have attracted considerable recent attention in the field of organic electronics. We have shown that the trap states in pentacene TFTs can be classified into three major groups: deep-trap states with a spatial extension of about 1.5 molecules (A), relatively shallow-trap states that extend over about 5 molecules (B), and shallower-trap states that extend over 6 to 20 molecules (C). These states correspond to deep and moderately shallow-trap states with binding energies 140 meV (A) and 22 meV (B). The most shallow in-gap states are reflected in a broad peak ranging from 5 to 15 meV (C). All these states are crucial for improving the performance of organic TFTs.

## 8.3 Conclusion

Experimental studies of transport properties of small molecule organic semiconductors (rubrene, pentacene, etc.) show that the mobility of the carriers in these materials is essentially determined by traps. The intrinsic mobility of hypothetical pure crystal without traps shows band-like monotonic decrease with temperature, as can be proved by Hall measurements probing only free carriers. On the other hand, freezing carriers on traps at low temperatures essentially changes the temperature dependence of mobility leading to nonmonotonic behavior. Complex behavior of intrinsic mobility makes detailed transport study of traps in this materials impossible and, hence, the crucial information about trap states has to be obtained from other kind of experiments.

In Section 8.2, we reviewed the method to study trapped states by the fine analysis of high-precision electron ESR spectra. High resolution of the method facilitates the measurement of trap distribution over the binding energy. We note that the energy of deep traps $\approx$ 140meV, found in the fine analysis of the ESR spectra, can be used for analysis of the temperature dependence of mobility in the framework of the trap and release model. This analysis shows that there is one deep-trap state per approximately one thousand of organic molecules.

## Acknowledgments

The studies reviewed in this chapter would have been impossible without inspiring collaboration and discussions with Prof. Tatsuo Hasegawa, Prof. Giulio De Filippis, Prof. Hiroyuki Matsui, Prof. Vittorio Cataudella, and Prof. Antonio de Candia. A.S.M. was supported by the ImPACT Program of the Council for Science, Technology and Innovation (Cabinet Office, Government of Japan).

## References


1 Shaw, J.M. and Seidler, P.F. (2001). *IBM J. Res. Dev.* 45:3.
2 Hoppe, H. and Sariciftci, N.S. (2004). *J. Mater. Res.* 19:1924.
3 Katz, H.E. (2004). *Chem. Mater.* 16:4748.
4 Forrest, S.R. (2004). *Nature* 428:911.
5 Gershenson, M.E., Podzorov, V. and Morpurgo, A.F. (2006). *Rev. Mod. Phys.* 78:973
6 Troisi, A. and Orlandi, G. (2005). *J. Phys. Chem. B* 109:1849.
7 Coropceanu, V., Cornil, J., da Silva Filho, D.A. (2007). *Chem. Rev.* 107:926.
8 Kenkre, V.M., Andersen, J.D., Dunlap, D.H. and Duke, C.B. (1989). *Phys. Rev. Lett.* 62:1165.
9 Hannewald, K. and Bobbert, P.A. (2004). *Appl. Phys. Lett.* 85:1535.
10 Zoli, M. (2002). *Phys. Rev. B* 66:012303.
11 Stojanovic, V.M. and Vanevic, M. (2008). *Phys. Rev. B* 78:214301.
12 Li, Y., Yi, Y., Coropceanu, V. and Bredas, J.L. (2012). *Phys. Rev. B* 85:245201.
13 Holstein, T. (1959). *Ann. Phys.* 8:343.
14 Podzorov, V., Menard, E., Borissov, A. et al. (2004). *Phys. Rev. Lett.* 93:086602.
15 Podzorov, V., Menard, E., Rogers, J.A. and Gershenson, M.E. (2005). *Phys. Rev. Lett.* 95:226601.
16 Minder, N.A., Ono, S., Chen, Z. et al. *Adv. Mater.* 24:503.
17 Jurchescu, O.D., Baas, J. and Palstra, T.T.M. (2004). *Appl. Phys. Lett.* 84:3061.
18 Sundar, V.C., Zaumseil, J., Podzorov, V. et al. (2004). *Science* 303:1644.
19 Hulea, I.N., Fratini, S., Xie, H. et al. (2006). *Nat. Mater.* 5:982.
20 Friedman, L. and Holstein, T. (1963). *Ann. Phys.* 21:494.
21 Friedman, L. (1964). *Phys. Rev.* 135:A233.
22 Chang, J.-F., Sirringhaus, H., Giles, M. et al. (2007). *Phys. Rev. B* 76:205204.
23 Emin, D. (1991). *Phys. Rev. B* 43:11720.
24 Austin, I.G and Mott, N.F. (1969). *Adv. Phys.* 50:757812.
25 Mishchenko, A.S., Nagaosa, N., De Filippis, G. et al. (2015). *Phys. Rev. Lett.* 114:146401.
26 Li, Z.Q., Podzorov, V., Sai, N. et al. (2007). *Phys. Rev. Lett.* 99:016403.
27 Takeya, J., Tsukagoshi, K., Aoyagi, Y. et al. (2005). *Jpn. J. Appl. Phys.* 44 Part 2:L1393.
28 Hoesterey, D.C. and Letson, G.M. (1963). *J. Phys. Chem. Solids* 24:1609.
29 Shockley, W. and Read, W.T. (1952). *Phys. Rev.* 87:835.
30 De Filippis, G., Cataudella, V., Mishchenko, A.S. et al. (2015). *Phys. Rev. Lett.* 114:086601.
31 Su, W.P., Schrieer, J.R. and Heeger, A.J. (1979). *Phys. Rev. Lett.* 42:1698.
32 Marchand, D.J.J., De Filippis, G., Cataudella, V. et al. (2010). *Phys. Rev. Lett.* 105:266605.
33 Anderson, P.W. (1977). *Nobel Lecture.*
34 Feher, G. (1959). *Phys. Rev.* 114:1219.
35 Altshuler, S.A. and Kozirev, B.M. (1964). *Electron Paramagnetic Resonance.* New York:Academic Press.
36 Slichter, C.P. (2010). *Principles of Magnetic Resonance.* Springer Series in Solid-State Sciences.
37 Bolton, J.R. (1967. *J. Chem. Phys.* 46:408.

38 Weil, J.A., Bolton, J.R. and Wertz, J.E. (1994). *Electron Paramagnetic Resonance: Elementary Theory and Practical Applications.* New York:Wiley-Interscience.
39 Lewis, I.C. and Singer, L.S. (1965). *J. Chem. Phys.* 43:2712.
40 Marumoto, K., Kuroda, S.-I., Takenobu, T. and Iwasa, Y. (2006). *Phys. Rev. Lett.*97:256603.
41 Marumoto, K., Muramatsu, Y., Nagano, Y. (2005). *J. Phys. Soc. Jpn.* 74:3066.
42 Matsui, H., Hasegawa, T., Tokura, Y. et al. (2008). *Phys. Rev. Lett.* 100:126601.
43 Matsui, H. and Hasegawa, T. (2009). *Mater. Res. Soc. Symp. Proc.* 1154:B08–04.
44 Marumoto, K., Arai, N., Goto, H. et al. (2011). *Phys. Rev. B* 83:075302.
45 Kubo, R. and Tomita, K. (1954). *J. Phys. Soc. Jpn.* 9:888.
46 Matsui, H., Mishchenko, A.S. and Hasegawa, T. (2010). *Phys. Rev. Lett.* 104:056602.
47 Mishchenko, A.S., Matsui, H. and Hasegawa, T. (2012). *Phys. Rev. B* 85:085211.
48 Horowitz, G., Hajlaoui, M.E. and Hajlaoui, R. (2000). *J. Appl. Phys.* 87:4456.
49 Lang, D.V., Chi, X., Siegrist, T. et al. (2004). *Phys. Rev. Lett.* 93:086802.
50 Tal, O., Rosenwaks, Y., Preezant, Y. et al. (2005). *Phys. Rev. Lett.* 95:256405.
51 Calhoun, M.F., Hsieh, C. and Podzorov, V. (2007). *Phys. Rev. Lett.* 98:096402.
52 Sueyoshi, T., Fukagawa, H., Ono, M. et al. (2009). *Appl. Phys. Lett.* 95:183303.
53 Xie, H., Alves, H. and Morpurgo, A.F. (2009). *Phys. Rev. B* 80:245305.
54 Kalb, W.L., Haas, S., Krellner, C. et al. (2010). *Phys. Rev. B* 81:155315.
55 Mishchenko, A.S., Prokof'ev, N.V., Sakamoto, A. and Svistunov. B.V. (2000). *Phys. Rev. B* 62:6317.
56 Mishchenko, A.S. (2012). Stochastic Optimization Method for Analytic Continuation. In:*Correlated Electrons: From Models to Materials*, (eds. E. Pavarini, W. Koch, F. Anders and M. Jarrel), 14.1–14.28. Julich: Forschungszentrum Julich.
57 Mishchenko, A.S. (2005). *Usp. Phys. Nauk* 175:925.
58 Mishchenko, A.S. and Nagaosa, N. (2006). *J. Phys. Soc. J.* 75:011003.
59 Vafayi, K. and Gunnarsson, O. (2007). *Phys. Rev. B* 76:035115.
60 Mishchenko, A.S.,. Nagaosa, N, Alvermann, A. et al. (2009). *Phys. Rev. B* 79:180301(R).
61 Berciu, M., Mishchenko, A.S. and Nagaosa, N. (2010). *Europhys. Lett.* 89:37007.
62 Goodvin, G.L., Covaci, L. and Berciu, M. (2009). *Phys. Rev. Lett.* 103:176402.
63 Cataudella, V., De Filippis, G., Martone, F. and Perroni, C.A. (2004). *Phys. Rev. B* 70:193105.
64 De Filippis, G., Cataudella, V., Marigliano Ramaglia, V. and Perroni, C.A. (2005). *Phys. Rev. B* 72:014307.
65 Ebrahimnegad, H. and Berciu, M. (2012). *Phys. Rev. B* 85:165117.
66 Fehske, H., Schneider, R. and Weisse, A. (eds.) (2008). *Computational Many-Particle Physics.* Berlin-Heidelberg:Springer.
67 Weisstein, E.W. (2020). MathWorld, Normal Sum Distribution. http://mathworld.wolfram.com/NormalSumDistribution.html (Accessed, 8 December 2020)
68 Jarrel, M. and Gubernatis, J.E. (1996). *Phys. Rep.* 269:133.
69 Tikhonov, A.N. and Arsenin, V.Y. (1977). *Solutions of Ill posed Problems.* Washington:Winston.
70 Perhttp://arxih.org/abs/math-ph/0302045 (Accessed, 8 December 2020)
71 Mishchenko, A.S., Prokof'ev, N.V. and Svistunov, B.V. (2001). *Phys. Rev. B.* 64:033101.

72 Mishchenko, A.S., Prokof'ev, N.V., Sakamoto, A. and Svistunov, B.V. (2001). *Int. J. Mod. Phys. B* 15:3940.

73 Mishchenko, A.S., Nagaosa, N., Prokof'ev, N.V. et al. (2002). *Phys. Rev. B* 66:020301(R).

74 Mishchenko, A.S., Nagaosa, N., Prokof'ev, N.V. et al. (2003). *Phys. Rev. Lett.* 91:236401.

75 Mishchenko, A.S. and Nagaosa, N. (2004). *Phys. Rev. Lett.* 93:036402.

76 Mishchenko, A.S. and Nagaosa, N. (2006). *Phys. Rev. B* 73:092502.

77 Mishchenko, A.S. and Nagaosa, N. (2006). *J. Phys. Chem. Solids* 67:259.

78 De Filippis, G., Cataudella, V., Mishchenko, A.S. et al. (2006). *Phys. Rev. Lett.* 96:136405.

79 Mishchenko, A.S. (2006). *Proceedings of the International School of Physics.* Enrico Fermi, Course CLXI, 177–206.

80 Mishchenko, A.S. and Nagaosa, N. (2007). Polarons in Complex Matter. In: *Material Science,* (ed. A.S. Alexandrov), 503–544. Springer.

81 Cataudella, V., De Filippis, G., Mishchenko, A.S. and Nagaosa, N. (2007). *Phys. Rev. Lett.* 99:226402.

82 Mishchenko, A.S. (2008). Diagrammatic Monte Carlo and Stochastic Optimization Methods for Complex Composite Objects in Macroscopic Baths. In:*Computational Many-Particle Physics* (ed. H. Fehske, R. Scheider and A. Weisse), 367–395. Berlin-Heidelberg:Springer.

83 Mishchenko, A.S., Nagaosa, N., Shen, Z.-X. et al. (2008). *Phys. Rev. Lett.* 100:166401.

84 Cataudella, V., De Filippis, G., Mishchenko, A.S. and Nagaosa, N. (2009). *J. Supercond. Nov. Magn.* 22:17.

85 Mishchenko, A.S. (2009). *Usp. Phys. Nauk* 179:1259.

86 Mishchenko, A.S. (2010). *Advances in Condensed Matter Physics.* 2010:306106.

87 Goodvin, G.L., Mishchenko, A.S. and Berciu, M. (2011). *Phys. Rev. Lett.* 107:076403.

88 Mishchenko, A.S, Nagaosa, N., Shen, K.M. et al. (2011). *Europhys. Lett.* 96:57007.

89 Hafermann, H., Brener, S., Rubtsov, A.N. et al. (2009). *J. Phys. Condens. Matter* 21:064248.

90 Hafermann, H., Katsnelson, M.I. and Lichtenstein, A.I. (2009). *Europhys. Lett.* 85:37006.

91 Gorelov, E., Karolak, M., Wehling, T.O. et al. (2010). *Phys. Rev. Lett.* 104:226401.

92 Gorelov, E., Koloren, J., Wehling, T. et al. (2010). *Phys. Rev. B* 82:085117.

93 Bao, Z. and Locklin, J. (2007). *Organic Field-Effect Transistors.* Boca Raton:CRC Press.

94 Parisse, P., Ottaviano, L., Delley, B. and Picozzi, S. (2007). *J. Phys. Condens. Matter* 19:106209.

95 Ohashi, N., Tomii, H., Matsubara, R. et al. (2007).*Appl. Phys. Lett.* 91:162105.

96 Hasegawa, T. and Takeya, J. (2009). *Sci. Tech. Adv. Mater.* 10:024314.

97 Yang, Y.S., Kim, S.H., Lee, J.-I. et al. (2002). *Appl. Phys. Lett.* 80:1595.

98 Lang, D.V., Chi, X., Siegrist, T. et al. (2004). *Phys. Rev. Lett.* 93:086802.

99 Goldman, C., Krellner, C., Pemstich, K.P. et al. (2006).*J. Appl. Phys.* 99:034507.

100 Matsushima, T., Yahiro, M. and Adachi, C. (2007). *Appl. Phys. Lett.* 91:103505.

101 Verlaak, S. and Heremans, P. (2007). *Phys. Rev. B* 75:115127.

**102** Kang, J.H., da Silva Filho, D., Bredas, J.-L. and Zhu, X.-Y. (2005). *Appl. Phys. Lett.* 86:152115.

**103** Veres, J., Ogier, S.D., Leeming, S.W. et al. (2003). *Adv. Funct. Mater.* 13:199.

**104** Houili, H., Picon, J.D., Zippiroli, L. and Bussac, M.N. (2006). *J. Appl. Phys.* 100:023702.

**105** Troisi, A. and Orlandi, G. (2006). *Phys. Rev. Lett.* 96:086601.

**106** Troisi, A. (2010). *Phys. Rev. B* 82:245202.

**107** Matsubara, R., Sakai, M., Kudo, K. et al. (2011). *Org. Electron.* 12:196.

# 9

# Second Harmonic Generation Spectroscopy

*Takaaki Manaka and Mitsumasa Iwamoto*

*Tokyo Institute of Technology, O-okayama, Meguro-ku, Tokyo, Japan*

**CHAPTER MENU**



## 9.1 Introduction

Performance of the organic devices such as organic field effect transistor (OFET), organic light emitting diode (OLED) and organic solar cell (OSC) has been rapidly improved due to recent development of the materials with large carrier mobility [1, 2] and high durability. Understanding of the operation mechanism of the devices is of great importance for the further improvement of performance, but is still insufficient. Recently, novel fabrication processes such as ink-jet printing and roll-to-roll flexographic printing have been attracting much attention. Understanding fundamental carrier processes is also helpful for the development of such state-of-the-art devices. In this chapter, we introduce novel techniques for the study of organic devices on the basis of nonlinear optical measurements, electric-field induced second harmonic generation (EFISHG). In particular, we have developed a method for visualizing the electric field and the transient carrier motion in organic semiconductor materials and devices.

For an inorganic semiconductor such as silicon, carrier density is governed by the impurity doping, and the carrier distribution is ruled by the Fermi-Dirac distribution statistics. In such devices, carrier density at a certain point is determined by the electric potential at that point. In such cases, direct observation of a potential distribution in the device is meaningful. Therefore, probe microscope techniques coupled with the Kelvin-surface potential measurement such as Kelvin-force microscope (KFM) have been used to evaluate the potential distribution in the channel of metal-oxide semiconductor (MOS) FET [3–6]. On the other hand, the conventional organic semiconductor, e.g. pentacene and phthalocyanine, initially has low-carrier density. The doping mechanism of an organic semiconductor differs from conventional inorganic semiconductors, and the precise control of the carrier density by a chemical doping process is quite difficult. Further, dielectric relaxation time ($\varepsilon/\sigma$: $\varepsilon$: dielectric constant, $\sigma$: conductivity) of the organic materials is quite long. In such situations, carrier distribution in the organic semiconductor is far from the thermal equilibrium, and thus differs from that predicted from the Fermi-Dirac distribution statistics. These issues motivate us to consider organic semiconductors as dielectric materials.

Information on the electric field rather than the electrical potential is important for those materials with a dielectric nature, because the electric field dominates the carrier transport phenomena in such materials whether the system obeys the Fermi-Dirac statistics or not. For instance, the basic equation that expresses a current flowing in materials, $J = en\mu E$ is still valid in materials with a dielectric nature, where $e$, $n$ and $\mu$ represent elementary charge, carrier density and mobility, respectively. Besides that, the injection properties are also governed by the electric field, and zero electric field at an injection point is an *a-priori* boundary condition for the space charge limited current (SCLC) [7]. Accordingly, it is worthwhile developing a technique which can directly evaluate an electric field distribution in materials and devices. EFISHG is one of the third-order nonlinear optical processes, and it generates second harmonic signal of the fundamental light under the presence of a DC electric field. On the basis of this technique, we proposed a novel technique for studying the device operation of the organic devices. The EFISHG measurements enable us to visualize the electric field distribution in the OFET channel with high spatial resolution [8, 9]. By coupling with the time-resolved technique, we successfully observed the carrier motion in the OFET [10–17] and the OLED [18–23] and the OPV [24–28]. This technique also opens a new way to study carrier behaviors in inorganic [29, 30] as well as organic devices.

## 9.2 Basics of the EFISHG

### 9.2.1 Macroscopic Origin of the SHG

Here, we summarize the simple theory of the EFISHG. Under the presence of the intense light such as pulsed laser, material is nonlinearly polarized, and the induced polarization is expressed as a power series of the electric field as

$$P = P_0 + \chi^{(1)}E(\omega) + \chi^{(2)}E(\omega)E(\omega) + \chi^{(3)}E(\omega)E(\omega)E(\omega) + \cdots. \tag{9.1}$$

where $P_0$ indicates the permanent dipole. $E(\omega)$ is the electric field of light. $\chi^{(1)}$, $\chi^{(2)}$ and $\chi^{(3)}$ are known as optical susceptibilities. The third- and forth-term represent the nonlinear response of the polarization to the electric field. We call $\chi^{(2)}$ and $\chi^{(3)}$ as second- and third-order nonlinear susceptibility, respectively. Here, as we assume the electric field of incident light as $E(\omega) = E_0e^{i\omega t}$, we can understand the polarization originating from $\chi^{(2)}$ and $\chi^{(3)}$ oscillates with a frequency of $2\omega$ and $3\omega$. Therefore, we will observe the second- and third-harmonic signal from the materials due to the contribution of $\chi^{(2)}$ and $\chi^{(3)}$. As the nonlinear optical susceptibilities is the tensor quantity, tensor components strongly depend on the symmetry of the material. If the material has inversion symmetry, second-order nonlinear optical susceptibility vanishes under the electric dipole approximation, and thus we cannot observe the SHG signal from such materials. However, under specific conditions, we can observe the SHG signal from the centrosymmetric materials. One of these conditions is the higher-order contribution beyond the electric-dipole approximation [31], such as the electric quadrupole [32, 33] and magnetic dipole contribution [34]. The other condition is the symmetry breaking owing to external stimuli such as the external static electric field. The latter contribution plays an important role to study the operation of the organic semiconductor device.

Under the presence of static electric field, the field may induce effective polarization in the material. These induced polarizations are the origin of the EFISHG. Depending on the type of induced polarization, we have to consider the various origins of the EFISHG. If the material consists of dipolar molecules, though these molecules orient randomly, static electric field is likely to align the dipolar molecules. Actually, this is the conventional EFISHG, and is used to estimate the second-order molecular hyperpolarizability of the dipolar molecules. In this measurement, the dipolar molecules are dissolved in a solvent, and the static electric field is applied to the solution in order to align the dipoles. By analyzing Maker fringe oscillation of the SHG intensity, we can evaluate the molecular hyperpolarizability [35, 36]. Besides that, the side-chain orientation in poled polymer is also studied by using the EFISHG technique [37]. Poled polymer is the important material for the optical application of polymers such as the electro-optical devices, where the dipolar side-chains of the polymer align with the electric field above the glass transition temperature of the polymers. After poling, the temperature will be decreased below the glass transition temperature to fix the side-chain orientation. We can observe the SHG signal from the polymers evidently indicating the broken inversion symmetry. In both cases, orientational polarization is the major origin of the SHG signal. On the other hand, there is no orientational polarization induced by the static electric field for the molecule with inversion symmetry such as phthalocyanine and pentacene. If the molecules possess inversion symmetry, electronic polarization mainly contributes to the induced polarization. The static electric field may distort the electron distribution of the molecule, and the effective polarization will be induced. In this sense, the centrosymmetry is broken by the static electric field, and we may observe the SHG signal. In other words, a centrosymmetric molecule initially has no second-order molecular hyperpolarizability, but possesses the effective hyperpolarizability under the static electric field.

Nonlinear optical polarization due to the EFISHG process is expressed as

$$P_i^{2\omega} = \chi_{ijk}^{(2)} E_j^{\omega} E_k^{\omega} + \chi_{ijkl}^{(3)} E_j^{0} E_k^{\omega} E_l^{\omega}. \tag{9.2}$$

where $E^0$ represents the static electric field. $\chi_{ijkl}^{(3)}$ is the nonlinear optical susceptibility tensor for the EFISHG, and the indices $ijkl$ refer to the Cartesian components of the electric fields. It should be noted that $P_i$ oscillates with the frequency of $2\omega$, and thus we call the SHG. For the EFISHG process, the second harmonic signal of the fundamental light is produced. However, the EFISHG is actually the third-order nonlinear optical process in a broad sense, because the third-order nonlinear optical susceptibility $\chi_{ijkl}^{(3)}$ describes the EFISHG process. As a result, the SHG intensity for the EFISHG is proportional to the external field as

$$I^{2\omega} \propto \left| \chi^{(3)} E(0) E(\omega) E(\omega) \right|^2 \tag{9.3}$$

Accordingly, the electric field in the organic semiconductor material can be evaluated by the EFISHG measurement.

### 9.2.2 Microscopic Description of the SHG

To understand the microscopic origin of the SHG, a quantum mechanical perturbation theory is used to derive explicit expression for the molecular hyperpolarizability. Since the molecular hyperpolarizability is originating from nonlinear interaction between electrons in molecule and the electric field of light, it depends on the material parameter such as the dipole transition moment and the resonant frequency. The nonlinear optical susceptibility for the EFISHG, $\gamma_{\mathrm{ijkl}}$, is expressed as

$$\gamma_{ijkl}^{DC} \propto \sum_{m,n,o} \left( \frac{\langle g|\hat{\mu}_i|o\rangle\langle o|\hat{\mu}_j|n\rangle\langle n|\hat{\mu}_k|m\rangle\langle m|\hat{\mu}_l|g\rangle}{(\omega_{og}-2\omega)(\omega_{ng}-2\omega)(\omega_{mg}-\omega)} + \frac{\langle g|\hat{\mu}_j|o\rangle\langle o|\hat{\mu}_i|n\rangle\langle n|\hat{\mu}_k|m\rangle\langle m|\hat{\mu}_l|g\rangle}{\omega_{og}(\omega_{ng}-2\omega)(\omega_{mg}-\omega)} \right.$$
$$\left. + \frac{\langle g|\hat{\mu}_j|o\rangle\langle o|\hat{\mu}_k|n\rangle\langle n|\hat{\mu}_i|m\rangle\langle m|\hat{\mu}_l|g\rangle}{\omega_{og}(\omega_{ng}+2\omega)(\omega_{mg}-\omega)} + \frac{\langle g|\hat{\mu}_j|o\rangle\langle o|\hat{\mu}_k|n\rangle\langle n|\hat{\mu}_l|m\rangle\langle m|\hat{\mu}_i|g\rangle}{\omega_{og}(\omega_{ng}+2\omega)(\omega_{mg}+\omega)} \right) \tag{9.4}$$

Under the electric dipole approximation, optical transition between identical parity states is forbidden. Namely, the transition dipole moment, $\langle n|\hat{\mu}|m\rangle$, has to take the finite value only when the parity of $|n\rangle$ state differs from that of $|m\rangle$ state. From the above equation, we can find that the resonance enhancement occurs when the SH photon energy is coincident with the energy difference between the ground state and the allowed excited state. This implies that the choice of the wavelength is quite important to effectively observe the EFISHG signal. Generally, SHG wavelength should be approached to the absorption peak to promote the resonance enhancement. For better understanding, the schematic images of the SHG process for to the EFISHG are illustrated in Figure 9.1(b). The diagram for the EFISHG process is depicted from the first term of Eq. (4). For reference, the THG process is illustrated in Figure 9.1(a). It should be noted that state $|n\rangle$ overlies $|p\rangle$ state in Figure 9.1(b), because one of the photon energies is considered to be zero for the EFISHG process. Considering the diagram, we can easily understand that the resonance enhancement may occur when the SH photon energy is coincident with the energy difference between the ground state and the allowed excited state for the EFISHG.

The macroscopic nonlinear optical susceptibilities are expressed using the ensemble-average of the molecular hyperpolarizabilities as $\chi_{ijkl}^{(3)} = N\langle T_{ijkl}^{\mu\nu\xi o}\rangle\gamma_{\mu\nu\xi o}$, where $N$ is

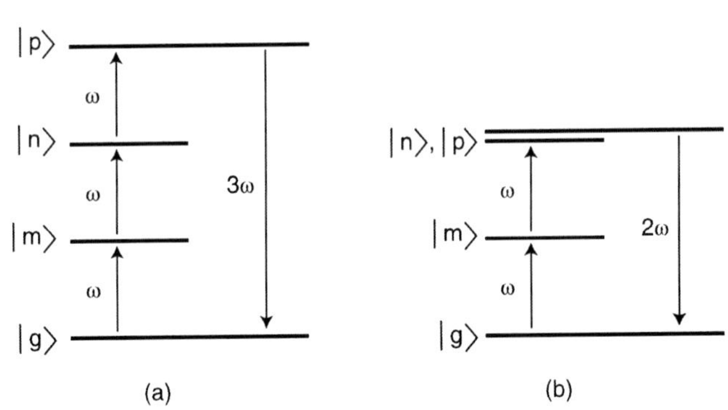


**Figure 9.1** Schematic images of the transition process in the THG and the EFISHG.

the volume density of molecules. $T_{ijkl}^{\mu\nu\xi o}$ describes the coordinate transformation between the molecular system and the laboratory system. And the transformation matrices are conventionally expressed as the Euler rotation matrices [38]. $\langle\rangle$ denotes an ensemble-average over the molecular orientations, and $\langle T_{ijkl}^{\mu\nu\xi o}\rangle$ is expressed explicitly using orientational distribution function $f(\phi, \beta, \varphi)$ as,

$$\left\langle T_{ijkl}^{\mu\nu\xi o}\right\rangle = \frac{\int_0^{2\pi}\int_0^{\pi}\int_0^{2\pi} T_{ijkl}^{\mu\nu\xi o} f(\phi,\beta,\varphi)\sin\beta d\phi d\beta d\varphi}{\int_0^{2\pi}\int_0^{\pi}\int_0^{2\pi} f(\phi,\beta,\varphi)\sin\beta d\phi d\beta d\varphi} \tag{9.5}$$

where $\phi$, $\beta$ and $\varphi$ represent the Euler angels between the molecular system and the laboratory system. It should be noted that complexity of the above calculation can be reduced depending on the structure of the films, that is, macroscopic symmetry of the film. If the film is amorphous or consists of randomly-oriented molecules, we can assume $f(\phi, \beta, \varphi) = 1$.

### 9.2.3 EFISHG Measurements

A schematic image of the experimental setup for the EFISHG measurement is shown in Figure 9.2(a). The light source for the EFISHG measurement is a femtosecond optical parametric amplifier (OPA: Coherent OPerA Solo) pumped by a Ti:sapphire regenerative amplifier system (Coherent: Libra). The pulse width and repetition rate of the system are 80 fs and 1 kHz, respectively. We can also use the Q-switched YAG laser with a pulse width of a few nanosecond, instead of a femtosecond laser system. The advantage of using the femtosecond laser system is to improve the signal-to-noise ratio of the measurement and to reduce measurement time. The wavelength of the fundamental light was chosen so that the SHG signal can be observed effectively from the devices. For instance, an appropriate wavelength for the pentacene FET device with a 500-nm-thick $SiO_2$ insulator is 1120 nm. The fundamental light from the OPA is focused onto a channel region of the OFET by using an objective lens with a long working distance (Mitutoyo: M Plan Apo SL20x, numerical aperture = 0.28, working

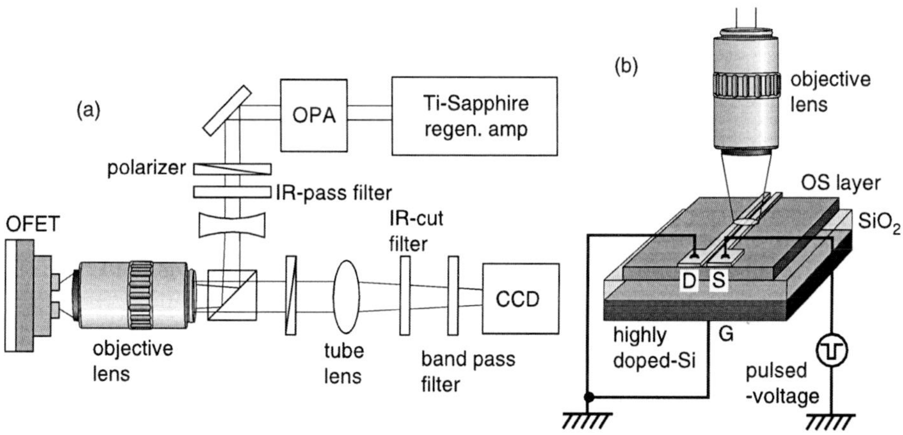


**Figure 9.2** (a) Schematic images of the optical setup of the SHG measurement. (b) Device structure and the electrical connection for the TRM-SHG measurement.

distance = 30.5 mm). The effective spot size of the fundamental light is approximately 150 μm so that the laser completely irradiates the OFET channel between the source and the drain electrodes (see Figure 9.2(b)). The SH light generated from the sample is filtered by the infrared-cut and interference filters to eliminate the fundamental light and other unnecessary light, and it is detected by a cooled charge-coupled device (CCD) camera (Andor Technology: DU971P-BV).

For visualizing carrier motion in the OFET, the EFISHG measurement is employed in combination with the time-resolved technique. Thus, we call this technique time-resolved microscopic SHG (TRM-SHG) measurement. A schematic image of the equipment setup is shown in Figure 9.3(a). The TRM-SHG measurement is similar to the pump-probe measurement in the laser spectroscopy. Electrical stimulus such as pulsed-voltage is applied to the devices instead of the pump beam excitation in the pump-probe measurement. During the TRM-SHG measurement for the OFET, pulsed-voltages are continuously applied to the source electrode (see Figure 9.2(b)), and the pulsed laser light from the OPA is impinged on the device after applying each voltage pulse. The voltage pulses are completely synchronized with the laser pulses. As shown in Figure 9.3(b), delay time is defined as the time difference between the rise-up edge of the voltage pulse and the laser pulse. Then, we can take snapshots of the SHG image from the device with different delay times. The delay time should be strictly controlled by using a precise delay generator to maintain accuracy of the TRM-SHG measurement. Temporal resolution is determined by the pulse-width of laser as well as the rise time of the voltage pulse, that is, the slew rate of the pulse generator and the high-speed amplifier. For the current setup, the temporal resolution is approximately 5 ns, when we use the high-speed pulse generator (Avtech: AV-1011B1-B).

For the electric field imaging and visualizing carrier motion, the devices used here were the top-contact OFETs. We chose pentacene and 6,13-Bis(triisopropylsilylethynyl) (TIPS)-pentacene known as p-type organic semiconductor materials. Both molecules have inversion symmetry, and intrinsically produce no SHG signal under electric dipole approximation. Heavily-doped Si-wafers coated with a 500-nm-thick $SiO_2$ layer are used as substrates. The pentacene layers, approximately 100 nm thick, were deposited on an $SiO_2$ surface. The process pressure during deposition of pentacene was kept at less than

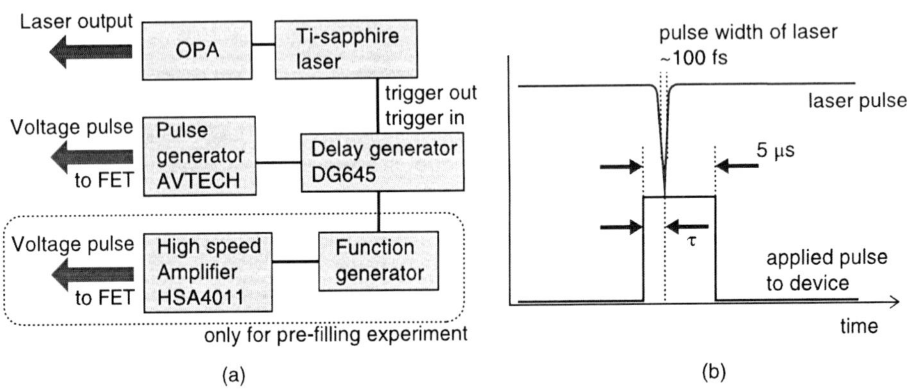


**Figure 9.3** Schematic images of the SHG setup transition process in the THG and the EFISHG.

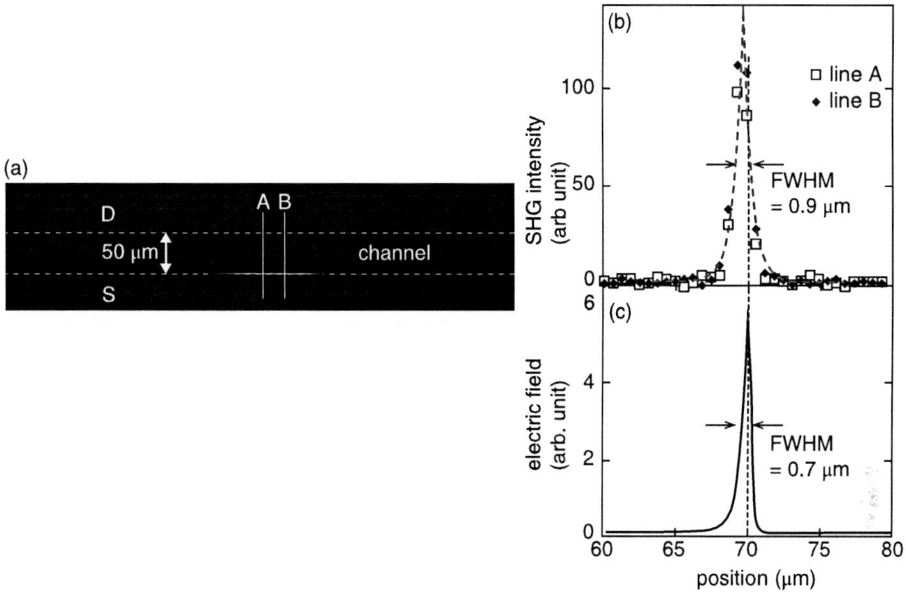


**Figure 9.4** (a) SHG image obtained from the channel of pentacene FET under the application of negative pulse to the source electrode. (b) SHG intensity profile across the channel by taking a line scan at point A and B. (c) Theoretically calculated in-plane component of the electric field distribution in the pentacene layer.

$1 \times 10^{-4}$ Pa, and the deposition rate was controlled at approximately 3 nm/min. The substrate temperature was not controlled during the evaporation. For the TIPS-pentacene layers, they were deposited onto the $SiO_2$ layer by the dip-coating method from a toluene solution with a concentration of 0.5 wt%. The dipping speed is 8.5 μm $s^{-1}$. After the deposition of organic semiconductor layers, gold layers with a thickness of 100 nm are deposited as the source and drain electrodes by vacuum evaporation. The designed channel length ($L$) and width ($W$) are 30 μm and 1.5 mm, respectively, for typical OFET measurements.

### 9.2.4 Evaluation of In-plane Electric Field in OFET

Figure 9.4(a) shows the SHG image obtained from the channel of pentacene FET under the application of negative pulse to the source electrode, i.e., in the off-state of the FET. As shown in Figure 9.4(a), a strong SHG emission was observed at an edge of source electrode. Negative voltage applied between the source and the gate electrodes produced a strong in-plane component of the electric field at the edge of the source electrode due to the edge effect, and it activated the SHG at the edge of the source electrode. Figure 9.4(b) indicates the SHG intensity profile across the channel by taking a line scan at point A and B in Figure 9.4(a). Figure 9.4(c) shows the in-plane component of the electric field distribution in the pentacene layer theoretically calculated by solving the Laplace equation. The sharpness of the SHG peak at the edge of source electrode is quantified in terms of the full width at half maximum (FWHM) values of the signal profile. The FWHM values of the SHG profile and the electric field distribution are estimated as 0.9 μm and

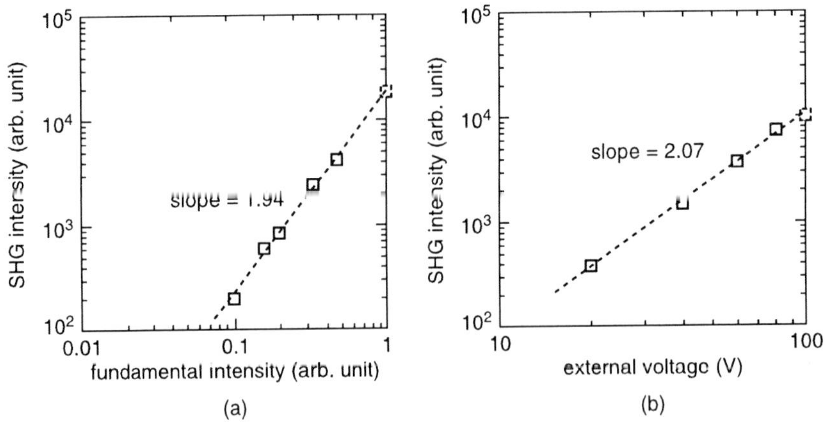


**Figure 9.5** (a) Fundamental power and (b) External voltage a dependence of the SHG intensity.

0.7 μm, respectively. The electron injection from Au electrode to the pentacene layer is prohibited, due to the large injection barrier for electron at the pentacene/Au interface. The energy difference between the lowest unoccupied molecular orbital (LUMO) of pentacene and the work function of Au electrode was reported as 2.7 eV [39]. Thus, the strong electric field remains around the edge of the electrode during voltage application, and it activated the SHG signal around the edge of the source electrode as shown. Figure 9.5(a) and 9.5(b) show log-log scale plot of the external voltage and the fundamental power dependence of the SHG intensity. Intensity of the fundamental light was changed using neutral density filters. As expressed in Eq. (3), SHG intensity will be proportional to the square of the static electric field and is the fourth power of the electric field of light. The latter means that the SHG intensity will be proportional to the square of the fundamental power. Figure 9.5(a) and 9.5(b) clearly indicates that SHG intensity is proportional to the square of both external voltage and fundamental intensity, implying that the SHG signal originates from the EFISHG process. The width of the SHG peak at the edge of the source electrode did not depend on the external voltage and the fundamental intensity. In principal, the electric field magnitude can be evaluated directly from the SHG intensity using Eq. (3). But actually, field evaluation from the SHG image is ambiguous, because the SHG intensity is influenced by many parameters such as laser power, the nonlinear optical susceptibility and refractive index of the materials. It is difficult to know the absolute value of all parameters. For that reason, the magnitude of the electric field was evaluated by comparing the square root of the SHG intensity with the electric field obtained by solving Laplace equation [40]. Laplace electric field distribution is uniquely and rigorously determined if the geometric information of the electrode configuration, the dielectric constant and the electric potential are provided.

### 9.2.5 Direct Imaging of Carrier Motion in OFET

According to our previous results, SHG intensity distribution of the on-state of the FET completely differed from that in the off-state [8]. This is due to the electric field contribution from the injected charge, i.e., space charge field. Gauss's law in Maxwell's

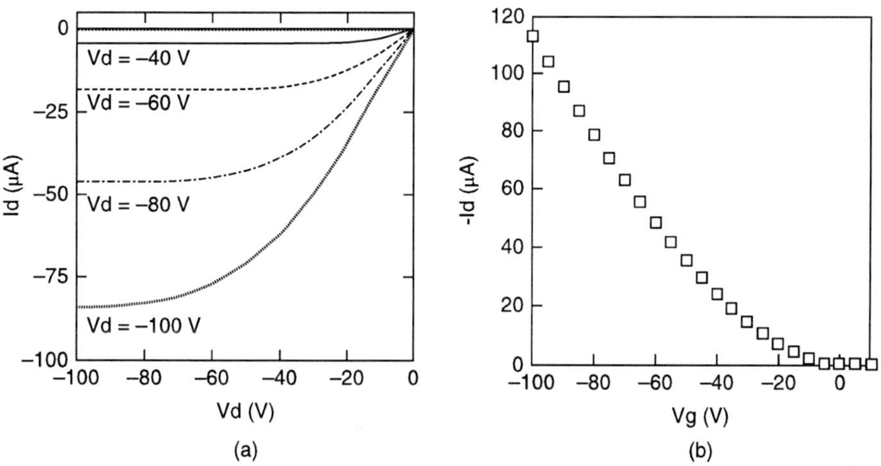


**Figure 9.6** (a) Output and (b) transfer characteristics of pentacene FET used in the SHG experiment.

electromagnetic theory denotes that the electric flux density originates from charge carriers. Thus, the electric field in the on-state is the sum of the electric field in the off-state and the electric field created by the charges. This motivated us to evaluate the carrier distribution in the device by analyzing the electric field distribution in the on-state. In particular, we can evaluate the transient carrier transport phenomena in the material by analyzing the transient change of the electric field distribution. This is the principle of the visualizing carrier motion by EFISHG measurements in combination with the time-resolved technique (TRM-SHG).

Figure 9.6(a) and 9.6(b) show the output and transfer characteristics of pentacene FET used in the SHG experiment, respectively. As shown in the figure, p-channel FET characteristics were observed in a manner often reported in the pentacene FET [41]. Field-effect mobility ($\mu_{FET}$) and threshold voltage ($V_{th}$) is evaluated as 0.10 $cm^2/Vs$ and −20 V, respectively. Possible reasons for such small mobility compared with reported values are due to the small grain size and the presence of a contact resistance. According to the atomic force microscope (AFM) measurement, grain size of the present film is approximately 500 nm. Grain size of deposited film depends on the substrate temperature, specifically, high temperature substrate is required to form large size domains [42].

Figure 9.7 (a) and 9.7 (b) show the transient changes in the SHG image from the pentacene FET channel under the positive and negative pulses application to the source electrode, respectively. As shown in Figure 9.7(a), the emission band of the SHG moves from the edge of the source electrode to the drain side. This indicates that the injected holes from the source electrode transport in the FET channel with an increase of delay time. Positive voltage applied to the Au electrode injects holes easily into pentacene because of the energy difference between Au work function and pentacene HOMO level. Transient carrier behavior can be analyzed theoretically by solving a 2D drift-diffusion equation. Based on the theoretical calculation, we previously reported that the transient peak position of the SHG intensity moves along the channel in a diffusion-like manner, corresponding to the square root of time dependence [12, 13]. This provides insight into the driving force of carrier. That is, the driving force is the electric field created by injected charges, not a source-drain field. It should be noted that the equivalent circuit

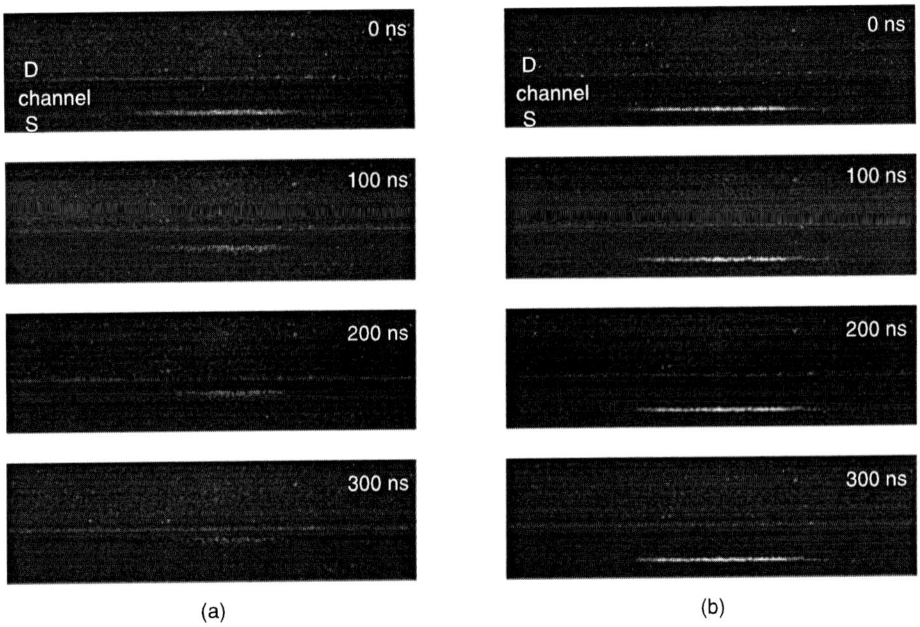


**Figure 9.7** Time-resolved SHG image under (a) positive and (b) negative voltage pulse application to the source electrode. Amplitude of voltrage pulses were 100 V.

model, based on the ladder connection of the resistance and capacitance, successfully demonstrated the square root of time dependence of the transient carrier transport [43]. By assuming that the electric field in the channel is expressed as $V_s/x$, we have derived a simple formula for the carrier mobility as

$$\mu = \frac{x^2}{2V_s}t \tag{9.6}$$

where $t$ denotes the time when we observe the front of the SHG emission band at position $x$ and $V_s$ is the source pulse voltage. Using this equation, carrier mobility of this sample is obtained as 0.17 $cm^2/Vs$. It is noteworthy that the spatial and temporal resolutions of the present system are approximately 0.5 μm and 5 ns depending on a laser and an electrical system, respectively.

## 9.3 Some Application of the TRM-SHG to the OFET

### 9.3.1 Trap Effect

A number of papers reported that optimizing the gate insulator could enhance the carrier mobility of the device [44–47]. Among these, control of the trap states at the gate insulator surface is important to improve mobility. In this section, we showed that visualizing carrier motion in the OFET channel is also effective in observing the trap states in the OFET. Here, we introduce the difference in the transient electric field distribution between the pentacene FET with and without PMMA thin layer on

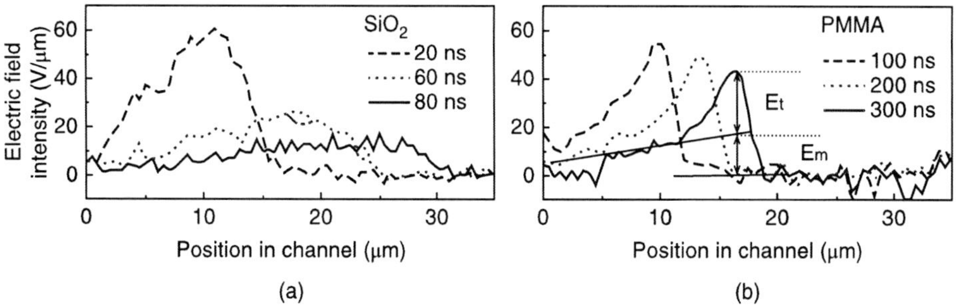


**Figure 9.8** Time-evolution of SHG distribution from the pentacene FET channel with different insulators.

the $SiO_2$ surface. Hereafter, the former and latter devices are respectively denoted as PMMA and $SiO_2$ device.

Figure 9.8(a) and 9.8(b) show the electric field distributions obtained by the TRM-SHG measurement in (a) the $SiO_2$ and (b) the PMMA device. For the PMMA device, the PMMA layer with a thickness of 50 nm was deposited on the $SiO_2$. The electric field distributions at three different delay times were plotted for the PMMA and $SiO_2$ device. We found, at a glance, that their shapes were significantly different for the samples with different insulators. In the PMMA device, we saw the sharp peaks at the front edge of field distribution. On the contrary, the profiles were broad and we did not see obvious sharp peaks at the front edge in the $SiO_2$ device. Further, the carrier velocity in the PMMA device is quite slow compared with the $SiO_2$ device (about 7 times smaller). We have supposed that the amplitude of the electric field at the peak position decayed in a power function with an exponent about -0.5 [12]. To justify this theoretical prediction, we plotted the time-dependence of the electric field at the peak position $E_x(x)$ for the same samples in a log-log scale in Figure 9.9(a). For the $SiO_2$ device, the experimental data are strongly consistent with the theoretical prediction. However, a single power function cannot fit the data for the PMMA device, indicating that a simple drift-diffusion treatment may not be available to describe the electric field decay in the PMMA device.

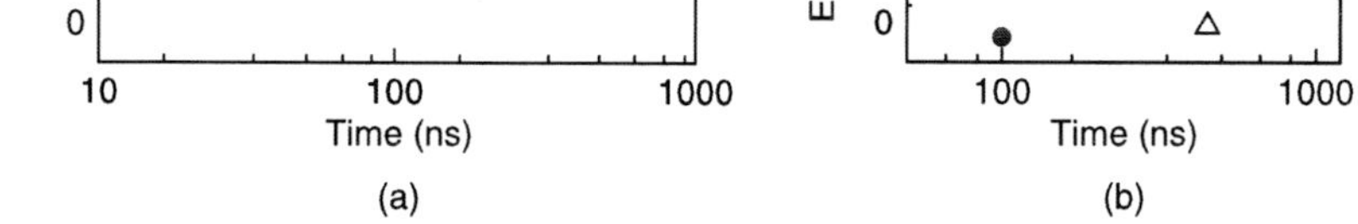


**Figure 9.9** (a) Time-evolution of the electric field at the peak position and (b) the electric field components $E_t$ and $E_m$.

When we carefully observed these electric field distributions measured in the PMMA device, we found that these distributions seem to be a superposition of two fields: one increases slightly without remarkable peak and contributes the linear-like region near the source electrode, which is quite similar to the field distribution observed in the $SiO_2$ device, another contributes a sharp peak and decays rapidly away from the front peak. The solid thin line in Figure 9.8(b) illustrated a possible partition, where we defined two values of the two fields at the peak position $x$, $E_t$ and $E_m$, respectively. Figure 9.9(b) replot the time decay of the electric field components $E_t$ and $E_m$. Interestingly, $E_m$ decreases monotonically, while $E_t$ increases dramatically and then saturates after a certain time. In addition, we also saw that the shape of the peak in the PMMA device seems to be invariant with the delay time and the source-gate voltages. Then, it is reasonable to consider that the carrier behavior observed in the PMMA device is attributed to the carrier traps at the pentacene/insulator interface. When we measure the threshold voltage of the device, the $SiO_2$ device often shows the large positive $V_{th}$, whereas the PMMA device shows the negative $V_{th}$. This implies that the number of empty traps in the PMMA device at $V_g$ =0 is larger than that in the $SiO_2$ device, because some trap sites are already filled in the $SiO_2$ device.

To confirm validity of the above argument, we carried out a new experiment called a "prefilling" experiment. In this experiment, before the application of the positive voltage to the source electrode, negative pulse is applied to the gate electrode to inject hole. Figure 9.10(a) illustrates the schematic diagram of the prefilling experiment. The negative gate voltage $V_g$ is applied to inject holes prior to the application of source voltage pulse. If the PMMA device has a high trap density, the injected carriers by applying the gate voltage would fill the vacant traps during the prefilling process. Namely, the transient response of the prefilled device should almost be similar to that of the $SiO_2$ device, because the number of vacant traps will be reduced.

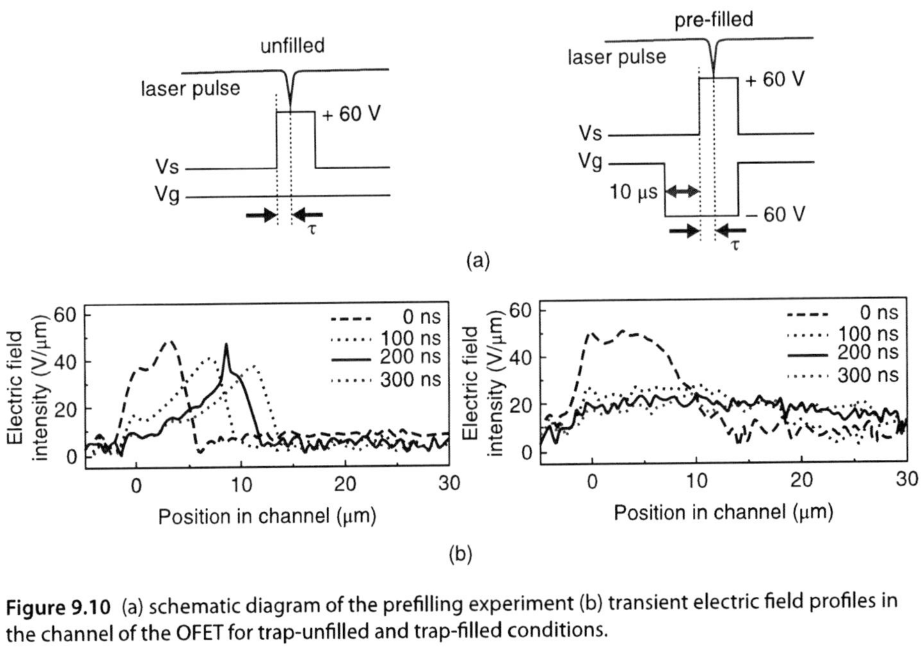


**Figure 9.10** (a) schematic diagram of the prefilling experiment (b) transient electric field profiles in the channel of the OFET for trap-unfilled and trap-filled conditions.

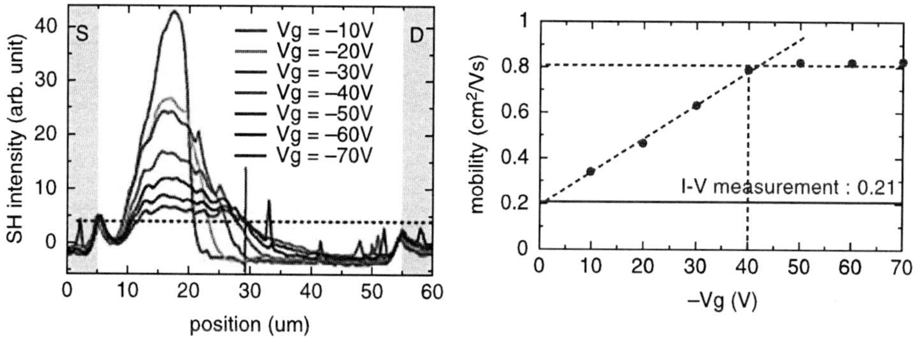


**Figure 9.11** SHG intensity profile with different gate voltages from $V_g$ =−10 V to $V_g$ =−70 V for prefilling experiment (b) prefilling voltage dependence of the carrier mobility.

Figure 9.10(b) shows the transient electric field profiles in the channel of the OFET for trap-unfilled and trap-filled conditions. We found that the sharp peak in the transient electric field completely disappeared in the trap-filled case and the whole profile was quite similar to those shown in Figure 9.8(a). We see that the mobility under the trap-filled condition was approximately 10 times greater than that of the unfilled condition. This is almost consistent with the mobility difference between bare $SiO_2$ and PMMA devices. We can evaluate the trap density from the prefilling experiment. Figure 9.11(a) shows the SHG intensity profile with different gate voltages from $V_g$ =−10 V to $V_g$ =−70 V. To take these data, delay time was

fixed at 50 ns. We see a single peak appeared at a position around 15 μm from the source electrode. This peak gradually decreases with the increase of the gate voltage $V_g$, accompanying the spread of the profile. These results suggest that the increase of $V_g$ results in the decrease of empty trapping sites. Consequently, carriers injected from the source can move smoothly under the prefilling condition. Figure 9.11(b) represents the prefilling voltage dependence of the carrier mobility estimated using Eq. (6). Interestingly, mobility monotonously increases in the region -10 V > $V_g$ > -40 V as $V_g$ decreases, and eventually saturates in the region $V_g$ < -40 V. This reasonably indicates that all empty traps are filled at the saturation voltage. Accordingly, trap density is evaluated from this saturation voltage as $1.4\times10^{12}/cm^2$.

### 9.3.2 Metal Electrode Dependence

For the OFET device operation, carrier transport process successively takes place after carrier injection. Thus, both injection and transport processes should be optimized to improve organic device performance. To do that, discriminating evaluation of the injection and transport processes is also important. From a single current-voltage measurement, injection and transport processes cannot be separately evaluated, because the contact resistance and channel resistance are connected in series in the equivalent circuit of OFET. Consequently, a field effect mobility that is evaluated from $I_{ds}$-$V_{gs}$ characteristics depend on the electrode metals. For instance, large carrier mobility is likely to be reported using large work function electrodes, suggesting that a close relationship exists between work function of electrode and the mobility. However, this is a strange situation, because the carrier motility essentially depends only on the semiconductor

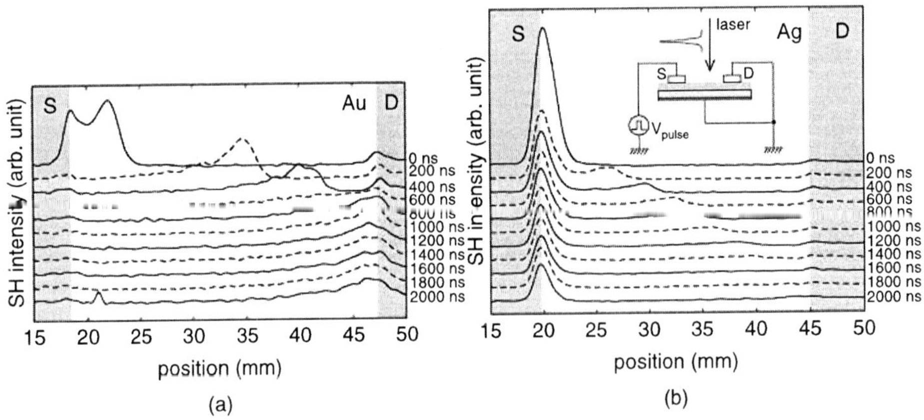


**Figure 9.12** Time evolution of the SHG intensity distribution along the FET channel for the device using (a) Au and (b) Ag top-electrodes. Positive voltage pulses with an amplitude of 100 V were applied to the source electrode.

material. In this section, TRM-SHG measurement is conducted to reveal the carrier injection and transport process in pentacene FET with different electrode, Au and Ag.

Figures 9.12(a) and 9.12(b) show time evolution of the SHG intensity distribution along the channel of the pentacene FET with (a) Au and (b) Ag source and drain electrodes. To inject holes, positive voltage with amplitude of 100 V was applied to the source electrode. As discussed above, the SHG peak motion in the device with Au electrode followed a diffusion-like manner. For devices using the Ag electrode, the slope in log-log scale of time evolution for the Ag device was 0.56, also indicating a diffusion-like manner, but peak moved slowly compared with the Au device. For the Au device, carrier front reached the drain electrode at a delay time of 600 ns, whereas it reached the drain electrode at approximately 2000 ns. This indicates that the time to reach steady state is approximately 2000 ns for the Ag device. Significant difference was found in the SHG intensity at the edge of the source electrode. As shown in Figure 9.12(b), the SHG signal for the Ag device remained at a steady state, whereas it completely disappeared for the Au device. These results imply that the in-plane electric field at the edge of Au electrode disappeared immediately after the hole injection. In fact, the decay rate for the Au device was less than 100 ns.

As mentioned, under the presence of the injected charges, the total electric field in the device is the sum of the electric field due to the external voltage created by the injected charges. Positive voltage applied to the Au source electrode injects holes easily into the pentacene because of the low injection barrier from the metal electrode to pentacene layer for the hole injection. In such cases, the electric field formed due to the external voltage is rapidly cancelled out by the electric field created by the injected charges. This also implies that the channel potential underneath the source electrode is approximately equivalent to the source potential for the Au electrode at a steady state. In other words, the SHG intensity at the electrode edge reflects the potential generated across the pentacene below the source electrode. In contrast, the SHG signal from the edge of the Ag electrode remained even after reaching a steady state, indicating that the electric field remained at the electrode edge even in a steady state. This is due to insufficient accumulation of carrier underneath the source electrode, and the electric field at the edge of the electrode is not fully cancelled out. This observation motivates us to compare the injection barriers between the Au and Ag electrode. The work functions of Au and Ag are 5.1 eV and 4.3 eV [48], respectively, given a pentacene with an energy gap of 2.5 eV and an ionization potential of 2.6 eV, the injection barriers for holes are 0.0 and 0.8 eV, correspondingly.

Using TRM-SHG results, mobilities of these samples were evaluated to be 0.11 and 0.06 $cm^2/Vs$ for the devices using Au and Ag electrodes, respectively. To evaluate the net mobility using the Ag electrode, a potential drop should explicitly be taken into account. Here, the potential drop represents the potential difference between the source electrode and the channel below the source electrode. Accordingly, the potential drop was estimated at approximately 8.4 V for the present case. Interestingly, the evaluated mobility of the device using the Au electrode reached almost identically to that using the Ag electrode if the measurement was performed under an application voltage of 60 V. This clearly indicates that the transport properties are identical between both different devices using the Au and Ag electrodes. Accordingly, we could show that though the carrier behavior was dependent on the work function of the metals, the evaluated carrier mobility in the channel region had no dependence on the work function of metals.

### 9.3.3 Anisotropic Carrier Transport

For the development of the organic devices, much effort has been paid to improve the carrier mobility of organic semiconductor materials. For organic molecular crystals, control of the intermolecular interaction is important to optimize the mobility. In this sense, one of the most effective ways is to use single crystalline organic semiconductors [49, 50]. Single crystals exhibit anisotropic carrier transport properties, such as a mobility anisotropy, owing to the anisotropic intermolecular interaction in the crystals such as charge transfer integrals between the adjacent molecules. The anisotropic transport property has been reported for pentacene [51] rubrene [52], 6,13-Bis(triisopropylsilylethynyl) (TIPS) pentacene, [53, 54] and dinaphtho-thiopheno-thiophenes (DNTT) [55] single crystals. For the current-voltage measurement, the radial-shaped electrode allows us to measure the angle dependence of the mobility in a single mounting [51, 55], though it requires relatively large single crystal. In this section, transient anisotropic carrier transport in a small area of the single crystal grains is shown to be directly observed by using the TRM-SHG measurement.

In the experiment, we used a round-shaped electrode, because the carriers are injected around the edge of the round-shaped electrode homogeneously, and the angular dependence of the carrier mobility is directly observed at once. Figures 9.13(b)–9.13(d) show the TRM-SHG images from the round-shaped Au electrode at various delay times. Positive voltage pulses with an amplitude of 100 V were applied to the electrode with respect to the gate electrode. At a delay time of 0 ns, the SHG signal is observed near the edge of the electrode, reflecting that the hole injection has just begun. With increasing delay time, the SHG distribution gradually spreads from the round-shaped electrode. It should be noted that the carriers spread smoothly from the round-shaped electrode despite the absence of the opposite electrode, for example, the drain electrode with respect to the source electrode. This also implies that the driving force of the carrier is the space charge field formed by the carriers injected from the round-shaped electrode.

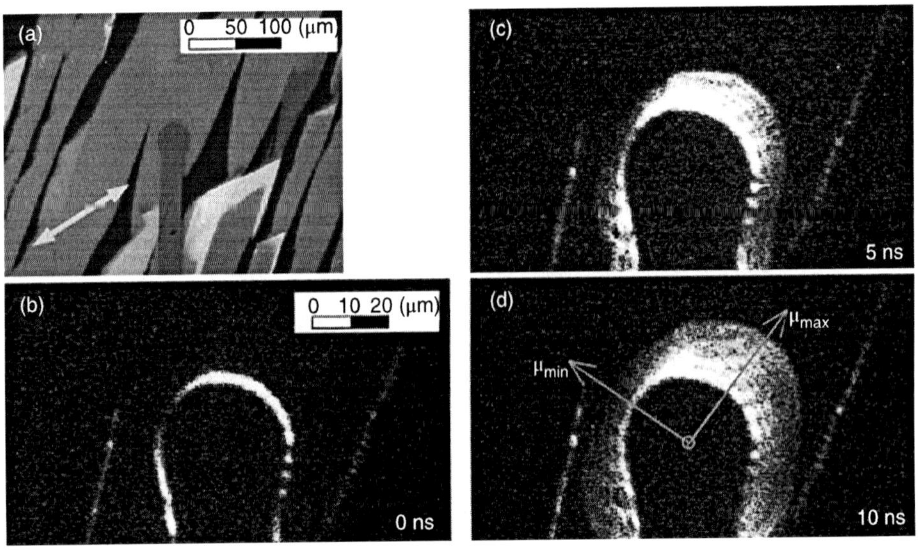


**Figure 9.13** (a) Polarized microscopic image of the round-shaped Au electrode. Extinction direction is represented using a double-sided arrow. Time evolution of the TRM-SHG images from the round-shaped electrode at delay times of (b) 0 ns, (c) 10 ns, and (d) 20 ns.

The most important result is that the shape of the SHG distribution is not a perfect circle, indicating that the carrier velocity strongly depends on the carrier transporting direction. In other words, the ellipse directly represents the anisotropic carrier transport property of the film. The carrier velocity, along the major axis of this ellipse, reaches its maximum value and that along the minor axis reaches its minimum. Thus, the maximum and minimum directions of the mobility can be uniquely determined, and the mobility anisotropy is evaluated as 4.5. The difference in the orientation of the extinction direction and of the direction of the highest mobility is approximately 20°. This difference is in agreement with the previously reported value [56].

## 9.4 Application of the TRM-SHG to OLED

Electric field in other organic devices, such as OLED and OSC can also be probed by the EFISHG measurement. However, we need to take into account the difference in the device structure between OLED, OSC and OFET. In the OFET, in-plane electric field governs the carrier transport in the device, whereas out-of-plane (longitudinal) electric field plays an important role in the sandwiched-devices such as OLED and OSC. We have to employ an oblique incidence of p-polarized light for probing electric field and carrier motion in the sandwiched-devices. Here, we briefly summarize the EFISHG measurement for OLED.

In double-layer OLED, electrons and holes are injected into the electron and hole transport layers from electrodes with appropriate work-functions, respectively. Injected carriers are transported in the device, and accumulate at the interface between the hole and electron transport layers owing to the Maxwell-Wagner effect. Some charges immediately recombine leading to electroluminescence, but some remain at the interface if

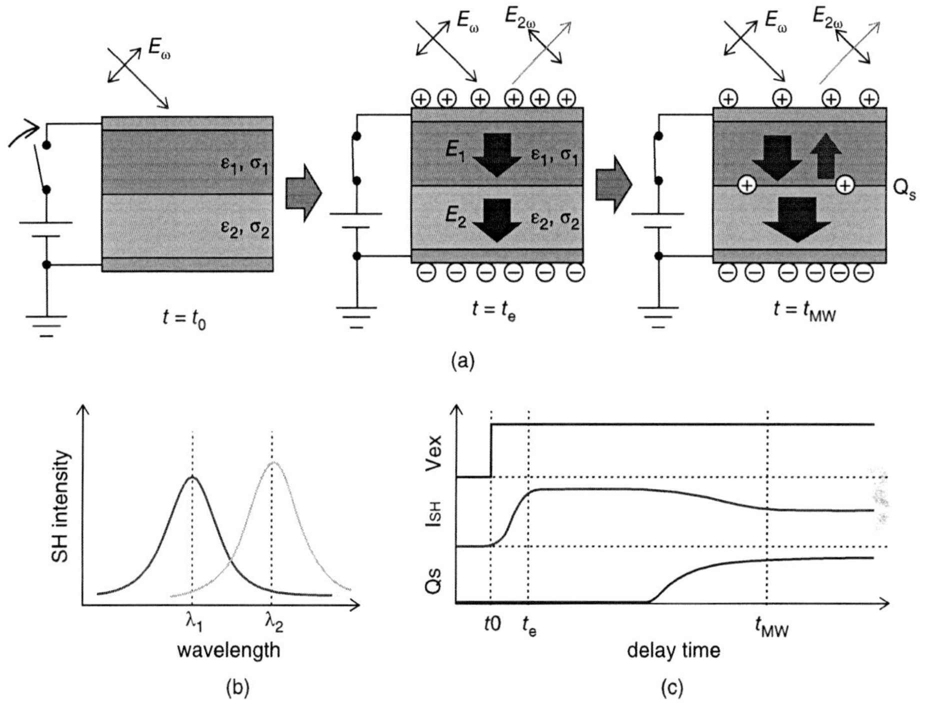


**Figure 9.14** (a) Schematic image of carrier behavior in OLED by applying pulse voltage. (b) Schematic image of the SHG spectrum of each layer of OLED. (c) Transient SHG response and the accumulated carriers with respect to the voltage pulse.

the carrier balance is not satisfied. SHG measurement is also effective to the OLED because the electric field governs the carrier transport in the device, and probing carrier behavior in constituent layers of the OLED is effective to analyze the device operation. Figure 9.14(a) represents the schematic image of carrier behavior in OLED by applying pulse voltage. In order to detect the electric field between electrodes, p-polarized fundamental light is incidental to the device, and p-polarized SHG light is detected. OLED was operated by the pulsed-voltage application. For the optical measurement, spectroscopic selectivity can be employed to analyze the carrier behavior in each layer separately. Figure 9.14(b) illustrates the SHG spectrum of double-layered structure. Each organic material has its own intrinsic optical spectrum. This allows us to probe the electric field in each layer of a double-layer structure separately by choosing the proper fundamental wavelength.

Here, we discuss the relationship between the transient SHG signal and carrier behavior in OLED under pulse voltage application. Figure 9.14(c) schematically represents the transient SHG response and the accumulated carriers with respect to the delay time. SHG intensity rapidly increases and saturates at $t = t_e$ due to the increase of electric field in the device by applying the pulsed voltage. At the same time, carrier injection starts from the electrodes, and the injected carriers are transported to the interface. During the transport process, constant SHG signal will be observed, because the average electric field in the transport layer keeps constant. After that, SHG intensity

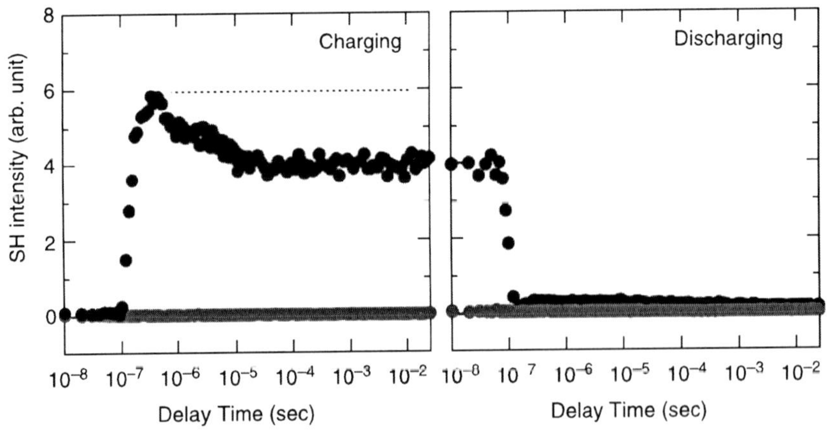


**Figure 9.15** Typical examples of the transient SHG signal in charging and discharging process of ITO/pentacene/ $\alpha$-NPD/$Alq_3$/Al structure.

begins to decrease, and again saturates at the Maxwell-Wagner time constant,$t = t_{MW}$, indicating the Maxwell-Wagner charging at the interface.

Figure 9.15 represents a typical example of the transient SHG signal from ITO/pentacene/$\alpha$-NPD/$Alq_3$/Al structure. Thicknesses of pentacene, $\alpha$-NPD and $Alq_3$ layers were 10 nm, 150 nm and 50 nm, respectively. Wavelength of the fundamental light was set to 820 nm to selectively probe the SHG signal from the $\alpha$-NPD layer. This implies that the SHG signal intensity corresponds to the electric field in an $\alpha$-NPD layer. As shown in Figure 9.15, SHG intensity increases immediately after applying the forward-bias voltage with a response time of approximately $10^{-7}$ s. This response time is in good agreement with the RC circuit time constant of a series capacitance of pentacene/$\alpha$-NPD/$Alq_3$ layers, and a lead resistance of ITO electrode. Then SHG signal decreased and saturated. This result evidently indicates that the electric field in $\alpha$-NPD decreased, because positive charges were accumulated at the $\alpha$-NPD/$Alq_3$ interface. By analyzing the SHG intensity, amplitude of the electric field in $\alpha$-NPD layer and the amount of the accumulated charge can be quantitatively obtained.

## 9.5 Conclusions

In this chapter, we briefly introduced our developed technique for visualizing electric field and carrier motion in organic semiconductor thin films on the basis of the microscopic EFISHG measurements. EFISHG is a nonlinear optical phenomenon which produces a second harmonic signal of the incident laser under the presence of the electric field. In this sense, EFISHG is nothing but a technique that observes the induced polarization. In combination with a time-resolved technique, we can visualize the carrier motion in the material, through the observation of a transient change of the electric field distribution. By using this technique, a variety of device parameter such as carrier mobility, trap density and the potential drop at the metal/organic interface is estimated. This technique is applicable not only to the OFET but also to a variety of the organic devices, such as OLED and OPV. The findings obtained by the EFI-SHG measurements will be useful for further development of organic device physics.

## Acknowledgement

Financial supports by the Grants-in-Aid for Scientific Research (No.22226007, 21686029, 24360118, 15H03971) from Japan Society for the Promotion of Science are greatly appreciated.

## References

1 Zhou, L., Wanga, A., Wu, S.C. et al. (2006). *Appl. Phys. Lett.* 88: 083502.
2 Lodha, A. and Singh, R. (2001). *IEEE Trans. Semi. Manu.* 14: 281.
3 Nonnenmacher, M., O'Boyle, M.P. and Wickramasinghe, H.K. (1991). *Appl. Phys. Lett.* 58: 2921.
4 Vatel, O. and Tanimoto, M. (1995). *J. Appl. Phys.* 77: 2358.
5 Kikukawa, A., Hosaka, S., and Imura, R. (1995). *Appl. Phys. Lett.* 66: 3510.
6 Mizutani, T., Arakawa, M., and Kishimoto, S. (1997). *IEEE Electr. Dev. Lett.* 18: 423.
7 Lampert, M.A. (1970). *Current Injection in Solids*. New York: Academic Press.
8 Manaka, T., Lim, E., Tamura, R. et al. (2006). *Appl. Phys. Lett.* 89: 072113.
9 Manaka, T., Nakao, M., Yamada, D. et al. (2006). *Opt. Express* 15: 15964.
10 Manaka, T., Lim, E., Tamura, R., and Iwamoto, M. (2007). *Nat. Photonics* 1: 581.
11 Manaka, T., Nakao, M., Lim, E., and Iwamoto, M. (2008). *Appl. Phys. Lett.* 92: 142106.
12 Manaka, T., Liu, F., Weis, M., and Iwamoto, M. (2008). *Phys. Rev. B* 78: 121302.
13 Manaka, T., Liu, F., Weis, M., and Iwamoto, M. (2009). *J. Phys. Chem. C* 113: 10279.
14 Nakao, M., Manaka, T., Weis, M. et al. (2009). *J. Appl. Phys.* 106: 14511.
15 Manaka, T., Liu, F., Weis, M., and Iwamoto, M. (2010). *J. Appl. Phys.* 107: 043712.
16 Manaka, T., Matsubara, K., Abe, K., and Iwamoto, M. (2013). *Appl. Phys. Express* 6: 101601.
17 Matsubara, K., Manaka, T., and Iwamoto, M. (2015). *Appl. Phys. Express* 8: 041601.
18 Taguchi, D., Inoue, S., Zhang, L. et al. (2010). *J. Phys. Chem. Lett.* 1: 803.
19 Taguchi, D., Zhang, L., Li, J. et al. (2011). *Jpn. J. Appl. Phys.* 50: 04DK08.
20 Sadakata, A., Taguchi, D., Yamamoto, T. et al. (2011). *J. Appl. Phys.* 110: 103707.
21 Sadakata, A., Osada, K., Taguchi, D. et al. (2012). *J. Appl. Phys.* 112: 083723.
22 Sadakata, A., Oda, Y., Taguchi, D. et al. (2013). *Jpn. J. Appl. Phys.* 52: 05DC03.
23 Taguchi, D., Shino, T., Chen, X. et al. (2011). *Appl. Phys. Lett.* 98: 133507.
24 Taguchi, D., Shino, T., Chen, X. et al. (2011). *J. Appl. Phys.* 110: 103717.
25 Taguchi, D., Shino, T., Zhang, L. et al. (2011). *Appl. Phys. Express* 4: 021602.
26 Chen, X., Taguchi, D., Lee, K. et al. (2011). *Chem. Phys. Lett.* 511: 491.
27 Chen, X., Taguchi, D., Weis, M. et al. (2012). *Jpn. J. Appl. Phys.* 51: 041605.
28 Chen, X., Taguchi, D., Lee, K. et al. (2013). *Jpn. J. Appl. Phys.* 52: 04CR05.
29 Ou-Yang, W., Manaka, T., Naitou, S. et al. (2012). *Jpn. J. Appl. Phys.* 51: 070209.
30 Katsuno, T., Manaka, T., Ishikawa, T. et al. (2014). *Appl. Phys. Lett.* 104: 252112.
31 Shen, Y.R. (1984). *The Principle of Nonlinear Optics*. New York: John Wiley & Sons.

32 Koos, D.A., Shannon, V.L., and Richmond, G.L. (1993). *Phys. Rev. B* 47: 4730.
33 Yamada, T., Hoshi, H., Manaka, T. et al. (1996). *Phys. Rev. B* 53: R13 314.
34 Koopmans, B., Janner, A.-M., and Jonkman, H.T. (1993). *Phys. Rev. Lett.* 71: 3569.
35 Levine, B.F. and Bethea, C.G. (1975). *J. Chem. Phys.* 63: 2666.
36 Bosshard, C., Knopfle, G., Pretre, P., and Gunter, P. (1992). *J. Appl. Phys.* 71: 1594.
37 Burland, D.M., Miller, R.D., and Walsh, C.A. (1994). *Chem. Rev.* 94: 31.
38 Rose, M.E. (1957). *Elementary Theory of Angular Momentum*. New York: Wiley
39 Karl, N. (1974). *Festköerperproblemes* 14: 261.
40 Yamada, D., Manaka, T,, Lim, E. et al. (2008). *J. Appl. Phys.* 104: 0745022.
41 Kitamura, M. and Arakawa, Y. (2008). *J. Phys. Condens. Matter* 20: 184011.
42 Knipp, D., Street, R.A., and Volkel, A.R. (2003). *Appl. Phys. Lett.* 82: 3907.
43 Weis, M., Manaka, T., and Iwamoto, M. (2009). *J. Appl. Phys.* 105: 024505.
44 Knipp, D., Street, R.A., Volkel, A., and Ho, J. (2003). *J. Appl. Phys.* 93: 347.
45 Chua, L.L., Zaumseil, J., Chang, J.F. et al. (2005). *Nature* 434: 194.
46 Kim, C., Facchetti, A., and Marks, T.J. (2007). *Science* 318: 76.
47 Yang, S.Y., Kim, S.H., Shin, K. et al. (2006). *Appl. Phys. Lett.* 88: 173507.
48 Michaelson, H.B. (1978). *IBM. J. Res. Dev.* 22: 72.
49 de Boer, R.W.I., Klapwijk, T.M., and Morpurgo, A.F. (2003). *Appl. Phys. Lett.* 83: 4345.
50 Takeya, J., Yamagishi, M., Tominari, Y. et al. (2007). *Appl. Phys. Lett.* 90: 102120.
51 Lee, J.Y., Roth, S., and Park, Y.W. (2006). *Appl. Phys. Lett.* 88: 252106.
52 Sundar, V.C., Zaumseil, J., Podzorov, V. et al. (2004). *Science* 303: 1644.
53 Ostroverkhova, O., Cooke, D.G., and Hegmann, F.A. (2006). *Appl. Phys. Lett.* 89: 192113.
54 Sele, C.W., Kjellander, B.K.C., Niesen, B. et al. (2006). *Adv. Mater.* 21: 4926.
55 Uno, M., Tominari, Y., Yamagishi, M. et al. (2009). *Appl. Phys. Lett.* 94: 223308.
56 Headrick, R.L., Wo, S., Sansoz, F., and Anthony, J.E. (2008). *Appl. Phys. Lett.* 92: 063302.

# 10

# Device Physics of Organic Field-effect Transistors

*Hiroyuki Matsui*

*Graduate School of Organic Materials Science, Yamagata University, Yamagata, Japan*

**CHAPTER MENU**



## 10.1 Organic Field-Effect Transistors (OFETs)

Organic semiconductors without doping are hardly conductive because of the wide bandgap, typically 2-3 eV, and low carrier density. When electron or hole carriers are doped, by contrast, the conductivity increases drastically by more than seven orders [1]. Organic field-effect transistors (OFETs) utilize a charge carrier accumulation at the metal-insulator-semiconductor (MIS) structure by the gate electric field, and act as electronic switches, logics or amplifiers [2, 3].

In this section, we first review the structural aspects of OFETs such as device structures, film morphology, and crystal structures in 10.1.1. Then, in 10.1.2, the basic operation principles of OFETs will be shown. In 10.1.3, we will discuss carrier traps which are one of the main factors in determining carrier mobility. In 10.1.4, five models for the charge transport in semiconductor channels will be reviewed. Finally, in 10.1.5, we will discuss the charge injection from the source/drain metal electrodes to the organic semiconducting channel.

### 10.1.1 Structure of OFETs

Organic field-effect transistors (OFETs) consist of at least four layers: (1) gate electrode layer, (2) gate insulator layer, (3) organic semiconductor layer, and (4) source/drain electrode layer. The structure of transistors stacked in order is called bottom-gate top-contact geometry (Figure 10.1(a)). Optional layers such as substrates, surface modification layers, and passivation layers are also added if necessary. Other geometries of top-gate bottom-contact, bottom-gate bottom-contact, and top-gate top-contact are also known (Figure 10.1(b)-(d)). The bottom-gate geometry has been widely used in fundamental studies because of the ease of device fabrication with $SiO_2$/Si substrates [4]. The bottom-contact geometry is preferred when source/drain electrodes are fabricated by photolithography or printing methods, otherwise, the organic semiconductor

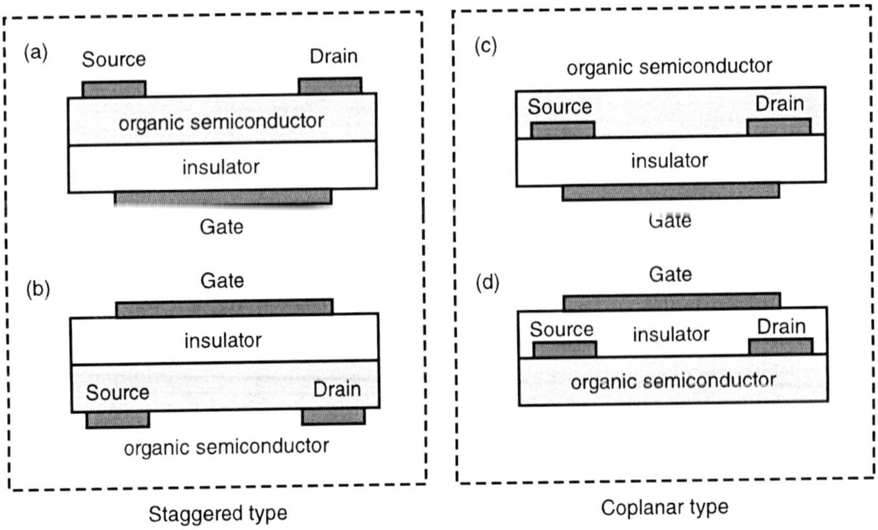


**Figure 10.1** Device structure of organic field-effect transistors. (a) Bottom-gate top-contact, (b) top-gate bottom-contact, (c) bottom-gate bottom contact, and (d) top-gate top-contact geometries.

layers may be damaged by photoresists or solvents. The four geometries can be further categorized into two groups: (1) staggered type, including bottom-gate top-contact and top-gate bottom-contact geometries, and (2) coplanar type, including bottom-gate bottom-contact and top-gate top-contact geometries. Recent studies have revealed that staggered-type OFETs are more suitable to the efficient charge injection from source/drain electrodes to semiconductor layers [5, 6].

An organic semiconductor layer is the most important part of OFETs. To obtain high carrier mobility, single-crystalline or polycrystalline semiconductor films are preferred to amorphous. In particular, some of the single-crystal OFETs made from rubrene or heteroacene-based molecules have exhibited quite high field-effect mobility more than 10 $cm^2/Vs$ [7–9]. Many X-ray diffraction (XRD) and scanning probe microscope (SPM) experiments have revealed that the crystallinity and the device performances have a strong correlation with each other (Figure 10.2) [10, 11]. Not only small molecules, polymer semiconducting films also exhibit XRD peaks, indicating a local periodicity in the polymer films [12–14].

Typical crystal structures of small molecule semiconductors suitable to OFETs are so-called *herringbone* and *brickwork* packings (Figure 10.3) [15, 16]. Since the intermolecular interactions in these packings form 2D conduction paths, they are suitable for the charge transport along the semiconductor/insulator interfaces and less sensitive to the in-plane direction of crystalline domains. Most semiconducting polymers used for OFETs have a lamellar structure shown in Figure 10.2(b) [12–14]. The lamellar structure also forms quasi-2D conduction paths. The mobility along polymer chains is typically 3–10 times higher than the mobility in the direction normal to the molecular planes [17, 18].

It is worth stating the uniqueness of OFETs compared to other organic devices such as organic light-emitting diodes (OLEDs) and organic photovoltaics (OPVs). One significant difference is that the current flow in OTFTs is usually parallel to substrates whereas

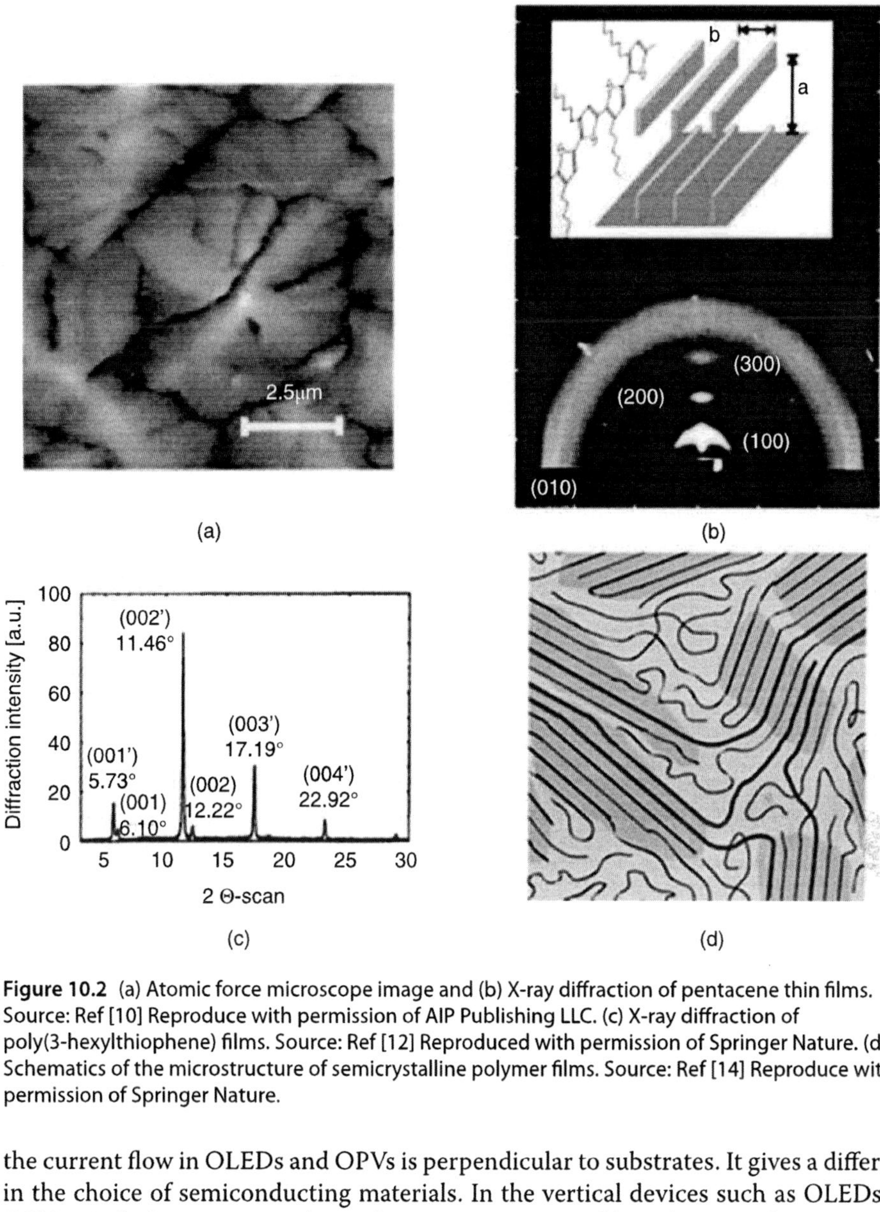


**Figure 10.2** (a) Atomic force microscope image and (b) X-ray diffraction of pentacene thin films. Source: Ref [10] Reproduce with permission of AIP Publishing LLC. (c) X-ray diffraction of poly(3-hexylthiophene) films. Source: Ref [12] Reproduced with permission of Springer Nature. (d) Schematics of the microstructure of semicrystalline polymer films. Source: Ref [14] Reproduce with permission of Springer Nature.

the current flow in OLEDs and OPVs is perpendicular to substrates. It gives a difference in the choice of semiconducting materials. In the vertical devices such as OLEDs and OPVs, pin-holes in semiconductor layers are serious problems because they cause leakage between top and bottom electrodes. For this reason, amorphous semiconductors which can form uniform films without pin holes are preferred in OLEDs and OPVs. In OFETs, by contrast, pin holes in semiconductor layers do not seriously matter. Since mobility is a more important factor for device performances, crystalline semiconductors which usually possess higher mobility than amorphous ones, are preferred in OFETs. The second important difference of OFETs from OLEDs and OPVs is that only one type of carriers, that is, electrons or holes, plays an important role in OFETs, whereas both type of carriers are important in OLEDs and OPVs. In other words, excitons do not exist

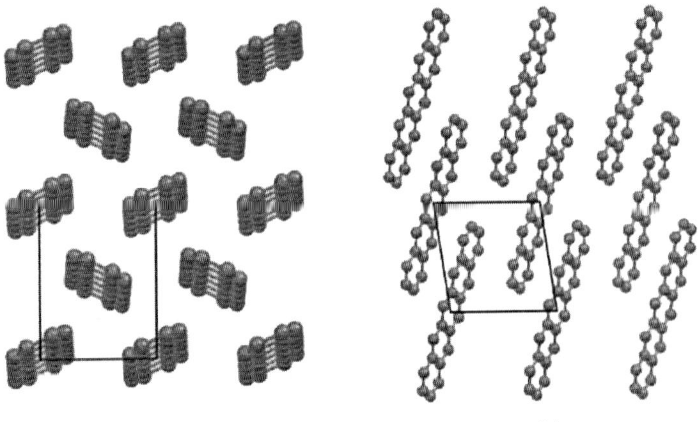


Figure 10.3 (a) Herringbone-type crystal structure of pentacene. Source: Based on Ref, [15]. (b) Brickwork-type crystal structure of 6,13-bis(triisopropylsilylethynyl)pentacene (TIPS-pentacene). Only $\pi$-conjugated cores are shown. Source: Based on Ref [16].

in OFETs except for light-emitting OFETs [19]. Lastly, the carrier density in OFETs can be much higher than those of OLEDs and OPVs because of the charge accumulation by gate electric field. The high carrier density results in high conductivity and high mobility in OFETs.

### 10.1.2 Operation Principles of OFETs

Intrinsic organic semiconductors have quite low carrier density. For example, a rough estimation of a carrier density $n = Dk_BT\exp(-E_g/2k_BT)$, $D = 10^{22}$ cm$^{-3}$ eV$^{-1}$, $E_g$ = 2 eV, $T$ = 300 K, and $\mu$ = 1 cm$^2$/Vs, gives a conductivity of $\sigma = en\mu \approx 10^{-15}$ S/cm or a sheet conductivity of $\sigma_{\text{sheet}} = \sigma d_s \approx 10^{-20}$ S for a film thickness of $d_s$ = 100 nm. Here $D$ is the density of states at the band edge, $E_g$ is the band gap energy, $k_B$ is the Boltzmann constant, $T$ is the temperature, and $\mu$ is the mobility. Field effect is a technique to accumulate charge carriers in such organic semiconductors by utilizing a metal-insulator-semiconductor (MIS) structure [20, 21]. For simplicity, we assume that the work function of gate electrodes is the same as the Fermi level of organic semiconductors and that the thickness of the gate dielectric layer is much larger than that of the charge accumulation layer. Then, a metal gate electrode, a gate dielectric layer and a semiconductor layer can be described as a flat-band diagram shown in Figure 10.4(a). If we apply a negative voltage $V_G$ at gate electrode, Gauss's law gives the sheet carrier density $en_{\text{sheet}} = \epsilon\epsilon_0 V_G/d = C_i|V_G|$ (Figure 10.4(b)). Assuming $C_i$ = 10 nF cm$^{-2}$ and $V_G = -10$ V, the sheet conductivity $\sigma_{\text{sheet}} = en_{\text{sheet}}\mu \approx 10^{-7}$ S. Thus, the conductivity of semiconductor channels in OFETs can be changed drastically by a gate electric field. In real devices, charge carriers start to accumulate at a non-zero threshold voltage $V_{\text{th}}$, and the sheet carrier density is described as $en_{\text{sheet}} = C_i|V_G - V_{\text{th}}|$ because of the presence of trap states [21, 22], the interfacial dipole moments [23], and the mismatch between the work function of gate electrodes and the Fermi level of organic semiconductors [24].

When a drain voltage $V_D$ is applied, the charge density at position $x$ is given by $en_{\text{sheet}}(x) = C_i|V_G - V_{\text{th}} - V(x)|$, where $x$ is the distance from the source electrode and

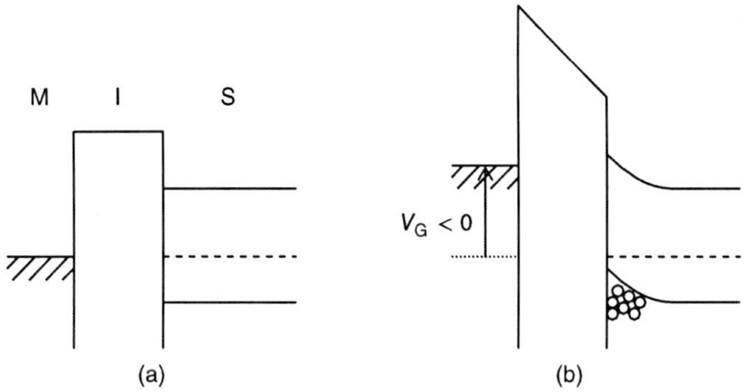


**Figure 10.4** Energy band diagram at (a) zero and (b) negative gate voltages.

$V(x)$ is the voltage of the semiconductor film at the position. Then, the current is

$$I_D = W\sigma_{sheet}(x)E(x) = WC_i\mu|V_G - V_{th} - V(x)|\frac{dV(x)}{dx} \quad (10.1)$$

and should not depend on $x$ because of the continuity equation. The boundary conditions are $V(0) = 0$ and $V(L) = V_D$, where $L$ is the distance between source and drain electrodes. Calculating the integrals $\int_0^L I_D dx = \int_0^L WC_i\mu|V_G - V_{th} - V|dV$, the drain current is finally given by

$$I_D = \frac{W}{L}C_i\mu V_D\left|V_G - V_{th} - \frac{V_D}{2}\right|. \quad (10.2)$$

When $|V_D| \ll |V_G - V_{th}|$, the drain current is proportional to the drain voltage $V_D$ as described by

$$I_D = \frac{W}{L}C_i\mu V_D(V_G - V_{th}). \quad (10.3)$$

This condition is called a linear regime, where the charge carriers are accumulated uniformly in the whole channel area (Figure 10.5(a)). The field effect mobility in the linear regime can be calculated by

$$\mu = \frac{L}{W}\frac{1}{C_iV_D}\frac{\partial I_D}{\partial V_G}. \quad (10.4)$$

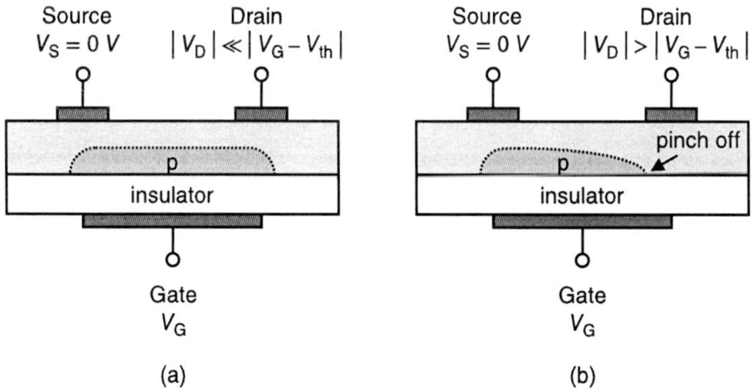


**Figure 10.5** Charge carrier distribution in (a) linear and (b) saturation regimes.

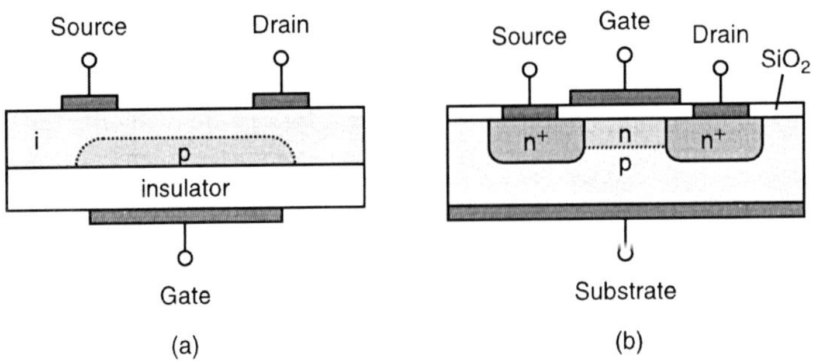


**Figure 10.6** Comparison of (a) OFETs and (b) silicon MOSFETs.

When $|V_D| > |V_G - V_{th}|$, on the other hand, charge carriers are not accumulated near the drain electrode (Figure 10.5(b)). This phenomenon is called pinch-off. Since the excess drain voltage over $|V_G - V_{th}|$ is applied at the pinch-off point, the drain current does not depend on the drain voltage in this regime. It is called a saturation regime. Substituting $V_D = V_G - V_{th}$ to Eq. (10.2) gives the drain current in the saturation regime

$$I_D = \frac{W}{2L} C_i \mu (V_G - V_{th})^2. \tag{10.5}$$

Then, the field effect mobility in the saturation regime can be calculated by

$$\mu = 2\frac{L}{W}\frac{1}{C_i}\left(\frac{\partial\sqrt{I_D}}{\partial V_G}\right)^2. \tag{10.6}$$

In addition to the mobility and threshold voltage, turn-on voltage $V_{on}$ and subthreshold slope $SS$ are important device parameters. The turn-on voltage $V_{on}$ is extracted from $\log(I_D) - V_G$ plot as the gate voltage where the drain current starts to increase steeply. This is actually the cross point of ideal drain current and the floor current due to semiconductor bulk conduction, surface leakage, or gate leakage. The subthreshold slope is defined as the inverse of the slope in $\log I_D - V_G$ plot. $SS = dV_G/d(\log I_D)$. $SS$ can be used for the simple estimation of deep trap density $N_t$ by [25]

$$SS = \frac{k_B T \ln 10}{e}\left(1 + \frac{e^2 N_t}{C_i}\right), \tag{10.7}$$

and small $SS$ is important in both analog and digital circuits. The effect of the metal/semiconductor contacts was ignored in the above discussion, while the contacts are actually very important and can limit the performance of the devices with a short channel length [26]. The effect of contacts will be discussed in detail in Section 10.1.5.

The operation mechanisms of OFETs are different from those of silicon metal-oxide-semiconductor (MOS) FETs in several points (Figure 10.6) [20]. Firstly, organic semiconductors are the intrinsic semiconductors which are not intentionally doped by chemical dopants. This is because introducing dopant impurities into organic semiconductors always cause structural disorders. Secondly, standard OFETs do not have the substrate potential that silicon transistors always have. The other side of organic semiconductor layers is usually air or insulating materials such as substrates and passivation layers.

### 10.1.3 Carrier Traps

In discussing the charge transport in organic semiconductors, traps are an inevitable issue because actual organic semiconductor layers are far from perfect single crystals. Traps can be defined as the localized electronic states between the valence and conduction bands (Figure 10.7) [22]. According to Anderson's theory, the localized in-gap states and delocalized band states have density of states in different energy ranges at 0 K [27]. In other words, we can define a threshold energy, that is, the so-called mobility edge, which separates the density of localized states from that of delocalized states. The name comes from the fact that, at 0 K, only the charge carriers in delocalized states can contribute to the conduction. At non-zero temperature, however, carriers in the localized trap states can also contribute to the conduction in two ways. One is a phonon-assisted tunneling: a carrier hops directly from one localized state at a certain energy to another localized state at a different energy [28]. Phonons need to be absorbed or emitted at that time to satisfy the energy conservation. It is also referred to as hopping transport. The other way is the trap-and-release process: a carrier travels from one localized state to another via delocalized states [21]. Since the probability that the carrier is released from the trap depends on the energy of the trap with respect to the mobility edge, we usually refer to the traps whose energy distance from the mobility edge is less than a few $k_B T$ as shallow traps, and refer to the traps whose energy distance from the mobility edge is more than a few $k_B T$ as deep traps [22].

Traps in organic semiconductors are considered to have a variety of origins. Figure 10.8 illustrates four types of traps in OFETs. First, impurity molecules which have energy levels inside bandgap can trap carriers. Such impurities can come from, for example, by-products of synthesis processes, contaminations, degradation, reactions with oxygen and water in the atmosphere [29, 30]. and adsorption of gas molecules [31]. The effect of impurities strongly depends on their energy levels, since the impurity levels within the valence and conduction bands do not trap carriers. Secondly, lattice defects which violate the periodicity of the semiconductor crystals also act as traps.

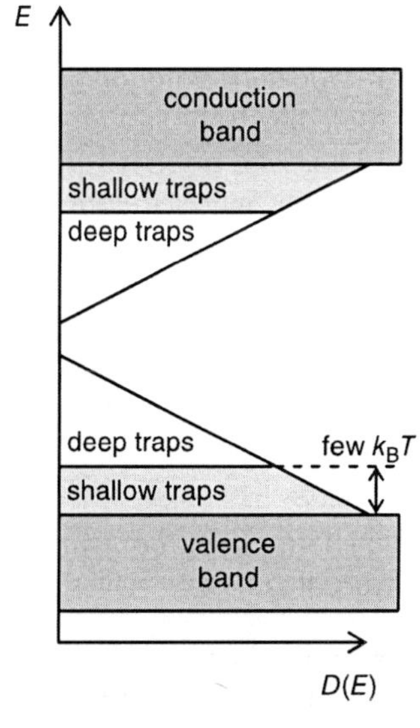


**Figure 10.7** Energy distribution of the density of trap and band states.

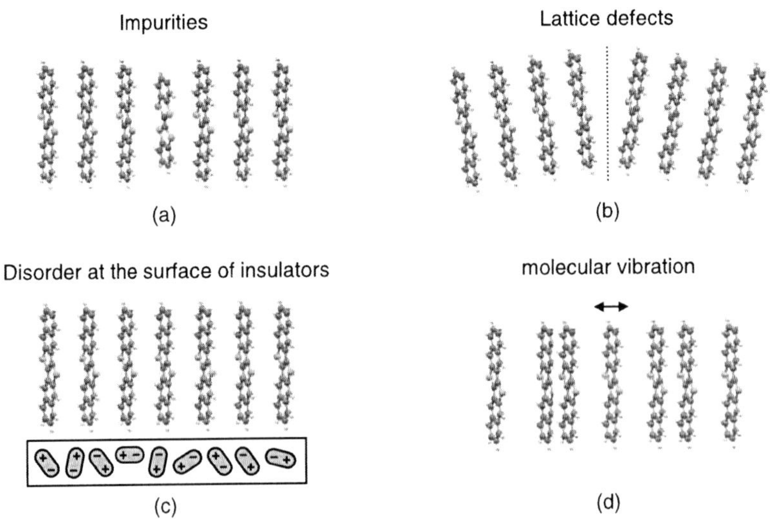


**Figure 10.8** Origins of traps in OFETs. (a) Chemical impurities. (b) Lattice defects. (c) Interaction with gate insulators. (d) Molecular vibration.

Such defects are created during crystal growth or film deposition processes. Grain boundaries are the most obvious defects in polycrystal films [32, 33], and there are also other traps which are more difficult to observe [34]. The third is the interactions with gate insulators because charges are accumulated on the semiconductor/insulator interfaces in OFETs [35]. The roughness, random dipole moments, and inhomogeneity of the surface of the gate insulators have been proposed as the causes of trap states. The last is traps due to molecular vibration, which are dynamic traps while the former three types are static traps [36–38]. Since the motion of molecules is far slower than the motion of electrons, the electronic states can be considered localized in the timescale of carrier conduction by the slow molecular vibration.

### 10.1.4 Transport Models in Channels

The mobility of OFETs have a wide variation of between $10^{-5}$ $cm^2/Vs$ and $10\,cm^2/Vs$ [1]. Accordingly, it is not a good idea to explain all OFETs using one single theoretical model. We need to choose one of several models which fit the OFETs under the issue. Here, we review five different theoretical models which are often used to explain charge transport in OFETs (Figure 10.9).

When the trap density in organic semiconductor crystals is low and electron-phonon coupling is not too high, charge carriers can behave like the Bloch wave in the periodic potential. This is the so-called *band model*. Electron-phonon coupling can be included as a *small polaron* picture [39]. The band model would be suitable for OFETs which exhibit mobility greater than $1\,cm^2/Vs$, and can explain the negative temperature dependence of mobility and Hall effect. When the trap density is intermediate, the temperature dependence of mobility becomes positive, and the *multiple trap-and-release (MTR) model* describes the system better than the band model [21]. This model includes three

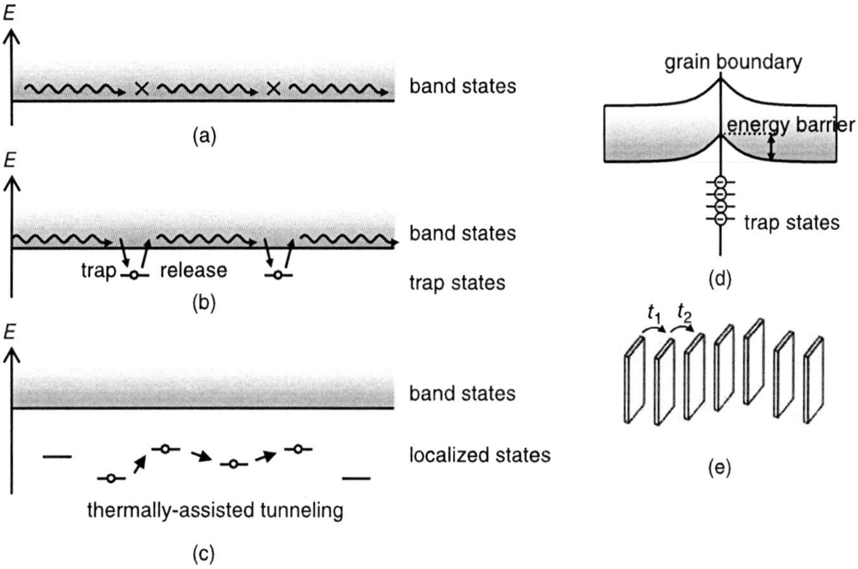


**Figure 10.9** Schematics of five transport models. (a) Band model. (b) Multiple trap-and-release (MTR) model. (c) Hopping model. (d) Grain boundary model. (e) Dynamic disorder model.

processes in the charge transport: a carrier is activated from a trap to delocalized states (releasing), travels through the delocalized states, and is captured by another trap (trapping). This model requires the presence of the delocalized states which should not be too far from the Fermi energy. When the system contains a lot of disorder and the trap density is extremely high, delocalized states do not exist or are too far from Fermi energy to contribute to the conduction. In such disordered systems, phonon-assisted tunneling from traps to traps is the major path of charge transport [28]. This is the so-called *hopping model.* This model would be the best to describe the charge transport in disordered amorphous films. When the systems are polycrystals, lattice defects are not distributed uniformly but concentrated at the grain boundaries [32, 33]. Such systems require another model, which takes into account any potential barriers at the grain boundaries. In the small polaron model above, the electron-phonon coupling is included as a perturbation. Recently, however, Troisi and co-workers have proposed that the electron-phonon coupling is so large in some organic systems that the perturbative treatment is invalid [36]. The relatively new model with a non-perturbative treatment is called the *dynamic disorder model.* Such a model has been studied on the basis of numerical calculations [36–38, 40].

#### 10.1.4.1 Band Transport Model

Band transport characteristics have been observed in high-quality single-crystal OFETs. One characteristic feature of the band transport is the negative temperature dependence of mobility. Since the mobility in the band model is given by

$$\mu = \frac{e\tau}{m^*} \tag{10.8}$$

and the relaxation time $\tau$ increases as temprature decreases, the mobility $\mu$ also increases as temperature decreases. Podzorov and co-workers reported negative

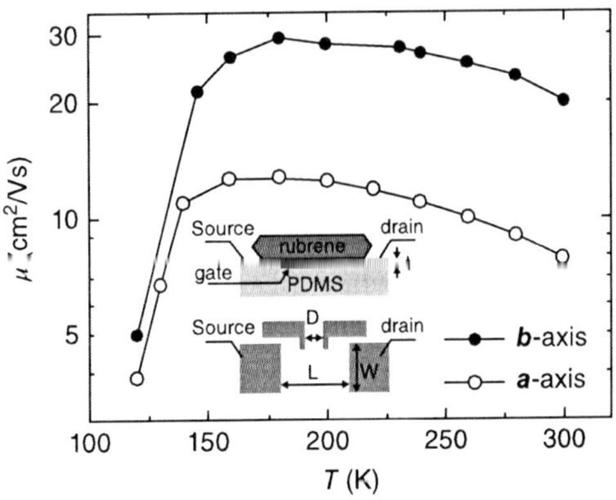


**Figure 10.10** Temperature dependence of field-effect mobility in rubrene single-crystal OFETs extracted from four-probe measurements. Source: Based on Ref. [7].

temperature dependence of mobility in air-gap rubrene single-crystal OFETs in 2004 (Figure 10.10) [7]. The use of air gap as a gate insulating layer can minimize the disorder at the semiconductor/insulator interfaces. The mobility increased when the temperature was cooled down in the range of 180– 00 K, whereas the mobility decreased when the temperature was cooled down further. The transport in the lower temperature range is considered trap-dominated. Recently, many other OFETs have been reported to exhibit negative temperature dependence of mobility in high temperature ranges [41].

The Hall effect is another characteristic phenomenon in band-transport systems. Podzorov and co-workers and Takeya and co-workers reported the Hall effect of accumulation charge carriers in single-crystal transistors [42, 43]. The Hall voltage, proportional to the magnetic field, was observed in the transverse direction (Figure 10.11(a)). By defining the Hall coefficient as $R_H = V_H/BI_D$, where $V_H$ is the Hall voltage and $B$ is

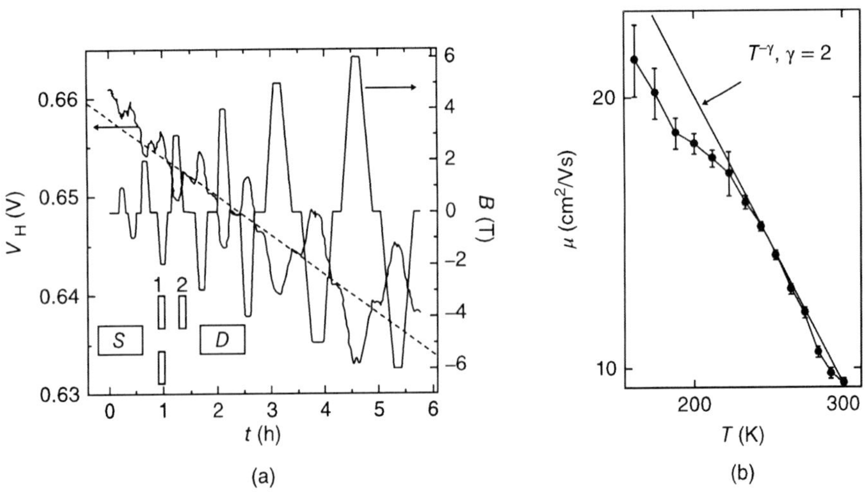


**Figure 10.11** Hall effect of rubrene single-crystal OFETs. (a) Magnetic field and Hall voltage. (b) Temperature dependence of Hall mobility. Source: Based on Ref. [42].

the magnetic field, we can calculate the Hall carrier density $n_H$ and the Hall mobility $\mu_H$ independently

$$n_H = \frac{1}{eR_H} = \frac{BI_D}{eV_H}, \tag{10.9}$$

$$\mu_H = R_H\sigma = \frac{V_H L}{BV_D W}. \tag{10.10}$$

The calculated Hall mobility is plotted in Figure 10.11(b). The mobility followed the power law $\mu \propto T^{-\gamma}$ with $\gamma \sim 2$, which is consistent with the calculation for tetracene [44]. Recently, the Hall effect of other OFETs, including polycrystalline OFETs and polymer OFETs, has also been reported [45–48].

The band dispersion of rubrene single crystals has been observed by angle-resolved ultraviolet photoelectron spectroscopy (ARUPS) by Machida and co-workers in 2010 (Figure 10.12) [49]. ARUPS is a powerful technique for clarifying the band dispersion of materials and obtaining the band width $W$ and the effective mass $m^*$ experimentally. Figure 10.12(b) shows the ARUPS spectra mapped on the $E$-$k_{||}$ plane. From the band dispersion, the band width and effective mass were estimated at 0.4 eV and 0.65 $m_0$ ($m_0$ is the mass of free electron), respectively. The ARUPS measurement has provided evidence for the existence of band states in organic semiconductors.

Theoretical studies of band mobility have been reported by Northrup and Kobayashi and co-workers [50, 51]. They calculated the effective mass $m^*$ by a band calculation, and the relaxation time $\tau$ by the acoustic deformation potential model

$$\tau = \frac{\hbar^3 B L_{\rm eff}}{\epsilon_{\rm ac}^2 k_B T m_d}. \tag{10.11}$$

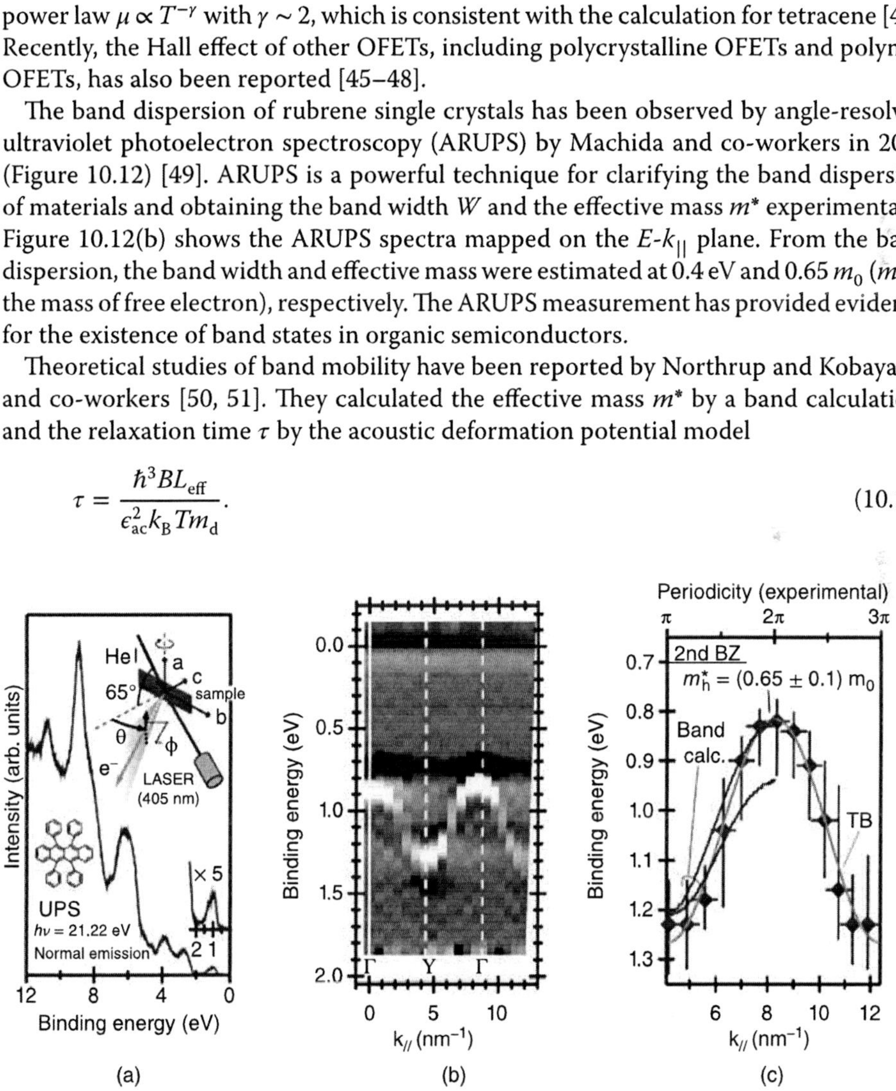


**Figure 10.12** (a) Angle-resolved ultraviolet photoelectron spectroscopy (ARUPS) of rubrene single crystals taken at normal emission angle. (b) ARUPS spectra mapped on the $E$-$k_{||}$ plane. (c) Comparison of $E$-$k_{||}$ diagram of the ARUPS main peaks, the band calculation, and the tight-binding model Source: Ref [49] Reproduced with permission of American Physical Society.

**Table 10.1** The mobility calculated by the acoustic deformation potential model.

| | Pentacene | Rubrene | $C_8$-BTBT |
|---|---|---|---|
| $L_{eff}$(Å) | 11.3 | 2.8 | 8.9 |
| $B$(meV/Å$^3$) | 89(17) | 180(9) | 93(8) |
| $\varepsilon_{ac}$(eV) | 1.74(7.90) | 2.17(7.69) | 1.49(6.71) |
| $\tau$(fs) | 43(0.41) | 19(0.08) | 28(0.12) |
| $\mu_a^{band}$(cm$^2$/Vs) | 58(0.55) | 51(0.21) | 36(0.15) |
| $\mu_b^{band}$(cm$^2$/Vs) | 44(0.42) | 18(0.07) | 11(0.04) |
| $\mu^{exp}$(cm$^2$/Vs) | 35 [a] | 40 [b] | 31[c] |

Source: Ref [51] Reproduce with permission of AIP Publishing LLC.

Here, $B$ is the elastic modulus, $L_{\text{eff}}$ is the effective thickness of the conduction channel, $\varepsilon_{\text{ac}}$ is the acoustic deformation potential, and $m_{\text{d}}$ is the density of states mass. Table 10.1 summarizes the calculated mobility and other parameters for three different materials. Although the calculation does not take defects into account and the mobility is rather too high compared to experimental results, the calculation should give the upper limit of the mobility for the organic semiconductor materials.

Electron-phonon coupling is treated perturbatively in the band model [39]. Although it is only valid when the coupling is relatively small, the perturbation treatment helps to simplify the calculation. The Hamiltonian including electron-phonon coupling is given by

$$H = \sum_{i,j} V_{i,j} a_i^\dagger a_j + \sum_{\mathbf{q},\lambda} \hbar\omega_{\mathbf{q},\lambda} \left( b_{\mathbf{q},\lambda}^\dagger b_{\mathbf{q},\lambda} + \frac{1}{2} \right) + \sum_{\mathbf{q},\lambda,i,j} \hbar\omega_{\mathbf{q},\lambda} g_{\mathbf{q},\lambda,i,j} (b_{\mathbf{q},\lambda}^\dagger + b_{-\mathbf{q},\lambda}^\dagger) a_i^\dagger a_j. \tag{10.12}$$

Here $a_i^\dagger$ and $a_i$ are the creation and annihilation operators for an electron at site $i$, and $b_{\mathbf{q},\lambda}^\dagger$ and $b_{\mathbf{q},\lambda}$ are the creation and annihilation operators for a phonon with wavevector $\mathbf{q}$ in the mode $\lambda$. The first term is the energy of electrons, the second is the energy of phonons, and the last is the electron-phonon coupling. Applying a canonical transformation akin to the small polaron transformation, it can be transformed into

$$\widetilde{H} = \sum_{i,j} \widetilde{V}_{i,j} a_i^\dagger a_j + \sum_{\mathbf{q},\lambda} \hbar\omega_{\mathbf{q},\lambda} \left( b_{\mathbf{q},\lambda}^\dagger b_{\mathbf{q},\lambda} + \frac{1}{2} \right). \tag{10.13}$$

This equation looks like the Hamiltonian of non-interacting electrons and phonons. Actually, the electron-phonon coupling is hidden in $\widetilde{V}_{i,j}$, which contains $b_{\mathbf{q},\lambda}^\dagger$ and $b_{\mathbf{q},\lambda}$. The transformation results in a *polaronic band*, describing the state of the charge carriers dressed by phonons.

#### 10.1.4.2 Multiple Trap and Release Model

In many organic thin-film transistors with polycrystalline or polymeric semiconductor films, the temperature dependence of mobility follows an activation law: $\mu \propto \exp(-E_A/k_BT)$, as seen in Figure 10.13 [52]. The activation energy $E_{\text{A}}$ is typical of between 0.05 – 0.2 eV. The temperature dependence can be explained by the multiple trap-and-release (MTR) model [21]. If the system has intermediate density of traps,

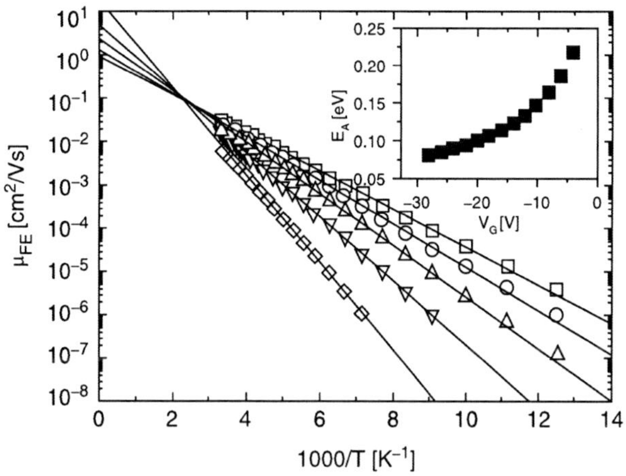


**Figure 10.13** Activation-type temperature dependence of mobility in pentacene thin-film transistors. Source: Based on Ref. [52].

some of the accumulated carriers are trapped and only the rest of carriers in delocalized states can contribute to the conduction. As a result, effective mobility is given by

$$\mu_{\text{eff}} = \mu_0 \frac{n_\text{f}}{n_\text{f} + n_\text{t}}. \tag{10.14}$$

where $\mu_0$ is the intrinsic mobility in delocalized states, $n_\text{f}$ the density of free carriers, and $n_\text{t}$ the density of trapped carriers. Since $n_\text{f}$ is approximately proportional to $\exp(-E_A/k_BT)$ and the temperature dependence of $\mu_0$ is relatively weak, the temperature dependence of $\mu_{\text{eff}}$ follows the activation law. More precisely, $n_\text{f}$ and $n_\text{t}$ can be calculated for a given density of states $D(E)$ as

$$n_\text{t} = \int_{-\infty}^{E_\text{C}} D(E)f(E)dE, \tag{10.15}$$

$$n_\text{f} = \int_{E_\text{C}}^{\infty} D(E)f(E)dE, \tag{10.16}$$

where $E_\text{C}$ is the energy at the bottom of the conduction band, that is the mobility edge, $f(E) = \{1 + \exp[(E - E_\text{f})/k_\text{B}T]\}^{-1}$ is the Fermi distribution function, and $E_\text{f}$ is the Fermi energy.

The methods to extract the density of trap states from the transistor characteristics have been reported so far from several groups. The simplest way was reported by Lang and co-workers [53]. They approximated as $E_\text{f} \sim E_\text{A}$, and estimated the density of states by

$$N(E) = \frac{C_\text{i}}{e}\left(\frac{dE_\text{A}}{dV_\text{G}}\right)^{-1}. \tag{10.17}$$

Other methods which are more complicated and more accurate have been reported by Horowitz et al. [54], Fortunato et al. [55], Grünewald et al. [56], and Kalb et al. [57, 58]. Kalb et al. compared the six methods shown in Figure 10.14(a) [59]. They have almost one-order variation in the density of trap states, and Lang's method underestimates the

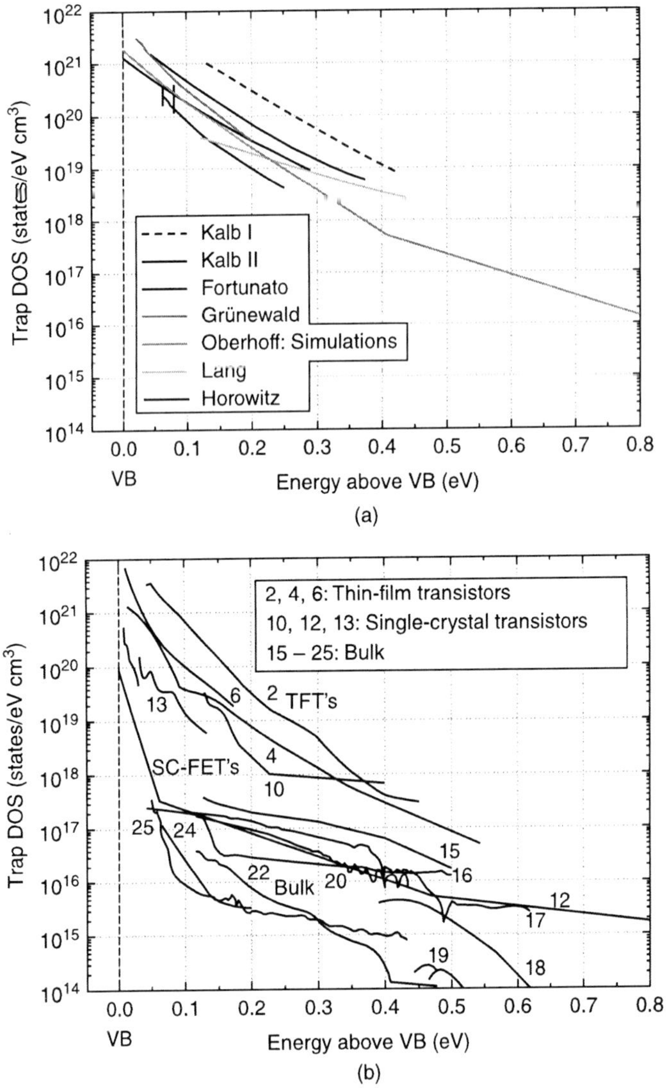


**Figure 10.14** (a) Density of trap states calculated in the different methods from the same data set of measured data for pentacene OFETs. (b) Density of trap states in small-molecule organic semiconductors of thin-film transistors, single-crystal transistors, and in the bulk of single crystals. Source: Ref [59] Reproduced with permission of American Physical Society.

slope of the density of states. These deviations need to be considered in the analyses of trap densities. Figure 10.14(b) compares the density of trap states in thin-film transistors, single-crystal transistors and the bulk of single crystals. As a whole, thin-film transistors exhibit the highest density of trap states, and the bulk of single crystals exhibit the lowest density. It indicates not only the grain boundaries in thin films but also the semiconductor/insulator interfaces in the OFET structure can increase trap states.

The microscopic picture of the MTR transport can be observed by field-induced electron spin resonance (FI-ESR) measurement [60–63]. ESR is a technique to detect unpaired electrons by the resonance of Zeeman splitting under a static magnetic

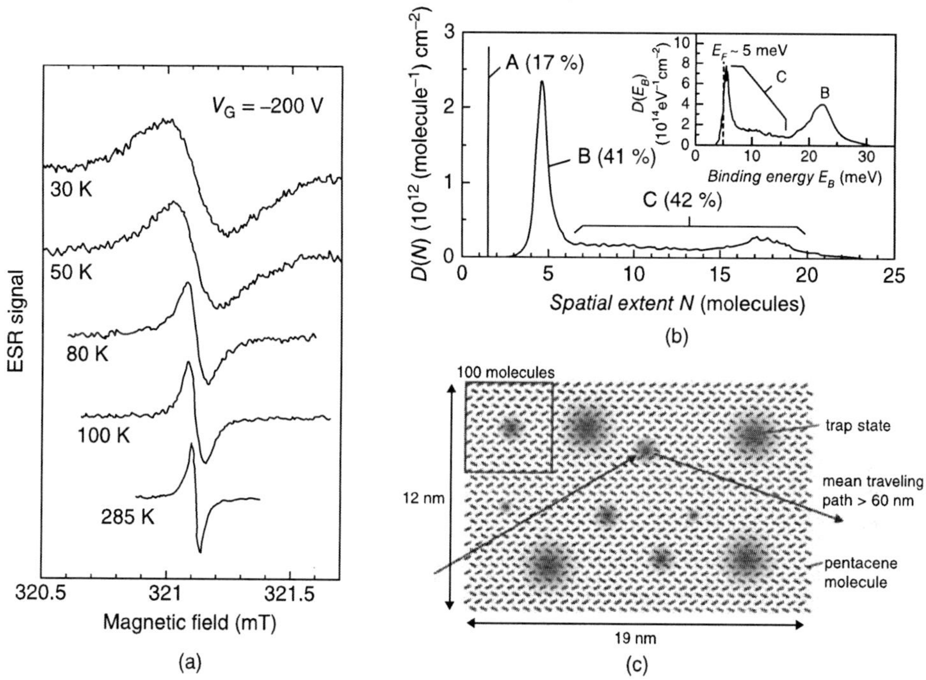


**Figure 10.15** (a) Motional narrowing of electron spin resonance (ESR) spectra in pentacene thin-film transistors. Source: Based on Ref. [61]. (b) Distribution of the spatial extent of trap states extracted from the ESR spectrum at 20 K. Source: Ref [62] Reproduced with permission of American Physical Society. (c) Schematics of the multiple trap-and-release (MTR) picture obtained from the ESR analyses.

field to microwave. FI-ESR detects the field-induced carriers in OFETs by performing ESR measurement with gate voltage applied [60]. Figure 10.15(a) shows the motional narrowing effect of the FI-ESR spectra in pentacene thin-film transistors [61, 63]. By analyzing the linewidth, the persistent time of the carriers in a trap site was estimated as 1–100 ns, depending upon temperature and gate voltage. The activation energy of the persistent time was found in the range of 5–21 meV, which indicates that shallow traps are limiting the carrier dynamics inside crystalline domains, while the macroscopic transport can be limited by grain boundaries or contact resistance. The FI-ESR measurement at low temperature, where all carriers are immobilized by traps, can provide more information about the trap states [62]. Figure 10.15(b) shows the spatial extent of trap states in pentacene thin-film transistors extracted from the low-temperature FI-ESR spectra. The distribution of the spatial extent in the range of 1–20 molecules reveals that the carriers are weakly bound to the trap sites. It is consistent with the small activation energy observed in motional narrowing. As a result of such FI-ESR analyses, a schematic picture of MTR in Figure 10.15(c) can be obtained.

#### 10.1.4.3 Hopping Model

When semiconductor films are extremely disordered like amorphous, the excitation of charge carriers into the delocalized states rarely occurs. Then, the most probable conduction path is the hopping between the localized (trap) states [28]. Once the charge carriers are strongly localized at each semiconductor molecules, intramolecular electron-phonon coupling becomes important [64]. Figure 10.16(a) illustrates the

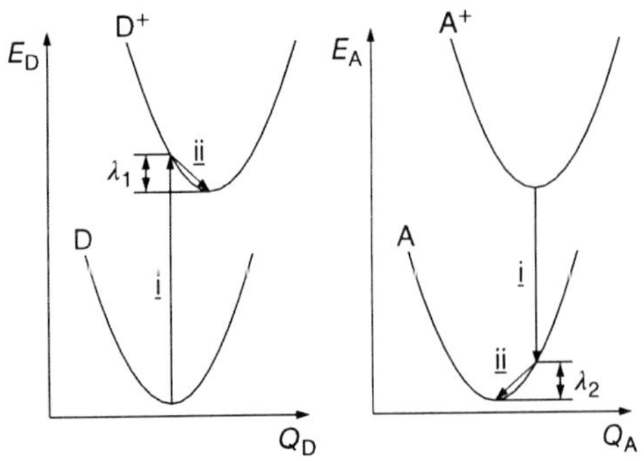


**Figure 10.16** Energy diagram of the hopping process, $D + A^+ \rightarrow D^+ + A$, between the two molecules referred to as D and A.

energy diagram on the hopping process, $D + A^+ \rightarrow D^+ + A$, between the two molecules referred to as D and A. The bottom axis denotes the coordinates of the atoms in each monomer. Since the electron transfer is much faster than the motion of nucleus, the hopping process can be divided into two steps: Step 1 is the simultaneous oxidization of D and reduction of $A^+$ at frozen nuclear geometries, and Step 2 is the relaxation of the nuclear geometries. As seen in the figure, Step 1 requires additional energy of $\lambda = \lambda_1 + \lambda_2$, which is called *reorganization energy*. Based on the Fermi's Golden rule and the reorganization energy, the transition probability per unit time can be expressed by

$$\mu = \frac{ed^2V^2}{\hbar k_B T}\left(\frac{\pi}{\lambda k_B T}\right)^{1/2} \exp\left(-\frac{\lambda}{4k_B T}\right). \tag{10.18}$$

Here $V$ is the transfer integral, and can be calculated by [65]

$$V = \frac{H_{12} - \frac{1}{2}(H_{11} + H_{22})S_{12}}{1 - S_{12}^2}, \tag{10.19}$$

$$H_{ij} = \langle \psi_i \mid H \mid \psi_j \rangle, \tag{10.20}$$

$$S_{ij} = \langle \psi_i \mid \psi_j \rangle. \tag{10.21}$$

Figure 10.17 summarizes the calculated parameters such as reorganization energy, transfer integral, and mobility for pentacene, rubrene, and $C_8$-BTBT [51]. The hopping mobility is found more than one order lower than the band mobility listed in Table 10.1. Actually, the mobility 3.51 cm$^2$/Vs for rubrene and 0.26 cm$^2$/Vs for $C_8$-BTBT is too low compared to the experimental results, whereas the mobility 1.90 cm$^2$/Vs for pentacene is rather reasonable. Thus, the hopping model is not suitable for the high-mobility crystalline materials such as rubrene and $C_8$-BTBT.

#### 10.1.4.4 Dynamic Disorder Model

In the small polaron model above in 10.1.4.1, the electron-phonon coupling is included as a perturbation. Recently, however, Troisi et al. proposed that the electron-phonon coupling is so large in some organic systems that the perturbative treatment is invalid [36]. Figure 10.18 indicates the fluctuation of the transfer integrals in pentacene thin films based on the molecular dynamics (MD) calculations at 300 K. The average and standard deviation of the transfer integrals were 455.3 ± 258.8 meV, −615.9

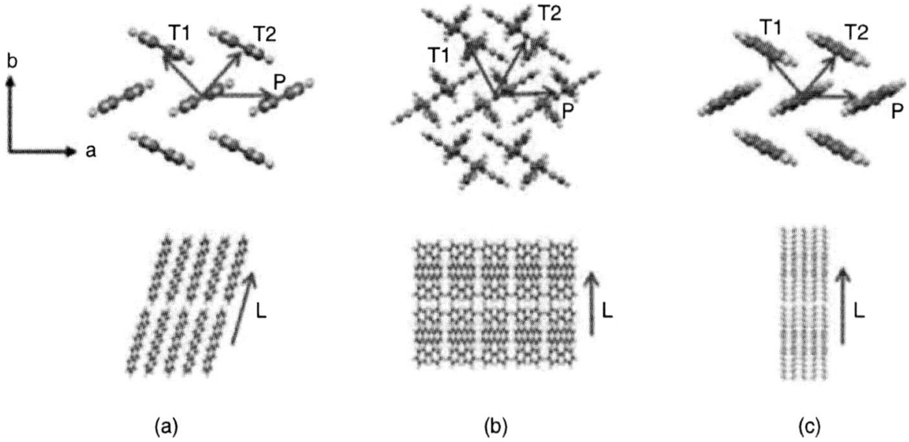


| | | Pentacene | Rubrene | $C_8$-BTBT |
|---|---|---|---|---|
| λ (meV) | | 92 | 152 | 242 |
| *t* (meV) | T1 | 81 | 21 | 26 |
| | T2 | 68 | 21 | 26 |
| | P | 39 | 107 | 67 |
| | L | 0 | 0 | 0 |
| *r* (Å) | T1 | 4.761 | 7.959 | 4.930 |
| | T2 | 5.214 | 7.959 | 4.930 |
| | P | 6.266 | 7.170 | 5.927 |
| | L | 14.530 | 13.866 | 29.180 |
| $\mu^{hop}$ (cm²/Vs) | T1 | 3.36 | 0.02 | 0.01 |
| | T2 | 2.01 | 0.02 | 0.01 |
| | P | 0.32 | 10.48 | 0.77 |
| | L | 0.00 | 0.00 | 0.00 |
| | ave. | 1.90 | 3.51 | 0.26 |

(d)

**Figure 10.17** Calculation of hopping mobility for (a) pentacene, (b) rubrene and (c) $C_8$-BTBT. (d) Summary of the calculated parameters. Source: Ref [51] Reproduced with permission of AIP Publishing LLC.

± 356.4 meV, and 983.3 ± 404.4 meV for the pairs A, B, and C, respectively. Results have revealed that the magnitude of the fluctuation of the transfer integrals is comparable to the transfer integrals themselves.

For the organic systems with large intermolecular vibrations, a relatively new model with a non-perturbative treatment called *dynamic disorder* has been developed by

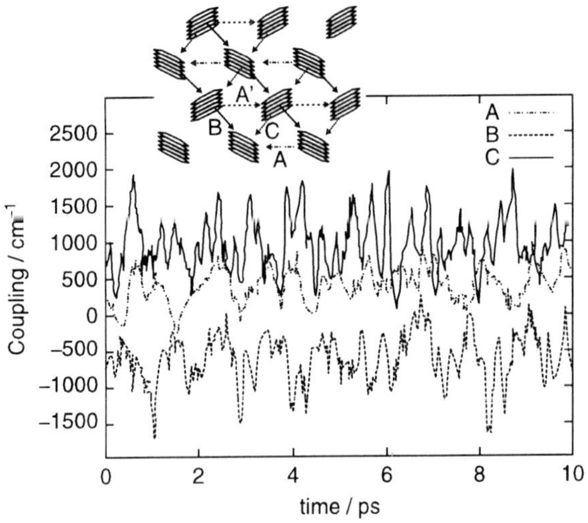


**Figure 10.18** The fluctuation of transfer integrals between three different pairs in pentacene thin films. Source: Ref [36] Reproduce with permission of American Chemical Society.

Troisi et al. and Ishii et al. [36–38, 40]. The model is based on the Hamiltonian for electrons:

$$H_e(t) = \sum_i \varepsilon_i(\Delta \mathbf{u}_i(t)) a_i^\dagger a_i - \sum_{i,j} \tau_{ij}(\Delta \mathbf{R}_i(t), \Delta \mathbf{R}_j(t))(a_i^\dagger a_j + a_j^\dagger a_i). \tag{10.22}$$

Here $\Delta \mathbf{u}_i(t)$ and $\Delta \mathbf{R}_i(t)$ are the coordinates for the intramolecular deformation and the displacement of the $i$th molecule, respectively, $\varepsilon_i$ is the energy of electron at the molecule, and $\tau_{ij}$ is the transfer integral between the $i$th and $j$th molecules. The dependence of $\varepsilon_i$ on $\Delta \mathbf{u}_i(t)$ represents the intramolecular Holstein electron-phonon coupling, associated with the reorganization energy in the small polaron theory. The dependence of $\tau_{ij}$ on $\Delta \mathbf{R}_i(t)$ and $\Delta \mathbf{R}_j(t)$ represents the intermolecular Peierls electron-phonon coupling. The time evolution of the molecular lattice dynamics, $\Delta \mathbf{u}_i(t)$ and $\Delta \mathbf{R}_i(t)$, is calculated classically. Figure 10.19(a) illustrates the energy diagram and molecular lattice coordinates in a pentacene crystal. To calculate mobility, a diffusion coefficient

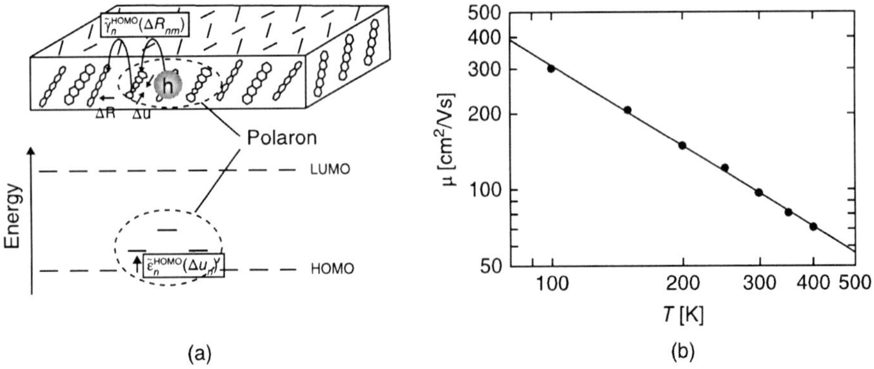


**Figure 10.19** (a) Schematic picture of the hole transport in pentacene crystal with Holstein and Peierls electron-phonon couplings. (b) The calculated temperature dependence of mobility. Source: Ref [40] Reproduce with permission of American Physical Society.

is calculated, for example, by

$$D = \lim_{t\to\infty} \frac{\langle\langle r_i^2 t\rangle - \langle r_i t\rangle^2\rangle}{2t}, \quad (10.23)$$

and then mobility is calculated by the Einstein's relation

$$\mu = \frac{eD}{k_B T}. \quad (10.24)$$

Figure 10.19(b) shows the calculated temperature dependence of mobility. Such a new theory with fewer approximations is expected to explain the charge transport of organic semiconductor in a wide mobility range.

#### 10.1.4.5 Grain Boundary Model

The semiconductor films are usually polycrystalline with a domain size of 0.1–10 μm in small molecular thin-film transistors fabricated by vacuum deposition (Figure 10.20(a)). Even polymer semiconductors such as PBTTT and F8BT have been reported to show polycrystal-like domains by the transmission electron microscopy (TEM) and the scanning transmission X-ray microscopy (STXM) (Figure 10.20(b)). Verlaak et al. classified the domain boundaries in organic semiconductor films into two types: those which contain scatter centers and those which contain trapping centers [33]. In other words, the energy level of carriers at the domain boundaries of the former type is higher than that within crystal domains, whereas the energy level at the domain boundaries of the latter type is lower. The former boundaries naturally form a potential barrier for the carriers. In the latter type, on the other hand, charge carriers are trapped at the boundaries until an equilibrium occupation has been reached. Then, the trapped charge carriers repel following carriers. Accordingly, a potential barrier is formed even in the latter type with trapping centers. Such potential barriers increase the channel resistance at boundaries, and a large potential drop is observed when source-drain voltage is applied [66]. Based on the diffusion theory, the effective mobility in a boundary-dominated regime is

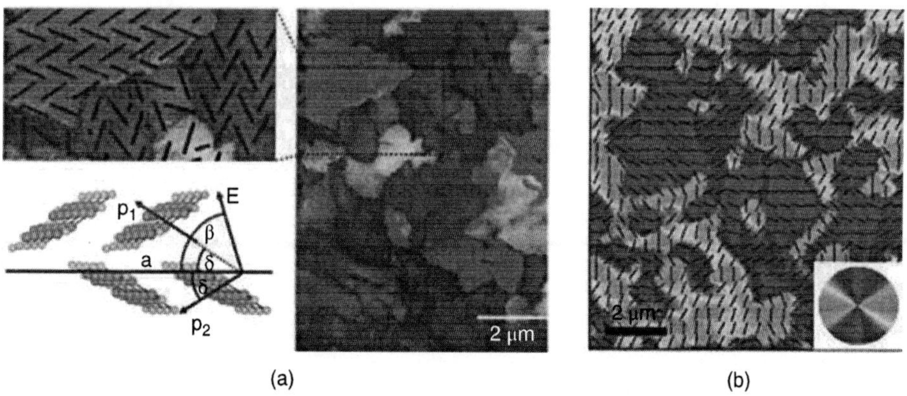


**Figure 10.20** Domain structure of (a) pentacene and (b) PBTTT thin films observed by scanning transmission X-ray microscopy (STXM). Source: Ref [32] Reproduce with permission of John Wiley & Sons.

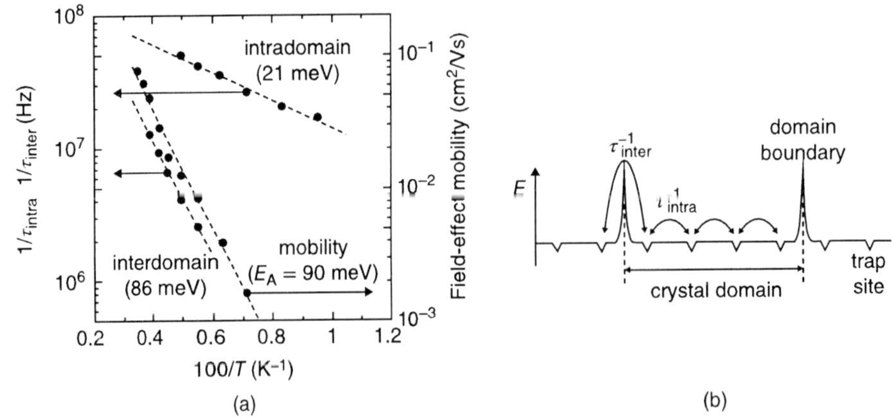


**Figure 10.21** (a) Persistent times at each trap site, $\tau_{intra}$, and at each domain, $\tau_{inter}$, estimated by field-induced electron spin resonance (FI-ESR) analysis. Activation energies are shown in parentheses. (b) Schematics of the intra- and interdomain carrier dynamics in polycrystalline OFETs. Source: Based on Ref. [63] 2012, American Physical Society.

given by

$$\mu = \frac{el}{2k_B T}\mu_{int}\sqrt{\frac{eN_A\phi_B}{2\epsilon_S}}\exp\left(-\frac{e\phi_B}{k_B T}\right). \tag{10.25}$$

Here $l$ is the average size of domains, $\mu_{int}$ is the mobility in domains, $N_A$ is the acceptor (donor) density, $\varphi_B$ is the barrier height, and $\varepsilon_S$ is the dielectric constant.

The intra- and interdomain carrier dynamics in polycrystalline organic thin-film transistors have been investigated independently by FI-ESR measurement [63]. Figure 10.21 shows the persistent times at each trap site, $\tau_{intra}$, and at each domain, $\tau_{inter}$, for PBTTT thin-film transistors estimated from the motional narrowing effect of the FI-ESR spectra. Temperature dependence indicates that the activation energy for the interdomain carrier hopping over domain boundaries is 86 meV, which corresponds to the barrier height, while those for the intradomain carrier hopping between trap sites is 21 meV. The field-effect mobility extracted from transfer curves exhibited an activation energy of 90 meV. These results indicate that the overall transport in the PBTTT transistors is limited by the domain boundaries. In addition, the ratio $\tau_{inter}/\tau_{intra}$ gives an average number of the intradomain hoppings required for traveling to the next domains.

### 10.1.5 Carrier Injection at Source and Drain Electrodes

The carrier injection at source and drain electrodes can be more important than the transport of channels in short-channel transistors, since the ratio of contact resistance to channel resistance increases as channel length decreases [26]. In coplanar-type transistors such as bottom-gate bottom-contact geometry, the contact resistance simply means the resistance at the semiconductor/metal interfaces (Figure 10.22(a)). Because the interface area of the metal electrodes and the charge accumulation layer in the semiconductor is so small, the coplanar-type transistors tend to have high contact resistance. In staggered-type transistors such as bottom-gate top-contact geometry,

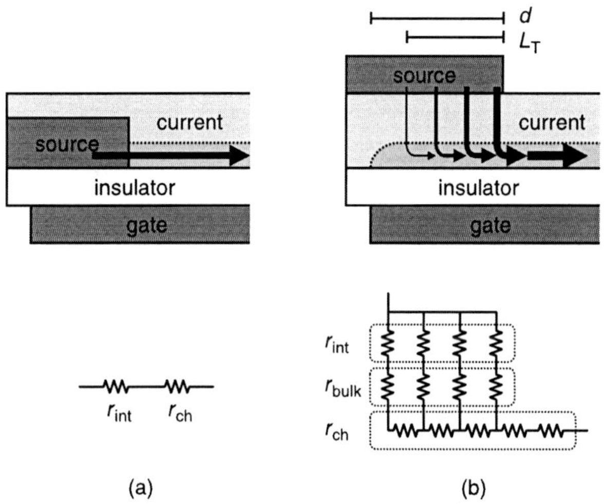


**Figure 10.22** Schematics and equivalent circuit model of the charge injection at source electrodes for (a) coplanar-type and (b) staggered-type transistors.

by contrast, the charge injection occurs in a large area at the semiconductor/metal interface (Figure 10.22(b)). The contact resistance in staggered-type transistors comprises two components: the resistance at the semiconductor/metal interfaces $r_{int}$ and the resistance for the vertical conduction in the bulk semiconductor $r_{bulk}$. Based on the current crowding model, which uses the equivalent circuit shown in Figure 10.22(b), the overall contact resistance $R_c$ [Ω] can be given by [6, 67]

$$R_c = \frac{\sqrt{r_c r_{ch}}}{W} \coth\left(\sqrt{\frac{r_{ch}}{r_c}} d\right) = \frac{r_{ch} L_T}{W} \coth\left(\frac{d}{L_T}\right). \tag{10.26}$$

Here $W$ is the channel width, $r_c = r_{int} + r_{bulk}$ [Ω·cm$^2$] is the sum of interfacial and bulk resistances per unit area, $r_{ch}$ [Ω] is the sheet resistance of the semiconductor channel, and $d$ [cm] is the length of overlap between source and gate electrodes. $L_T$ [cm] is the effective injection length defined as

$$L_T = \sqrt{\frac{r_c}{r_{ch}}}. \tag{10.27}$$

If the overlap length $d$ is large enough, $d \gg L_T$, Eq. (10.26) can be simplified to

$$R_c = \frac{\sqrt{r_c r_{ch}}}{W}. \tag{10.28}$$

Notice that the overall contact resistance $R_c$ has the contribution by the sheet resistance of semiconductor channel $r_{ch}$. Since the $r_{ch}$ generally depends on gate voltage, $R_c$ is also dependent on gate voltage even if $r_c$ is constant. We note that, in the above discussion, all resistances are assumed as Ohmic. The assumption is not always valid because, for example, the semiconductor/metal interfaces may be Schottky. In such cases, more complicated models are required.

To measure the contact resistance $R_c$ in OFETs, two methods called the transmission line method (TLM) and four-terminal measurement are widely used.

#### 10.1.5.1 Transmission Line Method (TLM)

In the transmission line method (TLM), the resistances of several transistors with a variety of channel lengths are measured at the same gate voltage (Figure 10.23(a)) [68]. Since the total resistance of an OFET, $R_{total}$, is given by

$$R_{total}(V_G) = r_{ch}(V_G)\frac{L}{W} + 2R_c(V_G) = \frac{L}{\mu C_i W(V_G - V_{th})} + 2R_c(V_G), \tag{10.29}$$

$2R_c$ can be estimated as the intercept of the extrapolated lines for the $R_{total}$ vs $L$ plot (Figure 10.23(b)). In many cases, contact resistance and channel resistance are evaluated in a normalized form, $R_cW$ [Ω·cm] and $R_{total}W$ [Ω·cm], since $R_c$ and $R_{total}$ are proportional to $1/W$. The normalization helps to compare resistances among devices with different channel width. In addition, the sheet resistance of the semiconductor channel can be estimated from the slope of the same plot as

$$r_{ch} = \frac{\partial(R_{total}W)}{\partial L}. \tag{10.30}$$

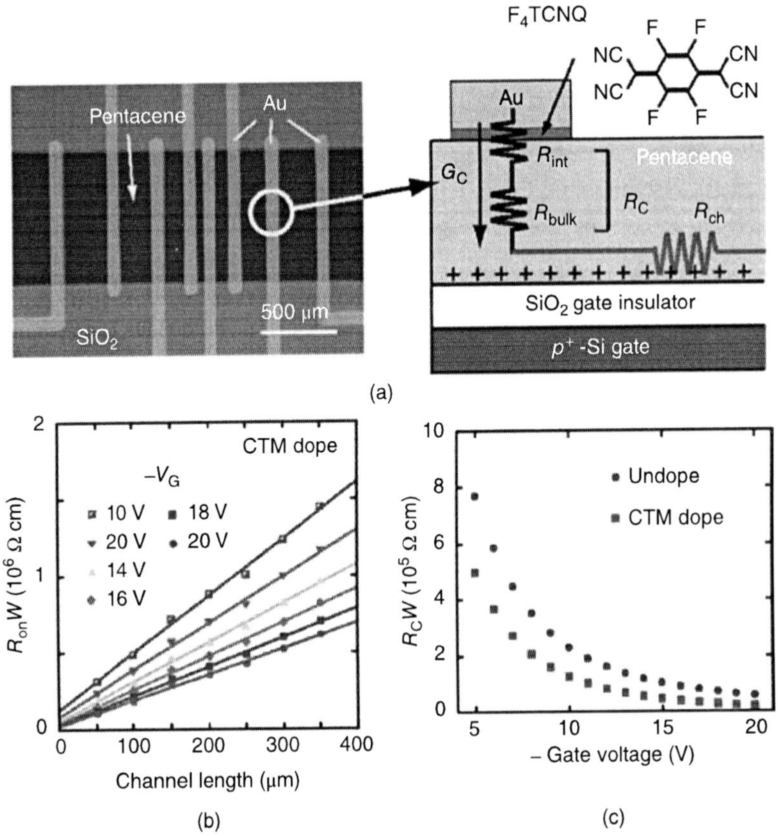


Figure 10.23 An example of transmission line method (TLM). Source: Ref [68] Reproduced with permission of AIP Publishing LLC.

Accordingly, the intrinsic mobility which excludes the effect of contact resistance can be calculated by

$$\mu_{\mathrm{int}} = \frac{1}{C_{\mathrm{i}}} \frac{\partial(1/r_{\mathrm{ch}})}{\partial V_{\mathrm{G}}}. \tag{10.31}$$

Figure 10.23(c) shows the extracted contact resistance for Au/pentacene and Au/$F_4$TCNQ/pentacene contacts by Minari et al. The $F_4$TCNQ layer was inserted in order to reduce the contact resistance by doping carriers in the vicinity of the electrode. In this type of device, the contact resistance is comparable to the channel resistance at a channel length of 50 μm.

Although the TLM is a simple and useful method, attention to the following assumptions need to be paid. Firstly, this method assumes that the semiconductor channels are so uniform that the channel resistance is proportional to the channel length. This is not always true, and needs to be checked by measuring several devices for each channel length. The second assumption is that the contacts are Ohmic and can be expressed as a constant resistance $R_{\mathrm{c}}$. However, the contacts of organic semiconductors and metals frequently exhibit non-linear characteristics like Schottky junctions. The easiest way to check the linearity is to measure the output ($I_{\mathrm{D}}$-$V_{\mathrm{D}}$) characteristics of the OFETs with a short-channel length. If the contacts have non-linear characteristics, non-linearity can be observed in the low $V_{\mathrm{D}}$ region in the output characteristics. Lastly, the measurement must be carried out in a linear regime with a low enough drain voltage.

#### 10.1.5.2 Four-Terminal Measurement

Four-terminal measurement is another popular method to measure contact resistance and intrinsic mobility. The OFETs for four-terminal measurements need at least two additional electrodes for measuring the potential inside the semiconductor channels. Figure 10.24(a) shows the image of the pentacene OFETs for four-terminal measurement [69]. The semiconductor layer (dark gray) was patterned by laser ablation. The electrodes labeled as 1 and 4 are drain and source electrodes, which are used to flow constant drain current $I_{\mathrm{D}}$. The electrodes labeled as 2 and 3 are used to measure the potential inside semiconductor channels by voltmeters. Employing voltmeters with high enough input impedance minimizes the current through electrodes 2 and 3, and hence the voltage drop at the two electrodes are negligible. In this way, the potential inside semiconductor channels is measured precisely. An additional two electrodes at the left side of Figure 10.24(a) can be used to check the uniformity of the semiconductor films. Assuming that the semiconductor film is uniform and the drain voltage is low, the potential profile $V(x)$ inside the channel should be linear (Figure 10.24(b)) and given by

$$V(x) = \frac{V_2 - V_3}{L_{23}}\left(x - \frac{L_{14}}{2}\right) + \frac{V_2 + V_3}{2}, \tag{10.32}$$

where $x$ is the distance from the electrode 4, and $V_2$ and $V_3$ are the voltages measured by the electrodes 2 and 3, respectively. Therefore, the voltage drops at electrode 1 (drain) and electrode 4 (source) can be estimated as

$$\Delta V_1 = V_1 - V(L_{14}) = V_1 - \frac{L_{14}(V_2 - V_3)}{2L_{23}} - \frac{V_2 + V_3}{2}, \tag{10.33}$$

$$\Delta V_4 = V(0) = -\frac{L_{14}(V_2 - V_3)}{2L_{23}} + \frac{V_2 + V_3}{2}. \tag{10.34}$$

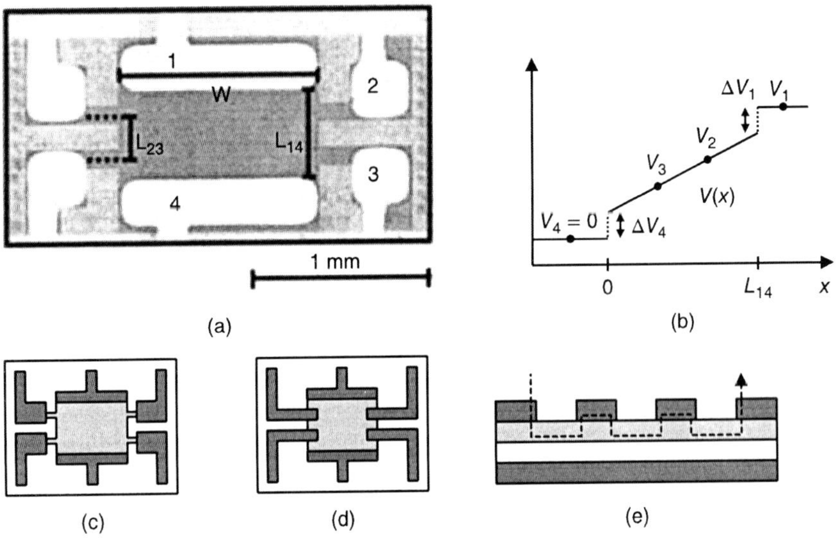


**Figure 10.24** (a) Image of OFET sample for four-terminal measurements [69]. (b) Potential profile between the source and drain electrodes. Schematics of (c) desirable and (d) undesirable device layouts for four-terminal measurement. Dark gray area indicates metal electrodes and light gray area indicates semiconductor layers. (e) Undesirable current path in the device structure (d).

Contact resistances are finally given by $R_{c,D} = \Delta V_1/I_D$ for drain electrodes and $R_{c,S} = \Delta V_4/I_D$ for source electrodes. Sheet conductivity of the channels $\sigma_{sheet}$ can be calculated as

$$\sigma_{sheet} = \frac{I_D}{V_2 - V_3}\frac{L_{23}}{W}, \tag{10.35}$$

and mobility can be calculated by Eq. (10.31).

For precise four-terminal measurements, the layouts of the electrodes and semiconductor layers are very important. Electrodes 2 and 3 should not overlap the channel area where the current flows (Figure 10.24(c)). If the electrodes overlap the channel area as shown in Figure 10.24(d), the current may flow inside the electrodes as illustrated in Figure 10.24(e) because the conductivity of metals is much higher than that of organic semiconductors. This undesirable current path results in the voltage drop at the interface of semiconductor and electrodes 2 (or 3).

#### 10.1.5.3 Effect of Contact Resistance on Apparent Mobility

Neglecting contact resistance can cause serious problems in the estimation of channel mobility. Uemura et al. have pointed out that both the underestimation and overestimation of mobility are possible if contact resistance is neglected [70]. The apparent mobility in linear regime is usually estimated by

$$\mu_{app} = \frac{L}{W}\frac{1}{C_i V_D}\frac{\partial I_D}{\partial V_G}. \tag{10.36}$$

On the other hand, the intrinsic mobility is given by

$$\mu_{int} = \frac{1}{C_i}\frac{\partial \sigma_{sheet}}{\partial V_G} = \frac{L}{W}\frac{1}{C_i}\frac{\partial}{\partial V_G}\left(\frac{I_D}{V_D - 2R_c I_D}\right) \tag{10.37}$$

Be aware that $R_c$ is dependent on $V_G$ as discussed in 10.1.5. Comparing Eqs. (10.36) and (10.37), we finally obtain the relation between $\mu_{app}$ and $\mu_{int}$:

$$\mu_{app} = \left(1 - \frac{2R_c I_D}{V_D}\right)^2 \mu_{int} \pm 2\frac{L}{W}\frac{I_D^2}{C_i V_D^2}\frac{\partial R_c}{\partial V_G} \tag{10.38}$$

Here, the sign of the second term is positive for p-type OFETs and negative for n-type OFETs. If we assume that the $R_c$ is independent of $V_G$, the second term becomes zero. Then, the apparent mobility is lower than the intrinsic mobility as is usually expected. If the $R_c$ is strongly dependent on $V_G$, however, the second term has a positive contribution to the apparent mobility. As a result, the apparent mobility can be higher than the intrinsic mobility.

Figure 10.25 shows an example of the overestimation of mobility. These DNTT thin-film transistors exhibited an unusual peak in the apparent mobility as a function of gate voltage (Figure 10.25(b)). Comparing the apparent mobility with those of TLM and

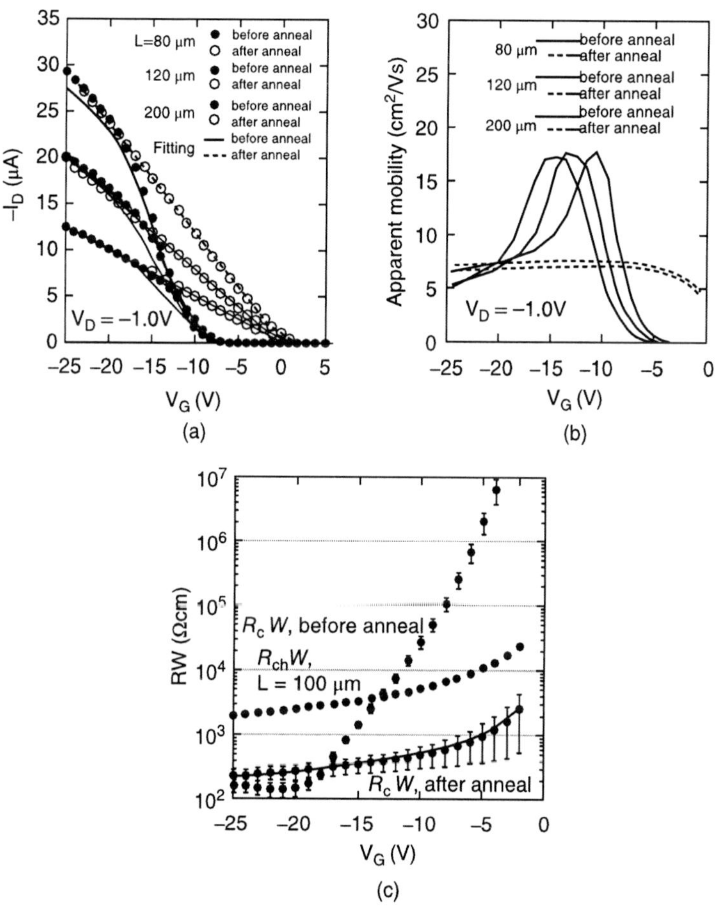


**Figure 10.25** (a) Transfer characteristics, (b) apparent mobility, and (c) gate voltage dependence of contact and channel resistances of DNTT thin-film transistors before and after annealing. Source: Ref [70] Reproduce with permission of John Wiley & Sons.

four-terminal measurement, they have revealed that the apparent mobility was overestimated because of the contact resistance. Actually, these devices exhibited extremely strong gate-voltage dependence of contact resistance as shown in Figure 10.25(c). The overestimation of the mobility occurred at the crossover of the contact resistance $R_c W$ and the channel resistance $R_{ch} W$. Thus, the TLM and four-terminal measurement are so important for the appropriate evaluation of mobility as well as contact resistance.

## References


1 Dimitrakopoulos, C.D. and Malenfant, P.R.L. (2002). *Adv. Mater.* 14: 99.
2 Crone, B., Dodabalapur, A., Lin, Y.-Y. et al. (2000). *Nature* 403: 521.
3 Someya, T., Kato, Y., Sekitani, T. et al. (2005). *Proc. Natl. Acad. Sci.* 102: 12321.
4 Kudo, K., Yamashina, M. and Moriizumi, T. (1984). *Jpn. J. Appl. Phys.* 23: 130.
5 Gundlach, D.J., Zhou, L., Nichols, J.A. et al. (2006). *J. Appl. Phys.* 100: 024509.
6 Richards, T.J. and Sirringhaus, H. (2007). *J. Appl. Phys.* 102 094510.
7 Podzorov, V., Menard, E., Borissov, A. et al. (2004). *Phys. Rev. Lett.* 93: 086602.
8 Minemawari, H., Yamada, T., Matsui, H. et al. (2011). *Nature* 475: 364.
9 Mitsui, C., Okamoto, T., Yamagishi, M. et al. (2014). *Adv. Mater.* 26: 4546.
10 Knipp, D., Street, R.A., Völkel, A., and Ho, J. (2003). *J. Appl. Phys.* 93: 347.
11 Di Carloa, A., Piacenza, F., Bolognesi, A. et al. (2005). *Appl. Phys. Lett.* 86: 263501.
12 Sirringhaus, H., Brown, P.J., Friend, R.H. et al. *Nature* 401: 685.
13 Salleo, A. (2007). *Mater. Today* 10: 38.
14 Noriega, R., Rivnay, J., Vandewal, K. et al. (2013). *Nat. Mater.* 12: 1038.
15 Yoshida, H. and Sato, N. (2008). *Phys. Rev. B* 77: 235205.
16 Mannsfeld, S.C.B., Tang, M.L., and Bao, Z. (2011). *Adv. Mater.* 23: 127.
17 Watanabe, S., Tanaka, H., Kuroda, S. et al. (2010). *Appl. Phys. Lett.* 96: 173302.
18 Soeda, J., Matsui, H., Okamoto, T. et al. (2014). *Adv. Mater.* 26: 6430.
19 Hepp, A., Heil, H., Weise, W. et al. (2003). *Phys. Rev. Lett.* 91: 157406.
20 Sze, S.M. (1981). *Physics of Semiconductor Devices*, 2e. New York: Wiley.
21 Horowitz, G. and Delannoy, P. (1991). *J. Appl. Phys.* 70: 469.
22 Gershenson, M.E., Podzorov, V., and Morpurgo, A.F. (2006). *Rev. Mod. Phys.* 78: 973.
23 Kobayashi, S., Nishikawa, T., Takenobu, T. et al. (2004). *Nat. Mater.* 3: 317.
24 Shiwaku, R., Yoshimura, Y., Takeda, Y. et al. (2015). *Appl. Phys. Lett.* 106: 053301.
25 Rolland, A., Richard, J., Kleider, J.P., and Mencaraglia, D. (1993). *J. Electrochem. Soc.* 140: 3679.
26 Ante, F., Kälblein, D., Zaki, T. et al. (2012). *Small* 8: 73.
27 Anderson, P.W. (1977). *Nobel Lecture.*
28 Vissenberg, M.C.J.M. and Matters, M. (1998). *Phys. Rev. B* 57: 57.
29 Knipp, D. and Northrup, J.E. (2009). *Adv. Mater.* 21: 2511.
30 De Angelis, F., Gaspari, M., Procopio, A. et al. (2009). *Chem. Phys. Lett.* 468: 193.
31 Cramer, T., Steinbrecher, T., Koslowski, T. et al. (2009). *Phys. Rev. B* 79: 155316.
32 McNeill, C. (2011). *Polym. Phys.* 49: 909.
33 Verlaak, S., Rolin, C., and Heremans, P. (2007). *J. Phys. Chem. B* 111: 139.
34 Kang, J.H., da Silva Filho, D., Bredas, J.-L. and Zhua, X.-Y. (2005). *Appl. Phys. Lett.* 86: 152115.
35 Sirringhaus, H. (2009). *Adv. Mater.* 21: 3859.

36 Troisi, A. and Orlandi, G. (2006). *J. Phys. Chem. A* 110: 4065.
37 Troisi, A. (2007). *Adv. Mater.* 19: 2000.
38 Troisi, A. (2010). *Adv. Polym. Sci.* 223: 259.
39 Hannewald, K., Stojanovic, V.M., Schellekens, J.M.T. et al. (2004). *Phys. Rev. B* 69: 075211.
40 Ishii, H., Honma, K., Kobayashi, N., and Hirose, K. (2012). *Phys. Rev. B* 85: 245206.
41 Kubo, T., Häusermann, R., Tsurumi, J. et al. (2016). *Nat. Comm.* 7: 11156.
42 Podzorov, V., Menard, E., Rogers, J.A., and Gershenson, M.E. (2005). *Phys. Rev. Lett.* 95: 226601.
43 Takeya, J., Tsukagoshi, K., Aoyagi, Y. et al. (2005). *Jpn. J. Appl. Phys.* 44: 1393.
44 Hannewald, K. and Bobbert, P.A. (2005). *Physics of Semiconductors*. New York: AIP.
45 Sekitani, T., Takamatsu, Y., Nakano, S. et al. (2006). *Appl. Phys. Lett.* 88: 253508.
46 Yamagishi, M., Soeda, J., Uemura, T. et al. (2010). *Phys. Rev. B* 81: 16130.
47 Uemura, T., Yamagishi, M., Soeda, J. et al. (2012). *Phys. Rev. B* 85: 035313.
48 Yamashita, Y., Tsurumi, J., Hinkel, F. et al. (2014). *Adv. Mater.* 26: 8169.
49 Machida, S., Nakayama, Y., Duhm, S. et al. (2010). *Phys. Rev. Lett.* 104: 156401.
50 Northrup, J.E. (2011). *Appl. Phys. Lett.* 99 062111.
51 Kobayashi, H., Kobayashi, N., Hosoi, S. et al. (2013). *J. Chem. Phys.* 139: 014707.
52 Meijer, E.J., Matters, M., Herwig, P.T. et al. (2000). *Appl. Phys. Lett.* 76: 3433.
53 Lang, D.V., Chi, X., Siegrist, T. et al. (2004). *Phys. Rev. Lett.* 93: 086802.
54 Horowitz, G., Hajlaoui, R., and Delannoy, P. (1995). *J. Phys. III* 5: 355.
55 Fortunato, G., Meakin, D.B., Migliorato, P., and Le Combers, P.G. (1988). *Philos. Mag. B* 57: 573.
56 Grünewald, M., Thomas, P., and Würtz, D. (1980). *Phys. Status Solidi B* 100: K139.
57 Kalb, W.L., Mattenberger, K., and Batlogg, B. (2008). *Phys. Rev. B* 78 035334.
58 Kalb, W.L. and Batlogg, B. (2010). *Phys. Rev. B* 81 035327.
59 Kalb, W.L., Haas, S., Krellner, C. et al. (210). *Phys. Rev. B* 81: 155315.
60 Marumoto, K., Muramatsu, Y., Nagano, Y. et al. (2005). *J. Phys. Soc. J.* 74: 3066.
61 Matsui, H., Hasegawa, T., Tokura, Y. et al. (2008). *Phys. Rev. Lett.* 100: 126601.
62 Matsui, H., Mishchenko, A.S., and Hasegawa, T. (2010). *Phys. Rev. Lett.* 104 056602.
63 Matsui, H., Kumaki, D., Takahashi, E. et al. (2012). *Phys. Rev. B* 85: 035308.
64 Bredas, J.-L., Beljonne, D., Coropceanu, V., and Cornil, J. (2004). *Chem. Rev.* 104: 4971.
65 Coropceanu, V., Cornil, J., da S. Filho, D.A. et al. (2007). *Chem. Rev.* 107: 926.
66 Matsubara, R., Ohashi, N., Sakai, M. et al. (2008). *Appl. Phys. Lett.* 92: 242108.
67 Chiang, C., Martin, S., Kanicki, J. et al. (1998). *Jpn. J. Appl. Phys.* 37: 5914.
68 Minari, T., Miyadera, T., Tsukagoshi, K. et al. (2007). *Appl. Phys. Lett.* 91: 053508.
69 Yagi, I., Tsukagoshi, K., and Aoyagi, Y. (2004). *Appl. Phys. Lett.* 84: 813.
70 Uemura, T., Rolin, C., Ke, T.-H. et al. (2016). *Adv. Mater.* 28: 151.

# 11

# Spontaneous Orientation Polarization in Organic Light-Emitting Diodes and its Influence on Charge Injection, Accumulation, and Degradation Properties

*Yutaka Noguchi, Hisao Ishii, Lars Jäger, Tobias D. Schmidt and Wolfgang Brütting*

CHAPTER MENU



## 11.1 Introduction

Spontaneous orientation polarization (SOP) in organic light-emitting diodes (OLEDs) can indirectly be traced back to the work of Berleb et al. in the year 2000, who concluded on the presence of fixed negative interfacial charge between tris-(8-hydroxyquinolate) aluminum ($Alq_3$) and 4,4′-bis[N-(1-naphthyl)-N-phenylamino]-biphenyl] ($\alpha$-NPD) layers in an archetypical OLED by using impedance spectroscopy [1]. The first direct observation of SOP, however, was reported by Ito et al. in 2002 [2]. In Kelvin probe measurements, they observed that the surface potential of an $Alq_3$ film deposited on an Au substrate grows linearly with a slope of 50 mV/nm as a function of the film thickness. The so-called giant surface potential (GSP) reaches 28 V at 560 nm but diminishes by light absorption of the $Alq_3$ film. Complementary studies using optical second harmonic generation (SHG) revealed that GSP originates from the spontaneous order of the permanent dipole moment (**p**) of $Alq_3$ [2, 3]. The average contribution of **p** to the polarization along the surface normal direction is expressed as the order parameter $\langle \cos\theta \rangle$ being about 0.05 for an $Alq_3$ film, where $\theta$ is the tilt angle of **p** with respect to the surface normal. Thus, the evaporated film of $Alq_3$ is only slightly ordered in the out-of-plane direction but is otherwise amorphous. However, this slight ordering can significantly affect the device properties, because the orientation polarization induces an electric field comparable to that of an operating device ($5 \times 10^5$ V/cm in the case of $Alq_3$).

Because of the photoinduced decay nature [2–5], GSP has not initially been considered as an important parameter in terms of device properties. People have believed that no influences remain in actual devices of OLEDs after vanishing GSP due to the absorption of the ambient light and emission from the device itself. Moreover, GSP has been considered as a unique property for a special class of materials, since there were only few reports on GSP of other materials [6, 7]. Research has focused on the photoinduced decay mechanism of GSP mainly motivated by the interests of fundamental material science rather than device physics [2, 4, 5, 8, 9].

On the other hand, the findings of Berleb and Brütting et al. clearly showed that negative interface charge governs the hole injection and accumulation properties [1, 10, 11]. Hole injection occurs at biases even lower than the built-in voltage ($V_{bi}$) of the device because of the electric field formed by the interface charge. The injected charges are captured at the interface until the fixed interface charge is compensated. In other words, the fixed interface charge determines the minimum amount of the accumulated charge during device operation. The interface charge density is typically as high as $-1$ mC/m$^2$, which corresponds to approximately some ten percent of the total accumulated charge density of the device under operation [12–15]. Though the charge injection characteristics below $V_{bi}$ were well explained by assuming the presence of the interface charge, its origin remained elusive.

Noguchi et al. have revealed that the interface charge and GSP have a common origin, namely the polarization charge due to SOP in the evaporated film [16, 17]. They pointed out that SOP of the $Alq_3$ film induces a constant charge of about 1.4 mC/m$^2$ on the film surface independent of its thickness and that this value is similar to the fixed interface charge density found at the $\alpha$-NPD/$Alq_3$ interface in OLEDs. They also found that GSP is not a unique property of $Alq_3$, but that several commonly used materials in OLEDs show a similar property. Moreover, orientation polarization is maintained in actual devices and thus induces fixed charges at the heterojunctions in these devices [16, 17]. Their results also suggest an alternative mechanism of the GSP decay which was under debate at that time [2, 4, 5, 8, 9]; the molecular order does not vanish due to the light absorption, but GSP decays due to the photo-generated carriers in the film.

Furthermore, researchers at Kodak had observed that the apparent interface charge density decreases proportionally to the loss of luminous efficiency in aged OLEDs. This behavior was discovered by Kondakov et al. in an $Alq_3$-based OLED [18], and the following studies revealed similar characteristics in several OLEDs incorporating polar films [19–22]. As the origin of the decrease in the interface charge density, the generation of hole traps (and trapped holes) and a change of the molecular orientation have been suggested [18–23]. Though the mechanism has not been fully understood, the apparent interface charge density often works as a sensitive probe for device degradation if the device contains a polar film. This is consistent with the fact that the emission zone of a multilayer OLED is confined near the interface due to nature of the charge accumulation [20].

Polar films are possibly included in common organic thin-film devices, since permanent dipole moments are inherent to many organic semiconductors. Spontaneous orientation polarization has not been considered as a significant factor to the device performance, though it may be used unintentionally. As Yokoyama et al. have pointed out, the molecular (transition dipole moment) orientation is quite common in "amorphous" organic semiconductor films and it impacts on the device performance [24–26]. In terms of device optimization, the orientation of molecules, including their permanent and transition dipole moments in the film, should be taken into account as well as other common material properties, such as the energy levels and charge carrier mobility. In this chapter, we present the relations between SOP and the charge carrier dynamics as well as degradation properties in OLEDs, particularly focusing on the interface properties.

## 11.2 Interface Charge Model

The interface charge model describes how SOP affects the charge injection and accumulation behavior [11, 27]. The interface charge model is based on simple electrostatics but explains the charge carrier dynamics in multilayer OLEDs fairly well. In the following discussions, organic semiconductors are assumed as an "insulator", i.e., no thermally generated carriers in the film.

We consider a thin film on an electrode where the film has a spontaneous polarization $\mathbf{P}_0$ (Figure 11.1(a)). At the film surface, the boundary condition

$$(\varepsilon\mathbf{E} + \mathbf{P}_0) \cdot \hat{\mathbf{n}} = 0, \tag{11.1}$$

is satisfied according to Gauss's law. Here, $\varepsilon$ is the dielectric constant of the film, $\mathbf{E}$ is the electric field in the film, $\hat{\mathbf{n}}$ is the unit vector of the surface normal direction. If the film

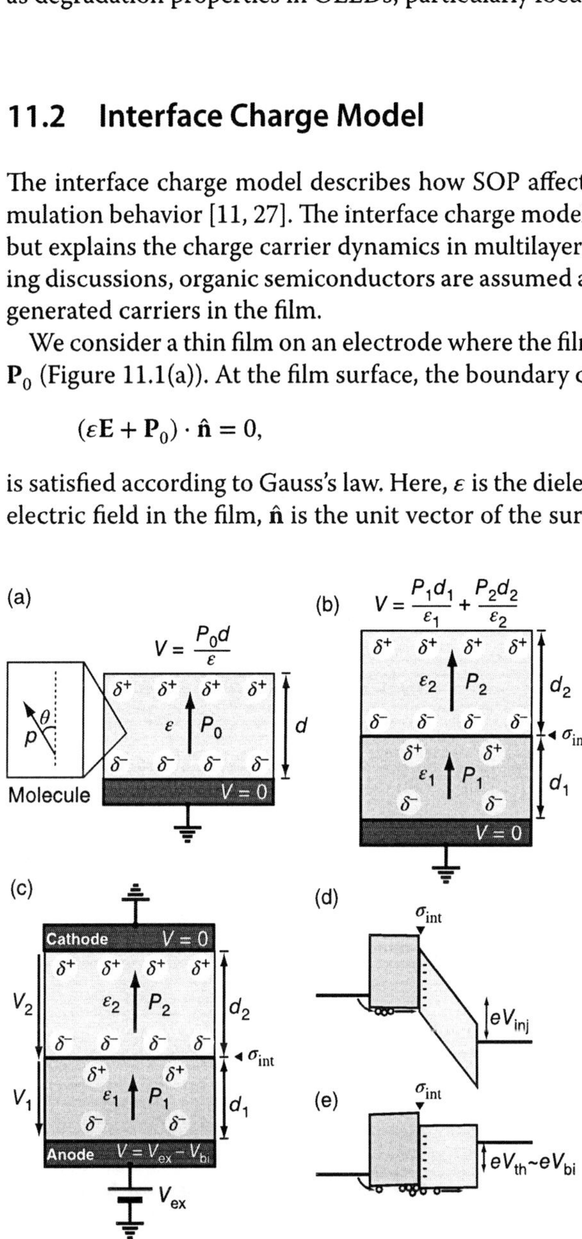


**Figure 11.1** (a)–(c) Schematic illustrations of the orientation polarization in single- and double- layer structures. (a) Orientation polarization and giant surface potential in a single-layer structure. (b) Orientation polarization and giant surface potential in a double-layer structure. (c) Orientation polarization and interface charge in a double-layer device. (d), (e) Schematic illustrations of energy diagram at the hole injection voltage (d) and at the threshold voltage of the actual current (e). Adapted from Ref [28].

is uniform, the potential at the film surface $V_s$ is

$$V_s = -\int_0^d \mathbf{E} \cdot \hat{\mathbf{n}} dz = -(\mathbf{E} \cdot \hat{\mathbf{n}})d = \frac{\mathbf{P}_0 \cdot \hat{\mathbf{n}}}{\varepsilon} d, \tag{11.2}$$

where $d$ is the film thickness. Therefore, the surface potential is proportional to the film thickness if $\mathbf{P}_0$ is constant. This property appears in GSP.

$\mathbf{P}_0$ is defined as $\sum_i \mathbf{p}_i/L^3$, where $\mathbf{p}_i$ is the permanent dipole moment of $i$-th molecule and $L^3$ is the volume of the film. If the film consists of a single component, the contribution of each molecule along the surface normal is given by

$$\mathbf{p}_i \cdot \hat{\mathbf{n}} = |\mathbf{p}| \cos\theta_i, \tag{11.3}$$

thus

$$\mathbf{P}_0 \cdot \hat{\mathbf{n}} = \frac{\sum_i |\mathbf{p}| \cos\theta_i}{L^3} = |\mathbf{p}| \langle \cos\theta \rangle n, \tag{11.4}$$

where $n$ is the density of the molecule, and $\langle \cos\theta \rangle$ is the average orientation angle of the permanent dipole moment with respect to the surface normal direction, namely

$$\langle \cos\theta \rangle = \frac{\sum_i \cos\theta_i}{nL^3}. \tag{11.5}$$

Note that $\sigma = |\mathbf{p}| \langle \cos\theta \rangle n$ corresponds to the polarization charge density induced on the film surface.

In a bilayer structure with different polarization, e.g., $\mathbf{P}_1$ and $\mathbf{P}_2$ (Figure 11.1(b)), GSP is observed as

$$V_s = \frac{\mathbf{P}_1 \cdot \hat{\mathbf{n}}}{\varepsilon_1} d_1 + \frac{\mathbf{P}_2 \cdot \hat{\mathbf{n}}}{\varepsilon_2} d_2, \tag{11.6}$$

where $d_1$ and $d_2$ are the film thicknesses of the first and second layer, respectively. The net polarization charge at the interface is easily obtained as

$$\sigma_{\text{int}} = (\mathbf{P}_1 - \mathbf{P}_2) \cdot \hat{\mathbf{n}} = \varepsilon_1 \frac{V_1}{d_1} - \varepsilon_2 \frac{V_2}{d_2}, \tag{11.7}$$

where $V_1/d_1$ and $V_2/d_2$ correspond to the GSP slope of each film. Note that $\sigma_{\text{int}}$ is independent of the film thicknesses.

Next, we consider the charge injection voltage and charge accumulation characteristics of a bilayer device, where two organic layers are sandwiched between bottom and top electrodes (Figure 11.1c). For simplicity, only hole injection and accumulation are assumed in the following case. The external voltage ($V_{\text{ex}}$) is applied to the bottom electrode with reference to the top electrode. The potential drop in the first and second layer is $V_1$ and $V_2$, respectively. The built-in voltage ($V_{\text{bi}}$) originating from the work function difference between two electrodes and interface dipole at the contacts should be taken into account [29], namely $V_{\text{ex}} - V_{\text{bi}} = V_1 + V_2$. The electric field in the first layer should be positive when the hole injection from the bottom electrode occurs. This condition is independent from the energy barrier height at the electrode/organic film contact. The hole injection voltage ($V_{inj}$) is thus given by $V_{ex}$ when $V_1/d_1 = 0$. The boundary condition at the organic heterojunction is

$$-\varepsilon_1 \frac{V_1}{d_1} - \mathbf{P}_1 \cdot \hat{\mathbf{n}} + \varepsilon_2 \frac{V_2}{d_2} + \mathbf{P}_2 \cdot \hat{\mathbf{n}} = 0. \tag{11.8}$$

Consequently

$$V_{inj} = V_{bi} + (\mathbf{P}_1 - \mathbf{P}_2) \cdot \hat{\mathbf{n}} \frac{d_2}{\varepsilon_2} = V_{bi} + \frac{\sigma_{int}}{\varepsilon_2} d_2. \tag{11.9}$$

Note that the hole injection voltage is proportional to the thickness of the second layer, if $\sigma_{int}$ is constant. This is the fixed charge model proposed by Berleb et al. [1, 11]

The injected holes into the first layer are accumulated at the organic heterointerface, if the electric field in the second layer is negative. Note that this charge accumulation occurs regardless of the energy barrier height at the interface. When the conductance of the first layer is sufficiently high, and the potential drop in the first layer is negligible small, the threshold voltage for hole injection into the second layer ($V_{th}$) equals to $V_{bi}$. The accumulated charge density at the interface at $V_{th}$ is given by

$$\sigma_{acc} = (V_{th} - V_{inj}) \frac{\varepsilon_2}{d_2} = -\sigma_{int}. \tag{11.10}$$

Therefore, the interface charge density can be estimated from capacitance-voltage ($C$-$V$) measurement of the bilayer device [11]. This simple model explains the device characteristics below $V_{th}$ and correlation between the interface charge density and GSP slope well [17].

OLEDs work at applied biases higher than $V_{bi}$. In this region, the interface charge is compensated by the injected counter charge. Thus, the electric field due to the interface charge no longer affects device operation at least from the viewpoint of electrostatics. However, importantly, the interface charge is the polarization charge but the accumulated charge is the real charge. There are excess ionized molecules with the opposite polarity to the interface charge at the heterointerface. The ionized molecules near the emission zone can work as an exciton quencher and induce a molecular decomposition [21, 30–32]. On the other hand, the accumulated charge defines the emission zone and enhances the charge carrier balance factor [12, 33]. Since the interface charge density is comparable to the maximum amount of the accumulation charge density under device operation [12–15], this excess charge accumulation should be taken into account for a detailed understanding of the device operation and degradation mechanisms. Moreover, we have to keep in mind the presence of the counter charge at the metal/organic film interface. The polarization charge may enhance the carrier injection from the electrode. We will discuss these issues in the following sections.

## 11.3 Interface Charge in Bilayer Devices

We show experimental results to verify that the origin of interface charge is SOP [17]. We employed polar and nonpolar molecules which are commonly used in OLEDs, where the molecular structure is shown in Figure 11.2a. As a model system, these molecules were deposited on the $\alpha$-NPD film on an indium-tin-oxide (ITO) coated glass substrate through a conventional vacuum evaporation technique.

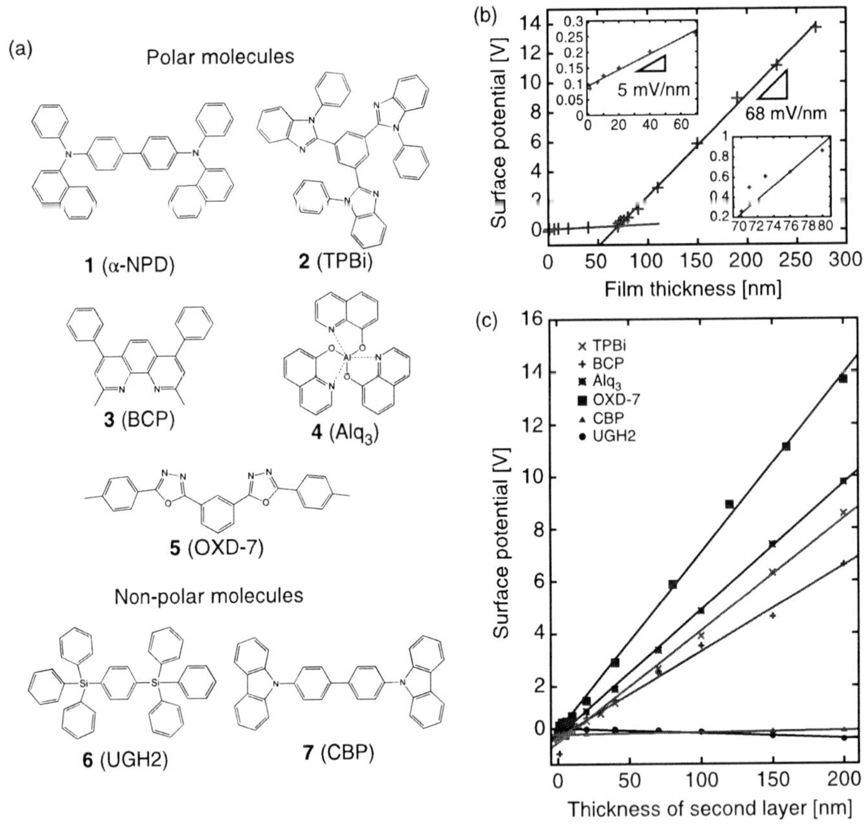


**Figure 11.2** (a) Molecular structure of the materials used in this study. (b) The surface potential of the OXD-7 and $\alpha$-NPD films on ITO as a function of the film thickness. The top inset shows the surface potential of the $\alpha$-NPD film. The bottom inset shows the potential jump at the $\alpha$-NPD/ OXD-7 interface. (c) Surface potential of various films on an $\alpha$-NPD layer. Adapted from Ref [17, 28].

Figure 11.2b shows the surface potential of a 2,2′-(1,3-phenylene)bis[5-(4-tert-butylphenyl)-1,3,4-oxadiazole] (OXD-7) film on $\alpha$-NPD/ITO as a function of film thickness, where the potential reference is the ITO electrode. A clear GSP behavior is observed for the OXD-7 film; the surface potential is proportional to the film thickness with a slope of 68 mV/nm. The GSP slope is about 1.4 times larger than that of $Alq_3$ (48 mV/nm). This value corresponds to 3.4 V for a 50-nm-film, which is a typical thickness in an actual OLED. Since the typical driving voltage of OLEDs is less than 10 V, GSP is significant for device operation. On the other hand, the $\alpha$-NPD film also shows weak GSP with a slope of ~5 mV/nm (inset of Figure 11.2b). Substituting these results into Eq. (11.7), the net polarization charge at the interface is estimated to be $-1.6\ mC/m^2$. Here, the dielectric constant of $\alpha$-NPD and OXD-7 was 3.3 and 2.9, respectively, which was evaluated from the capacitance measurement. Similar GSP characteristics were observed for the other polar molecules; $Alq_3$, bathocuproine (BCP), and 2,2′,2″-(1,3,5-Benzinetriyl)-tris(1-phenyl-1-H-benzimidazole) (TPBi), while not for nonpolar molecules such as 1,4-bis(triphenylsilyl)benzene (UGH2) and 4,4′-bis(N-carbazolyl)-1,1′-biphenyl (CBP). The results are summarized in Figure 11.2c.

The interface charge density was evaluated for the bilayer devices by displacement current measurement (DCM) [27, 34, 35]. DCM is a type of capacitance-voltage ($C$-$V$) measurement that uses a triangular wave voltage and measures all current responses including the actual current ($i_{act}$). The DCM current ($i_{dcm}$) at a quasi-static state is then given by

$$i_{dcm} = i_{act} + \frac{\partial D}{\partial t} \simeq i_{act} + C_{app}\frac{dV}{dt}. \tag{11.11}$$

DCM evaluates not only the apparent capacitance ($C_{app}$), but also the trapped charge density and resistance to charge injection. Details of this technique have been published in Ref. [27].

Figure 11.3a represents a typical DCM curve of ITO/$\alpha$-NPD/OXD-7/Al device. The sweep rate of the triangular wave voltage ($dV/dt$) was 1 V/s, which is slow enough to observe $C_{app}$ at a quasi-static limit. A constant displacement current corresponding to the total geometric capacitance of the device was observed from −6 to −4.5 V. The result indicates no mobile carriers in the device, i.e., the depletion state. The displacement current then gradually increases from −4.5 V ($V_{inj}$) to −3.7 V ($V_{acc}$). The interpretation of this intermediate state has not been well established. A possible explanation is that the injected holes from ITO gradually penetrates into the $\alpha$-NPD layer by filling the mid-gap states and finally reach the $\alpha$-NPD/OXD-7 interface [36]. A constant displacement current is observed again from $V_{acc}$ to 2.5 V ($V_{th}$), indicating that the injected holes are accumulated at the $\alpha$-NPD/OXD-7 interface, i.e., the accumulation state. Finally, the actual current rapidly increases at voltages higher than $V_{th}$ due to hole and electron injection into the OXD-7 layer.

Figure 11.3b shows $V_{inj}$ as a function of OXD-7-film thickness. $V_{inj}$ is proportional to the film thickness with a negative slope. From Eq. (11.9), the presence of negative interface charge is suggested, and the density is estimated as −2.7 mC/m$^2$, where the dielectric constant is 2.9. Similar characteristics were observed for bilayer devices using polar films of $Alq_3$, BCP, and TPBi as the second layer. On the other hand, $V_{inj} \simeq V_{th}$ was observed for devices with nonpolar films of UGH2 (Figure 11.3c) and CBP, indicating that no interface charge resides in the device [17].

Figure 11.4 shows the relationship between the polarization charge density estimated from the GSP slope and the interface charge density estimated from the DCM curves.

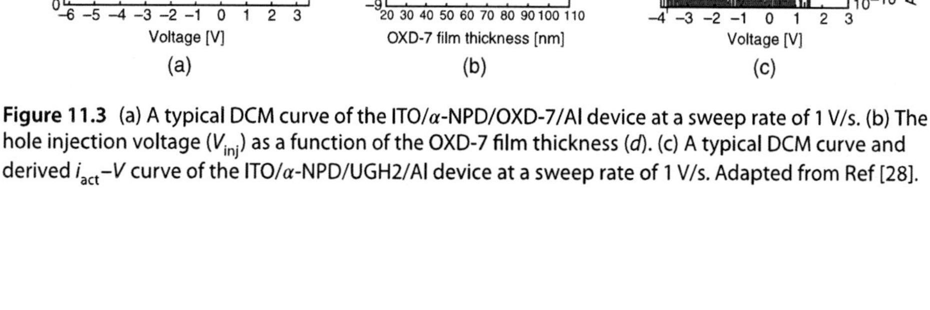


**Figure 11.3** (a) A typical DCM curve of the ITO/$\alpha$-NPD/OXD-7/Al device at a sweep rate of 1 V/s. (b) The hole injection voltage ($V_{inj}$) as a function of the OXD-7 film thickness ($d$). (c) A typical DCM curve and derived $i_{act}$–$V$ curve of the ITO/$\alpha$-NPD/UGH2/Al device at a sweep rate of 1 V/s. Adapted from Ref [28].

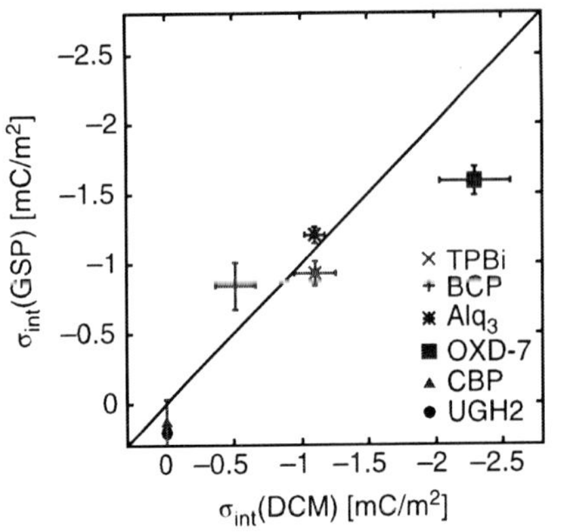


**Figure 11.4** Comparison between the KP and DCM results. Interface charge density estimated from the GSP slope and DCM curves. The solid line with slope 1 is shown. Adapted from Ref [17].

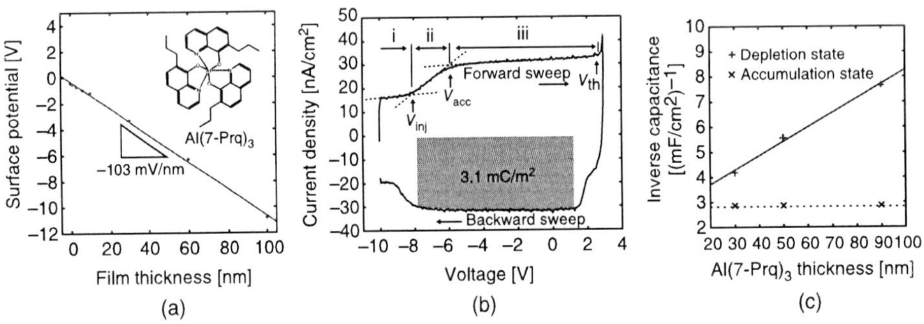


**Figure 11.5** (a)The surface potential of the $Al(7\text{-}Prq)_3$ film on ITO as a function of the film thickness. Molecular structure of $Al(7\text{-}Prq)_3$ is also shown. (b) A typical DCM curve of ITO/$\alpha$-NPD/$Al(7\text{-}Prq)_3$/Ca/Al device at a sweep rate of 1 V/s. (i) Depletion, (ii) intermediate, and (iii) accumulation states are observed in the forward sweep. (c) Inverse of the apparent capacitance at the depletion and accumulation states as a function of the $Al(7\text{-}Prq)_3$ film thickness. The inverse capacitance at the accumulation state is independent of the film thickness, whereas that at the depletion state is proportional. Adapted from Ref [37].

Data points are located around the line with slope 1, indicating that the origin of the interface charge is the polarization charge due to SOP of the second layer.

Though most GSP films including $Alq_3$ show positive GSP (Figure 11.2c), negative GSP was also found in an evaporated film of an $Alq_3$ derivative. We confirmed that the interface charge model is valid for the device incorporating the negative GSP film. Figure 11.5a shows the surface potential of the vacuum evaporated film of an $Alq_3$ derivative, i.e., $Al(7\text{-}Prq)_3$, on the ITO substrate [37]. The negative surface potential grows linearly as a function of the film thickness with a slope of −103 mV/nm, which is approximately two times larger than that of $Alq_3$, but the opposite polarity. This negative GSP in $Al(7\text{-}Prq)_3$ film was discovered by Isoshima et al. [38]. Since not only the molecular structure but also the permanent dipole moment and molecular orbitals of $Al(7\text{-}Prq)_3$ are similar to those of $Alq_3$ [37], it is surprising that a small change in the ligand sphere of the molecule, i.e., attachment of propyl group, induces significant changes in GSP. This result suggests a possible mechanism for the formation of SOP. Isoshima et al. proposed an "asymmetric dice model", in which the statistics of stable positions of the molecule on the surface determine the molecular orientation rather than a dipole-dipole interaction [38].

Figure 11.5b shows the DCM curve of an Al(7-Prq)$_3$-based bilayer device, where the sweep rate is 1 V/s. Depletion, intermediate, and accumulation states clearly appear in the DCM curve [(i), (ii), and (iii) in Figure 11.5b, respectively] [27]. Because of the negative GSP, polarity of the interface charge in this device should be positive, consequently, electron accumulation occurs at the biases lower than the onset of the actual current (2.4 V, $V_{th}$ in Figure 11.5b). Figure 11.5c shows the inverse of $C_{app}$ at the depletion and accumulation states of the bilayer device with varying Al(7-Prq)$_3$ film thickness. Though $1/C_{app}$ at the depletion state linearly increases with the Al(7-Prq)$_3$ film thickness, the value at the accumulation state is constant, indicating that the $\alpha$-NPD layer functions as a capacitor in this regime. Therefore, the accumulated carrier type is electrons, which is consistent with the interface charge model. The density of the positive interface charge is estimated to be 3.1 mC/m$^2$ by integrating the current during the discharging process (Figure 11.5b).

GSP has already been found for several polar molecules including common electron transporters and emitters. Interface charge commonly exists in actual devices, and one should analyze the device operation properties taking this fact into account. Moreover, it is worth mentioning that the average orientation of the permanent dipole moment ($\langle \cos\theta \rangle$) in the polar films ever found is only about 0.05. There is considerable potential for enhancement of the orientation polarization, though the mechanism to build-up GSP has been under debate. Recently, the orientation of transition dipole moment has attracted much attention because its horizontal orientation boosts the light outcoupling efficiency and charge transport property [24, 25]. Similarly, controlling SOP based on the molecular design would be the next important issue to improve device performance and exploit innovative functions of organic semiconductors.

## 11.4 Charge Injection Property

The interface charge dominates the charge injection voltage and minimum amount of accumulation charge density of a multilayer device under operation. Since the origin of the interface charge is SOP, there is the counter-charge with opposite polarity at the other interface of the film, e.g., the organic film/cathode interface in the case of the above-mentioned bilayer devices. The polarization charge at the interface can modify the interface properties, such as the energy level alignment and electronic structure [39, 40], and consequently the charge injection efficiency. In this section, we show the electron injection property of bilayer devices incorporating a polar film at the cathode side [37].

We compared the device properties of ITO/$\alpha$-NPD/Al(7-Prq)$_3$/Ca/Al and ITO/$\alpha$-NPD/Alq$_3$/Ca/Al in order to reveal the impact of orientation polarization. Figure 11.6 shows the current-density–voltage–luminance ($J$–$V$–$L$) characteristics of Alq$_3$ and Al(7-Prq)$_3$-based devices. The conductance of the Al(7-Prq)$_3$ device is remarkably low, which indicates low charge carrier mobilities of the Al(7-Prq)$_3$ film and a high resistance to the charge injection at the interfaces, i.e., $\alpha$-NPD/Al(7-Prq)$_3$ for holes and Al(7-Prq)$_3$/Ca for electrons. Though charge carrier mobilities of the Al(7-Prq)$_3$ film have not been examined, low mobilities are likely because an overlap of molecular orbitals, e.g., the highest occupied molecular orbital (HOMO) and lowest unoccupied molecular orbital (LUMO), between neighboring molecules may be

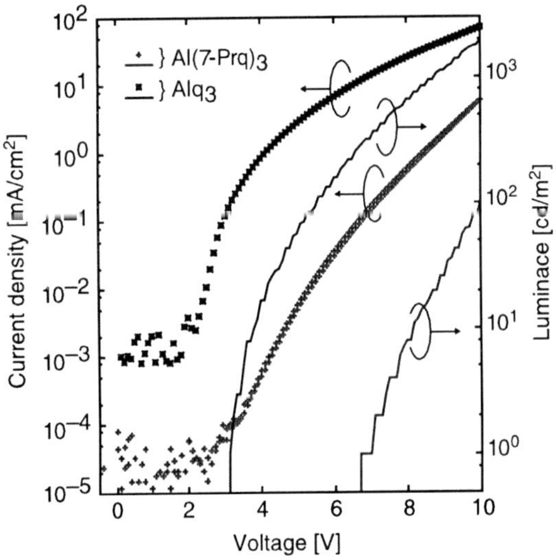


**Figure 11.6** *J–V–L* characteristics of ITO/$\alpha$-NPD/Al(7-Prq)$_3$/Ca/Al and ITO/$\alpha$-NPD/Alq$_3$/Ca/Al devices. Adapted from Ref [37].

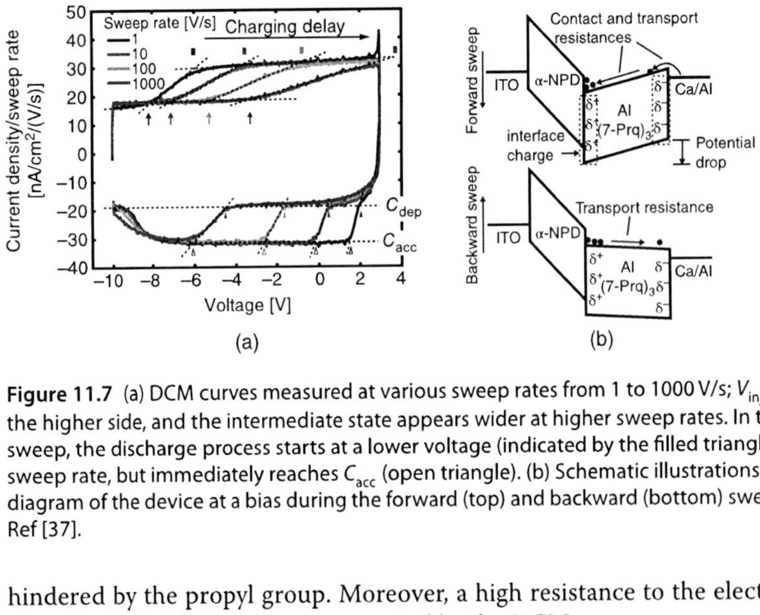


**Figure 11.7** (a) DCM curves measured at various sweep rates from 1 to 1000 V/s; $V_{inj}$ (arrows) shifts to the higher side, and the intermediate state appears wider at higher sweep rates. In the backward sweep, the discharge process starts at a lower voltage (indicated by the filled triangle) with increasing sweep rate, but immediately reaches $C_{acc}$ (open triangle). (b) Schematic illustrations of the energy diagram of the device at a bias during the forward (top) and backward (bottom) sweeps. Adapted from Ref [37].

hindered by the propyl group. Moreover, a high resistance to the electron injection at the Al(7-Prq)$_3$/Ca interface is suggested by the DCM curves.

Figure 11.7a shows the DCM curves at various sweep rates from 1 to 1000 V/s. The vertical axis shows the current density divided by the sweep rate, which corresponds to $C_{app}$. The curves in the forward sweep are significantly distorted with an increase in the sweep rate of the applied voltage; the increase in $C_{app}$ is suppressed at higher sweep rates. This result indicates a long RC time constant ($\tau_{RC}$) for carrier accumulation at the hetero interface of the test device [27, 41, 42]. Because the carrier type in this bias range is electrons, $\tau_{RC}$ can be attributed to the high resistance for electron injection and transport in the Al(7-Prq)$_3$ layer, as suggested by the *J–V* characteristics (Figure 11.6). The high resistance results in a potential drop in the Al(7-Prq)$_3$ layer in the forward sweep (Figure 11.7b, top). In the backward sweep, the observed displacement current is lower (in absolute value) than that corresponding to $C_{dep}$ at biases greater than a certain voltage (indicated by the filled triangles in Figure 11.7a). This is because the charge injection continues after the direction of the bias sweep is changed (Figure 11.7b, top). The extraction process only starts after the potential drop across the Al(7-Prq)$_3$ layer vanishes at the voltage indicated by the filled triangles in Figure 11.7a. At lower voltages, the negative electric field is formed in the Al(7-Prq)$_3$ layer, and the accumulated charge is extracted from the interface (Figure 11.7b, bottom) Interestingly, during the extraction process, $C_{app}$ rapidly increases to the value of the $\alpha$-NPD layer ($C_{acc}$), compared to the injection process (indicated by open triangles in Figure 11.7a). These results suggest that $\tau_{RC}$ of the extraction process is lower than that of the injection process. The asymmetric $\tau_{RC}$ in the forward and backward sweeps can be attributed to the contact resistance, e.g., when the injection barrier for electrons is significant at the Al(7-Prq)$_3$/Ca interface.

If we simply assume that the energy barrier for electron injection is the difference between the LUMO level and work function of Ca, no significant difference is observed between the Al(7-Prq)$_3$/Ca and Alq$_3$/Ca interfaces [37]. The origin of the high-contact resistance can be attributed to the negative polarization charge at the Al(7-Prq)$_3$/Ca interface, which is the counterpart of the positive polarization charge at the $\alpha$-NPD/Al(7-Prq)$_3$ interface. The presence of the negative charge can impede electron injection from the cathode to the Al(7-Prq)$_3$ layer. Similarly, at the $\alpha$-NPD/Al(7-Prq)$_3$ interface, hole injection to the Al(7-Prq)$_3$ layer can be suppressed. In contrast, the use of Alq$_3$ instead of Al(7-Prq)$_3$ at the cathode side can enhance electron injection due to the positive polarization charge at the cathode interface.

This speculation is supported by the observation of high-sensitivity ultraviolet photoemission spectroscopy (UPS) of the polar films. Kinjo et al. reported the direct observation of significantly relaxed negative carriers (anions) at the polar Alq$_3$ film surface [40]. They showed that the electron detachment energy of the anion is about 1 eV larger than the electron affinity of Alq$_3$ measured by inverse photoemission. This significant relaxation energy leads to the good electron injection nature of Alq$_3$. Moreover, Nakayama et al. reported a similar phenomenon at the organic heterojunctions in the bilayer systems of polar or non-polar films on an $\alpha$-NPD layer [39]. A UPS study revealed that the electron transfer is unlikely to occur at the heterojunctions, if polar films such as TPBi and OXD-7 are used as overlayers (in the case of TPBi and OXD-7, the *negative* interface charge exists at the heterojunction). These results suggest that charge transfer can be suppressed by the polarization charge.

The polarity of SOP can play an important role for the efficient charge injection, and a film with a positive orientation polarization, which corresponds to a positive polarization charge at the film surface, could be used as an electron injection layer (EIL). The use of Alq$_3$, BCP, TPBi, and OXD-7 layers as EILs is therefore a reasonable choice in terms of the polarity of the film.

## 11.5 Degradation Property

The apparent interface charge density works as a sensitive probe for device degradation; it decreases proportionally to the loss of luminous efficiency. This relation was first found by Kondakov et al. in an Alq$_3$-based OLED [18], and the following studies revealed similar behavior in other device structures incorporating polar films [19–22]. Because the interface charge adjacent to the emission layer (EML) confines the emission zone of the device due to the charge accumulation nature [20], it seems reasonable that the apparent interface charge density is sensitive to the changes (generation of charge traps, change of orientation polarization and so on) in the emission zone. In this section, we show degradation mechanisms of a phosphorescent OLED incorporating a polar film as an electron transport layer (ETL), studied by a combination of time resolved electroluminescence spectroscopy (TRELS) and DCM [22].

In order to analyze the drop in luminance during the electrical degradation, one has to define the measured quantity. The external quantum efficiency (EQE) of OLEDs is given by four independent factors [43, 44]:

$$\eta_{ext} = \gamma \eta_r q_{eff} \eta_{out}. \tag{11.12}$$

Therein $\gamma$ represents the charge carrier balance, $\eta_r$ is the ratio for creation of radiative excitons, $q_{eff}$ is the effective radiative quantum efficiency, and $\eta_{out}$ is the light outcoupling efficiency of the device. $\eta_r$ is quantum mechanically determined from the spin statistics of two-particles system forming pairs with discrete values for their total spin. The formation probability is 0.25 and 0.75 for singlet ($S = 0$) and triplet ($S = 1$) excitons, respectively. Generally, $\eta_r$ becomes unity for phosphorescent emitters where both species can emit light [45], while 0.25 is the upper limit for fluorescent emitters. However, it is possible to exceed the limit for fluorescent emitters via triplet-triplet annihilation (TTA) [46] or thermally activated delayed fluorescence (TADF) [47, 48]. $q_{eff}$ depends on the competition between radiative and non-radiative processes, which are modified by the Purcell effect taking into account the microcavity-like structure of an OLED [49]. $\eta_{out}$ strongly depends on the refractive indices and thicknesses of all used organic and inorganic layers and the orientation of transition dipole moments of the emitting species [25].

Among the four factors in $\eta_{ext}$, $\gamma$ and $q_{eff}$ are most likely to degrade during device operation, the latter corresponding to an increase in the non-radiative decay rate ($\Gamma_{nr}$). Here, we assume triplet-polaron qenching (TPQ) and trap-assisted recombination (TAR) as the mechanisms of the non-radiative decay induced by device degradation [21, 31, 32]. $q_{eff}$ is defined as the ratio of the radiative exciton decay rate ($\Gamma_r$) to the total decay rate, that is

$$q_{eff} = \frac{F\Gamma_r}{F\Gamma_r + \Gamma_{nr}}. \tag{11.13}$$

Here, $F$ is the Purcell factor, and the non-radiative decay rate $\Gamma_{nr}$ can be split into a pristine contribution $\Gamma_{nr,0}$ and a term growing proportional to the creation of traps

$$\Gamma_{nr} = \Gamma_{nr,0} + \Gamma_{TPQ} n_{tr}, \tag{11.14}$$

where $\Gamma_{TPQ}$ is the triplet-polaron-quenching rate, and $n_{tr}$ is the trapped charge density.

On the other hand, $\gamma$ is defined as the ratio of the recombination current ($j_r$) to the total current ($j_{tot}$); $\gamma = j_r/j_{tot}$, corresponding to $\gamma = 1 - j_{nr}/j_{tot}$, where $j_{nr}$ is the non-radiative recombination current. Note that the charges no longer forming excitons due to TAR account for $j_{nr}$. Following the simplified treatment by Coehoorn et al. [50] and Kuik et al. [51] and neglecting the variation of the current over the width $d_r$ of the recombination zone, $j_{tot}$ and $j_{nr}$ can be written as

$$j_{tot} = 2\frac{e^2}{\varepsilon}\mu d_r n^2, \text{ and } j_{nr} = \frac{e^2}{\varepsilon}\mu n_{tr} d_r n. \tag{11.15}$$

Therein, equal mobilities ($\mu$) as well as carrier densities of electron and holes $n$ are assumed and, overall, bimolecular recombination is assumed to be dominant compared to TAR. From these equations, one obtains

$$\gamma = 1 - \frac{n_{\mathrm{tr}}}{2n}, \tag{11.16}$$

with $n$ depending on the applied current.

According to the above model, the total decay of luminance with respect to the initial luminance ($L/L_0$) induced by trapped charges is

$$\begin{aligned} \frac{L}{L_0} &= \frac{q_{\mathrm{eff}}}{q_{\mathrm{eff},0}}\frac{\gamma}{\gamma_0} \\ &= \frac{F\Gamma_{\mathrm{r}} + \Gamma_{\mathrm{nr},0}}{F\Gamma_{\mathrm{r}} + \Gamma_{\mathrm{nr},0} + \Gamma_{\mathrm{TPQ}} n_{\mathrm{tr}}}\left(1 - \frac{n_{\mathrm{tr}}}{2n}\right) \\ &= \frac{1}{1 + \tau_0 \Gamma_{\mathrm{TPQ}} n_{\mathrm{tr}}}\left(1 - \frac{n_{\mathrm{tr}}}{2n}\right), \end{aligned} \tag{11.17}$$

where, $\tau_0 = (F\Gamma_{\mathrm{r}} + \Gamma_{\mathrm{nr},0})^{-1}$ is the excited states lifetime at the pristine state and $n_{\mathrm{tr}} = 0$ is assumed in the pristine device. Using $\tau_0$, the excited states lifetime ($\tau$) during degradation is expressed as

$$\tau = \frac{\tau_0}{1 + \tau_0 \Gamma_{\mathrm{TPQ}} n_{\mathrm{tr}}}. \tag{11.18}$$

Time-resolved electroluminescent spectroscopy evaluates the exciton lifetime in the device [52, 53]. In addition, the trapped charge density is required for checking the validity of the above-mentioned model. DCM was used in this study for this purpose as it is a powerful tool to evaluate the trapped charge density [27]. In a simplified model of a bilayer device (Figure 11.1c), where the capacitance of the first and second layer is $C_1$ and $C_2$, respectively, the displacement current $i_{\mathrm{dis}}$ is expressed as

$$i_{\mathrm{dis}} = \frac{C_2}{C_1 + C_2}\frac{d\sigma_{\mathrm{inj}}}{dt} + C_{\mathrm{dep}}\frac{dV}{dt}. \tag{11.19}$$

Here, $\sigma_{\mathrm{inj}}$ is the injected charge density at a biasing voltage $V$, and we assumed that only hole injection occurs and the injected holes accumulate at the organic heterointerface. $C_{\mathrm{dep}} = C_1C_2/(C_1 + C_2)$ is the total geometrical capacitance at the depletion state. Therefore

$$C_{\mathrm{app}} = \frac{C_2}{C_1 + C_2}\frac{d\sigma_{\mathrm{inj}}}{dV} + C_{\mathrm{dep}} \tag{11.20}$$

is obtained. The accumulated charge density is

$$\sigma_{\mathrm{inj}} = \int_{V_{\mathrm{inj}}}^{V} (C_{\mathrm{app}} - C_{\mathrm{dep}})\frac{C_1 + C_2}{C_2} dV'. \tag{11.21}$$

If part of the injected charge is trapped at the interface in the forward sweep, the extracted amount of charge in the backward sweep reduces to $\sigma_{\mathrm{inj}} - \sigma_{\mathrm{tr}}$, where $\sigma_{\mathrm{tr}}$ is the trapped charge density at the interface. Thus, $\sigma_{\mathrm{tr}}$ can be estimated by integrating $i_{\mathrm{dis}}$ in the forward and backward sweeps. Moreover, the injection voltage in the subsequent sweeps ($V'_{\mathrm{inj}}$) shifts to the higher side due to the electric field formed by the trapped

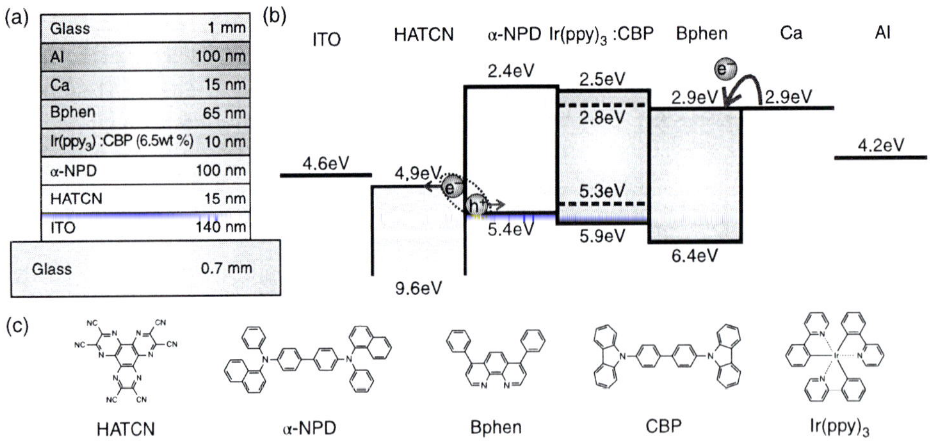


**Figure 11.8** (a) OLED stack layout under investigation. Strictly speaking, the encapsulation glass at the top of the device is not directly lying on the cathode of the OLED and therewith a small amount of the inert gas atmosphere of the glovebox is captured between the cathode and the glass lid. (b) Schematic energy diagram of the HOMO and LUMO levels of the used organic materials [54–57]. (c) Molecular structure of the used organic materials. Adapted from Ref [22].

charge. Then, the trapped charge density, $\sigma_{tr}$, is also given by the differences of the injected charge amount between the two voltage sweeps

$$\sigma_{tr} = \frac{C_1 + C_2}{C_2}\left[\int_{V_{inj}}^{V} (C_{app} - C_{dep})dV' - \int_{V'_{inj}}^{V} (C'_{app} - C_{dep})dV'\right], \tag{11.22}$$

where $C'_{app}$ indicates the apparent capacitance at the steady state (trap-filled state). Therefore, the trapped charge density is evaluated by integrating DCM curves measured for multiple sweep cycles. This method is available for devices with more than two layers, if the traps are located at a certain interface. For accurate estimation of the trapped charge density, charge detrapping treatment, e.g. intense white light irradiation under applying a reverse bias, is necessary prior to the measurement of DCM curves.

Figures 11.8a and 11.8b show the stack layout and the energy diagram of the device under investigation, respectively. The molecular structure of the materials used in this study is presented in Figure 11.8c. Hexa-azatriphenylene-hexanitrile (HATCN) is used as a hole injection layer (HIL) for reducing the contact resistance to hole injection. The emission layer consists of the matrix of CBP doped with 6.5 wt.% of a typical green phosphorescent emitter, fac-tris(2-phenylpyridine)iridium ($Ir(ppy)_3$). 4,7-Diphenyl-1,10-phenanthroline (Bphen) is employed as an ETL. Because of SOP of the Bphen layer, the negative interface charge exists between EML and ETL. A short RC-time constant of about 50 ns is achieved by reducing the pixel size to 1 mm$^2$. This low latency device structure allows for an analysis of the excited states lifetimes of the phosphorescent emitter by TRELS.

An accelerated aging protocol under constant current conditions (1 mA/cm$^2$) was performed. Figure 11.9a shows the aging curves of the device. The luminance decays exponentially, indicating the creation of non-radiative recombination centers inside the EML [50, 58]. After 15 hours of degradation, the luminance dropped to 40% of the initial

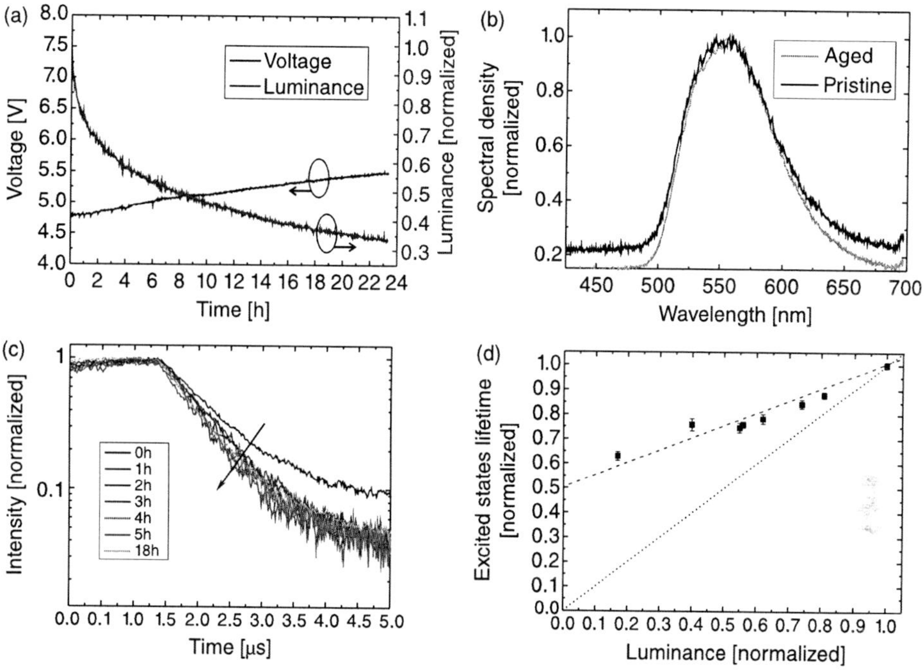


**Figure 11.9** (a) Degradation behavior of the OLED under investigation under constant current conditions (1 mA/cm$^2$) resulting in an initial luminance value around 500 cd/m$^2$. (b) Emission spectra of the pristine device and after 18 h electrical stressing. (c) Excited states lifetime investigations via TRELS during the degradation process. The excited states lifetime (mono-exponential fits) is subsequently decreased with aging time. (d) Excited states lifetime (dots) normalized by the pristine value of 0.81 μs as a function of the normalized luminance during degradation. The data can be fitted with a linear function (red dashed line) with a slope of 0.5. The one-to-one correlation is shown as black dotted line. Adapted from Ref [22].

value while the required voltage increased by 0.5 V (10% of the initial value, 4.8 V). These results suggest that the device degradation mainly occurred in the EML.

Figure 11.9c shows the excited state lifetime recorded after a short electrical pulse with a height of 6 V and a duration of 25 μs at various aging times. The lifetime decreased from the initial value of 0.81 μs to 0.61 μs after 18 h electrical aging, while the emission spectrum did not change on this timescale (Figure 11.9b). These results can be explained by the creation of quenching centers, which increases the non-radiative decay rate of the excited dye molecules, while the radiative rate is assumed to stay constant for this system.

Figure 11.9d shows the relation between the excited lifetime and luminance normalized by the initial value. A slope of 0.5 is obtained, indicating that the decay in the excited states lifetime explains only one half of the total luminance loss in this system. The other half can be attributed to the change of the charge carrier balance as expected from Eq. 11.17.

Figure 11.10a and 11.10b shows the DCM curves between −2 V and 3 V of the first and the second voltage sweep at different aging times, respectively. As a detrapping process, we used white light from an incandescent electric lamp combined with the application of

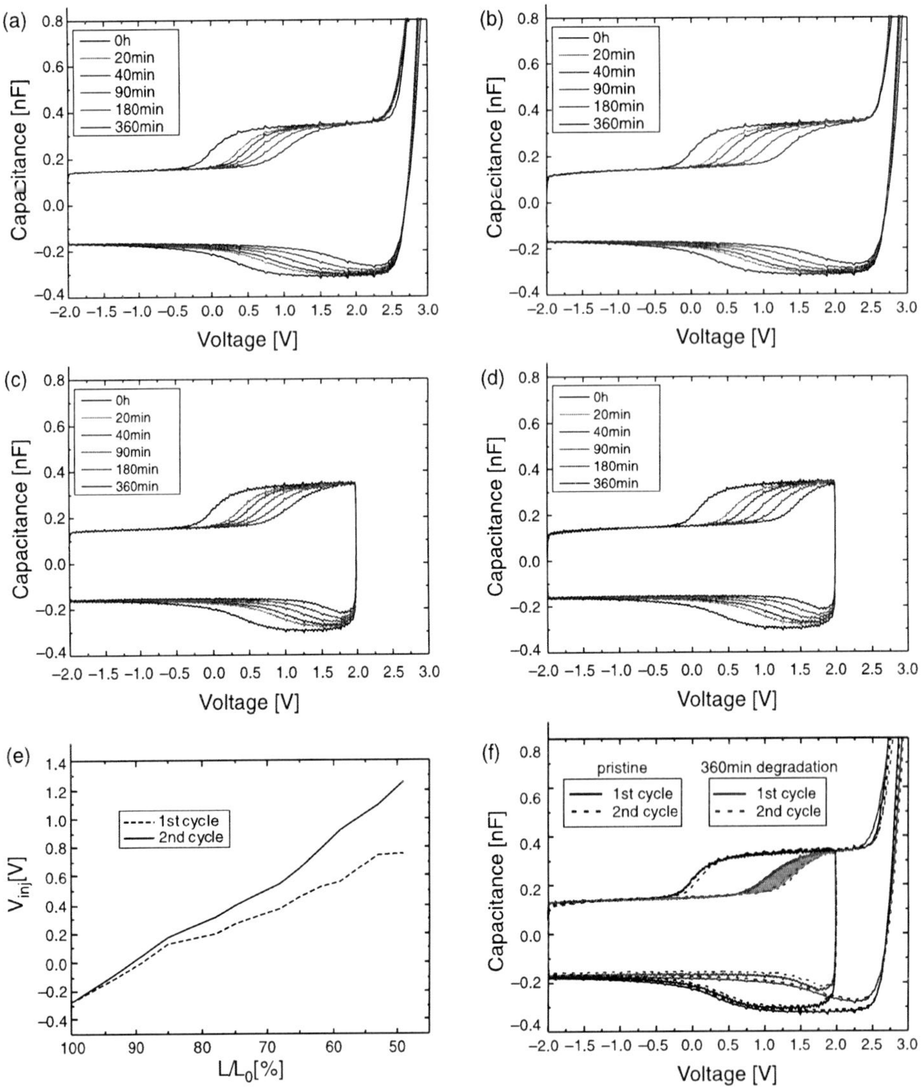


**Figure 11.10** DCM (100 V/s) responses for different degradation steps. (a) and (b) show the measurements in a voltage range of −2 V to 3 V of the first and the second sweep cycle, respectively. (c) and (d) show the measurement in a voltage range of −2 V to 2 V of the first and the second sweep cycle, respectively. Due to the identical behavior of the measurements for both voltage ranges, one can conclude that the trap states are positively charged due to the hole injection only. (e) The relation between the injection voltage and the normalized luminance. The slope for the second sweep is steeper than that for the first sweep. (f) The DCM curves of the first and second voltage sweep (100 V/s) for the pristine and aged (360 min) devices in the voltage range of −2 V to 2 V and −2 V to 3 V. Differences between the first and the second sweep are hardly seen in the DCM curves of the pristine device, whereas clearly seen in those of the aged device. These features are independent of the measured voltage range. The gray region correlates to the trapped charge density (see Eq. [11.22]). Adapted from Ref [22].

a reverse bias of −2 V for 20 s before the first sweep of DCM. The hole injection voltage ($V_{inj}$) shifts to the higher side with aging time while $V_{th}$ stays constant, indicating that the apparent interface charge density decreased with device aging. Moreover, the difference of $V_{inj}$ of the first and second sweep is proportional to the normalized luminance loss (Figure 11.10e) as pointed out by Kondakov et al. [18]. Here, the slope for the second sweep is steeper than that for the first sweep. The results indicate that the polarity of the trapped charge is positive, and the trapped charge density is also proportional to the luminance loss. Note that the rechargeable trap states alone cannot explain all changes in the DCM curves during degradation, because it is not possible to achieve a full recovery of the hole injection voltage by the detrapping process. One possible explanation is the creation of deep-trap states which cannot be released by the mentioned detrapping process or, what is more likely, the orientation polarization of the ETL is changed by electrical ageing [23].

In order to examine the location of the trapped states, DCM curves between −2 V and 2 V were also measured (Figures 11.10c and 11.10d). Within this voltage range, only hole injection occurs and the injected holes are accumulated at the EML/ETL interface due to the presence of the negative interface charge. Nevertheless, $V_{inj}$ of the second sweep shifts to the higher side compared to the first sweep, which is identical to the DCM curves between −2 V and 3 V (Figure 11.10f). Since there are no carriers in the ETL at bias voltages lower than 2 V and the transport properties of the HTL exhibit almost no degradation (confirmed by impedance spectroscopy measurements [not shown]), the trap states are likely to exist near the emission zone, i.e., the EML/ETL interface. We estimated the trapped charge density by integrating DCM curves according to Eq. 11.22 as shown in Figure 11.10f. Similar hole traps were also found in aged devices, such as a $\alpha$-NPD/$Alq_3$-based bilayer OLED [36] and a three-layer OLED doped with a TADF emitter [20].

Kondakov et al. pointed out that cleavage of exocyclic C-N bonds in CBP or $\alpha$-NPD is induced by the interaction of the electrical and optical processes in (near) the emission zone under device operation [31]. It is likely that the decomposed species act as hole traps and exciton quencher through TPQ in the aged devices and they exist only in (or near) the emission zone. Since the interface charge keeps a significant amount of counter charge during device operation, the emission zone could be confined near the interface, and degradation is likely to occur at this position. In order to suppress the creation of exciton quenchers, Nakanotani et al. have proposed OLEDs with a broader emission zone, which results in a longer lifetime [59]. Note that since the interface charge also leads to positive effects on the device properties such as a better charge injection efficiency, one needs to design a device structure to optimize the total device performance.

Figure 11.11a shows the relation between the exciton lifetime and trapped charge density. The trapped charge density was estimated from the DCM curves by assuming a constant trap distribution over 2 nm from the EML/ETL interface. Since, the excited state lifetime modified by the TPQ process is described in Eq. 11.18, $\Gamma_{TPQ}$ can be determined by a curve fitting method. The best fit curve was obtained with $\Gamma_{TPQ}$ of $(0.23 \pm 0.03) \times 10^{-13}$ $cm^3/s$, which is in excellent agreement with published results [60, 61]. Hence, TPQ is a probable cause for the decrease of the effective radiative quantum efficiency. However, this is insufficient to explain the total luminance loss as shown in Figure 11.9d.

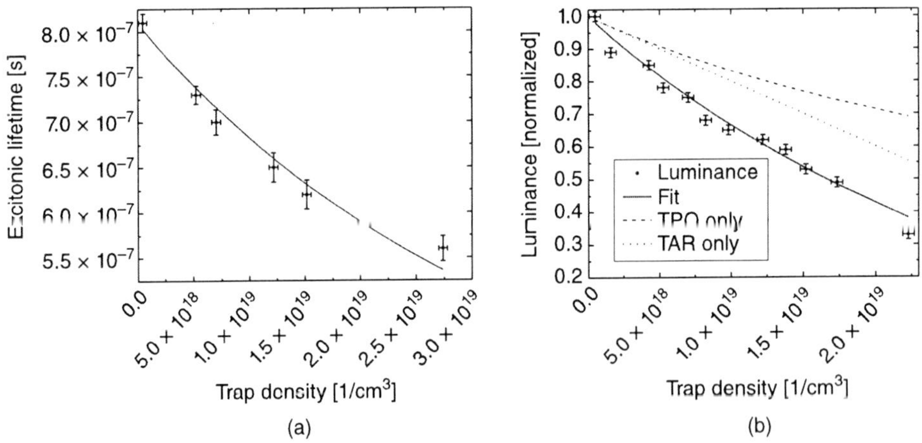


**Figure 11.11** (a) Dots: Excited states lifetime from TRELS experiments as a function of the trapped charge density determined via DCM investigations. Solid line: Fit of the experimental data using the modified TPQ model (Eq. [11.18]). The fit describes the measured date in an excellent way and results in a reasonable value of the TPQ rate of $(0.23 \pm 0.03) \times 10^{-13}$ cm$^3$/s. (b) Normalized luminance as a function of trapped charge density (dots). The red line represents the fit including the individual contributions of TPQ (blue dashed line) and TAR (brown dotted line), respectively. Adapted from Ref [22].

In addition to the TPQ process, the trapped charge influences the charge carrier balance as shown in Eq. (11.16). Figure 11.11b shows the relation between the normalized luminance and trapped charge density. The fit curve is obtained using Eq. 11.18, where the fit parameter is the charge carrier density due to the constant current flow through the device. In Figure 11.11b, the individual contributions of TPQ and TAR are also displayed. The fit describes the data in a reasonable way, however, the extracted charge carrier density seems to be overestimated ($n \sim 10^{19}$ cm$^{-3}$ at the applied current density). This might be caused by the simplified assumptions regarding the description of the charge carrier balance (equal mobilities and carrier densities of electrons and holes as well as bimolecular recombination has been assumed to be dominant compared to TAR). Though the degradation mechanisms in an archetypical OLED have not been fully understood, it is conceivable that the creation of hole traps in (or near) the emission zone play a key role for decreasing radiative quantum efficiency and charge carrier balance.

## 11.6 Conclusions

In this chapter, we have presented SOP in evaporated films of organic semiconductors, and its influence on the properties of OLEDs. The polarization charge induced by SOP appears at the organic heterointerface in the device and works as an "interface charge". Depending on the polarity of the interface charge, corresponding to the direction of the spontaneous polarization, either electron or hole injection into the device occurs even at voltages lower than the built-in voltage. The charge injection voltage is then proportional to the interface charge density and the thickness of the polar film. These primarily injected carriers are accumulated at the heterointerface until they compensate the interface charge. In other words, the interface charge density defines the minimum amount of the accumulated charge during device operation. The typical interface charge density is as much as some ten percent of the maximum amount of the total charge density during device operation. One needs to analyze the electrical and optical properties of the device by taking into account the presence of the excess charge. For instance, the apparent interface charge density decreases proportionally to the loss of luminous efficiency in an aged device. This phenomenon is reasonable because the interface charge confines the emission zone near the heterointerface due to the charge accumulation nature, so that chemical reactions between excitons and ionic species can be activated. Moreover, the counterpart of the interface charge is located at the organic film/cathode interface, if the polar film is used as the electron injection layer. The presence of the positive polarization charge can enhance the electron injection efficiency. The use of positive GSP materials as EIL is therefore a reasonable choice in terms of the polarity of the film.

Spontaneous orientation polarization has already been found for several polar molecules including common electron transporters and emitters. The results can probably be extended to many other polar molecules of organic semiconductors. Though SOP has not been considered as a significant factor to the device performance, in terms of device optimization, the orientational order of permanent dipole moments in the film should be taken into account as a unique material property, such as the energy levels and charge carrier mobilities. Moreover, it is worth mentioning that the average orientation of the permanent dipole moment in the polar films ever found is only about 0.05. There is considerable potential for enhancement of the orientation polarization, though the mechanism to build up GSP needs to be clarified. Controling SOP based on the molecular design would be an important issue to improve device performance and exploit innovative functions of organic semiconductors.

## Acknowledgement

Y.N. and H.I. would like to thank Yasuo Nakayama (Tokyo University of Science), Takashi Isoshima (RIKEN), Eisuke Ito (RIKEN), and students of Ishii group in Chiba University who contributed to the works reviewed in this chapter. This research is granted by the Japan Society for the Promotion of Science (JSPS) through the "Funding Program for World-Leading Innovative R&D on Science and Technology (FIRST Program)," initiated by the Council for Science and Technology Policy (CSTP), the Global-COE Project of Chiba University (Advanced School for Organic Electronics), and KAKENHI (Grant Nos. 21245042, 22750167, 25288114). The work of T.D.S., J.L. and W.B. was funded by the German Federal Ministry of Education and Research (BMBF) under the Contract No. 13N12240 "OLYMP" and by the Deutsche Forschungsgemeinschaft (Contract Nos. Br 1728/13-1 and Br 1728/15-1). T.D.S. also acknowledges support by JSPS summer program 2014. L.J. is a member of the Graduate School "Ressourcenstrategische Konzepte für zukunftsfähige Energiesysteme" (Institute of Materials Resource Management, University of Augsburg).

## References


1 Berleb, S., Brütting, W. and Paasch, G. (2000). *Org. Electron.* 1: 41.
2 Ito, E., Washizu, Y., Hayashi, N. et al. (2002). *J. Appl. Phys.* 92: 7306.
3 Manaka, T., Yoshizaki, K. and Iwamoto, M. (2006). *Curr. Appl. Phys.* 6: 877.
4 Yoshizaki, K., Manaka, T. and Iwamoto, M. (2004). *J. Appl. Phys.* 97: 023703.
5 Ozasa, K., Ito, H., Maeda, M. and Hara, M. (2011). *Appl. Phys. Lett.* 98: 013301.
6 Hayashi, N., Imai, K., Suzuki, T. et al. (2004). *Proc. Int. Symp. Super-Functional. Org. Devices IPAP Conf.* 69.
7 Kröger, M., Hamwi, S., Meyer, J. et al. (2007). *Phys. Rev. B Condens. Matter Mater. Phys.* 75: 1.
8 Sugi, K., Ishii, H., Kimura, Y. et al. (2004). *Thin Solid Films* 464–465: 412.
9 Kajimoto, N., Manaka, T. and Iwamoto, M. (2006). *Chem. Phys. Lett.* 430: 340.
10 Brütting, W., Riel, H., Beierlein, T. and Riess, W. (2001). *J. Appl. Phys.* 89: 1704.
11 Brütting, W., Berleb, S. and Mückl, A.G. (2001). *Org. Electron.* 2: 1.
12 Ruhstaller, B., Carter, S.A., Barth, S. et al. (2001). *J. Appl. Phys.* 89: 4575.
13 Matsumura, M., Ito, A. and Miyamae, Y. (1999). *Appl. Phys. Lett.* 75: 1042.
14 Rohlfing, F., Yamada, T. and Tsutsui, T. (1999). *J. Appl. Phys.* 86: 4978.
15 Taguchi, D., Inoue, S., Zhang, L. et al. (2010). *J. Phys. Chem. Lett.* 1: 803.
16 Noguchi, Y., Sato, N., Tanaka, Y. et al. (2008). *Appl. Phys. Lett.* 92 : 203306.
17 Noguchi, Y., Miyazaki, Y., Tanaka, Y. et al. (2012). *J. Appl. Phys.* 111: 114508.
18 Kondakov, D.Y., Sandifer, J.R., Tang, C.W. and Young, R.H. (2003). *J. Appl. Phys.* 93: 1108.
19 Jarikov, V.V. and Kondakov, D.Y. (2009). *J. Appl. Phys.* 105: 034905.
20 Noguchi, Y., Kim, H.j., Ishino, R. et al. (2015). *Org. Electron.* 17: 184.
21 Scholz, S., Kondakov, D., Lüssem, B. and Leo, K. (2015). *Chem. Rev.* 115: 8449.
22 Schmidt, T.D., Jäger, L., Noguchi, Y. et al. (2015). *J. Appl. Phys.* 117: 215502.
23 Miyamae, T., Takada, N., Yoshioka, T. et al. (2014). *Chem. Phys. Lett.* 616–617: 86.
24 Yokoyama, D. (2011). *J. Mater. Chem.* 21: 19187.
25 Brütting, W., Frischeisen, J., Schmidt, T.D. et al. (2013). *Phys. Status Solidi (a)* 210: 44.
26 Frischeisen, J., Yokoyama, D., Endo, A. et al. (2011). *Org. Electron* 12: 809.
27 Noguchi, Y., Tanaka, Y., Miyazaki, Y. et al. (2012). Displacement Current Measurement for Exploring Charge Carrier Dynamics in Organic Semiconductor Devices. In: *Physics of Organic Semiconductors* (ed. W. Brütting), 119–154. Wiley-VCH, Chapter 5.
28 Noguchi, Y., Nakayama, Y. and Ishii, H. (2015). *J. Vac. Soc. Japan* 58: 37.
29 Ishii, H., Sugiyama, K., Ito, E. and Seki, K. (1999). *Adv. Mater.* 11: 605.
30 Young, R.H., Tang, C.W. and Marchetti, A.P. (2002). *Appl. Phys. Lett.* 80: 874.
31 Kondakov, D.Y., Lenhart, W.C. and Nichols, W.F. (2007). *J. Appl. Phys.*, 101: 024512.
32 Murawski, C., Leo, K. and Gather, M.C. (2013). *Adv. Mater.* 25: 6801.
33 Malliaras, G.G. and Scott, J.C. (1999). *J. Appl. Phys.* 85: 7426.
34 Egusa, S., Miura, A., Gemma, N. and Azuma, M. (1994). *Jpn. J. Appl. Phys.* 33 (Part 1, No. 5A): 2741.
35 Ogawa, S., Kimura, Y., Ishii, H. and Niwano, M. (2003). *Jpn. J. Appl. Phys.* 42 (Part 2, No. 10B): L1275.
36 Noguchi, Y., Tamura, T., Kim, H.J. and Ishii, H. (2012). *J. Photon. Energy* 2: 021214.

37 Noguchi, Y., Lim, H., Isoshima, T. et al. (2013). *Appl. Phys. Lett.* 102: 203306.
38 Isoshima, T., Okabayashi, Y., Ito, E. et al. (2013). *Org. Electron.* 14: 1988.
39 Nakayama, Y., Machida, S., Miyazaki, Y., Nishi, T. et al. (2012). *Org. Electro.* 13: 2850.
40 Kinjo, H., Lim, H., Sato, T., Noguchi, Y. et al. (2016). *Appl. Phys. Express* 9: 021601.
41 Noguchi, Y., Sato, N., Miyazaki, Y. et al. (2010). *Jpn. J. Appl. Phys.* 49 (1 Part 2): 01AA01.
42 Tanaka, Y., Noguchi, Y., Kraus, M. et al. (2011). *Org. Electron.* 12: 1560.
43 Tsutsui, T., Aminaka, E., Lin, C.P. and Kim, D.U. (1997). *Philos. Trans. R. Soc.* 355: 801.
44 Tsutsui, T. and Takada, N. (2013). *Jpn. J. Appl. Phys.*, 52 (11R): 110001.
45 Baldo, M.A., O'Brien, D., You, Y. et al. (1998). *Nature* 395: 151.
46 Kondakov, D.Y., Pawlik, T.D., Hatwar, T.K. and Spindler, J.P. (2009). *J. Appl. Phys.* 106: 124510.
47 Uoyama, H., Goushi, K., Shizu, K. et al. (2012). *Nature* 492: 234.
48 Adachi, C. (2014). *Jpn. J. Appl. Phys.* 53: 060101.
49 Purcell, E.M., Torrey, H.C. and Pound, R.V. (1946). *Phys. Rev.* 69: 37.
50 Coehoorn, R., Van Eersel, H., Bobbert, P.A. and Janssen, R.A.J. (2015). *Adv. Funct. Mater.* 25: 2024.
51 Kuik, M., Koster, L.J.A., Wetzelaer, G.A.H. and Blom, P.W.M. (2011). *Phys. Rev. Lett.* 107: 1.
52 Schmidt, T.D., Setz, D.S., Flämmich, M. et al. (2012). *Appl. Phys. Lett.* 101: 2.
53 Popovic, Z.D., Aziz, H., Hu, N.X. et al. (2001). *J. Appl. Phys.* 89: 4673.
54 Zhou, Y., Shim, J.W., Fuentes-Hernandez, C. et al. (2012). *Phys. Chem. Chem. Phys.*, 14: 12014.
55 Broeker, B. (2010). *Electronic and structural properties of interfaces between electron donor & acceptor molecules and conductive electrodes*. Ph. D. thesis. Humboldt-University, Berlin.
56 Lee, D.H., Liu, Y.P., Lee, K.H. et al. (2010). *Org. Electron.* 11: 427.
57 Chin, B.D., Choi, Y., Baek, H.I. and Lee, C. (2009). *Proc. SPIE* 7415: 741510.
58 Fery, C., Racine, B., Vaufrey, D. et al. (2005). *Appl. Phys. Lett.* 87: 1–3.
59 Nakanotani, H., Masui, K., Nishide, J. et al. (2013). *Sci. Rep.* 3 (Lt 50): 2127.
60 Kalinowski, J., Stampor, W., Mezyk, J. et al. (2002). *Phys. Rev.* 66: 1–15.
61 Reineke, S., Walzer, K. and Leo, K. (2007). *Phys. Rev. B Condens. Matter Mater. Phys.* 75: 1–13.

# 12

# Advanced Molecular Design for Organic Light Emitting Diode Emitters Based on Horizontal Molecular Orientation and Thermally Activated Delayed Fluorescence

*Li Zhao, DaeHyeon Kim, Jean-Charles Ribierre, Takeshi Komino and Chihaya Adachi*

*Center for Organic Photonics and Electronics Research, Kyushu University*

CHAPTER MENU



## 12.1 Introduction

Since the first demonstration of efficient thin-film heterojunction organic light-emitting diodes (OLEDs) by Tang and VanSlyke in 1987 [1], extensive efforts have been devoted to developing the technology. Now, OLEDs are recognized to have a unique combination of features, including flexibility, light weight, and high internal quantum efficiencies reaching close to 100%. Thus, extensive mass production of OLEDs has begun in the flat panel display industry.

OLEDs are current-driven devices similar to conventional inorganic LEDs; however, light emission from OLEDs derives from electrically generated excitonic-states of luminescent molecules. Figure 12.1 shows the basic principles of the emission mechanism in OLEDs. Holes and electrons are injected into an organic layer from the anode and cathode, respectively, when a voltage is applied. The injected holes and electrons are transported in the organic layer through a hopping process [2] and recombine to generate molecular excitons. Finally, light emission is obtained via transitions from the excited state to a ground state.

The external quantum efficiency (EQE, $\eta_{ext}$) of an OLED is defined as the number of photons ($n_p$) emitted from the OLED for each injected charge carrier ($n_e$). This can be expressed by the following equation:

$$\eta_{ext} = \frac{n_p}{n_e} = \gamma \eta_{ST} \Phi_{PL} \eta_{out}, \quad (12.1)$$

where $\gamma$ represents the charge balance factor (ideally $\gamma = 1$), $\eta_{ST}$ is the fraction of radiative excitons, $\Phi_{PL}$ is the photoluminescence (PL) quantum yield, and $\eta_{out}$ is the outcoupling efficiency. Intensive studies have been performed to improve each of these factors and, in turn, the overall efficiency of OLEDs.

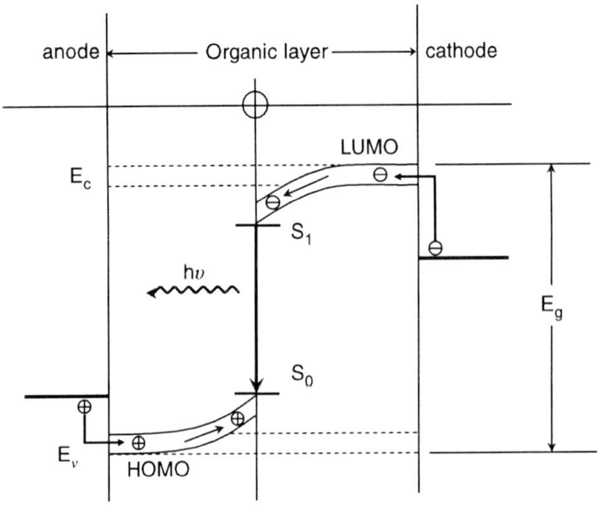


**Figure 12.1** Schematic diagram of emission mechanism in OLEDs.

First, it is crucial to have balanced charge injection, corresponding to $\gamma = 1$. As a highly transparent anode indium-tin-oxide (ITO) is commonly used, which has a relatively high work function (−5.0 eV) [3, 4]. On the cathode side, low work function metals are used, such as Al or Mg/Ag alloys [5, 6]. The process of carrier injection from electrodes to the organic layers plays an important role in optimization of the carrier balance in OLEDs. Efficient injection of carriers requires low energetic barriers or space charge layers to induce tunneling carrier injection [7]. Careful selection of the charge transport material or highly n- and p-doped transport layers [8–11] can lead to small ohmic losses and nearly flat band edges. Furthermore, multilayer structures help to confine carriers and facilitate well-balanced charge injection [12].

Second, the exciton formation process strongly influences the factor $\eta_{ST}$. Injected electrons and holes form excited-states (excitons) that can decay radiatively. The spin statistics of the electrically generated excitons leads to a branching ratio of singlet:triplet = 1:3. Singlet excitons are antisymmetric spin state with a total spin quantum number S = 0. Quantum mechanical laws allow singlet excitons to transition to the ground state, leading to fluorescence, typically with a nanosecond transient lifetime. Conversely, triplet states have an even symmetry with a total spin quantum number S = 1. Although the transition from a triplet state to the singlet ground state is quantum mechanically forbidden, in fact, heavy metal complexes show room temperature phosphorescence with a transient lifetime in the range of microseconds to seconds owing to spin-orbit coupling (SOC). The multiplicities of angular momentum states (i.e., $m_S = 0$ for $S = 0$ and $m_S = -1, 0,$ and $1$ for $S = 1$) and the random nature of spin production in OLEDs, means that statistically, only one quarter of generated excitons can contribute to fluorescence (i.e., from singlet states). The other 75% of excitons might contribute to phosphorescence (from triplet states) in the OLEDs. Thus, injected electrons and holes may be expected to form three times as many triplet states as singlet states; a result that has been verified both theoretically and experimentally [13–15]. Currently, many phosphorescent metal complexes are been used in phosphorescent OLEDs. These devices typically

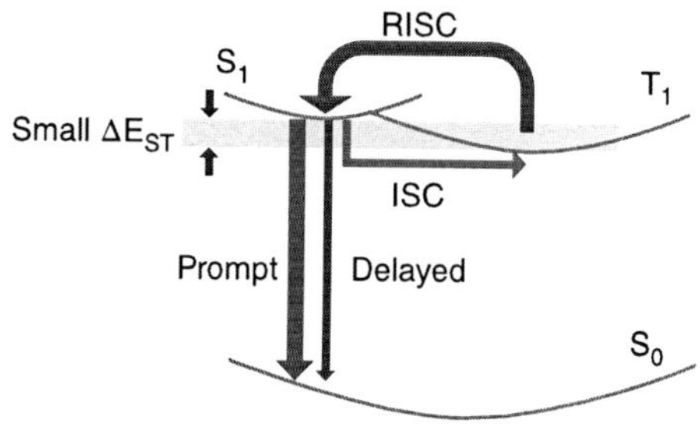


**Figure 12.2** Schematic illustration of TADF process.

feature metal complexes containing precious metals such as iridium, platinum, osmium, and gold. Unlike conventional fluorescent devices, phosphorescent OLEDs can generate light from both triplet and singlet excitons, allowing $\eta_{int} = \eta_{ST}\Phi_{PL}$ to reach values of nearly 100% [16].

Recently, it has been demonstrated that thermally activated delayed fluorescence (TADF) is another promising pathway in OLED devices whereby triplet excitons are converted into a singlet state, resulting in delayed fluorescence [17–20]. TADF occurs through two successive processes, as shown by the green arrows in Figure 12.2. The TADF process is composed of reverse intersystem crossing (RISC) from the lowest energy triplet state ($T_1$) to the lowest excited singlet state ($S_1$) and subsequent radiative decay from $S_1$ to the ground state ($S_0$). RISC is most effective when the energy difference between $S_1$ and $T_1$ ($\Delta E_{ST}$) is small compared with thermal energy at room temperature. Molecular orbital theory reveals that a small $\Delta E_{ST}$ can be achieved by spatially separating the highest occupied molecular orbital (HOMO) and the lowest unoccupied molecular orbital (LUMO) distributions [21]. Previous studies have provided clear evidence that TADF materials can achieve high efficiencies comparable to those of phosphorescent materials [17]. However, unlike phosphorescent materials, TADF materials do not contain rare and precious metals. This advantage makes TADF materials an excellent candidate for lowering the cost of OLED applications.

Third, the last factor involved in the overall device efficiency is $\eta_{ext}$, which is related to the outcoupling efficiency. Only a fraction of generated light is able to escape from the OLED device structure. In a simple model based on ray optics, the light outcoupling efficiency is given by [22]:

$$\eta_{out} = \frac{1}{2n^2}, \tag{12.2}$$

where $n$ represents the average refractive index of the organic layers. Considering that typically n = 1.6–1.8, for organic semiconducting thin films, the power extracted from an OLED is approximately 20%.

An excited molecule can lose energy to different optical channels in a multiple thin-film OLED structure (Figure 12.3). Viewed from the emitter positions, the cone of light which escapes from the OLED has an angle of about 30° with respect to the surface normal and typically amounts to approximately 20% of the total energy of the light output. The low output value is attributed losses to substrate modes from total internal reflection of light at the glass–air interface. At high emission angles, light cannot even reach the glass substrate, but is waveguided in the organic layers (including the transparent ITO electrode). Finally, the emitter can couple to the evanescent field of surface plasmon polaritons (SPPs) at the interface between the metal electrode and

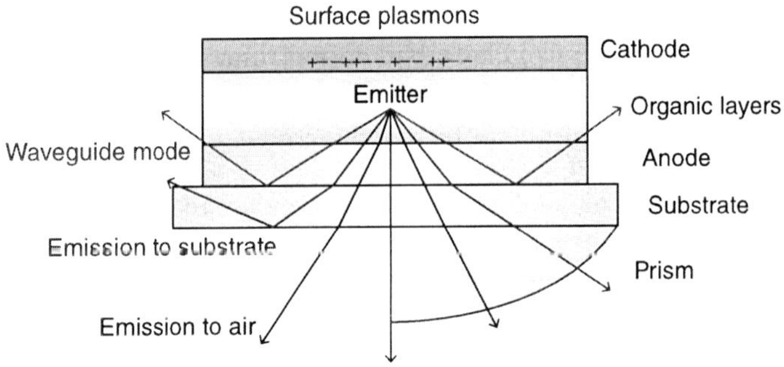


**Figure 12.3** Schematic illustration of OLED showing different optical loss channels.

the organic layers. SPPs are electromagnetic surface waves whose field intensities are trapped and decay exponentially, perpendicular to the interface.

In the method developed by Wasey and Barnes [22], the mode of an incoherent ensemble of dipole emitters embedded in a planar thin-film structure is described by a plane-wave expansion of the electric field under appropriate consideration of the electromagnetic boundary conditions. Simulation results based on this model, indicate that the greatest contribution to the light loss is related to coupling of exciton energy to SPPs, amounting to more than 40% of the total light output.

Researchers have developed several approaches to improve light outcoupling efficiency of OLEDs. If the active pixel area is not too large, light captured in the glass substrate can be fully extracted by attaching an index-matched macroscopic lens. To release energy trapped in waveguided modes, an internal scattering structure could, for example, be placed between the glass substrate and the ITO layer. To reduce energy coupling to SPP modes, one approach is to reduce the excitation of the SPP modes in OLEDs. It should be emphasized that horizontal dipoles couple to various optical channels, whereas vertical dipoles dissipate their energy almost exclusively to SPPs. Thus, perfectly horizontally oriented dipoles would only weakly couple to SPPs [23]. The EQE of OLEDs with a perfect horizontal orientation of emitters can be enhanced by a factor of approximately 1.5 compared with that of OLEDs with randomly oriented emitters, owing to considerably reduced coupling to SPPs when vertical dipoles are absent [24].

Recently, fluorescent [25], phosphorescent [26, 27], and TADF [28] OLEDs with horizontally-oriented guest emitters in host materials have been demonstrated to have external electroluminescent quantum efficiency (EQE) values over the theoretical maximum value predicted for an isotropic orientation of emitters. For instance, EQE values greater than 35% have recently been reported for green phosphorescent OLEDs containing horizontally-oriented emitters [29]. The enhancement of EQE in these devices can be explained by improved light outcoupling efficiency owing to horizontal orientation of the light-emitting transition dipoles. In addition, it should be mentioned that the horizontal molecular orientations in OLEDs can affect the charge transport properties of organic layers and can reduce coupling to surface plasmons at the organic/electrode interfaces [30, 31]. Previous studies have also investigated the dependence on molecular structure for a wide range of OLED materials [32]. On the basis of these results, guidelines for molecular design have been proposed to achieve horizontal molecular orientation in vapor-deposited OLEDs. While compact and bulky emitters show isotropic molecular orientation, light-emitting molecules with linear or planar structures tend to be horizontally oriented, even in an isotropic organic host matrix. The mechanism of these orientational processes is now well understood and many examples have confirmed that light outcoupling and EQE of vapor-deposited OLEDs can be greatly enhanced by horizontally orienting the light-emitting transition dipoles. In the following sections, we discuss the importance of molecular orientation for TADF-OLEDs and solution processed OLEDs.

## 12.2 Molecular Orientation in TADF OLEDs

TADF emitters have a small $\Delta E_{ST}$ that enables the formation of singlet excitons through thermal upconversion of triplet excitons, even at room temperature. Thus, devices based on these materials could ultimately achieve EQE values equal to those of phosphorescence-based OLEDs [33]. Considering fluorescence and phosphorescence-based OLEDs as first and second generations, respectively, TADF emitters can be classified as a third-generation emitter for highly efficient OLEDs. Recently, the best EQE of 31.2% has been demonstrated in green TADF OLEDs [34]. Furthermore, the EQE can be greatly improved by orienting the transition dipoles horizontally to the substrate surface. Such preferential dipole orientation is achievable by controlling the molecular orientation.

Since the discovery of preferential molecular orientation in oligofluorene vapor-deposited thin films in 2004 [35], there has been considerable research interest focused on molecular orientation in thin glassy films made from small molecules. It is now well recognized that vapor-deposited glasses, particularly those based on long linear-shaped molecules, can possess anisotropic molecular orientation; although these systems were previously thought to be isotropic. Early studies in the 1960s were conducted from the viewpoint of basic molecular science to examine the glass transitions of such films. Over the following decades, a large body of research has focused on molecular motion in vapor-deposited glasses during deposition. It has been shown that a deposition temperature ($T_{\text{deposition}}$) much lower than the glass transition temperature can affect the surface mobility [36], giving rise to molecules lying flat on the thin film surface. With regard to device applications, such horizontal orientations have been found in many vapor-deposited glasses used in OLEDs [24, 32]. Thus, this method of orienting molecules might enable enhanced intensity of electroluminescence outcoupling from devices.

Here, we show that cis-BOX2 is a promising TADF molecule, which shows a highly ordered molecular orientation. Three types of green TADF emitters, namely 4CzIPN, PXZ-TRZ, and *cis*-BOX2 doped into CBP or mCBP host matrices were examined at a concentration of 6 wt%, in the fabrication of 15-nm-thick thin films on glass substrates at 200 K by vacuum vapor deposition. To test the impact of molecular orientation on light outcoupling, a *cis*-BOX2-based OLED was also fabricated with the structure ITO (100 nm) / TAPC (50 nm) / 15 wt%-*cis*-BOX2:CBP (5 nm) / 6 wt%-*cis*-BOX2:CBP (25 nm) / PPT (40 nm) / LiF (0.8 nm) / Al (100 nm). The molecular orientation was evaluated by the following equation: $S = (k_e - k_o) / (k_e + 2k_o)$, where the extinction coefficient values were measured at the peak wavelength. This equation gives values of

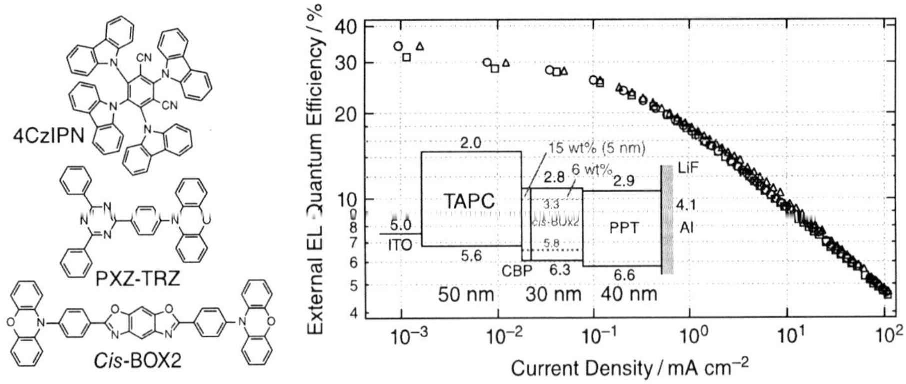


**Figure 12.4** Schematic illustration of OLED showing different optical loss channels.

S = −0.5 if the molecules are parallel to the substrate surface, S = 0 if the molecules are randomly oriented, and S = 1 if the molecules are aligned perpendicular to the substrate surface [32]. Angle-dependent PL measurements indicated that the *S* values for 4CzIPN, PXZ-TRZ, and *cis*-BOX2 doped into the mCBP host were −0.18, −0.31, and −0.44, respectively [37, 38]. Notably, at $T_{\text{deposition}} = 300$ K, the respective *S* values were −0.12, 0.05, and −0.4. This result is reasonable because *cis*-BOX2 has a longer linear structure compared with those of 4CzIPN and PXZ-TRZ (Figure 12.4). By replacing the mCBP host matrix with CBP, the *S* value of *cis*-BOX2 could be further improved to −0.50, meaning that all *cis*-BOX2 molecules were completely horizontally oriented [38]. Figure 12.4 shows the EQE–current density characteristics of a couple of *cis*-BOX2-based OLEDs. The maximum EQE of 33.4 ± 2.0% was obtained at a low current density of approximately $10^{-3}$ A cm$^{-2}$ and this value is the best reported among green TADF-based OLEDs [33]. Even compared with typical phosphorescence-based OLEDs, which have an EQE of ~20%, the obtained EQE is very high. We can attribute the high EQE to enhancement of the light outcoupling efficiency through the strong horizontal orientation of the emitter in the film. In fact, a photonic mode density simulation suggested the light outcoupling efficiency to be 36.0%–38.5%, which was almost the same as the experimentally obtained value of approximately 40%. Thus, the combination of TADF with control over the molecular orientation can be used to achieve highly efficient OLEDs.

## 12.3 Molecular Orientation in Solution Processed OLEDs

Empirical guidelines for molecular design have been established to achieve good horizontal molecular orientation in vapor-deposited OLEDs. Although compact and bulky emitters show isotropic molecular orientations, light-emitting molecules with linear or planar structures tend to be horizontally oriented, even in an isotropic organic host matrix. The mechanism of the orientational processes is now well understood. Many examples have validated the great enhancement of light outcoupling and EQE possible for vapor-deposited OLEDs by inducing horizontal orientation of the light-emitting transition dipoles. Although solution-processed light-emitting polymer films often

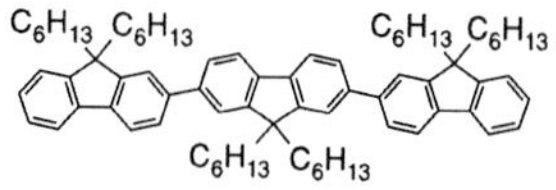


Terfluorene (C6) (d = 2.4 nm)

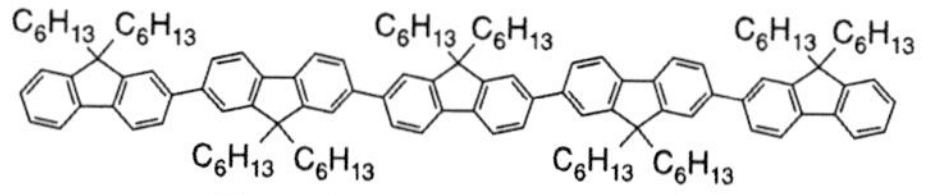


Pentafluorene(d = 4.0 nm)

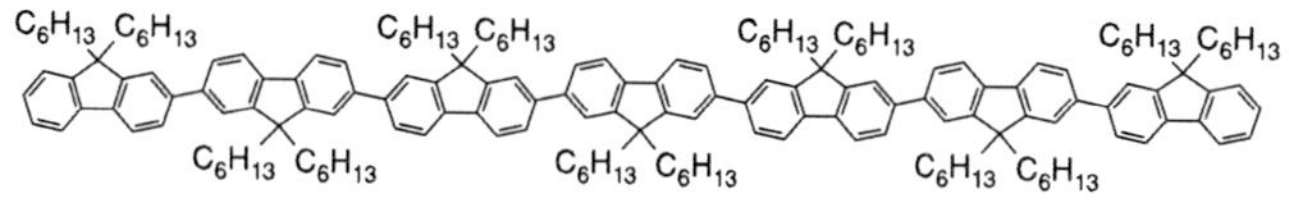


Heptafluorene(d = 5.7 nm)

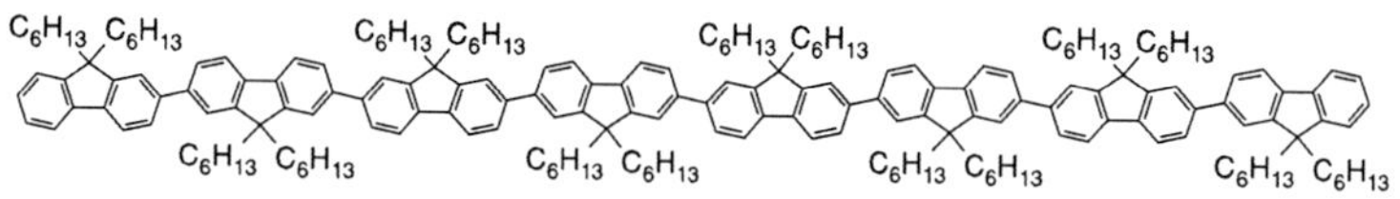


Octafluorene(d = 6.6 nm)

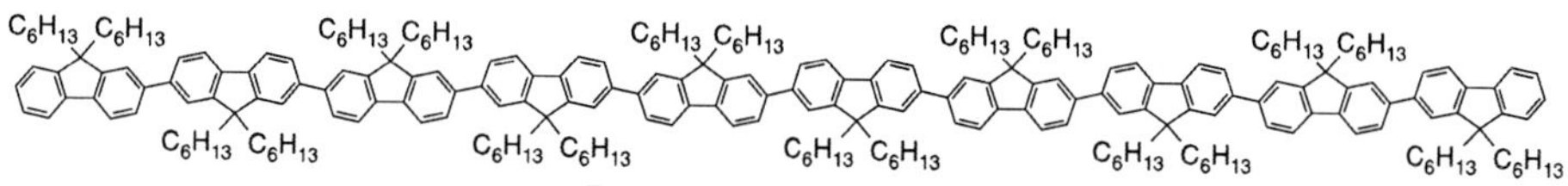


Decafluorene(d = 8.3 nm)

**Figure 12.5** Chemical structures of host molecules and oligofluorenes used in this study.

show a horizontal orientation of polymer chains [39–42], spin-coated glassy organic thin films based on small organic molecules are usually reported to be isotropic [43].

In this section, molecular orientation in thin films of several solution-processable oligofluorene derivatives dispersed into various host systems are discussed. The blends were deposited by spin-coating to form glassy thin films. The chemical structures of the oligofluorenes and host molecules are shown in Figure 12.5.

First, optical constants of a heptafluorene neat film were found to exhibit a small uniaxial optical anisotropy, implying that the molecules were not completely randomly oriented [44]. It is also important to note that the direction of transition dipoles in oligofluorenes lies parallel to the long axis of the molecule [45, 46]. To quantify the degree of molecular orientation, the orientation order parameter (S) was estimated [32]. From our experimental data, we determined an S value of −0.01 for the case of a terfluorene neat film, indicating a random distribution of the emitting dipoles. The S value was calculated to be approximately −0.1 for the pentafluorene, octafluorene, and decafluorene neat films. The S value of heptafluorene neat films has previously been reported to be −0.05 [44]. These results demonstrate that the pentafluorene, heptafluorene, octafluorene, and decafluorene molecules in neat films are only very slightly oriented parallel to the substrate plane.

Variable angle spectroscopic ellipsometer (VASE) measurements are not suitable for providing accurate information on the molecular orientation of oligofluorene emitters dispersed in host materials. Instead, angular dependence of the emission intensity of the blends is a useful tool for examining the orientation of emissive guest dipoles

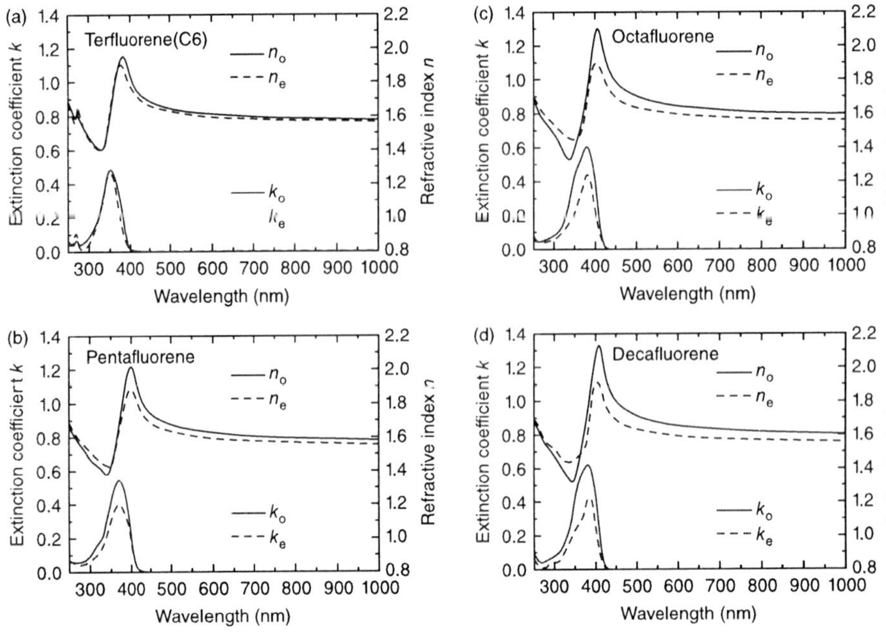


**Figure 12.6** Optical constants (k and n) of (a) terfluorene, (b) pentafluorene, (c) octafluorene. and (d) decafluorene neat films measured by variable angle spectroscopic ellipsometry. In the linear absorption region below 420 nm, the ordinary (in-plane) refractive index $n_o$ and extinction coefficient $k_o$ are slightly higher than the extraordinary (out-of-plane) index $n_e$ and coefficient $k_e$.

quantitatively. The results obtained for terfluorene, heptafluorene, octafluorene, and decafluorene are shown in Figure 12.6. For each CBP blend film, the angular-dependent PL characteristics were simulated for various S values.

Figure 12.7 summarizes the S values as a function of the oligomer length for the four different concentrations in the CBP matrix. The oligomer length is described here by the number of fluorene units contained in the different oligofluorenes. Note that the molecular lengths of the oligofluorenes are indicated in Figure 12.5 and vary from 2.4 nm for the terfluorenes to 8.3 nm for the decafluorene. All oligofluorenes in the CBP blends feature a negative orientation order parameter S value, implying that these light-emitting molecules are horizontally oriented in the host matrix. Notably, in the results obtained for terfluorene, pentafluorene, and heptafluorene, the absolute value of S was found to increase slightly with the number of fluorene units. However, the absolute values of S in the octafluorene blends were smaller than those determined for the heptafluorene and decafluorene blends. The results shown in Figure 12.8 indicate that longer molecular lengths do not necessarily give the best horizontal molecular orientation in spin-coated organic films, in contrast to observations for vacuum-deposited thin films. We also note a trend for lower dopant concentrations to provide slightly better orientation. On the basis of these results, we note that the S values measured for the spin-coated doped films seem to saturate at approximately −0.4 although values of −0.5 can often be achieved in thermally evaporated thin films [47].

Clearly, the method of depositing the molecules plays a critical role in determining the molecular orientation in the film. The densities of spin-coated organic thin films

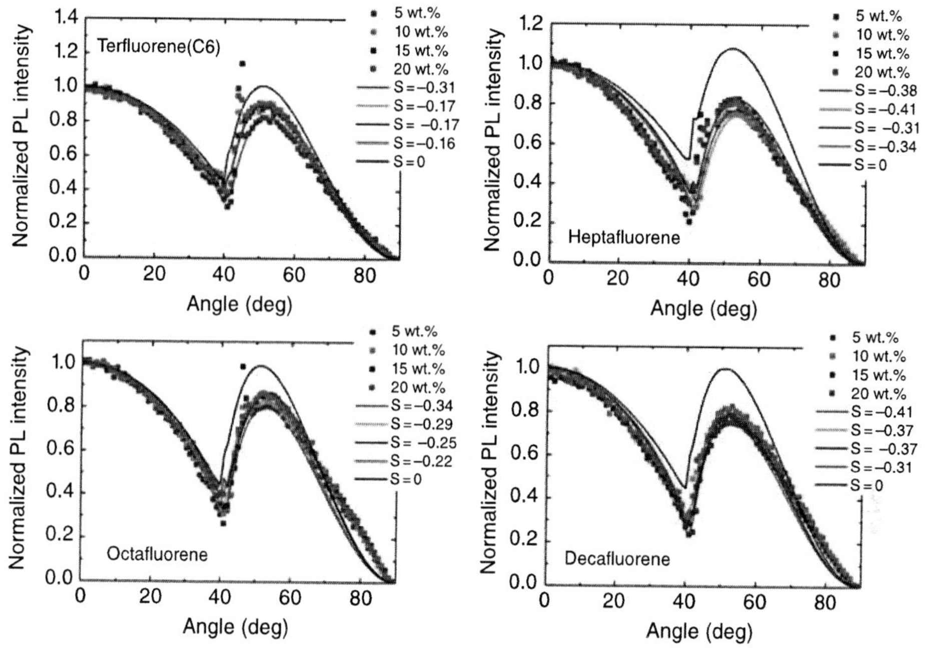


**Figure 12.7** PL intensity in TM mode as a function of emission angle in 15–20 nm thick CBP:terfluorene, CBP:heptafluorene, CBP:octafluorene, and CBP:decafluorene blend films. Concentrations of oligofluorenes in the CBP host matrix were 5, 10, 15, and 20 wt.%. Symbols correspond to experimental data and solid lines show simulation for different S values

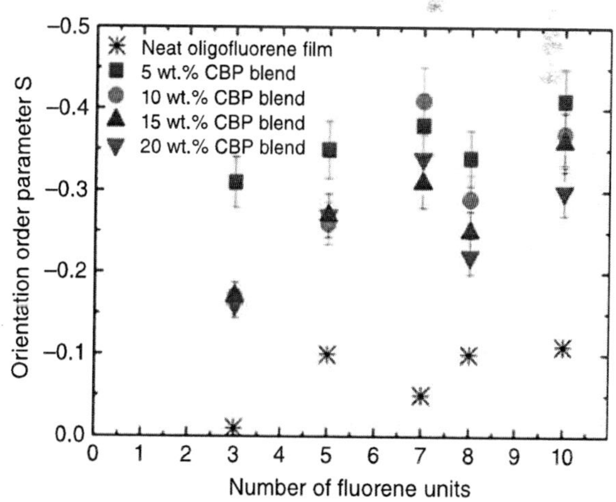


**Figure 12.8** Orientation order parameter S as a function of the oligomer length at four different doping concentrations in the spin-coated glassy doped films characterized by angle-dependent photoluminescence measurements. The S values determined by VASE in neat films are also displayed. The oligomer length is indicated in this figure by the number of fluorene units contained in the oligofluorenes.

have been shown to be lower than those of vacuum-deposited films based on the same material [43]. Layer-by-layer vacuum-deposition allows molecules to reorganize smoothly during the deposition process to minimize the surface energy of the substrate [32]. Weak Van der Waals intermolecular interactions in glassy organic films can induce horizontal molecular orientation of the emitters in the films. However, in spin-coating, the deposition of emitters is controlled by microscopic solution dynamics rather than nanoscopic molecular dynamics at the thin film surface. This environment yields a slightly more random orientation than that achievable in vapor-deposited films. In the case of oligofluorene doping of a CBP host, the number of molecular conformations presumably depends on the oligomer length. Although octafluorene and decafluorene molecules show a good degree of horizontal molecular orientation in CBP blend films, it is plausible that these molecules cannot perfectly align parallel to the substrate during film formation by spin-coating because of their greater molecular flexibility.

In short, the orientation mechanism occurring during spin-coating processes remains unclear; however, it has been experimentally demonstrated that highly oriented emitters can be obtained with the use of guest-host systems processed by spin-coating. Further studies involving computer simulations to clarify the role of the fluid mechanical properties during spin-coating are still needed to fully understand the mechanism of these molecular orientational processes.

## References

1 Tang, C.W. and VanSlyke, S.A. (1987). *Appl. Phys. Lett.* 51: 913.
2 Bässler, H. (1993). *Phys. Status Solidi B* 175: 15.
3 Wu, C.C., Wu, C.I., Sturm, J.C. and Kahn, A. (1997). *Appl. Phys. Lett.* 70: 1348.
4 Kim, J.S., Granström, M., Friend, R.H. et al. (1998). *J. Appl. Phys.* 84: 6859.
5 Kido, J. and Matsumoto, T. (1998). *Appl. Phys. Lett.* 73: 2866.
6 Oyamada, T., Maeda, C., Sasabe, H. and Adachi, C. (2003). *Jpn. J. Appl. Phys.* 42: L1535.
7 Walzer, K., Maennig, B., Pfeiffer, M. and Leo, K. (2007). *Chem. Rev.* 107: 1233.
8 Reineke, S., Lindner, F., Schwartz, G. et al. (2009). *Nature* 459: 234.
9 He, G.-F., Pfeiffer, M., Leo, K. et al. (2004). *Appl. Phys. Lett.* 85: 3911.
10 Blochwitz, J., Pfeiffer, M., Fritz, T. and Leo, K. (1998). *Appl. Phys. Lett.* 73: 729.
11 Huang, J.-S., Pfeiffer, M., Werner, A. et al. (2002). *Appl. Phys. Lett.* 80: 139.
12 Adachi, C., Tokito, S., Tsutsui, T. and Saito, S. (1988). *Jpn. J. Appl. Phys.* 27: 2.
13 Baldo, M.A., O'Brien, D.F., Thompson, M.E. and Forrest, S.R. (1999). *Phys. Rev. B* 60: 14422.
14 Baldo, M.A., O'Brien, D.F., You, Y. et al. (1998). *Nature* 395: 151.
15 Baldo, M.A., Lamansky, S., Burrows, P.E. et al. (1999). *Appl. Phys. Lett.* 75: 4.
16 Adachi, C., Baldo, M.A., Thompson, M.E. and Forrest, S.R. (2001). *J. Appl. Phys.* 90: 5048.
17 Uoyama, H., Goushi, K., Shizu, K. et al. (2012). *Nature* 492: 234.
18 Zhang, Q., Li, J., Shizu, K. et al. (2012). *J. Am. Chem. Soc.* 134: 14706.
19 Sato, K., Shizu, K., Yoshimura, K. et al. (2013). *Phys. Rev. Lett.* 110: 247401.
20 Zhang, Q., Li, B., Huang, S. et al. (2014). *Nat. Photon.* 8: 326.
21 Kaji, H., Suzuki, H., Fukushima, T. et al. (2015). *Nature Comm.* 6: 8476.
22 Wasey, J.A.E. and Barnes, W. (2000). *J. Mod. Opt.* 47: 725.
23 Smith, L.H., Wasey, J.A.E., Samuel, I.D.W. and Barnes, W.L. (2005). *Adv. Funct. Mater.* 15: 1839.
24 Brütting, W., Frischeisen, J., Schmidt, T.D. et al. (2013). *Phys. Status Solidi A* 210: 44.
25 Yokoyama, D., Sakaguchi, A., Suzuki, M. and Adachi, C. (2009). *Appl. Phys. Lett.* 95: 243303.

26 Kim, S.Y., Jeong, W.I., Mayr, C. et al. (2013). *Adv. Funct. Mater.* 23: 3896.
27 Kim, K.H., Moon, C.K., Lee, J.H. et al. (2014). *Adv. Mater.* 26: 3844.
28 Komino, T., Tanaka, H. and Adachi, C. (2014). *Chem. Mater.* 26: 3665.
29 Kim, K.H., Lee, S., Moon, C.K. et al. (2014). *Nature Comm.* 5: 4769.
30 Frischeisen, J., Yokoyama, D., Adachi, C. and Brütting, W. (2010). *Appl. Phys. Lett.* 96: 073302.
31 Frischeisen, J., Yokoyama, D., Endo, A. et al. (2011). *Org. Electron.* 12: 809.
32 Yokoyama, D. (2011). *J. Mater. Chem.* 21: 19187.
33 Uoyama, H., Goushi, K., Shizu, K. et al. (2012). *Nature* 492: 234.
34 Lee, D.R., Kim, B.S., Lee, C.W. et al. (2015). *ACS Appl. Mater. Int.* 7: 9625.
35 Lin, H.W., Lin, C.L., Chang, H.H. et al. (2004). *J. Appl. Phys.* 95: 881.
36 Dalal, S.S., Walters, D.M., Lyubimov, I. et al. (2015). *Proc. Natl. Acad. Sci. USA* 112: 4227.
37 Komino, T., Tanaka, H. and Adachi, C. (2014). *Chem. Mater.* 26: 3665.
38 Komino, T., Sagara, Y., Tanaka, H. et al. (2016). *Appl. Phys. Lett.* 108: 241106.
39 Ramsdale, C.M. and Greenham, N.C. (2002). *Adv. Mater.* 14: 202–215.
40 Quiles, M.C., Alonso, M.I., Bradley, D. and Richter, L.J. (2014). *Adv. Funct. Mater.* 24: 2116.
41 Kim, J.S., Ho, P.K.H., Greenham, N.C. and Friend, R.H. (2000). *J. Appl. Phys.* 88: 1073.
42 Mehes, G., Pan, C., Bencheikh, F. et al. (2016). *ACS Macro Lett.* 5: 781.
43 Shibata, M., Sakai, Y. and Yokoyama, D. (2015). *J. Mater. Chem. C* 3: 11178.
44 Zhao, L., Komino, T., Inoue, M. et al. (2015). *Appl. Phys. Lett.* 106: 063301.
45 Schumacher, S., Ruseckas, A., Montgomery, N.A. et al. (2009). *J. Chem. Phys.* 131: 154906.
46 Fortrie, R., Anemian, R., Stephan, O. et al. (2007). *J. Phys. Chem. C* 111: 2270.
47 Komino, T., Sagara, Y., Tanaka, H. et al. (2016). *Appl. Phys. Lett.* 108: 241106.

# 13

# Organic Field Effect Transistors Integrated Circuits

*Mayumi Uno*

*Osaka Research Institute of Industrial Science and Technology (ORIST), Osaka, Japan*

**CHAPTER MENU**



## 13.1 Introduction

Organic semiconductors are important candidates as materials of active layers of transistors to be applied in flexible electronics, and various semiconductor materials have been proposed [1–16]. Flexible electronics technology allows us to manufacture electric components even on plastic substrates to fit arbitrary shapes of objects, so that various kinds of innovative flexible or wearable applications can be possible for the oncoming smart society of Internet of Things. Organic semiconductor materials have attractive features of both flexibility and electric conductivity, owing to their weak van der Waals interaction between $\pi$-conjugated molecules. They have a great potential to be applied to new and attractive applications that will not be realized using conventional silicon technologies. Recently, high carrier mobility exceeding $10\,cm^2/Vs$ has been reported for newly-synthesized organic materials from several groups [10–15], and developments for realizing practical applications have also advanced.

Organic transistors are crucial devices as building blocks of logic components to be built in flexible devices. Organic devices have great advantages of high flexibility, light weight, and an easy fabrication process of printable and low-temperature conditions, so that many attractive applications have been proposed and developed such as active-matrix elements for plastic displays [17–22], sensor arrays [23–26], low-cost radio-frequency identification (RFID) tags [27–29], and logic components [30–39].

In this chapter, recent developments in organic circuits based on organic semiconductor transistors are described. Elementary circuits are explained in Section 13.2, and advancements of raising the performance of each organic transistor to be built in logic circuits are described in Section 13.3. Then, in Section 13.4, recent reports on integrated organic circuits are described.

## 13.2 Organic Fundamental Circuits

### 13.2.1 Inverter for Logic Components

In logic components, one of the most fundamental elements is an inverter which outputs inverse signals of its input signals, i.e., when the applied input is low, the output becomes high and vice versa, working as a NOT gate. Figure 13.1 shows examples of its circuit diagrams especially reported for organic transistors. Inverters can be constructed using only one single unipolar transistor connected with a resistance, as shown in Figure 13.1a. When a signal "low" is applied to the input $V_{in}$ in the circuit shown in Figure 13.1a, the p-type transistor is switched to on-state, so that the output voltage $V_{out}$ becomes almost the same as the power supply voltage $V_{DD}$. To realize this operation, the resistivity of the resistance must be sufficiently high compared to the impedance of the on-state in the p-type transistor. When a "high" signal is applied to $V_{in}$, the p-type transistor turns to off-state, so that the charges accumulated at $V_{out}$ are discharged through the resistance and $V_{out}$ turns to the "low" state. The impedance of off-state of the p-type transistors should be sufficiently higher than the resistivity of the resistance. In this

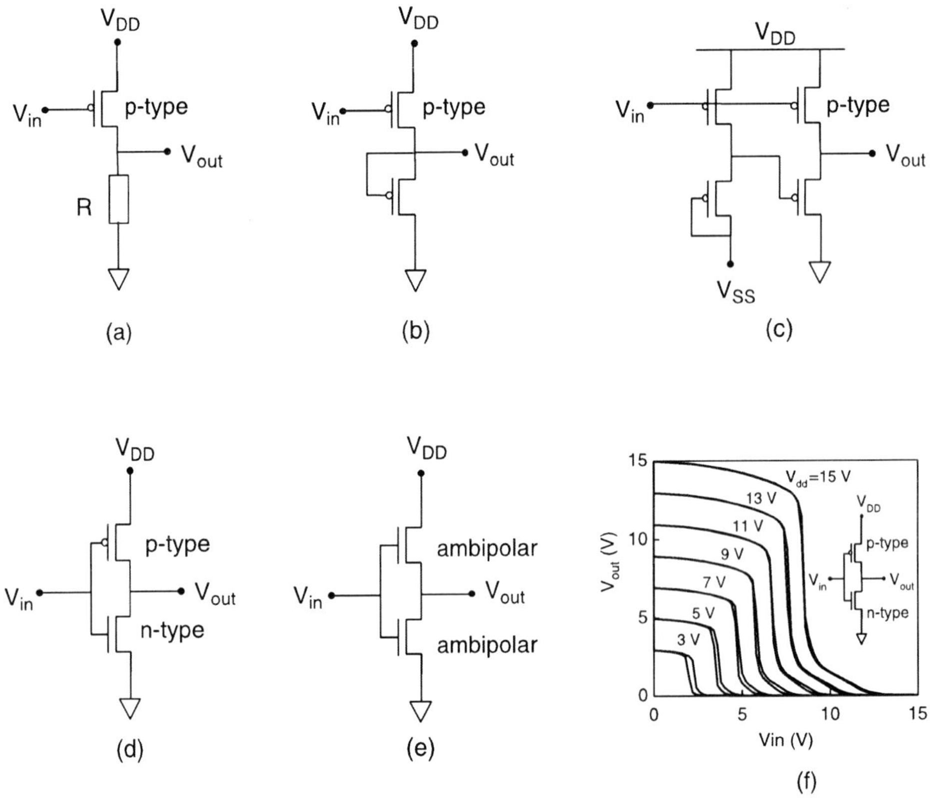


**Figure 13.1** Examples of inverter circuits consisted of (a) a p-type transistor connected with a resistance, (b) p-type transistors (c) a pseudo-CMOS inverter with only p-type transistors, (d) a p-type and n-type transistors, and (e) ambipolar transistors. (f). Source: Part (f) from Ref [35] Reproduced with permission of John Wiley & Sons.

configuration, the penetrating current inevitably occurs, and the output voltages cannot produce full-range voltages, which are serious drawbacks to constructing logic circuits.

To suppress the penetrating current drastically, complementary circuits consisting of p-type and n-type transistors, as shown in Figure 13.1d, have an advantage of low power consumption, so that advancement of silicon technology has been based on complementary metal-oxide-semiconductor (CMOS) circuits. In organic semiconductors, stable n-type operation in ambient conditions has been a challenging issue because electron charge injections to organic semiconductors are suppressed with rather shallow lowest unoccupied molecular orbitals (LUMOs). So, much efforts have been done to develop circuits with only using p-type transistors, as shown in Figures 13.1b and 13.1c. Figure 13.1c shows an example of a pseudo-CMOS inverter proposed as an alternative approach to obtain full-range output voltage operations [30, 31]. Using this circuit, almost full-range output voltages can be obtained when appropriate voltages are applied to $V_{SS}$, lowering the voltage levels of the outputs at the first stage. Although the circuit design requires double numbers of transistors and tuning of $V_{SS}$, it is advantageous to construct circuits only using a semiconductor for p-type operations in cases where the carrier mobility of p-type is extremely high.

It is nonetheless crucial to adopt a complementary configuration as shown in Figure 13.1d, for logic circuits to enable low power consumption and full-range operation. Especially for the applications such as RFID tags in which the battery power is limited, logic circuits must operate with low power consumption. So, new material developments for n-type operations are crucial for practical organic complementary circuits. Some organic semiconductors were proposed to be operated as n-channels, in which LUMO levels are deeper than around 4.0 eV to promote the carrier injections from metals to semiconductors [6, 7].

Pioneering works on organic complementary circuits have been reported by several groups since the 1990s [32, 33]. Dodabalapur et al. demonstrated that organic complementary inverter operates using copper phthalocyanine etc. for p-channel, and naphthalene tetracarboxylic dianhydride (NTCDA) materials for n-channel operations, although the carrier mobility was in the order of $10^{-2}$ $cm^2/Vs$. Low voltage operations of 3 V for complementary circuits were proposed by Klauk, et al. using pentacene and hexadecafluorocopperphthalocyanine ($F_{16}CuPc$) transistors with extremely thin gate insulators. The differences between the carrier mobility of p-channel and n-channel transistors are compensated by tuning the channel widths and lengths for each transistor. There are many possibilities for the combinations of organic semiconductor materials for p-type and n-type operations, however, they should satisfy certain conditions to be implemented in complementary circuits: sufficient high on/off ratio, considerably low threshold voltages compared to the voltage power supply, and sufficient high carrier mobility for circuit operations. Recently, high carrier mobilities in n-channel operations have been reported for newly-synthesized materials, even when measured in ambient conditions [15, 16, 34]. Using these new materials, circuit operations were demonstrated based on solution-processed p-type and n-type organic transistors with short channels [35, 36], to be mentioned in Section 13.3.3.

Another approach is to adopt ambipolar transistors instead of p-type or n-type transistors, as shown in Figure 13.1e [37]. It is a unique feature of organic semiconductors as essentially intrinsic semiconductors that they can be operated in both p-type and n-type polarities in certain combinations of molecular designs and contact electrodes.

Although threshold voltages and the balance of the impedance between the both operations have to be severely controlled, it is one alternative approach to constructing organic logic circuits.

Figure 13.1f shows an example of the output characteristics of organic complementary inverters depicted in Figure 13.1d. When the input voltages are high, the output voltages become low, and vice versa, so that it certainly works as a NOT gate.

### 13.2.2 Logic NAND and NOR Gates

Other basic logic components are AND gate and OR gate. Logic NAND and NOR gates can be constructed, for example from complementary circuits, by assembling two p-type and two n-type transistors, as shown in Figure 13.2 with their truth tables. These are essential building blocks used in digital logic circuits, such as flip-flop circuits, counters, and shift registers. In the case of organic complementary circuits in which the carrier mobility of p-type operations is superior to that of n-type operations, it is preferable to use NOR gates from the point of downscaling device sizes and increasing the device density, because the device space required for serially connected n-type transistors to produce a conductivity comparable to that of parallel-connected p-type transistors becomes large, reflecting the large channel widths. Utilizing these NAND, NOR, and NOT components previously mentioned, more advanced logical components such as flip-flop circuits, shift registers, code generators, to be functioned in logic circuits.

### 13.2.3 Active Matrix Elements

Although it is a slightly different topic from 'integrated circuits', active-matrix element is mentioned as another important electric component, which is expected to be used in display devices or sensor arrays. It is necessary to attach active matrix elements to each pixel of a video display to ensure sufficient high responsibility and to suppress

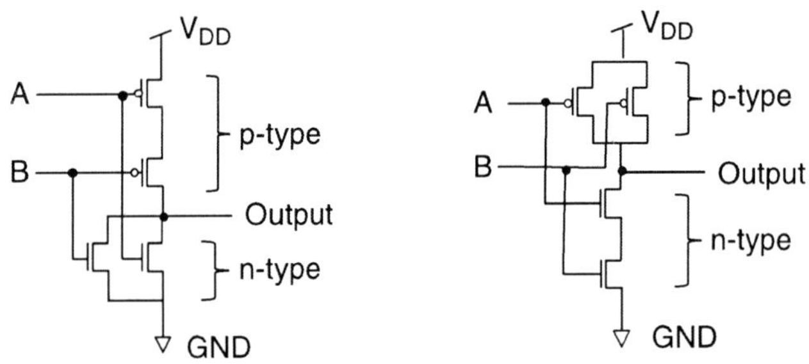


| A | B | NOR |
|---|---|---|
| 0 | 0 | 1 |
| 0 | 1 | 0 |
| 1 | 0 | 0 |
| 1 | 1 | 0 |

(a)

| A | B | NAND |
|---|---|---|
| 0 | 0 | 1 |
| 0 | 1 | 1 |
| 1 | 0 | 1 |
| 1 | 1 | 0 |

(b)

**Figure 13.2** Circuit diagrams with truth tables for (a) NOR and (b) NAND logic gates.

crosstalk between each pixel. The elements can be constructed using unipolar transistors, thus organic semiconductors with dominant p-type operations can be applied. Circuit designs of the active-matrix elements depend on the required specifications and the characteristics of the display devices, for example, whether the lightening elements are current-driven or voltage-driven. As a simple example, Figure 13.3a shows an element consisted of one transistor and one capacitor (1T1C), which can be used to drive each pixel in liquid crystal displays (LCDs). When gate voltages are applied to one particular gate line and transistors in the pixels turn on, the impedance of the transistor becomes sufficiently low, so that the voltages applied to the liquid crystal $V_{LC}$ become almost the same as those in the data line. Since the alignments of liquid crystal materials depend on the applied voltage, the pixel colors can be controlled to black or white depending on the applied voltage $V_{LC}$. The other transistors connected in the same data line were all switched off, so the only designated pixel was charged or discharged to control the liquid crystal lightening. Storage capacitor Cs are provided to keep the charge accumulated in $V_{LC}$ even when the gate line becomes off state. The carrier mobility of the transistors must be high enough to charge the pixel voltage within one gate time. The sweeping frequency of gate lines is ordinary 60Hz, and if possible, higher sweeping frequency is adopted to suppress flickering.

In the case of current-driven-type display like OLEDs, the voltages should be more strictly controlled; otherwise, the brightness in each pixel varies a lot. In these cases, pixel circuits consisted of two transistors and one capacitor (2T1C) as shown in Figure 13.3b [20], or more complicated designs are adopted according to the required specifications.

Much research has been devoted to drive flexible displays with organic active matrix elements. [17–22] E-papers and video displays have been successfully demonstrated on plastic substrates switched by organic transistors. Developments of practical devices are ongoing at organic electronics companies. An example of applications to OLED display is shown in Figure 13.4 [20]. In addition to applications to displays, active matrices can also be utilized to switch arrayed devices such as sensor arrays [4, 23–26]. Figure 13.5 shows an example of switching the array pixels of a temperature sensor [24].

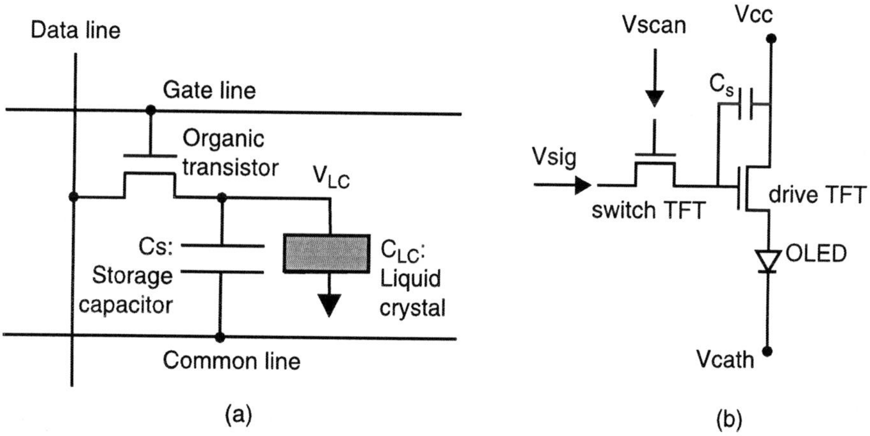


**Figure 13.3** Circuit diagrams for (a) 1T1C and (b) 2T1C pixel elements. Source: Part (b) from Ref [20] Reproduced with permission of John Wiley & Sons.

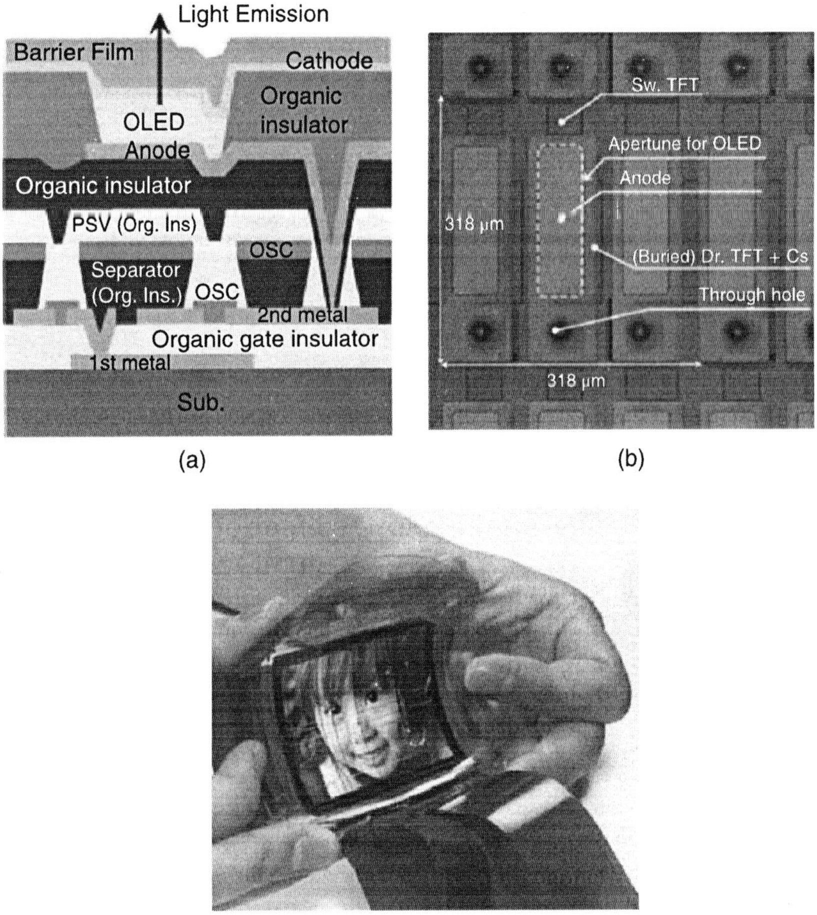


(c)

**Figure 13.4** (a) An example of a pixel structure of the top-emission OTFT-driven OLED display, (b) optical image of a pixel, and (c) photograph of OLED display driven by the active matrix. Source: Ref [20] Reproduced with permission of John Wiley & Sons.

## 13.3 High Performance Organic Transistors Applicable to Flexible Logic Circuits

In order to realize commercial devices which can produce an electric function, a large number of transistors should work at the same time, so that the operating speed of one transistor must be reasonably high. Therefore, organic transistors with short channels and high mobility are strongly required, because the cut-off frequency $f_c$ of a transistor is described as

$$f_c = \frac{\mu_{eff} V_D}{2\pi L^2}\left(\frac{c_i WL}{c_i WL + C_{para}}\right) \tag{13.1}$$

in the linear regime, where $L$ and $W$ are the channel length and the width of the transistor, respectively, $\mu_{eff}$ is the effective carrier mobility including the effects of contact resistance, $C_{para}$ is gate parasitic capacitance, and $c_i WL$ corresponds to the

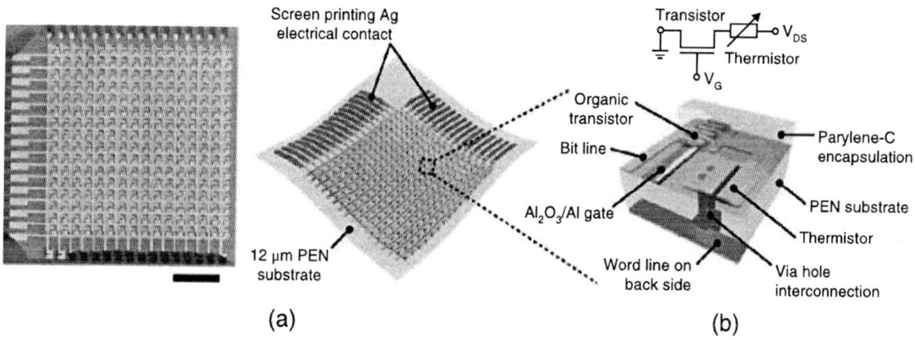


**Figure 13.5** Temperature sensor arrays and an active matrix to read their signals. Source: From Ref [24] Reproduce with permission of John Wiley & Sons.

channel capacitance. In the saturation regime, $V_D$ is replaced by $V_G$. From Eq. (13.1), it is obvious that short-channel high-mobility transistors are strongly important to raise up the maximum operational speed of organic transistors. To realize high carrier mobility in a short-channel device, it is crucial not only to utilize high-mobility organic semiconductor materials, but also to reduce the contact resistance between the organic materials and the contact electrodes, which has been a challenging issue in organic transistors. Short-channel and high-mobility transistors are also important in the aspect of the circuit density, as we can recognize the great advances in silicon technologies along with the down-sizing technique and integrated density of the circuits. In this section, attempts to realize high-performance organic transistors to be built in organic circuits are described.

### 13.3.1 Reducing the Contact Resistance

In attempts to shorten channel lengths, reducing the contact resistances have been challenging issues for organic transistors. In the early stage of development, the contact resistances reported for organic transistors were in the order of being as large as a few kΩcm or even higher [38]. Challenges to reduce the contact resistances have energetically been made [38–46]. As for the device geometry, top-contact configurations provide reliable electric contacts with organic semiconductors so that reproducible circuits are actually formed, using the metal electrodes evaporated through shadow masks. Though stable and rather low contact resistances are easily realized in the top-contact geometry [38, 40], the fabrications of top-contact short-channel devices have been challenging before the orthogonal lithography was proposed [47, 48], because conventional organic semiconductors are damaged by solutions and chemicals used for common photolithography and wet etching processes. So, the photolithography process to fabricate short-channel transistors was more often employed in bottom-contact transistors, because the patterning processes of electrodes are finished before the semiconductor deposition. However, a major concern for the bottom-contact OFETs with a polycrystalline semiconductor film is the molecular disorder occurring in the vicinity of the metal electrodes [43]. To suppress this phenomenon, electrode surface modifications on gold films by thiol-based self-assembled monolayers [44] or UV/$O_3$ treatment [45] are effective to reduce the contact resistance.

But there is still a challenge for bottom-contact configurations to be put into practical use, because these devices suffer from irreproducibility and poor long-term reliability due to chemical instability of most Au-thiolate SAMs [43]. To overcome this problem, a method using orthogonal fluorinated photoresist was proposed [47, 48], which enables the making of fine patterns without dissolution of organic semiconductors even in top-contact geometry. Nakayama. et al. reported a micro-patterning process based on photolithography with wet-etching of gold electrodes deposited directly on pristine organic films without any adverse effect of residual photoresist. Since the wet-etching of gold is governed by iodide/iodine redox reaction, it is essential to use organic semiconductor materials that are robust to the reaction [46]. Employing a newly developed high-mobility organic semiconductor materials with tuned ionization potential, low contact resistance of 123 Ωcm was reported for $C_{10}$-DNBDT, and 1.2 kΩcm for n-type operations [35]. The issue of contact resistances of organic semiconductors is to be overcome in the near future to raise transistor performances and integrate electric functional devices.

### 13.3.2 Downscaling the Channel Sizes and Vertical Transistors

This subsection describes the attempts to shorten the channel length beyond a few micrometers, or to sub-microns. A conventional method to fabricate contact electrodes with very short channels is electron beam (EB) lithography. Polymer transistors based on poly(3-hexylthiophene) were fabricated with the channel lengths of below 500 nm in bottom-contact configurations using EB lithography, demonstrating high-seed operation of 2 MHz with the carrier mobility of 0.2 $cm^2/Vs$ at the applied voltages of 10 V [49]. Since the EB lithography process suffers from its low throughput, alternative fabrication methods have been proposed to shorten the channel lengths to sub-micrometers [50–52]. For example, liquid solutions can be patterned to the order of one micrometer, utilizing the parent-repellent treatment method to control the surface energy, so the micro-patterning of electrodes can be made possible [50].

As an alternative approach to downscale the channel length, vertical transistors were also proposed [52–56]. In the vertical transistors, the channel length direction corresponds to the height direction of the structures, so that the short-channel devices can be easily fabricated beyond the in-plane patterning resolution. Figure 13.6 shows an example of the structure and characteristics of reported vertical transistors. Even when using photolithography technique, the short channels with sub-micrometer size can be fabricated, and high-speed operations were demonstrated [55].

### 13.3.3 High-Speed Organic Transistors

As transistors are active components, we should not only discuss their static characteristics but also their dynamic responses. From Eq. (13.1), the cutoff frequency is proportional to the carrier mobility and the applied voltages, and inversely proportional to the square of the channel lengths, so that high mobility transistors with short-channel lengths are desired. Furthermore, it is noteworthy that the gate parasitic capacitance is also a crucial factor to decide the operational speed of transistors.

A lot of effort has been made to examine and raise the response speeds of organic transistors, combining the abovementioned techniques, and the reported response speed

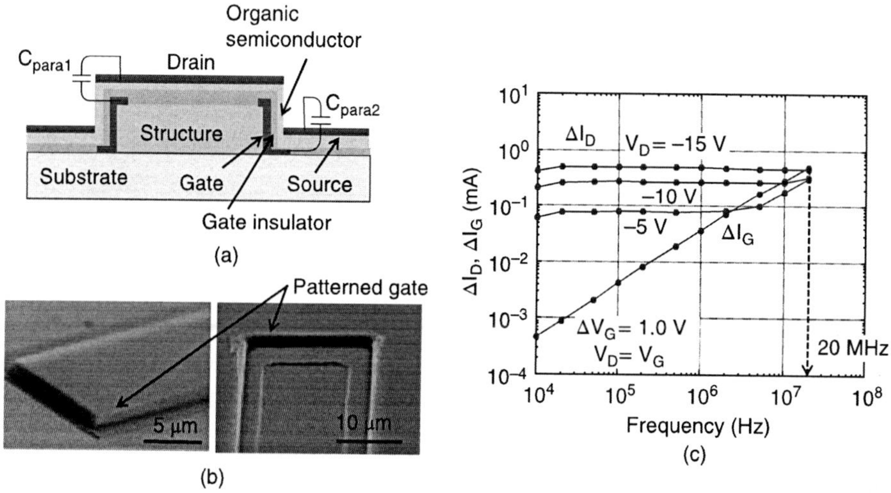


**Figure 13.6** (a) Schematic of a cross-sectional view of an example of a vertical transistor, (b) SEM image of the organic channel part of the structure, and (c) results of dynamic measurements in response to the applied AC gate voltages. Source: Reproduced from Ref [55] Reproduced with permission of Elsevier.

for organic transistors have progressed in recent years [47, 49, 55, 57–60]. The cutoff frequency of the unity gain condition was evaluated for organic transistors, measuring drain-current and gate-current amplitude in response to gate voltage modulations [47, 49, 57]. Wagner et al. reported the cutoff frequency of 2 MHz for polymer transistors with channel lengths of below 500 nm, fabricated by EB lithography, as previously mentioned [47]. Organic transistors based on small molecules were also examined and were shown to have cutoff frequencies above 20 MHz [47, 55, 57]. It is meaningful for organic transistors to exceed their response speed of more than 13.56 MHz, the HF band frequency of RFID tag communications, which means various applications related to identification tags are expected to be realized.

When a transistor is implemented in inverters for logic circuits, the delay time per single gate stage has to be considered to estimate the response time of total circuits. The delay time for a single inverter can be evaluated using ring oscillators which consist of odd numbers of inverters. The outputs of each invertor are connected to the inputs of the next stage invertors, and the output of the last stage is fed back to the input of the first stage invertor, so that the output of the ring oscillator oscillates with the time reflecting any time delays in every gate stage [20, 35, 58–60]. Along with the progress of new materials with high carrier mobility, the time for single gate was shortened to the order of below or reaching 1 μs [58, 60].

## 13.4 Integrated Organic Circuits

To realize practical organic integrated circuits, a large number of transistors should be operated at once. For example, a 64-bit shift-register to operate radio frequency identification (RFID) tags includes more than 2000 transistors, and each transistor should work much faster than the clock speed to drive the next-stage circuits, typically to be at least several MHz for one transistor to operate an entire circuit with a clock speed of a few tens of kHz. Thus, as we mentioned, it is important to obtain reasonable high speed for each transistor, and excellent reproducibility and a high yield are absolute necessities for all transistors, otherwise, the total electric functions of the circuits would not work. This section described recent progress in the integration of organic transistors to be applied to functional electric devices such as RFID tags and readout circuits of sensor devices.

### 13.4.1 RFID Tag Applications

Radio-frequency identification (RFID) tags are one prospective application because organic device features of low-cost, light-weight, and high flexibility are suitable for this application [27–29, 35]. Moreover, multi-functional devices with logic circuits, sensors, and any other devices can easily be integrated with low cost when using organic devices. The RFID transponder includes an antenna, a rectifier to produce DC voltages from AC signals of readers, and organic digital circuits for data transfer. Cantatore reported multi-bit organic transponders using the read-out frequency of 13.56 MHz (HF band), in which a 64-bit code generator was demonstrated, including as many as 1938 p-channel transistors at a clock frequency of 300 Hz. Organic 8-bit RFID transponders comprising 294 transistors were also reported by Myny et al. [28] using 25 μm thick polyethylene naphthalate (PEN) foils, as shown in Figure 13.6. The transistors were based on p-type pentacene transistors with the channel lengths of 2 μm. From the oscillating frequency of the fabricated 19-stage ring oscillators, the single stage delays were estimated to be below 1 μs.

In RFID tag applications, the transponder part of a coupling antenna and a rectifier should operate at 13. 56 MHz, but logic blocks are not necessarily required to operate at the same speed, as long as the clock rates are within the standard depending on the specifications. Shift registers are key components used for the transfer and the reading of data. Although their circuits can be designed in many different ways, they basically consist of latches that function as flip-flop circuits to store the data temporarily, and transfer them to the next latch in every clock cycle. When the clock signal shifts its state, the state in the input is reflected to the outputs of "Q", which means that the D flip-flop circuit can function as a temporary storage for part of the multi-bit data generator in synchronization with the clock signals. Figure 13.7 shows an example of output characteristics of a D-flip flop circuits consisted of 26 organic transistors on plastic substrates, based on solution-processed p-type and n-type transistors [36]. In the paper, the yield of each organic transistor was estimated to be above 99.5 %, from the yield of a D-flip flop circuit block of 88.8 %.

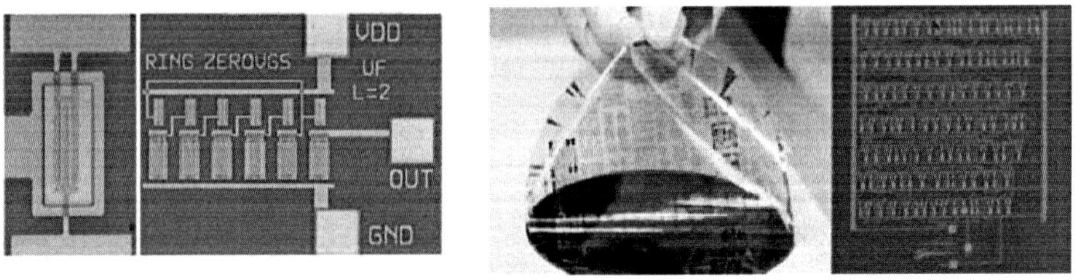


**Figure 13.7** An example of integrated organic circuits for RFID tag applications. Source: Ref. [28] with kind permission.

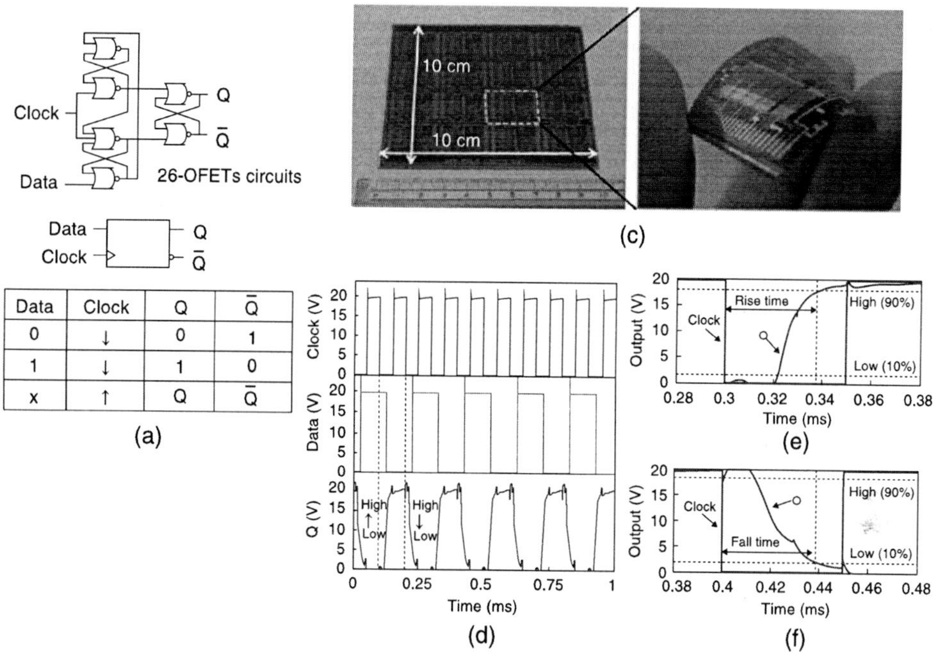


**Figure 13.8** D-flip flop circuits integrated on a plastic substrate with high yield. Source: Ref [36] Reproduced with permission of John Wiley & Sons.

Continued efforts are important to realize practical logic devices and enhance performance of the organic circuits. Ideally, in the tag applications of organic circuits, code generators are connected to organic memories and/or sensors integrated on the same substrates, so that sensing or item-level identification monitoring will be possible with low cost using flexible platforms.

### 13.4.2 Sensor Readout Circuits

Organic semiconductor devices are a good match to be applied to flexible sensors because of their high flexibility. It is necessary for sensing devices that their analog output signals are converted to digital signals so that the sensor information can be read in read-out systems. In some cases, the sensor signals are feeble so ideally, sensors should be connected with amplifiers and analog-to-digital (A/D) converters. So far, several research groups have made an effort to fabricate A/D converters based on organic field-effect transistors (OFETs) [65–69]. It is also demonstrated that flexible temperature sensor signals were converted to digital bits using a 2-bit parallel A/D converter [68]. We expect that these works will facilitate the development of organic electronics for future ubiquitous sensor systems.

## 13.5 Conclusions

Recent advancements in organic circuits based on organic semiconductor transistors are described. Elementary digital logic circuits and elements for active matrix devices are explained, as well as the requirements for one transistor to be integrated in organic circuits are described. Along with the progress of newly-developed high-mobility materials, the performances obtained in organic integrated circuits have been advanced remarkably. These active research and developments on organic integrated circuits offer flexible platforms which can be attached on arbitrary shapes of various objects, and contribute to the progress of flexible electronics for coming smart society.

## References


1 Gamier, F., Horowitz, G., Peng, X., and Fichou, D. (1990). *Adv. Mater.* 2: 592.
2 Horowitz, G. (1998). *Adv. Mater.* 10: 365.
3 Crone, B., Dodabalapur, A., Lin, Y.-Y. et al. (2000). *Nature* 403: 521.
4 Someya, T., Sekitani, T., Iba, S. et al. (2004). *Proc. Natl. Acad. Sci USA* 101: 9966.
5 Anthony, J.E. (2008). *Angew. Chem.* 47: 452.
6 Jones, B.A., Ahrens, M.J., and Yoon, M.-H. (2004). *Angew. Chemie Int. Ed.* 43: 6363.
7 Yan, H., Zheng, Y., Blache, R. et al. (2008). *Adv. Mater.* 20: 3393.
8 Usta, H., Facchetti, A., and Marks, T.J. (2011). *Acc. Chem. Res.* 44: 501.
9 Kang, M.J., Doi, I., Mori, H. et al. (2011). *Adv. Mater.* 23: 1222.
10 Minemawari, H., Yamada, T., Matsui, H. et al. (2011). *Nature* 475: 364.
11 Soeda, J., Uemura, T., Okamoto, T. et al. (2013). *Appl. Phys. Express* 6: 076503.
12 Mitsui, C., Okamoto, T., Yamagishi, M. et al. (2014). *Adv. Mater.* 26: 4546.
13 Iino, H., Usui, T., and Hanna, J. (2015). *Nat. Commun.* 6: 6828.
14 Park, J.-I., Chung, J.W., Kim, J.-Y. et al. (2015). *J. Am. Chem. Soc.* 137: 12175.
15 Gsänger, M., Oh, J.H., Könemann, M. et al. (2010). *Angew. Chem. Int. Ed.* 49: 740.
16 Mamada, M., Shima, H., Yoneda, Y. et al. (2015). *Chem. Mater.* 27: 2928.
17 Gelink, G.H., Huitema, H.E.A., van Veenendaal, E. et al. (2004). *Nat. Mater.* 3: 106.
18 Forrest, S.R. (2004). *Nature* 427: 911.
19 Sekitani, T., Nakajima, H., Maeda, H. et al. (2009). *Nat. Mater.* 8: 494.
20 Gelink, G., Heremans, P., Nomoto, K., and Anthopoulos, T.D. (2010). *Adv. Mater.* 22: 3778.
21 McCarthy, M.A., Liu, B., Donoghue, E.P. et al. (2011). *Science* 332: 570.
22 Sheraw, C.D., Zhou, L., Huang, J.R. et al. (2002). *Appl. Phys. Lett.* 80: 1088.
23 Sekitani, T., Zschieschang, U., Klauk, H., and Someya, T. (2010). *Nat. Mater.* 9: 1015.
24 Ren, X., Pei, K., Zhang, Z. et al. (2016). *Adv. Mater.* 28: 4832.
25 Mannsfeld, S.C.B., Tee, B.C.-K., Stoltenberg, R.M. et al. (2011). *Nat. Mater.* 9: 859.
26 Wu, X. and Huang, J. (2016). *J. Emerg. Sel. Topics Circuits. Syst.* 99: 1.
27 Cantatore, E., Geuns, T.C.T., Gelink, G.H. et al. (2007). *J. Solid-St. Circ.* 42: 84.
28 Myny, K., Steudel, S., Smout, S. et al. (2010). *Org. Electron.* 11: 1176.
29 Tinivella, R., Camarchia, V., Pirola, M. et al. (2011). *Org. Electron.* 12: 1328.
30 Fukuda, K., Sekitani, T., Yokota, T. et al. (2011). *Electron Device Lett.* 32: 1448.
31 Yokota, T., Sekitani, T., Tokuhara, T. et al. (2012). *Trans. Electron Devices* 59: 3434.
32 Dodabalapur, A., Laquindanum, J., Katz, H.E., and Bao, Z. (1996). *Appl. Phys. Lett.* 69: 4227.
33 Klauk, H., Zschieschang, U., Pflaum, J., and Halik, M. (2007). *Nature* 445: 745.
34 Kempa, H., Hambsch, M., Reuter, K. et al. (2011). *Trans. Electron Devices* 58: 2765.
35 Uno, M., Kanaoka, Y., Cha, B.-S. et al. (2015). *Adv. Electron. Mater.* 1: 1500178.

36 Uno, M., Isahaya, N., Cha, B.-S. et al. (2017). *Adv. Electron. Mater.* 3: 1600410.
37 Baeg, K.J., Caironi, M., and Noh, Y.Y. (2013). *Adv. Mater.* 25: 4210.
38 Gundlach, D.J., Zhou, L., Nichols, J.A. et al. (2006). *J. Appl. Phys.* 100: 024509.
39 Zaumseil, J., Baldwin, K.W., and Rogers, J.A. (2003). *J. Appl. Phys.* 93 (10): 6117.
40 Darmawan, P., Minari, T., Xu, Y. et al. (2012). *Adv. Funct. Mater.* 22: 4577.
41 Natali, D. and Caironi, M. (2012). *Adv. Funct. Mater.* 24: 1357.
42 Wada, H., Shibata, K., Bando, Y., and Mori, T. (2008). *J. Mater. Chem.* 18: 4165.
43 Bock, C., Pham, D.V., Kunze, U. et al. (2006). *J. Appl. Phys.* 100: 114517.
44 Kitamura, M., Kuzumoto, Y., Kang, W. et al. (2010). *Appl. Phys. Lett.* 97: 033306.
45 Stadlober, B., Haas, U., Gold, H. et al. (2007). *Adv. Funct. Mater.* 17: 2687.
46 Nakayama, K., Uno, M., Uemura, T. et al. (2014). *Adv. Mater. Interfaces* 1: 1300124.
47 Lee, J.K., Chatzichristidi, M., Zakhidov, A.A. et al. (2008). *J. Am. Chem. Soc.* 130: 11564.
48 Zakhidov, A.A., Lee, J.K., DeFranco, J.A. et al. (2011). *Chem. Sci.* 2: 1178.
49 Wagner, V., Wöbkenberg, P., Hoppe, A., and Seekampppl, J. (2006). *Appl. Phys. Lett.* 89: 243515.
50 Noh, Y.-Y., Zhao, N., Caironi, M., and Sirringhaus, H. (2007). *Nature Nano.* 2: 784–789.
51 Stutzmann, N., Friend, R.H., and Sirringhaus, H. (2003). *Science* 299: 1881.
52 Parashkov, R., Becker, E., Hartmann, S. et al. (2003). *Appl. Phys. Lett.* 82: 4579.
53 Takano, T., Yamauchi, H., Iizukal, M. et al. (2009). *Appl. Phys. Express* 2: 071501.
54 Nakahara, R., Uno, M., Uemura, T. et al. (2012). *Adv. Mater.* 24: 5212.
55 Uno, M., Cha, B.S., Kanaoka, Y., and Takeya, J. (2015). *Org. Electron.* 20: 119.
56 Li, J., Zhao, C., Wang, Q. et al. (2012). *Org. Electron.* 500: 4628.
57 Kitamura, M. and Arakawa, Y. (2011). *Jpn. J. Appl. Phys.* 50: 01BC01.
58 Ante, F., Kälblein, D., Zaki, T. et al. (2012). *Small* 8: 73.
59 Singh, S., Mohapatra, S.K., Sharma, A. et al. (2013). *Appl. Phys. Lett.* 102: 153303.
60 Baeg, K.J., Khim, D., Kim, D.-Y. et al. (2010). *Polym. Phys.* 49: 62.
61 Crone, B.K., Dodabalapur, A., Sarpeshkar, R. et al. (2001). *J. Appl. Phys.* 89: 5125.
62 Schwartz, D.E. and Ng, T.N. (2013). *Electron Device Lett.* 34: 271.
63 Abdinia, S., Ke, T.-H., Ameys, M. et al. (2015). *J. Disp. Technol.* 11: 564.
64 Takeda, Y., Hayasaka, K., Shiwaku, R. et al. (2016). *Sci. Rep.* 6: 25714.
65 Marien, H , Steyaert, M.S.J., van Veenendaal, E. and Heremans, P (2011). *J. Solid-State Circuits* 46: 276.
66 Nausieda, I., Ryu, K.K., He, D.D. et al. (2011). *Trans. Electron Devices* 58: 865.
67 Xiongl, W., Zschieschang, U., Klauk, H., and Murmann, B. (2010). *ISSCC Dig. of Tech. Papers* 7: 134.
68 Nakayama, K. (2016). *Org. Electron.* 36: 148.
69 Maruen, H., Steyaert, M.S.J., van Veenendaal, E., and Heremans, P. (2010). *Solid-State Circuits. Society* 46: 276.

# 14

# Naphthobisthiadiazole-Based Semiconducting Polymers for High-Efficiency Organic Photovoltaics

*Itaru Osaka*[1] *and Kazuo Takimiya*[2]

[1] *Graduate School of Advanced Science and Technology, Hiroshima University*
[2] *RIKEN Center for Emergent Matter Science*

**CHAPTER MENU**



## 14.1 Introduction

Organic photovoltaics (OPVs) is one of the most widely studied areas in organic electronics [1]. Similar to other organic devices, OPVs can be fabricated by solution-processes, which allow for low-cost and low-environmental-impact production [2, 3]. They also provide light-weight, flexible, and semitransparent solar cells which, in turn, can be installed on curved surfaces, vertical wall surfaces, or windows. These features clearly differentiate OPVs from conventional silicon photovoltaic technologies.

Improving the power conversion efficiency (PCE) is a critical issue for OPVs. A large number of studies on the development of new materials, in particular, p-type (donor) semiconducting polymers have been made to resolve this issue in the past decade [4–9]. Since the short-circuit density ($J_{SC}$) and the open-circuit voltage ($V_{OC}$) of the photovoltaic cell are largely dependent on the optical bandgap ($E_g$) of the material and the energy difference between the highest occupied molecular orbital (HOMO) of the p-type semiconducting polymer and the lowest unoccupied molecular orbital (LUMO) energy levels of the n-type semiconducting material, respectively, the control of the electronic structure is of particular importance in designing semiconducting polymers [10, 11]. In parallel, the photovoltaic property is also dependent on the transport of the generated charges upon exciton dissociation at the p/n interface and, therefore, the crystallinity and backbone orientation of the polymers must be carefully controlled [12, 13].

2,1,3-benzo[*c*][1,2,5]thiadiazole
(**BTz**)

naphtho[1,2-*c*:5,6-*c*′]bis[1,2,5]thiadiazole
(**NTz**)

**Figure 14.1** Chemical structures of benzo[*c*][1,2,5]thiadiazole (BTz) and naphtho[1,2-*c*:5,6-*c*′]bis[1,2,5]thiadiazole (NTz).

Donor–acceptor (D–A) semiconducting polymers, consisting of electron-rich (donor) and electron-deficient (acceptor) building units, is a state-of-the-art material system that greatly functions in OPVs [7, 8]. As the HOMO and LUMO energy levels are roughly determined by the orbital mixing of the donor and acceptor units, one can easily tune the electronic structure of the semiconduting polymers by carefully designing the donor and acceptor units [14]. Furthermore, the polymer backbones can have stronger intermolecular interactions likely due to the local dipole caused by the D–A intramolecular interaction, and thus can form more highly ordered structures relative to the classical polythiophene-type semiconducting polymers [15].

This chapter focuses on a particular D–A semiconducting polymer system based on an emerging acceptor unit, naphtho[1,2-*c*:5,6-*c*′]bis[1, 2, 5]thiadiazole (NTz) [16]. NTz is a heterocycle in which two benzo[*c*][1, 2, 5]thiadiazole (BTz) are fused in a kinked manner (Figure 14.1). Recently, semiconducting polymers based on this acceptor unit have been shown to demonstrate great performances in OPVs due to their ability to provide a highly crystalline structure and fascinating electronic properties. The electronic structures, ordering structures, OPV performances, and structure–property relationships especially drawn through the works by our group will be discussed here. We hope that this chapter will provide useful insights into the creation of new high performance semiconducting polymers for OPVs.

## 14.2 Semiconducting Polymers Based on Naphthobisthiadiazole

NTz was first reported in 1991, in which Mataka and co-workers showed the preparation of NTz and related compounds by using the reaction of sulfur nitride and various naphthols [16]. In 2011, Huang, Cao, and co-workers reported the synthesis of an NTz based D–A polymer, PBDT-DTNT (Figure 14.2), where benzodithiophene was employed as the donor unit [17]. The solar cells using PBDT-DTNT were initially reported to show PCEs of ~6%, which improved to more than 7% by optimization [18, 19]. They then synthesized a series of PBDT-DTNT derivatives and NTz-based polymers having various fused aromatics such as fluorene (PF-C12NT), carbazole (PCz-C12NT), benzodithiophene (PBDT-C12NT), and indacenodithiophene (PIDT-C12NT) (Figure 14.2) [20]. Shortly after the first report of PBDT-DTNT, we also reported a series of NTz-based semiconducting polymers [21–27], which will be discussed below. Several other groups

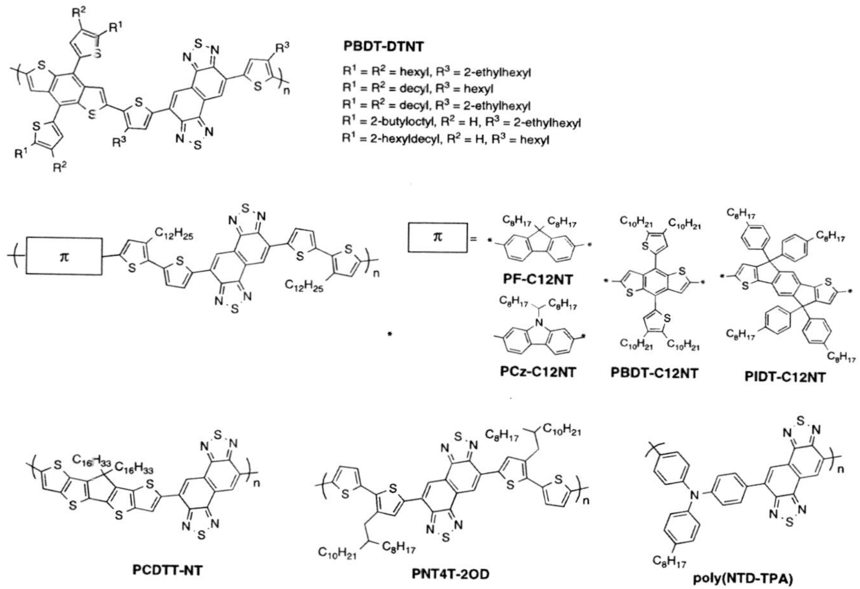


Figure 14.2 NTz-based D–A polymers.

also reported NTz-based polymers with cyclopentadithienothiophene (PCDTT-NT) [28], quaterthiophene having 2-octyldodecyl groups (PNT4T-2OD) [29], and triarylamine (poly(NTD-TPA)) [30] (Figure 14.2).

## 14.3 Quaterthiophene–NTz Polymer: Comparison with the Benzothiadiazole Analogue

In this section, we discuss the structure–property relationships of an NTz polymer having a quaterthiophene moiety with the 2-decyltetradecyl groups as the donor unit (PNTz4T) in comparison with the BTz analogue (PBTz4T) (Figure 14.3a) [21]. Figure 14.3b shows the UV–vis absorption spectra of the polymers in thin film. PNTz4T exhibited an absorption maximum ($\lambda_{max}$) at 725 nm, which was red-shifted for about 60 nm from that of PBTz4T ($\lambda_{max}$ = 662 nm). Furthermore, in PNTz4T, the longer wavelength region of the absorption peak at $\lambda_{max}$ appears sharper than that of PBTz4T, indicating better molecular ordering, likely due to its rigid π-extended structure. The absorption onset ($\lambda_{edge}$) in the thin film for PNTz4T was determined to be 805 nm,

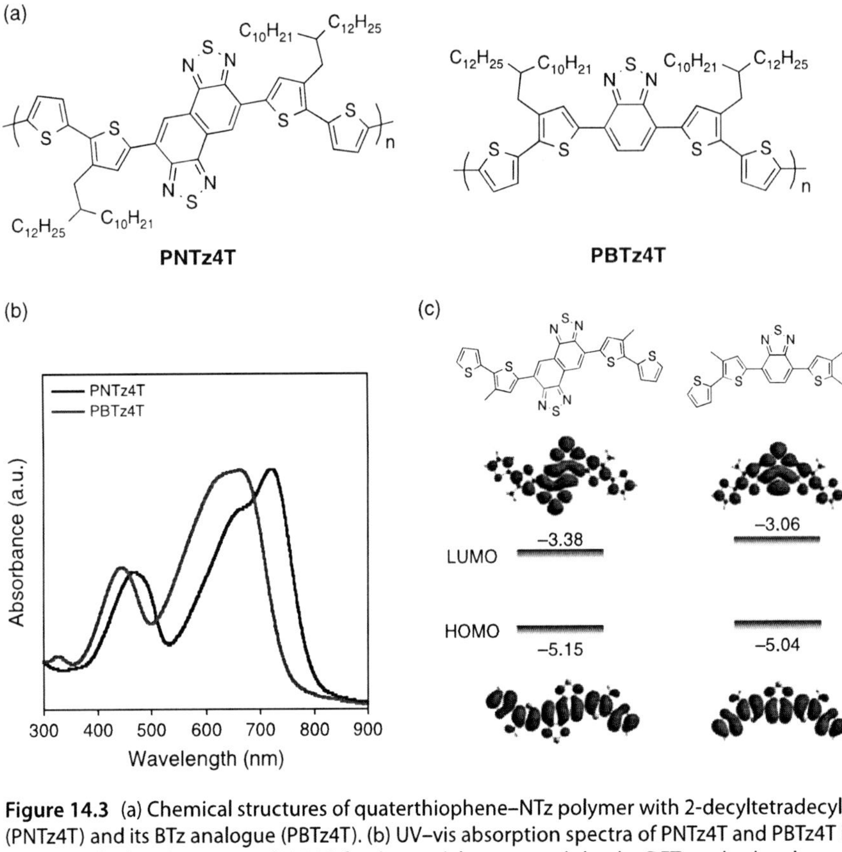


**Figure 14.3** (a) Chemical structures of quaterthiophene–NTz polymer with 2-decyltetradecyl groups (PNTz4T) and its BTz analogue (PBTz4T). (b) UV–vis absorption spectra of PNTz4T and PBTz4T in thin film. (c) Calculated HOMO and LUMO for the model compounds by the DFT method at the B3LYP/6-31g(d) level.

which corresponds to the optical bandgap ($E_g$) of 1.54 eV, whereas $E_g$ for PBTz4T was about 0.1 eV larger (1.65 eV). The HOMO energy level ($E_{HOMO}$) of PNTz4T was estimated to be −5.16 eV by cyclic voltammetry, which is 0.09 eV deeper than that of PBTz4T (−5.07 eV). On the other hand, the LUMO energy level ($E_{LUMO}$) for PNTz4T was −3.77 eV, which was 0.24 eV deeper than that of PBTz4T (−3.53 eV). The larger difference in $E_{LUMO}$s than in $E_{HOMO}$s resulted in a smaller HOMO–LUMO gap for PNTz4T than for PBTz4T. This suggests that the replacement of BTz with NTz has a greater influence on the LUMO than the HOMO. This is consistent with the calculated $E_{HOMO}$ and $E_{LUMO}$ of the model compounds (Figure 14.3c). Charge carrier mobilities were evaluated by using transistor devices with the annealed (200° C) polymer thin films on the Si/$SiO_2$ substrate surface-treated with 1*H*,1*H*,2*H*,2*H*-perfluorodecyltriethoxysilane (FDTS). PNTz4T provided mobilities as high as 0.56 $cm^2$ $V^{-1}$ $s^{-1}$, which was about one order of magnitude higher than that of PBTz4T (~0.05 $cm^2$ $V^{-1}$ $s^{-1}$), indicating the high potential of the NTz unit.

Solar cells with a conventional structure (ITO/PEDOT:PSS/(polymer/$PC_{61}BM$)/LiF/Al) were fabricated by spin-coating the polymer/$PC_{61}BM$ solutions in DCB (Figure 14.4a). *J–V* curves of the cells with the optimal polymer to $PC_{61}BM$ (p:n) weight ratio (1:1.5 and 1:1 for the PNTz4T and PBTz4T systems, respectively) under one sun of simulated AM 1.5G solar irradiation (100 mW/$cm^2$) are displayed in Figure 14.4b. While the PBTz4T cell showed a power conversion efficiency (PCE) of 2.6% ($J_{SC}$ = 5.6 mA $cm^{-2}$, $V_{OC}$ = 0.74 V, FF = 0.63), the PNTz4T cell showed a PCE of 6.3% ($J_{SC}$ = 12.0 mA $cm^{-2}$, $V_{OC}$ = 0.76 V, FF = 0.69). One of the reasons for the higher $J_{SC}$ in the PNTz4T cell than in the PBTz4T cell would be the wider absorption range.

In order to understand the higher performances observed in transistor and solar cells for PNTz4T relative to PBTz4T, the ordering structures in the thin film were investigated by the grazing incidence X-ray diffraction (GIXD) studies [31]. Two-dimensional (2D) diffraction images of PNTz4T and PBTz4T films on the FDTS-modified $SiO_2$ substrate, which reflect the ordering structure in the transistors, are shown in Figure 14.5a. In PNTz4T, diffractions assignable to the lamellar ($q \approx 0.25$ Å$^{-1}$) and the π–π stacking

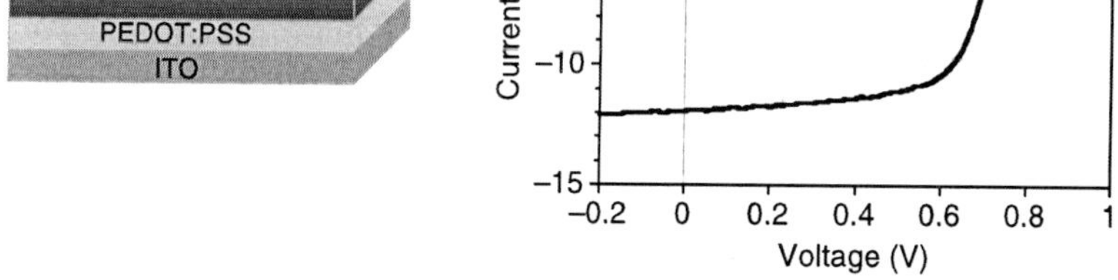


**Figure 14.4** (a) Schematic illustration of the solar cell with the conventional structure. (b) *J–V* characteristics of the cells that used PNTz4T and PBTz4T. The p:n weight ratios were 1:1.5 and 1:1 for the PNTz4T and PBTz4T systems, respectively.

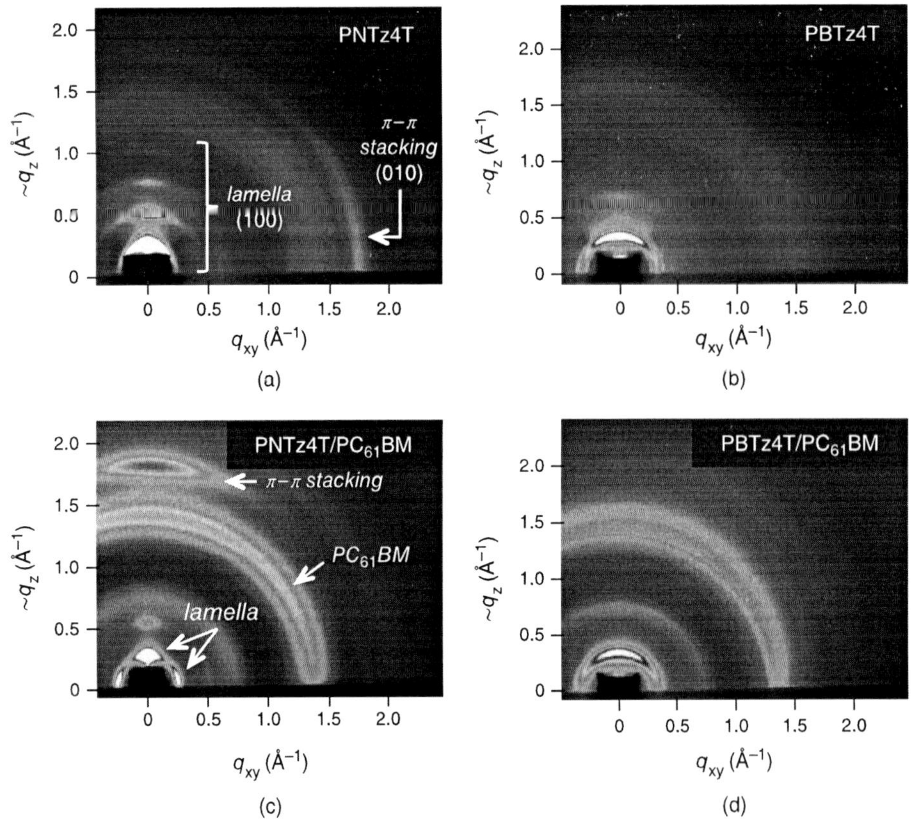


**Figure 14.5** 2D GIXD images of the PNTz4T thin film (a), PBTz4T thin film (b), PNTz4T/$PC_{61}$BM blend film (c), and PBTz4T/$PC_{61}$BM blend film (d).

structures ($q \approx 1.7$ Å$^{-1}$) appeared on the quasi $q_z$ ($\sim q_z$) and the $q_{xy}$ axes, respectively, indicating the predominant edge-on orientation on the substrate surface (Figure 14.6), though there are some misoriented fractions. The π–π stacking distance ($d_\pi$) of PNTz4T determined by the in-plane XRD pattern was 3.5 Å, which is very narrow for semiconducting polymers, and thus rationalizes the high mobilities observed well for PNTz4T. In contrast, PBTz4T showed largely arcing diffraction corresponding to the lamellar structure, indicative of the randomly oriented crystallite. PBTz4T did not show clear diffraction corresponding to the π–π stacking structure, indicating that the crystallinity is much less than that of PNTz4T. The large difference in the ordering structure between PNTz4T and PBTz4T is in good agreement with the fact that the transistor performances are quite distinct.

Polymer/$PC_{61}$BM blend films on the PEDOT:PSS-coated ITO glass substrate were also subjected to the GIXD measurements to investigate the ordering structure in the solar cells (Figure 14.5b). Interestingly, it was found that PNTz4T mainly oriented in a face-on manner (Figure 14.6), as the diffraction corresponding to the π–π stacking appeared on the $\sim q_z$ axis. Such a drastic change of the orientation by blending with $PC_{61}$BM has also been seen in other polymer systems [32]. It should also be noted that

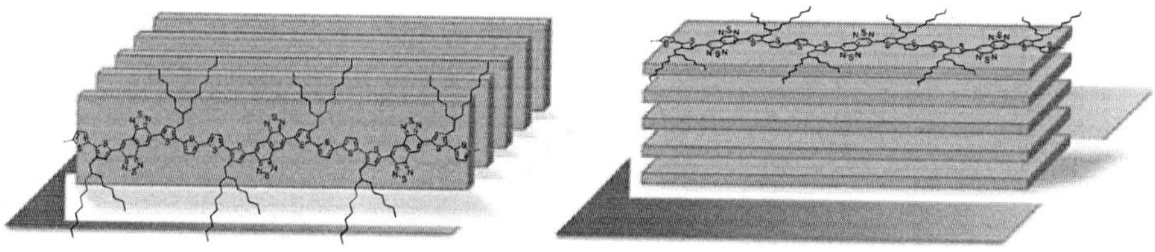


**Figure 14.6** Schematic illustration of typical orientation motifs of the polymer backbone. "Edge-on" with the polymer backbones stand on the substrate (left) and "face-on" with the polymer backbones lie flat on the substrate (right).

PNTz4T preserved the narrow $d_\pi$ of 3.5 Å in the blend film. In the meantime, PBTz4T provided much less crystalline feature in the blend film, where there was no π–π stacking diffraction, as seen in the polymer-only film. The predominant face-on orientation as well as the highly crystalline structure, and the narrow $d_\pi$ of PNTz4T in the blend film are most likely the main reasons that the PNTz4T cell shows higher PCE than the PBTz4T cell.

NTz is a more highly extended fused ring compared to BTz, and thus it is fairly acceptable that PNTz4T forms a more highly ordered structure as compared to PBTz4T. However, the difference in crystallinity, especially the π–π stacking, between the present two polymers in both the polymer-only film and the blend film was markedly large. We speculated that this marked difference originated in the difference of symmetry between the NTz and BTz unit. As shown by the single-crystal X-ray analysis of the model compounds, NTz2T and BTz2T (Figure 14.7a), the sulfur atom in the thiophene ring always points to the benzene substructure of the NTz or BTz unit. Thus, NTz with a centrosymmetrical structure ($C_{2h}$ symmetry) affords an *anti* arrangement of the thiophene rings that sandwich NTz, whereas BTz with an axisymmetrical structure ($C_{2v}$ symmetry) affords a *syn* arrangement of the neighboring thiophenes. Based on these arrangements, PNTz4T can give a more linear-shaped backbone as compared to PBTz4T, which gives a "wavy" shape (Figure 14.7b). In addition, in PNTz4T, the alkyl side chains are always in the *anti* placement, whereas in PBTz4T, the side chains are in both the *anti* and *syn* arrangements. The backbone shape and side chain arrangement often largely affect the packing structure, and thus these structural features lead to the highly ordered packing structure in PNTz4T than in PBTz4T [33–37].

## 14.4 Naphthodithiophene–NTz Polymer: Importance of the Backbone Orientation

Incorporation of π-extended heteroaromatic rings into the polymer backbone is beneficial to enhance the intermolecular interaction and thereby the crystallinity [15, 38–40]. As a family of thienoacenes [41], naphthodithiophenes have been of great interest as a core unit for organic semiconductors [42–47]. Naphtho[1,2-*b*:5,6-*b*′]dithiophene (NDT3) is the most promising naphthodithiophenes for semiconducting polymers, as NDT3-based polymers exhibited highly crystalline structures and yielded high hole mobility of ~0.8 $cm^2\ V^{-1}\ s^{-1}$ compared to the other isomers [34, 48, 49]. It was thus

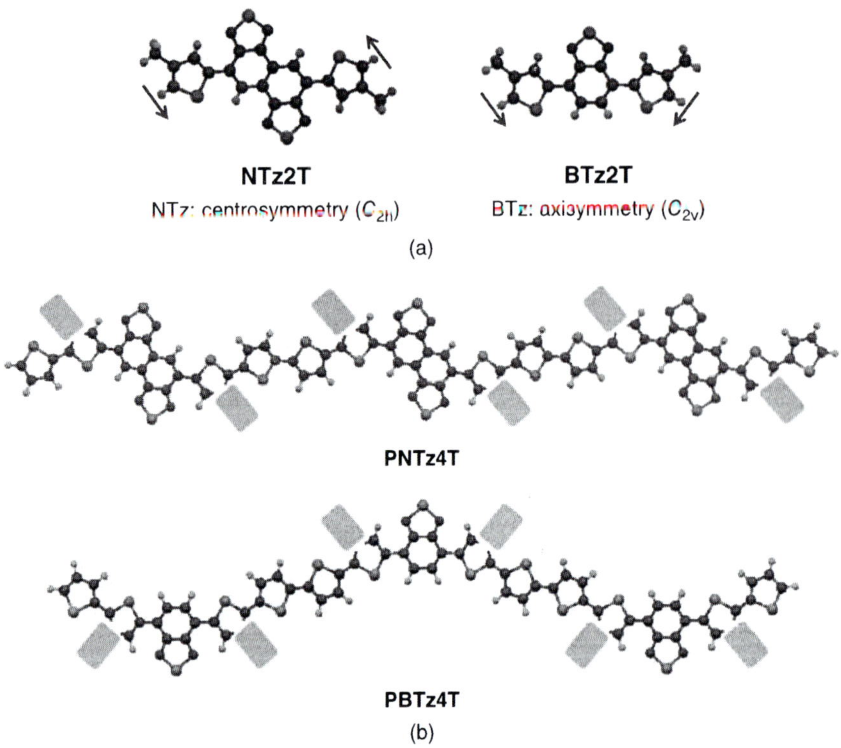


**Figure 14.7** (a) Molecular structures of the model compound using NTz (NTz2T) and BTz (BTz2T) determined by single-crystal X-ray analysis. (b) Optimized backbone structures for PNTz4T and PBTz4T determined by the DFT calculation at B3LYP/6-31g(d) level.

expected that the incorporation of NDT3 in combination with NTz in the backbone can lead to high performance semiconducting polymers.

Figure 14.8a depicts the chemical structure of NTz–NDT3 polymers (PNNTs) [22, 24]. As PNNT-DT had limited solubility in solvents such as chlorobenzene (CB) or dichlorobenzene (DCB) despite the fact that it had long branched alkyl groups (2-decyltetradecyl; DT) on the thiophene rings, PNNT-12HD and -12DT having additional linear alkyl groups (dodecyl) on the NDT3 ring at the 5,10-positions were also synthesized. PNNT-12HD and PNNT-12DT had indeed greatly improved solutibily in those solvents. The $E_{HOMO}$ determined by the photoemission yield spectroscopy for PNNT-DT was −5.25 eV, and that for both PNNT-12HD and -12DT were almost identical, −5.22 eV. This implies that the inductive effect of the alkyl groups at the 5,10-positions of the NDT3 core is quite small, which may partially be understood by the fact that the HOMO coefficient is not located at the 5,10-positions of the NDT3 core. This insensitivity is particularly important because a rise in $E_{HOMO}$ can cause a decrease in $V_{OC}$. Figure 14.8b shows the absorption spectra of PNNT-DT and -12HD. In both solution and films, these polymers showed similar spectra with absorption maxima at 660–670 nm. The absorption onsets in the film were found to be at 750–760 nm, which correspond to $E_g$s of 1.64–1.69 eV, indicating that the electronic structure was almost unchanged by the introduction of the alkyl groups.

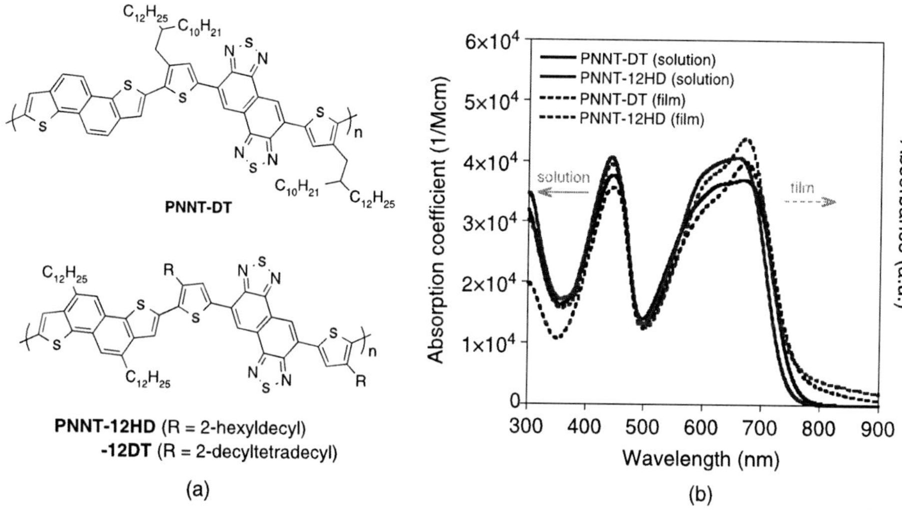


**Figure 14.8** (a) Chemical structures of naphthodithiophene–NTz polymers. (b) UV-vis absorption spectra of PNNT-DT and PNNT-12HD in chlorobenzene solution and in film.

The charge transport properties of the polymers were investigated using the "hole-only" device (ITO/PEDOT:PSS/polymer or (polymer/$PC_{61}$BM)/$MoO_x$/Ag), with which the out-of-plane charge carrier mobility can be estimated. The charge carrier mobilities ($\mu_{SCLC}$) evaluated by the space-charge-limited current model were of the $10^{-3}$ $cm^2$ $V^{-1}$ $s^{-1}$ order for both PNNT-12HD and -12DT in both polymer-only and polymer/$PC_{61}$BM blend films, whereas those of PNNT-DT were one order of magnitude lower (of the $10^{-4}$ $cm^2$ $V^{-1}$ $s^{-1}$ order). Thus, charge transport in the out-of-plane direction was enhanced by the alkylation.

The photovoltaic properties were investigated using a conventional cell (ITO/PEDOT:PSS/(polymer/PCBM)/LiF/Al). Figure 14.9a shows the $J$–$V$ curves of the polymer/$PC_{61}$BM cells with the thickness of 170–180 nm, in which the p:n ratio was 1:1 for the PNNDT-12DT and -DT systems and 1:2 for the PNNT-12HD system. PNNT-DT cell gave $J_{SC}$ of 12.3 mA $cm^{-2}$, whereas PNNT-12HD and -12DT cells gave slightly lower $J_{SC}$s of 10.3 and 7.3 mA $cm^{-2}$, respectively. Meanwhile, similar $V_{OC}$s of 0.84–0.86 V were obtained for all the cells, which is consistent with the almost identical $E_{HOMO}$s. Higher FFs of >0.65 were obtained in the cells using PNNT-12HD and -12DT than that in the cell cells using PNNT-DT (0.54). This is likely due to the efficient charge transport, as shown above. As a result, the PNNT-DT cell exhibited PCE of as high as 5.5%, and the PNNT-12HD and -12DT cells exhibited PCEs of 5.9% and 4.1%, respectively. Interestingly, PCEs of the PNNT-12HD cell improved when thicker active layers were used (Figure 14.9b). As the active layer thickness increased to ca. 300 nm, $J_{SC}$ increased to 14.7 mA $cm^{-2}$, while FF was reduced but remained relatively high (i.e., >0.6). As a result, PCE reached 7.5% for the cell that used $PC_{61}$BM with the thickness of 290 nm. Furthermore, when $PC_{71}$BM was used instead of $PC_{61}$BM, the PNNT-12HD cell with the thickness of 300 nm exhibited a high PCE of 8.2% ($J_{SC}$ = 15.6 mA $cm^{-2}$, $V_{OC}$ = 0.82 V, FF = 0.64).

To correlate these performances, the ordering structure of the polymers was confirmed by GIXD studies (Figure 14.10a). PNNT-DT was found to orient in an edge-on

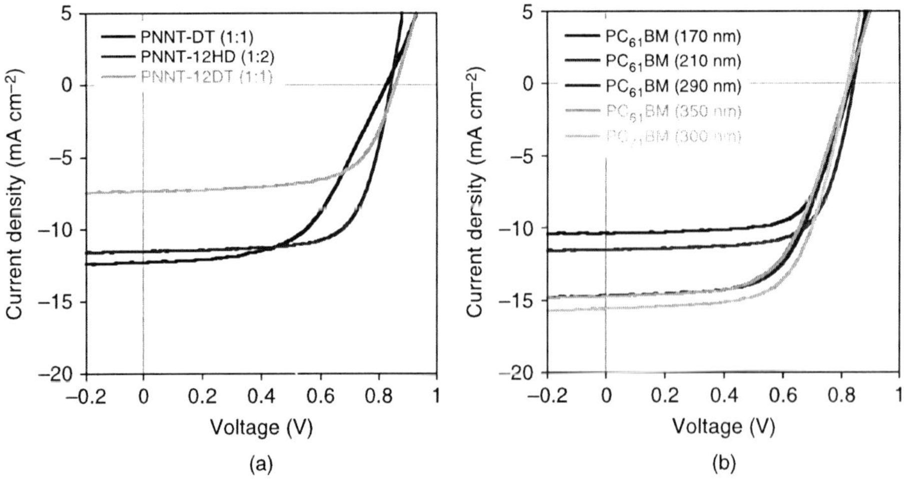


**Figure 14.9** (a) *J–V* curves of the polymer/$PC_{61}$BM-based cells using PNNT-DT, PNNT-12HD, and PNNT-12DT with the active-layer thicknesses of 170–180 nm. The p:n ratios are shown in the legend. (b) *J–V* curves of the PNNT-12HD cells (p:n = 1:2) with different active-layer thicknesses. The PCBMs and thicknesses are shown in the legend.

manner in the polymer-only film, as the diffractions corresponding to the lamellar and the $\pi$–$\pi$ stacking structure predominantly appeared along the $\sim q_z$ axis ($q = 0.30$ Å$^{-1}$) and the $q_{xy}$ axis ($q = 1.83$ Å$^{-1}$), respectively (Figure 14.10a). The lamellar $d$ spacing ($d_l$) and $d_\pi$ were 21.1 and 3.43 Å, respectively, for PNNT-DT. We note that, in sharp contrast, PNNT-12HD and -12DT preferentially adopted the face-on orientation, as the $\pi$–$\pi$ stacking diffraction appeared along the $\sim q_z$ axis (Figures 14.10b and 14.10c). The larger $d_l$ values for PNNT-12HD (24.1 Å) and -12DT (28.1 Å) relative to PNNT-DT were observed. The increased $d_l$ for PNNT-12DT, relative to PNNT-DT with the same alkyl chain on the thiophene rings, implies the decreased degree of side-chain interdigitation by the alkylation (Figure 14.11). Though the $d_\pi$ values for PNNT-12HD and -12DT (3.51–3.55 Å) were slightly greater than that for PNNT-DT, these are still narrow for semiconducting polymers. We speculated that the drastically changed backbone orientation was as a consequence of weakened intermolecular interactions [50, 51], as evidenced by the larger values of $d_l$ and $d_\pi$. This most likely originates in the increased

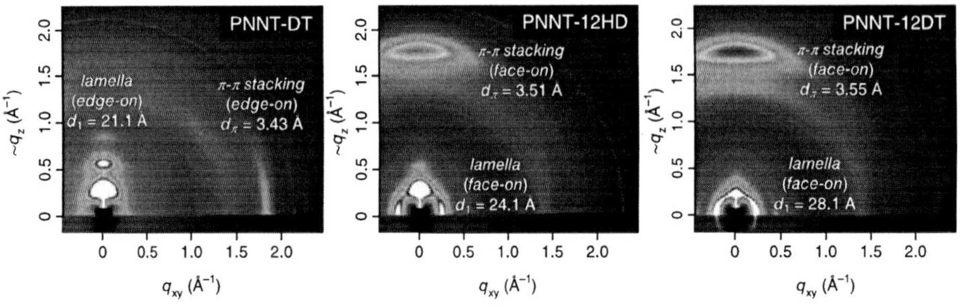


**Figure 14.10** 2D GIXD images of the polymer-only film for PNNT-DT (a), PNNT-12HD (b), and PNNT-12DT (c).

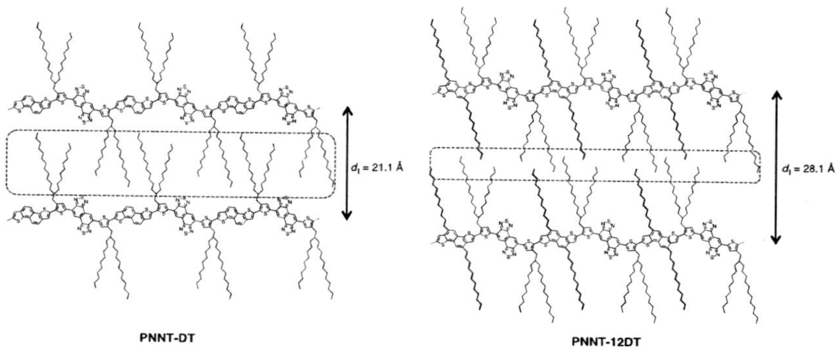


**Figure 14.11** Schematic illustration of the polymer structure predicted by the GIXD study. Whereas PNNT-DT allows strong side chain interdigitation, PNNT-12DT with the additional alkyl groups shows weak interdigitation.

side-chain density by the addition of the alkyl groups on the NDT3 unit, which hinders the side-chain interdigitation (Figure 14.11). Strong π–π intermolecular interaction should drive the polymer backbones to self-assemble each other while preventing them to interact with the substrate, which is likely to give rise to the edge-on orientation. Strong side-chain interdigitation can make the polymer lamellae stack each other, which would contribute to the edge-on orientation whole through the film thickness. However, when the π–π intermolecular interaction becomes weaker, the polymer backbone may, in part, interact with the substrate. In addition, when the side-chain interdigitation becomes weaker, the polymer lamellae cannot stack each other strongly, which may destroy the inter-lamellar ordering. These could drive the polymer backbone into the face-on orientation. It should also be noted that when PNNT-12HD and -12DT were blended with $PC_{61}BM$, both $d_l$ and $d_\pi$ were unchanged, indicating that the polymer ordering was mostly preserved. The face-on orientation and the crystalline π–π stacking of PNNT-12HD and -12DT are consistent with the high out-of-plane mobility compared with PNNT-DT. Presumably, this would again contribute to the greatly improved FF of the solar cells, even with the thicker active layer. It is interesting to note that even though PNNT-DT also formed the face-on orientation in the presence of $PC_{61}BM$, the $\mu_{SCLC}$ and FF were significantly lower than for PNNT-12HD and -12DT. This suggests that control of the primary ordering structure could be important to achieve high efficiencies.

## 14.5 Optimization of PNTz4T Cells: Distribution of Backbone Orientation vs Cell Structure

As PNTz4T was found to have great potential in solar cells, we further optimized the PNTz4T cell to improve efficiency [25]. First, we optimized the active layer thickness in a conventional cell using $PC_{61}BM$ with the p:n ratio of 1:2. A PCE of 6.6% with a $J_{SC}$ of 12.1 mA cm$^{-2}$, a $V_{OC}$ of 0.75 V and an FF of 0.73 were obtained with an active layer thickness of 150 nm. This observed FF is fairly high considering the active layer is 1.5 to 2 times thicker than that of typical PSCs (70–100 nm). We then fabricated cells with even thicker active layers. Similarly to previous studies using thicker active layers [24, 50, 52–54], the value of $J_{SC}$ increased with increasing thickness, whereas $V_{OC}$ decreased slightly with increasing thickness, but the change was relatively small. Though the FF showed a gradual decrease, it remained moderately high (0.67 at 300 nm). As a result, the overall PCE reached 8.7% with a $J_{SC}$ of 17.7 mA cm$^{-2}$ and $V_{OC}$ of 0.74 V at 300 nm (Figure 14.12a). The external quantum efficiency (EQE) was generally high, at 70% in the polymer absorption ranges of 300–500 nm and 600–800 nm (Figure 14.12b). In the conventional cell using $PC_{71}BM$, a large $J_{SC}$ was obtained compared with the cell using $PC_{61}BM$ (Figure 14.12a), which is due to the increased absorption and thus the high EQE in the range of 500–600 nm (Figure 14.12b), as commonly observed for PSCs. The highest PCE for the $PC_{71}BM$ cells was 8.9% ($J_{SC}$ = 18.9 mA cm$^{-2}$, $V_{OC}$ = 0.71 V, FF = 0.66) with the 290-nm-thick active layer.

Then, we used cells with an inverted structure ITO/ZnO/(PNTz4T/$PC_{61}BM$ or $PC_{71}BM$)/$MoO_x$/Ag). Interestingly, the inverted cells demonstrated high photovoltaic performances compared to the conventional cells, in particular in terms of $J_{SC}$ and FF. Notably, the PCE reached 10.1% ($J_{SC}$ = 19.4 mA cm$^{-2}$, $V_{OC}$ = 0.71 V, FF = 0.73), with an

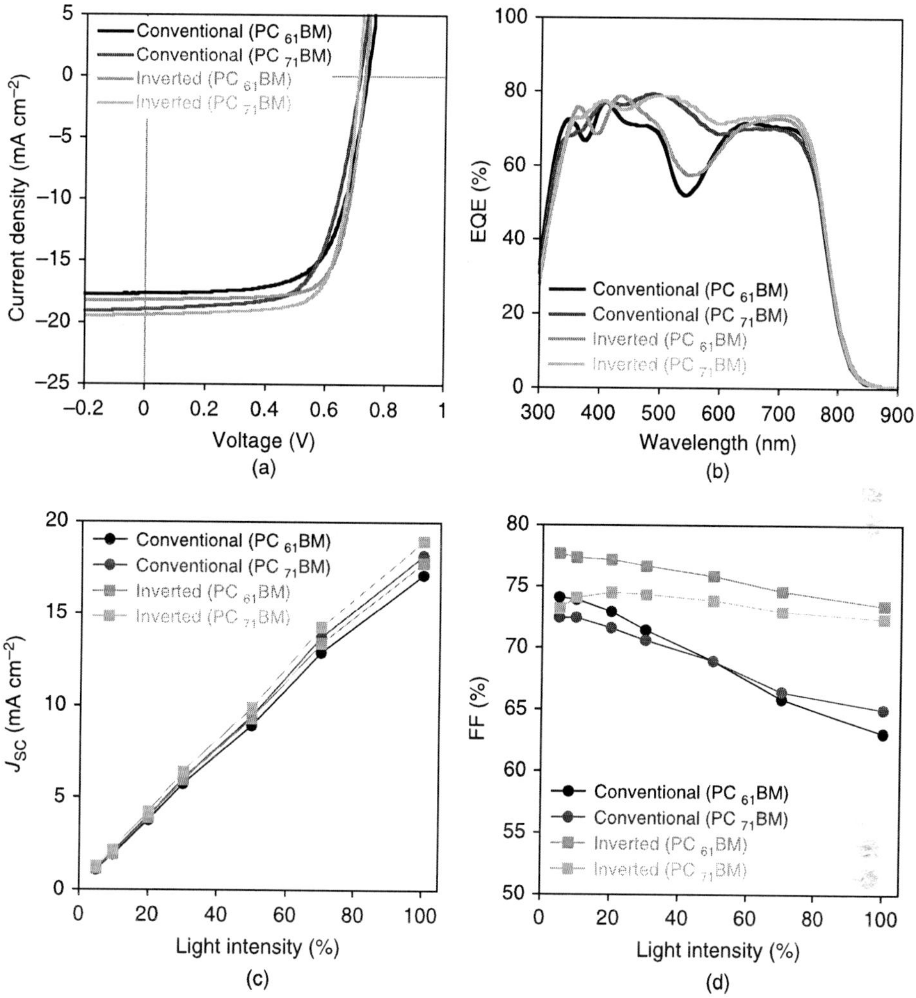


**Figure 14.12** Photovoltaic characteristics of PNTz4T-based cells with conventional and inverted architectures. (a,b) *J–V* curves (a) and EQE spectra (b) of the best cells. (c,d) $J_{SC}$ (c) and FF (d) as a function of light intensity. Source: Ref [25] Reproduced with permission of Springer Nature.

average of 9.8%, for the inverted $PC_{71}BM$ cell with an active layer thickness of ~290 nm, which is one of the highest PCEs observed in a single-junction cell. It is interesting to note that PCEs close to 10% were also observed for the inverted $PC_{61}BM$ cell with a thickness of ca. 280 nm (PCE = 9.8% (average 9.6%), $J_{SC}$ = 18.2 mA cm$^{-2}$, $V_{OC}$ = 0.73 V, FF = 0.74).

To investigate the reason for the higher FF in the inverted cell rather than in conventional cells, we conducted a qualitative study on the difference in charge recombination between both device architectures by plotting $J_{SC}$ and FF as a function of light intensity (Figure 14.12c,d) [55]. It is clear that for both conventional and inverted $PC_{61}BM$ and $PC_{71}BM$ cells, $J_{SC}$ increased linearly as the light intensity increased, wherein the number of free carriers increases. The decrease in FF with increasing light intensity was milder in the inverted cells than in conventional cells. The difference in the decrease in FF implies that bimolecular recombination is reduced in the inverted cell compared with the conventional cell, which may be one of the reasons for the higher FF in the inverted cell of this system.

The polymer microstructures in the blend films were studied by the GIXD measurements and the pole figure analysis [31]. Figures 14.13a and 14.13b display the 2D GIXD images of PNTz4T/$PC_{61}$BM blend films spun on ITO/PEDOT:PSS and ITO/ZnO substrates (film thickness, ~250 nm). In both cases, a diffraction corresponding to the π–π stacking, (010), appeared only along the ~$q_z$ axis, suggesting that there is a large population of polymer crystallites with the face-on orientation, as discussed above. Nevertheless, diffractions corresponding to the lamellar structure, (100), appeared along

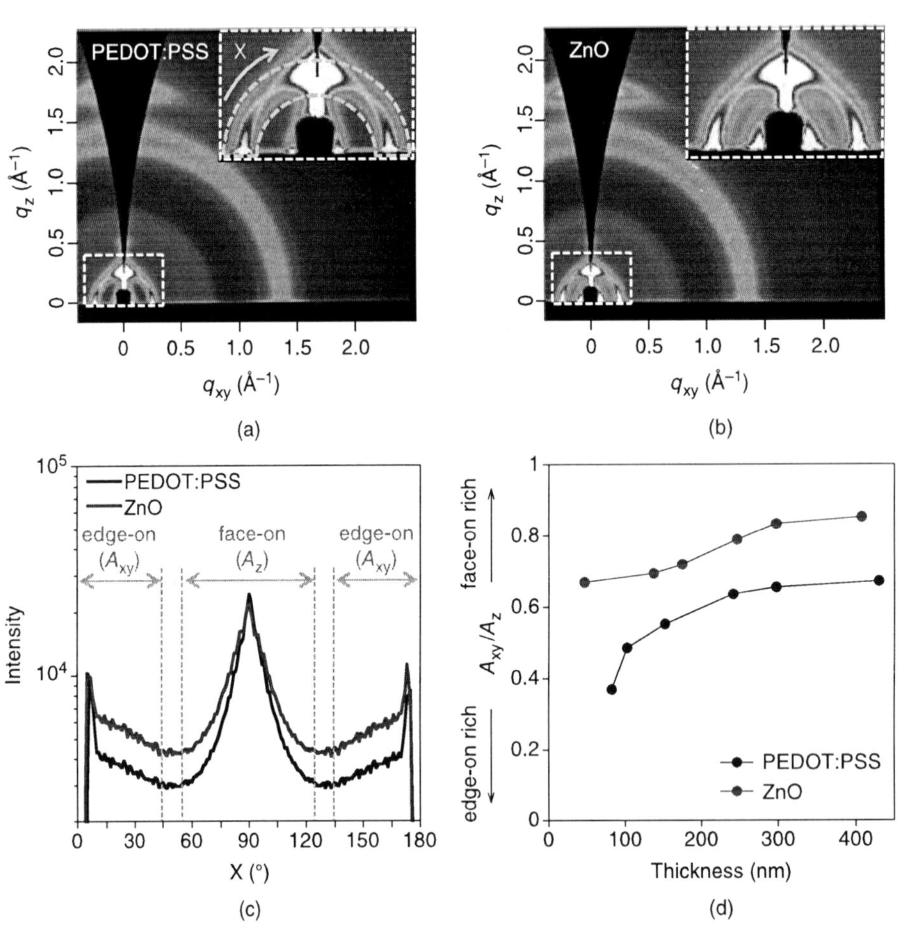


**Figure 14.13** GIXD data for PNTz4T/$PC_{61}$BM blend films. (a) 2D GIXD image of the blend film on the ITO/PEDOT:PSS substrate (thickness, 241 nm). (b) 2D GIXD image of the blend film on the ITO/ZnO substrate (thickness, 246 nm). Insets: close-up of the lamellar, (100), diffraction region. (c) Pole figures extracted from the lamellar diffraction for the blend films on both ITO/PEDOT:PSS and ITO/ZnO substrates. Definitions of the polar angle ($\chi$) range corresponding to the edge-on ($A_z$) and face-on ($A_{xy}$) crystallites are shown. (d) Dependence of $A_{xy}/A_z$ on film thickness, where $A_{xy}/A_z$ is the ratio of the face-on to edge-on orientation. Source: Ref [25] Reproduced with permission of Springer Nature.

both the $\sim q_z$ and $q_{xy}$ axes, indicating that edge-on and face-on crystallites co-exist in the film. Figure 14.13c shows the pole figures extracted from the lamellar diffraction of PNTz4T in the 2D GIXD images as shown in Figurse 14.13a and 14.13b (see insets for a close-up of the lamellar diffraction). We defined the areas integrated with polar angle $\chi$ ranges of 0–45° and 135–180° ($A_{xy}$) and 55–125° ($A_z$) as those corresponding to the fractions of face-on and edge-on crystallites, respectively. It is interesting to note that the ratio of $A_{xy}$ to $A_z$ ($A_{xy}/A_z$) for the blend film on the ITO/ZnO substrate was 0.79, which was higher than that for the blend film on the ITO/PEDOT:PSS substrate, namely 0.64. This means that the population of the face-on crystallite is larger in the inverted cells than in the conventional cells. We also carried out the pole figure analysis with different thicknesses ranging from ~50 nm to 400 nm [56], then plotted $A_{xy}/A_z$ as a function of film thickness (Figure 14.13d). For all thicknesses, $A_{xy}/A_z$ was greater for the ITO/ZnO substrate than for the ITO/PEDOT:PSS substrate. Notably, $A_{xy}/A_z$ increased gradually with increasing film thickness in both cases; that is, the population of the face-on crystallites increased with film thickness. This suggests that the face-on to edge-on ratio is not distributed evenly along the film thickness. One can assume that the orientation in the interfacial layers at the bottom (PEDOT:PSS or ZnO) and top (air), as well as the thickness of the interfacial layer, is independent of total film thickness. Thus, this increase in the face-on crystallite population mainly occurs at the bulk, and arises from the increase in the bulk volume in thicker films. This means that, in all the films with any thickness, the edge-on crystallites are abundant either at the film–bottom or film–air interface. In regioregular poly(3-hexylthiophene) films, it has been reported previously that edge-on crystallites exist at the film–bottom layer interface and that face-on crystallites exist in the bulk and at the film–air interface. Thus, in the case of PNTz4T, a class of polythiophene-based polymer, it is natural to consider that the edge-on crystallites are abundant at the film–bottom layer interface and that the face-on crystallites are abundant in the bulk through the film–air interface, which corresponds to the film–top layer in the cells, regardless of the substrate (Figure 14.14).

In conventional cells, where the generated holes flow towards the bottom PEDOT:PSS layer through the edge-on-rich region, such distribution of the polymer orientation should be detrimental to vertical hole transport, resulting in inefficient hole collection. In contrast, in the inverted cells, where the holes flow towards the top $MoO_x$ layer through the face-on-rich region, this polymer orientation distribution would facilitate vertical hole transport, leading to efficient hole collection. Together with the different ratio of the face-on crystallite between the films on the PEDOT:PSS and ZnO substrates, this model shown in Figure 14.14 is in good agreement with the high $J_{SC}$ and FF, as well as reduced charge recombination, in the inverted cell compared with the conventional cell.

## 14.6 Thiophene, Thiazolothiazole–NTz Polymers: Higly Thermally Stabe Solar Cells

The reliability of the solar cells is of particular importance as well as the efficiency [57–61]. In this section, we show that the polymers based on thiophene, thiazolothiazole (TzTz), and NTz (PTzNTzs, Figure 14.15a) gives solar cells with siginificantly high thermal stability [27]. Initially, we expected that incorporation of the NTz unit into the backbone of a thiophene–TzTz polymer (PTzBT-BOHD, Figure 14.15a) [50], which

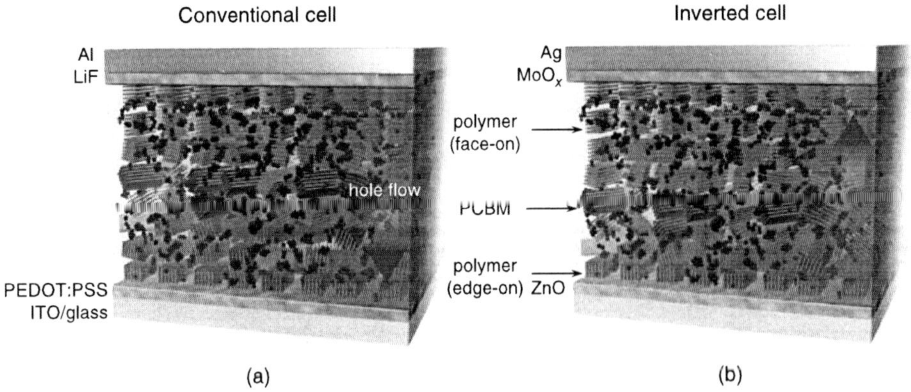


**Figure 14.14** Schematic illustrations of PNTz4T/$PC_{61}BM$ blend films in the cells. (a) Conventional cell with PEDOT:PSS as the bottom and LiF as the top interlayer. (b) Inverted cell with ZnO as the bottom and $MoO_x$ as the top interlayer. The population of face-on crystallite is larger in the inverted cell than in the conventional cell. In both cases, the population of edge-on crystallites is large at the bottom interface and the population of face-on crystallites is large in the bulk through the top interface. Note that the amount of PCBM shown is markedly reduced compared with real cells, and the distribution of the orientation is exaggerated in order to better visualize the polymer orientation. Source: Ref [25] Reproduced with permission of Springer Nature.

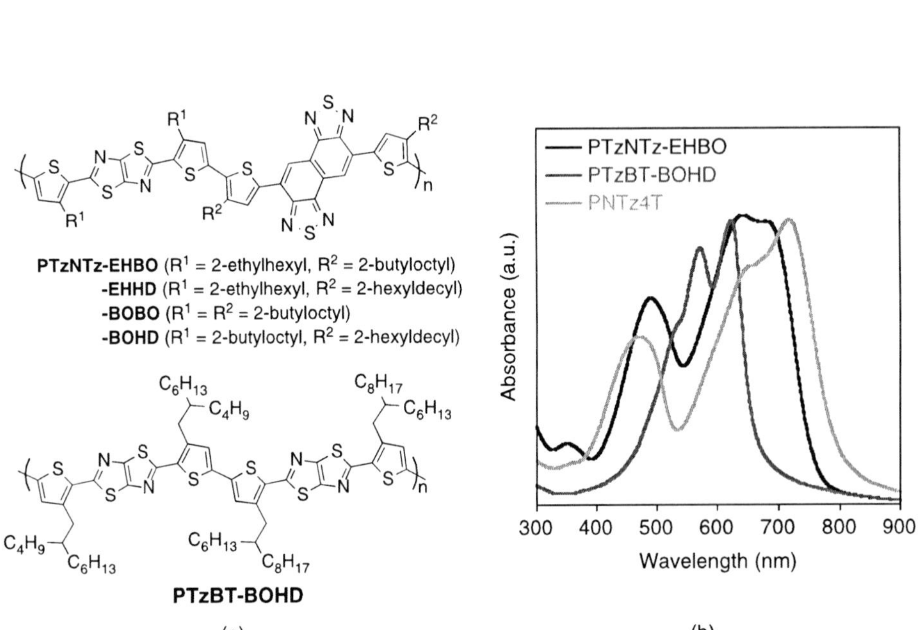


**Figure 14.15** (a) Chemical structures of the polymers based on thiophene, TzTz, and NTz (PTzNTzs) and the thiophene–TzTz polymer (PTzBT-BOHD). (b) UV-vis absorption spectra of PTzNTz-EHBO, PTzBT-BOHD, and PNTz4T in thin film.

possesses a relatively wide $E_g$ (~1.8 eV) and a deep $E_{HOMO}$ (−5.2–5.3 eV), can lead to a polymer system with a narrow $E_g$ and a deep $E_{HOMO}$ and thereby high $J_{SC}$ and $V_{OC}$ at the same time, resulting in high PCEs.

PTzNTzs showed $E_{HOMO}$s of −5.40–5.41 eV and $E_{LUMO}$s of −3.44–3.46 eV as evaluated by CV. Indeed, the $E_{HOMO}$s of PTzNTzs were deeper than those of both PTzBT-BOHD (−5.31 eV) and PNTz4T (−5.16 eV). The absorption maximum ($\lambda_{max}$) was ca. 680nm (Figure 14.15b), which was red-shifted from that of PTzBTs by 60nm. The absorption edge ($\lambda_{edge}$) was determined to be ca. 790 nm, which corresponds to the optical band gap ($E_g$) of 1.57–1.58 eV. These values were about 0.2 eV smaller than that of PTzBT-BOHD and slightly larger than that of PNTz4T (1.54 eV).

2D GIXD images of the polymer/$PC_{71}BM$ blend films gave a clear π–π stacking diffraction along the $q_z$ axis only for PTzNTz-EHBO with short alkyl groups (Figure 14.16a). The PTzNTz-BOBO blend film also showed a π–π stacking diffraction, but the intensity was quite weak. In contrast, when 1,8-diiodooctane (DIO) was used for the film fabrication as the solvent additive, all the polymers exhibited a diffraction corresponding to the π–π stacking of face-on crystallites (Figure 14.16b), indicating that the ordering structure was enhanced particularly for PTzNTz-EHHD, -BOBO, and -BOHD. The different behavior between PTzNTz-EHBO and the others in the DIO-aided films can be explained as follows. In the case of PTzNTz-EHHD, -BOBO, and -BOHD, with the higher solubility and thus weaker aggregation nature, the addition of DIO may slow down the crystallization of $PC_{71}BM$, during which time, the polymers crystallize as typically seen for many polymers [62].

Solar cells with an inverted structure were used to investigate the photovoltaic properties of the polymers. The optimal polymer to $PC_{71}BM$ weight ratio was 1:1.5 for all polymers. $J$–$V$ curves of the cells are displayed in Figure 14.17a. All the cells showed similar $V_{OC}$ of 0.84–0.85 V. Though $E_{HOMO}$ of PTzNTzs was deeper than that of PTzBT by 0.1 eV, $V_{OC}$s of the PTzNTzs cells were slightly decreased by 0.03–0.04 V. The origin of the energetic loss is yet unknown. Nevertheless, these $V_{OC}$s were higher than

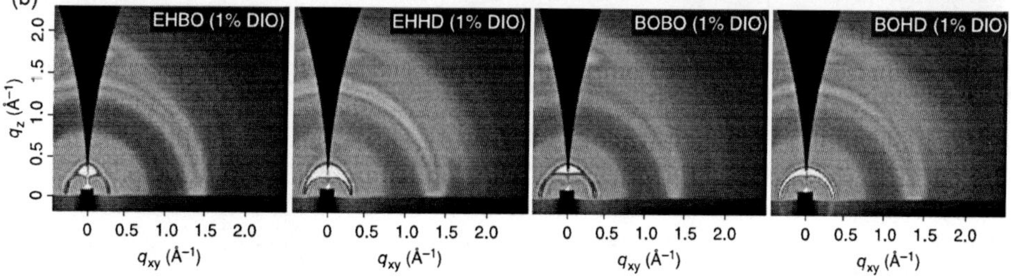


**Figure 14.16** 2D GIXD images of PTzNTz/$PC_{71}BM$ blend films (c), and DIO (1%)-aided PTzNTz/$PC_{71}BM$ blend films (d). The alkyl groups are shown at the right top of the image.

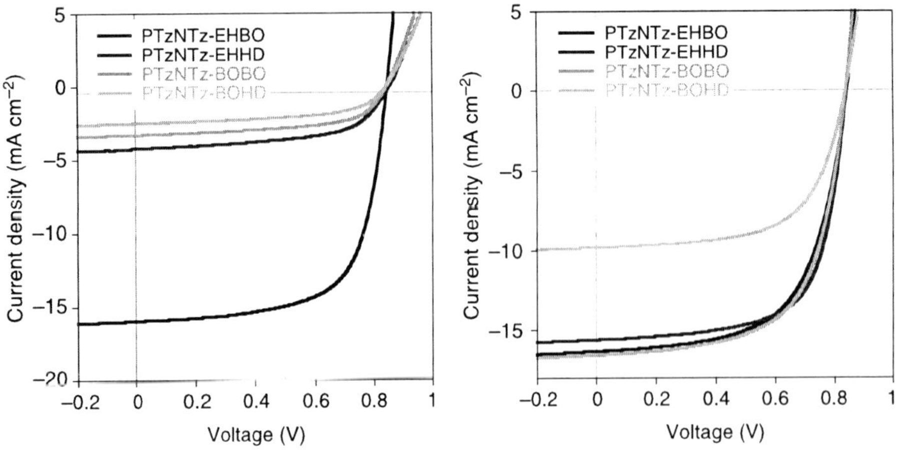


**Figure 14.17** *J–V* curves of the solar cells based on PTzNTzs. (a) The active layer was spun from the CB solution. (c) The active layer was spun from the CB/DIO (1v/v%) solution.

that of the PNTz4T cell. The PTzNTz-EHBO cells gave relatively high $J_{SC}$ of 16.0 mA cm$^{-2}$ compared to that of the PTzBT cells (~13 mA cm$^{-2}$), which is consistent with the narrower $E_g$. Interestingly, the cells that used PTzNTz-EHHD, -BOBO, and -BOHD exhibited significantly lower $J_{SC}$ of around 2–4 mA cm$^{-2}$. As a result, PTzNTz-EHBO cells exhibited PCEs of up to 9.0% ($J_{SC}$ = 16.0 mA cm$^{-2}$, $V_{OC}$ = 0.84 V, FF = 0.67), and those of the other polymer cells were 1.2–2.1%. Note that PCEs of the cells using PTzNTz-EHHD, -BOBO and -BOHD were markedly improved when the active layer was fabricated using 1% of DIO as the solvent additive (Figure 14.17b). In particular, the DIO-aided cells that used PTzNTz-EHHD and -BOBO showed significant increase in $J_{SC}$ to 16.3 and 15.6 mA cm$^{-2}$, respectively, which were similar to the value of the cells that used PTzNTz-EHBO. Thus, PCEs of the cells were greatly improved with the association of DIO: 2.1% to 8.8% for PTzNTz-EHHD, 1.7% to 8.8% for PTzNTz-BOBO, and 1.2% to 5.2% for PTzNTz-BOHD. PCE of the PTzNTz-EHBO-based cells fabricated with DIO was mostly similar to those without DIO. The improved performances for the cells using these polymers are in good agreement with the changes observed in the 2D GIXD images.

We then tested the thermal stability of the cells that used PTzNTz-EHBO fabricated with and without DIO and PTzNTz-BOBO fabricated with DIO in comparison with the cells using PTzBT-BOHD and PNTz4T. There have been several procedures established for testing device stability based on the consensus reached by the consortia of the international summit on OPV stability (ISOS) [63, 64]. In this work, we chose to test our cells in the dark at 85° C, which refers to category ISOS-D-3. To avoid degradation regarding the encapsulation such as the adhesive material, we tested the cells without encapsulation and stored them on a hotplate in a glovebox. Figure 14.18a shows the change of PCE as a function of the storage time. PCE of the PTzNTz-EHBO cell decreased gradually from 8.6% to 7.5% after 500 hours, which corresponds to a 13% drop from initial PCE. The DIO-aided PTzNTz-EHBO cell showed similar behavior with a PCE drop of 13%, where PCE decreased from 8.6% to 7.5%, suggesting that DIO does not affect the device stability under these conditions for the PTzNTz system. The DIO-aided

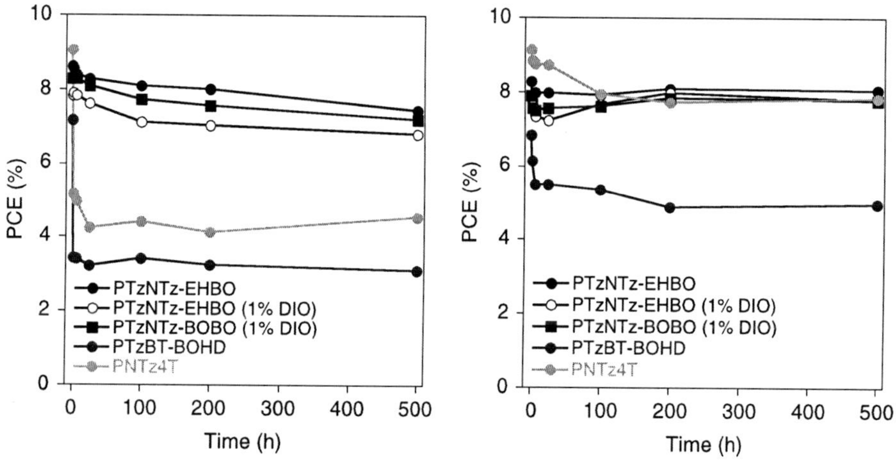


**Figure 14.18** Change of PCE for the cells using PTzNTz-EHBO fabricated without and with DIO (1 v/v%), PTzNTz-BOBO fabricated with DIO (1 v/v%), PTzBT-BOHD, and PNTz4T in combination with $PC_{71}BM$ under the storage for 500 hours at 85 °C in the glovebox. $MoO_x$ (a) and $WO_x$ (b) were used as the hole transport layer of the cells.

PTzNTz-BOBO cell also showed similar behavior with a PCE drop of 13%, where PCE decreased from 8.3% to 7.2%, implying that the effect of the side chain is negligible. In sharp contrast, PCE of the PTzBT-BOHD cell showed a significant drop at the initial stage; PCE decreased from 7.2% to 3.4% after one hour, though it showed only a slight degradation thereafter, resulting in an overall drop of 57% after 500 hours. The PNTz4T cell also showed significant degradation of 49% from the initial value (PCE decreased from 9.0% to 4.6%).

Interestingly, when the hole transport layer, $MoO_x$, was replaced with $WO_x$, the thermal stability of the cells was greatly improved (Figure 14.18b). All the PTzNTzs cells tested here with the initial PCE of 8.3%, 7.9%, and 7.9% for PTzNTz-EHBO, PTzNTz-EHBO with DIO, and PTzNTz-BOBO with DIO, respectively, showed negligible degradation after 500 hours. The drop of PCE for the PNTz4T cell was improved to 14%, whereas that for PTzBT-BOHD cell was improved but was still 30% after 500 hours. To the best of our knowledge, such high stability has never been reported for PSCs with high PCEs of ~9% under standardized conditions. These results suggest that the introduction of NTz into the polymer backbone is advantageous for the thermal stability of the solar cells as well as PCE.

## 14.7 Summary

We have described the great potential of NTz-based polymers by showing the basic properties through solar cell applications. With the strong electron-deficient nature and rigid π-extented skelton, NTz offers narrow bandgaps, deep HOMO energy levels, and highly crystalline order to the polymers when used as the building unit. It was also demonstrated that, importantly, the orientation of the polymer backbone against the substrate can be controled by the careful design of the side chain placement. The solar cells, based on the NTz-based polymers, exhibited significantly high PCEs of ~10%. In addition, such high PCEs can be achieved with thick active layers of around 300 nm, which is about 2–3 times thicker than typical polymer-based solar cells. Furthermore, the solar cells, based on the NTz-based polymers, show unprecedentedly high thermal stability. These unconventional features observed in the NTz-based polymers are of particular importance for commercialization. We believe that these results will pave the way for further development of the semiconducting polymers for solar cells.

## References


1 Peet, J., Heeger, A.J. and Bazan, G.C. (2009). *Acc. Chem. Res.* 42: 1700.
2 Brabec, C., Scherf, U. and Dyakonov, V. (2014). *Organic photovoltaics: Materials, Device Physics, and Manufacturing Technologies*, 2e. Weinheim: Wiley-VCH.
3 Günes, S., Neugebauer, H. and Sariciftci, N.S. (2007). *Chem. Rev.* 107: 1324.
4 Liang, Y. and Yu, L. (2010). *Acc. Chem. Res.* 43: 1227.
5 InganÄs, O., Zhang, F. and Andersson, M.R. (2009). *Acc. Chem. Res.* 42: 1731.
6 Chen, J. and Cao, Y. (2009). *Acc. Chem. Res.* 42: 1709.
7 Boudreault, P.-L.T., Najari, A. and Leclerc, M. (2011). *Chem. Mater.* 23: 456.
8 Facchetti, A. (2011). *Chem. Mater.* 23: 733.
9 Beaujuge, P.M. and Fréchet, J.M.J. (2011). *J. Am. Chem. Soc.* 133: 20009.
10 Thompson, B.C. and Fréchet, J.M.J. (2008). *Angew. Chem. Int. Ed.* 47: 58.
11 Scharber, M.C., Mühlbacher, D., Koppe, M. et al. (2006). *Adv. Mater.* 18: 789.
12 Huang, Y., Kramer, E.J., Heeger, A.J. and Bazan, G. C. (2014). *Chem. Rev.* 114: 7006.
13 Osaka, I. and Takimiya, K. (2015). *Polymer* 59: A1.
14 Takimiya, K., Osaka, I. and Nakano, M. (2013). *Chem. Mater.* 26: 587.
15 Osaka, I., Sauvé, G., Zhang, R. et al. (2007). *Adv. Mater.* 19: 4160.
16 Mataka, S., Takahashi, K., Ikezaki, Y. et al. (1991). *Bull. Chem. Soc. Jpn.* 64: 68.
17 Wang, M., Hu, X., Liu, P. et al. (2011). *J. Am. Chem. Soc.* 133: 9638.
18 Yang, T., Wang, M., Duan, C. et al. (2012). *Energy Environ. Sci.* 5: 8208.
19 Liu, P., Dong, S., Liu, F. et al. (2015). *Adv. Funct. Mater.* 25: 6458.
20 Liu, L.-Q., Zhang, G.-C., Liu, P. et al. (2014). *Chem. Asian J.* 9: 2104.
21 Osaka, I., Shimawaki, M., Mori, H. et al. (2012). *J. Am. Chem. Soc.* 134: 3498.
22 Osaka, I., Abe, T., Shimawaki, M. et al. (2012). *ACS Macro Lett.* 1: 437.
23 Kawashima, K., Miyazaki, E., Shimawaki, M. et al. (2013). *Polym. Chem.* 4: 5224.
24 Osaka, I., Kakara, T., Takemura, N. et al. (2013). *J. Am. Chem. Soc.* 135: 8834.
25 Vohra, V., Kawashima, K., Kakara, T. et al. (2015). *Nat. Photon.* 9: 403.
26 Cheng, S.-W., Chiou, D.-Y., Tsai, C.-E. (2015). *Adv. Funct. Mater.* 25: 6131.
27 Saito, M., Osaka, I., Suzuki, Y. et al. (2015). *Sci. Rep.* 5: 14202.
28 Zhong, H., Han, Y., Shaw, J. et al. (2015). *Macromolecules* 48: 5605.
29 Liu, Y., Zhao, J., Li, Z. et al. (2014). *Nat. Commun.* 5: 5293.
30 Yasuda, T., Shinohara, Y., Kusagaki, Y. et al. (2015). *Polymer* 58: 139.
31 Rivnay, J., Mannsfeld, S.C.B., Miller, C.E. et al. (2012). *Chem. Rev.* 112: 5488.
32 Osaka, I., Saito, M., Mori, H. et al. (2012). *Adv. Mater.* 24: 425.
33 Rieger, R., Beckmann, D., Mavrinskiy, A. et al. (2010). *Chem. Mater.* 22: 5314.
34 Osaka, I., Abe, T., Shinamura, S. and Takimiya, K. (2011). *J. Am. Chem. Soc.* 133: 6852.

35 McCullough, R.D., Tristram-Nagle, S., Williams, S.P. et al. (1993). *J. Am. Chem. Soc.* 115: 4910.
36 Zhao, N., Botton, G., Zhu, S. et al. (2004). *Macromolecules* 37: 8307.
37 Kline, R.J., McGehee, M.D., Kadnikova, E.N. et al. (2005). *Macromolecules* 38: 3312.
38 McCulloch, I., Heeney, M., Bailey, C. et al. (2006). *Nat. Mater.* 5: 328.
39 Mühlbacher, D., Scharber, M., Morana, M. et al. (2006). *Adv. Mater.* 18: 2884.
40 Liang, Y., Wu, Y., Feng, D. et al. (2009). *J. Am. Chem. Soc.* 131: 56.
41 Takimiya, K., Shinamura, S., Osaka, I. and Miyazaki, E. (2011). *Adv. Mater.* 23: 4347.
42 Umeda, R., Fukuda, H., Miki, K. et al. (2009). *C. R. Chim.* 12: 378.
43 Tilak, B.D. (1951). *Proc. Indian Acad. Sci., Sect.n A* 33: 71.
44 Shinamura, S., Miyazaki, E. and Takimiya, K. (2010). *J. Org. Chem.* 75: 1228.
45 Shinamura, S., Osaka, I., Miyazaki, E. et al. (2011). *J. Am. Chem. Soc.* 133: 5024.
46 Loser, S., Miyauchi, H., Hennek, J.W. et al. (2012). *Chem. Commun.* 48: 8511.
47 Loser, S., Bruns, C.J., Miyauchi, H. et al. (2011). *J. Am. Chem. Soc.* 133: 8142.
48 Osaka, I., Abe, T., Shinamura, S. et al. (2010). *J. Am. Chem. Soc.* 132: 5000.
49 Osaka, I., Shinamura, S., Abe, T. and Takimiya, K. (2013). *J. Mater. Chem. C* 1: 1297.
50 Osaka, I., Saito, M., Koganezawa, T. and Takimiya, K. (2014). *Adv. Mater.* 26: 331.
51 Zhang, X., Richter, L.J., DeLongchamp, D.M. et al. (2011). *J. Am. Chem. Soc.* 110902124705042.
52 Li, W., Hendriks, K.H., Roelofs, W.S.C. et al. (2013). *Adv. Mater.* 25: 3182.
53 Peet, J., Wen, L., Byrne, P. et al. (2011). *Appl. Phys. Lett.* 98: 043301.
54 Price, S.C., Stuart, A.C., Yang, L. et al. (2011). *J. Am. Chem. Soc.* 133: 4625.
55 Stuart, A.C., Tumbleston, J.R., Zhou, H. et al. (2013). *J. Am. Chem. Soc.* 135: 1806.
56 Duong, D.T., Toney, M.F. and Salleo, A. (2012). *Phys. Rev. B* 86: 205205.
57 Brabec, C.J., Gowrisanker, S., Halls, J.J.M. et al. (2010). *Adv. Mater.* 22: 3839.
58 Peters, C.H., Sachs-Quintana, I.T., Kastrop, J.P. et al. (2011). *Adv. Energy Mater.* 1: 491.
59 Peters, C.H., Sachs-Quintana, I.T., Mateker, W.R. et al. (2012). *Adv. Mater.* 24: 663.
60 Neugebauer, H., Brabec, C., Hummelen, J.C. and Sariciftci, N.S. (2000). *Sol. Energy Mater. Sol. Cells* 61: 35.
61 Kesters, J., Verstappen, P., Raymakers, J. et al. (2015). *Chem. Mater.* 27: 1332.
62 Lou, S.J., Szarko, J.M., Xu, T. et al. (2011). *J. Am. Chem. Soc.* 111205073209001.
63 Reese, M.O., Gevorgyan, S.A., Jørgensen, M. et al. (2011). *Sol. Energy Mater. Sol. Cells* 95: 1253.
64 Tanenbaum, D.M., Hermenau, M., Voroshazi, E. et al. (2012). *RSC Advances* 2: 882.

# 15

# Plasmonics for Light-Emitting and Photovoltaic Devices

*Koichi Okamoto*

*Department of Physics and Electronics, Osaka Prefecture University, Japan*

**CHAPTER MENU**



## 15.1 Optical Properties of the Surface Plasmon Resonance

Surface plasmon (SP), which is the quantum of the plasma oscillation of free electrons at metal/dielectric interface, can couple to electromagnetic wave and forms special mode called surface plasmon polariton (SPP) at the interface. SPP offers the unique ability to localize and enhance electromagnetic fields and it brings novel optical properties and functions to materials. The technique controlling and utilizing SPP is called "plasmonics" and has attracted much attention with the recent rapid advance of nanotechnology[1–3].

There are two types of modes of the SPP, namely, the propagating and localized modes. The electric field distribution of the SPP propagating mode was obtained by the three-dimensional (3D) finite-difference time-domain (3D-FDTD) simulations and shown in Figure 15.1a. The arrows express the lines of electric force generated by the fluctuations of the electron density by the plasma oscillations. The compressional wave of the electron density propagates the metal surface alone. The SPP can interact with light wave because it includes transverse wave of the charge fluctuation at the metal surface, while the compressional wave of the SP has only the longitudinal wave.

The wave vector ($k_{SP}$) of the propagating SPP parallel to the interface can be written with the following equation [1]

$$k_{SP}(\omega) = \frac{\omega}{c}\sqrt{\frac{\varepsilon_m(\omega)\varepsilon_d(\omega)}{\varepsilon_m(\omega)+\varepsilon_d(\omega)}} \tag{15.1}$$

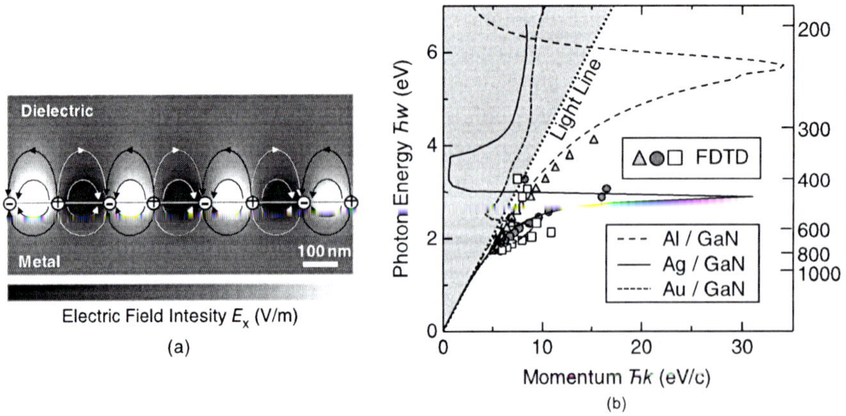


**Figure 15.1** (a) Electric field distribution of the SPP propagating mode calculated by the FDTD simulations. (b) Dispersion diagram of the SPP propagating mode.

where the relative permittivity of the metal and the dielectric materials are $\varepsilon_{\mathrm{m}}$ and $\varepsilon_{\mathrm{d}}$, respectively. $\omega$ and $C$ are the angular frequency and the light velocity in vacuum, respectively. This equation is called the dispersion equation of the SP and was derived from the boundary conditions of Maxwell's formula. The dispersion diagram, which is the relationship between $k_{\mathrm{SP}}$ and $\omega$ was obtained by Eq. 15.1 and shown in Figure 15.1b for Al/GaN, Ag/GaN, and Au/GaN interfaces, respectively. The permittivity of GaN was defined as 6 without wavelength dispersion. The angular frequency ($\omega$) and the wave vector ($k$) can be converted to the photon energy ($\hbar\omega$) and the momentum ($\hbar k$) by multiplying with the Dirac constant ($\hbar$). The black line is the light line given by $k_d = \sqrt{\varepsilon_d}\omega/c$ where $k_{\mathrm{d}}$ is the wave vector of the light wave propagated in the dielectric material, which can exist only at the inside area of this line shown as the shaded region in Figure 15.1b. On the other hand, the dispersion of SP exists outside of the light line because $k_{\mathrm{SP}}$ can be larger than $k_{\mathrm{d}}$ when $\varepsilon_{\mathrm{m}} + \varepsilon_{\mathrm{d}} < 0$. That is the most important property of the SP, so a negative value of the permittivity of a metal is necessary to generate the SPPs. The large $k_{\mathrm{SP}}$ means that the wavelength was compressed and the velocity was slowed down at the metal surface. This property enables the SPP to propagate into smaller space. Eq. 15.1 suggested that $k_{\mathrm{SP}}$ becomes infinity if the value of $\varepsilon_{\mathrm{m}} + \varepsilon_{\mathrm{d}}$ is close to zero. However, the dispersion diagrams of the SPP in Figure 15.1b have maximum values because the permittivity of a metal is complex number as $\varepsilon_{\mathrm{m}} = \varepsilon_{\mathrm{m}}' + \varepsilon_{\mathrm{m}}'' i$. The value of $\omega$ at the maximum of $k_{\mathrm{SP}}$ is called the SP resonance frequency ($\omega_{\mathrm{SP}}$) and shown in Figure 1(b) as dashed lines. The SP resonance energy (wavelength) are given as 5.7 eV (220 nm), 2.9 eV (430 nm), and 2.3 eV (540 nm) for Al/GaN, Ag/GaN, and Au/GaN interfaces, respectively. This suggests that Al, Ag, and Au must be suitable to use UV, blue, and red regions, respectively.

The electric field distribution of the SPP propagating mode obtained by the 3D-FDTD simulations was shown in Figure 15.2a. The polarizability ($\alpha$) of the metal sphere is given by the following equation if the radius ($r$) is much smaller than the wavelength.

$$\alpha = 4\pi r^3 \frac{\varepsilon_m - \varepsilon_d}{\varepsilon_m + 2\varepsilon_d} \tag{15.2}$$

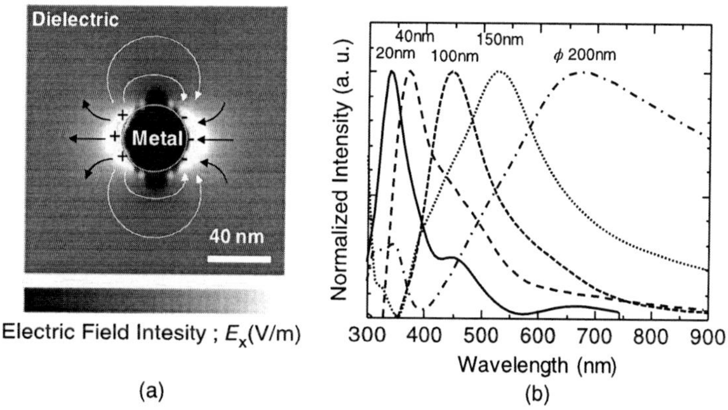


**Figure 15.2** (a) Electric field distribution of the localized SP mode around the Ag nanoparticle-calculated by the FDTD simulations. (b) The FDTD simulations of the resonance spectra around the Ag nanoparticles with various diameters.

Therefore $\alpha$ reaches to a maximum value at the condition of $2\varepsilon_m + \varepsilon_d = 0$. This condition is called the Fröhlich condition and the frequency under this condition is called the localized SP resonance frequency ($\omega_{LSP}$). In many cases, $\omega_{LSP}$ is lower than $\omega_{SP}$. If the sphere size becomes larger and closer to the wavelength, the electrostatic approximation is broken and the resonance peak of the localized SP mode is red-shifted and broadened by the time delay effect. In that case, the resonance spectra of the localized mode can be calculated by the Mie theory or the simulations of the electromagnetic analyses. Figure 15.2b shows the resonance spectra, that is the electrical field of the localized SP mode generated around an Ag nano-sphere of various diameters as calculated by the electromagnetic analysis with the FDTD method. It is possible to control the SP-photon resonance spectra by making an Ag nano-sphere with various diameters from 20 nm to 200 nm. The resonance peak wavelengths were changeable for wider wavelength regions from UV to near-infrared (IR).

## 15.2 High-Efficiency Light Emissions using Plasmonics

One futuristic application of plasmonics is the development of super bright light-emitting devices. Solid-state light-emitters have been expected to eventually replace traditional fluorescent tubes as new illumination sources. For example, InGaN-based quantum wells (QWs) provide bright light sources of UV, blue, green, red, and white lights [4]. Also, organic LEDs emitting in the visible spectrum provide stable sources of light for displays and illumination sources at a significantly lower cost than semiconductors. However, their efficiencies are still substantially lower than those of fluorescent lights. The most important requirement of these inorganic and organic LEDs is to increase emission efficiencies. The SP coupling technique is one of the most effective methods to increase these efficiencies. The idea of SP-enhanced light emission has been proposed since 1990, and it has been applied to increase emission efficiencies of several materials which include InGaN QWs. Gontijo et al. reported the coupling of the emission from InGaN QW into the SPP on silver thin film by using

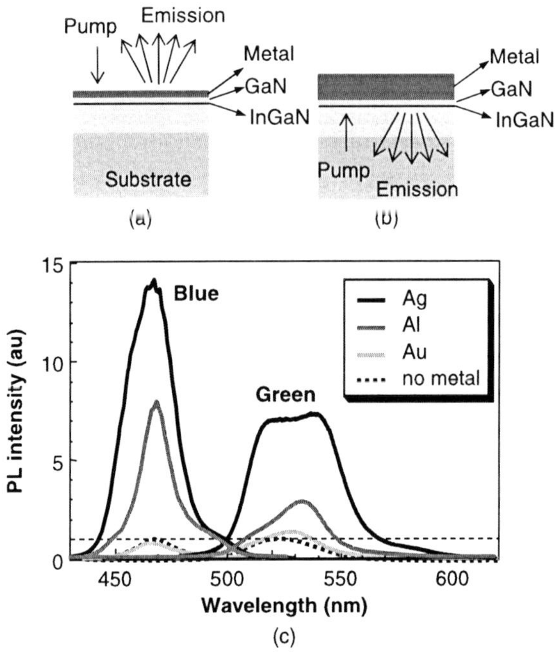


**Figure 15.3** Sample structure and experimental configuration of previous study reported in Ref. [6, 7] (a) and this study (b). (c) SP enhanced blue and green emissions from InGaN/GaN QWs coated with Ag, Al or Au.

the configuration shown in Figure 15.3a [5]. Unfortunately, they found that the PL intensities dramatically decreased with the SP coupling. By using the same sample structure, Neogi et al. confirmed that the recombination rate in an InGaN/GaN QW could be significantly enhanced by the time-resolved PL measurement [6]. However, in these early studies, light could not be extracted efficiently from the metal surface, and the SP coupling has been thought to be a negative factor for light-emitting materials.

Recently, we have reported for the first time that large photoluminescence (PL) increases from InGaN/GaN QW material coated with metal layers by using the configuration shown in Figure 15.3b [7]. InGaN/GaN, quantum well (QW) wafers were grown on a (0001) oriented sapphire substrate by a metal-organic chemical vapor deposition (MOCVD). The QW heterostructure consists of a GaN (4 μm) buffer layer, an InGaN QW (3 nm) and a GaN cap layer (10 nm). Silver, aluminum, or gold layers (50 nm) were deposited on top of the surfaces of these wafers by a high vacuum thermal evaporation. Photoluminescence (PL) measurements were performed by exciting the QW with a 406 nm diode laser and detecting the emission with a multi-channel spectrometer.

Figure 15.3c shows typical PL spectra from an InGaN/GaN QW separated from metal layers by 10 nm GaN spacers. Huge enhancements of PL spectra of both blue and green emissions were observed from the Ag- or Al-coated samples. These PL enhancements should be attributed to the SP coupling. No such enhancements were obtained from samples coated with Au, as its well-known plasmon resonance occurs only at longer wavelengths. The difference between our configuration and that of previous reports [5, 6] is the direction of photo-pumping and detection. In this study, we photo-pump and detect emission from the back-side of the samples through the transparent substrate by polishing the bottom surface (Figure 15.3b). By employing such back-side access to the QWs, we can avoid an absorption loss at the metal layer and obtain an effective light extraction from SPP at the interface. Thus, we can use very thick metal layers. This should also be a very important factor to obtain a huge enhanced light emission. If the metal layer is thinner than the penetration depth of SPP, other SPP mode is generated at the air/metal interface of the opposite side of the metal layer. These SPP modes couple each other and form symmetric and anti-symmetric mode of the SPPs. This should modify the SP frequency and coupling condition and make the light extraction very difficult. The thick metal layer is also useful to avoid the oxidation of the silver surface. Metal oxidation changes the surface roughness and SP mode. But the oxidation is typically generated only at air/metal interface and not at the metal/semiconductor interface. In this study, SPPs at the metal/semiconductor interface contribute to light emission enhancement. The thickness of metal films (50 nm) is large enough to ensure that metal oxidation at air/metal interface does not influence the metal/semiconductor interface. It is a very simple solution but the back-side access is the most important trick which enabled us to obtain light enhancements by the SP coupling for the first time.

## 15.3 Mechanism for the SP Coupled Emissions

We propose a possible mechanism of the QW–SP coupling and the light extraction shown in Figure 15.4 [7, 8]. Electron-hole pairs in the QW couple to plasma oscillation of electrons at the metal/semiconductor interface when the energies of electron-hole pairs and of the SP frequency are similar. Then, an electron-hole recombination may produce SPPs instead of photons or phonons, and this new recombination path increases the recombination rate and internal quantum efficiency. If the metal surface is perfectly flat, the SPP energy would be thermally dissipated. By providing roughness or nanostructure of the metal layers, the SPP energy can be extracted as light. Such roughness allows SPPs of high momentum to scatter, lose momentum, and couple to radiative photon. In order to obtain the high photon extraction efficiencies, the few tens of nanometer sized structures at the metal surfaces were obtained by controlling the evaporation conditions[9].

In order to evaluate the SP coupling mechanism we proposed, we employed the 3D-FDTD simulation [10]. To perform the 3D-FDTD simulations, we used "Poynting

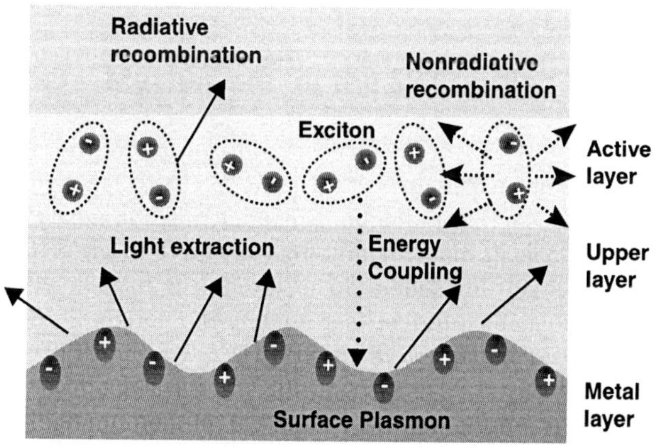


**Figure 15.4** Schematic diagram of the enhanced light emission efficiency by the exciton-SP coupling.

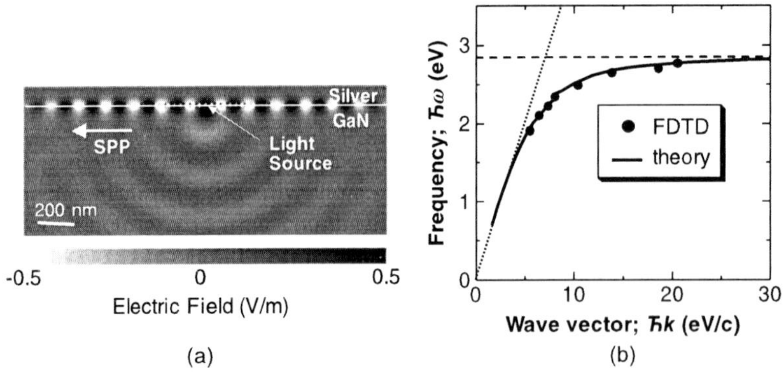


**Figure 15.5** Generation and propagation of SPP modes from the point light source located at the silver/GaN interface calculated by the 3D-FDTD simulation. (b) Dispersion curve of the SPP modes at silver/GaN interface by the calculated values by the 3D-FDTD simulations and the theoretical values with dielectric constants.

for optics" (Fujitsu Co.) which is known to be very suitable to simulate SP modes. Figure 15.5a shows the calculated spatial distribution of the electric field around the metal/semiconductor interface. The clear SPP mode appeared and propagated within the interface by the point light source located at the interface. A polarized plane wave with a 525 nm wavelength and 1 V/m amplitude was used as a point light source which is an assumption of an electron-hole pair. This result suggests that the SPP mode can be generated easily by direct energy transfer from electron-hole pairs without any special structures. Usually, some special configurations are necessary to generate SPP mode such as a grating coupler or an attenuated total reflection setting to satisfy a phase-matching condition between SPPs and photons. However, if the light source is located near the metal/dielectric interface within the wavelength scale, the SPP mode can be generated regardless of the phase matching condition. This calculation supports our proposed SP coupling model.

To verify the 3D-FDTD simulations, we calculated the SPP modes with various wavelengths and figured the dispersion curve of the SPP mode in Figure 15.5b. The dotted line and dashed line are the light line and SP frequency, respectively. The calculated values were in good agreement with the theoretical values given by Eq. 15.1 as the dispersion curve drawn with the solid line.

We also obtained much evidence that suggests the existence of the SP coupling.

(1) We found that the enhanced PL intensities decrease exponentially with increasing GaN spacer thickness. This should be reasonable because the SPP is strongly confined at the interface.
(2) We found that the obtained wavelength dependence of the PL enhanced ratios is clearly correlated to the dispersion curve of the SPPs calculated with the dielectric functions.
(3) We found that the internal quantum efficiencies were actually increased by measuring the temperature dependence of PL intensities.
(4) We also found that the spontaneous emission rates were also dramatically increased by the time-resolved PL measurement. These facts give support to our proposed coupling mechanism.

## 15.4 Quantum Efficiencies and Spontaneous Emission Rates

The external quantum efficiency ($\eta_{ext}$) of light emission from an LED is given by the light extraction efficiency ($C_{ext}$) and internal quantum efficiency ($\eta_{int}$). In turn, $\eta_{int}$ is determined by the ratio of the radiative (($k_{rad}$) and nonradiative ($k_{non}$) recombination rates of carriers.

$$\eta_{int}(\omega) = \frac{k_{rad}(\omega)}{k_{rad}(\omega) + k_{non}(\omega)} \tag{15.3}$$

Often, $k_{non}$ is faster than $k_{rad}$ at room temperature, resulting in modest $\eta_{int}$. There are three methods to increase $\eta_{ext}$; (1) increase $C_{ext}$, (2) decrease $k_{non}$, or (3) increase $k_{rad}$. Much effort has also been placed on reducing $k_{non}$ by growing higher quality crystals, but dramatic enhancements of $\eta_{ext}$ have so far been elusive. On the other hand, there have been very few studies focusing on increasing $k_{rad}$, though that could prove to be most effective for the development of high $\eta_{ext}$ light emitters. The SP coupling can provide a very rare technique to increase $k_{rad}$.

Under the existence of the SP coupling, the original and enhanced internal quantum efficiency of emission can be described as follows

$$\eta^{*}_{int}(\omega) = \frac{k_{rad}(\omega) + C'_{ex}(\omega)k_{SPC}(\omega)}{k_{rad}(\omega) + k_{non}(\omega) + k_{SPC}(\omega)} \tag{15.4}$$

where $k_{SPC}(\omega)$ is the SP coupling rate and should be very fast because the density of states of SP modes is much greater than that of the electron-hole pairs in the QW. $C'_{ext}(\omega)$ is the probability of photon extraction from the SPs energy. $C'_{ext}(\omega)$ is decided by the ratio of light scattering and dumping of the SPP mode through non-radiative loss. $C'_{ext}(\omega)$ should depend on the roughness and nano-structure of the metal surface. We succeeded in controlling the grain structure within nano-sizes, and achieved almost $C'_{ext}(\omega) = 1$.

If the SP coupling rate $k_{SPC}$ is faster than $k_{rad}$ and $k_{non}$, internal quantum efficiency should be increased. We can measure the spontaneous emission rate $k_{PL} = k_{rad} + k_{non}$ by using the time-resolved PL (TRPL) measurements [8]. By the SP coupling, the spontaneous emission rate should be increased to $k^{*}_{PL} = k_{rad} + k_{non} + k_{SPC}$. The observed $k_{PL}$ and $k^{*}_{PL}$ values of the sample used in Figure 15.1 were plotted against wavelength in Figure 15.3. To perform the PRPL measurements, frequency doubled beams of a mode-locked $Al_2O_3$:Ti laser (Spectra Physics, Tsunami) was used to excite the QW. The pulse width, wavelength, and repetition rate were 1.5 ps, 400 nm, and 80 MHz, respectively. A Hamamatsu Photonics C5680 streak camera was used as the detector. The emission rates with silver were much faster than those of the uncoated sample and strongly depend on wavelength. The difference becomes dramatically greater at the shorter wavelength region. In this shorter wavelength region, the PL enhancement was also much more effective and the SP coupling rates should be very fast. The spontaneous emission rate into the SPP mode (SP coupling rate) depend on the density of states of the SPP by Fermi's golden rule [5, 6]. The density of states of the SPP mode is proportional to $dk/d\omega$ which can be obtained by the dispersion curve. $dk/d\omega$ is also plotted in Figure 15.3 as the sold line [8].

The SP coupling rate should be almost equal to the PL decay rate with Ag layers because those values were much greater than the values of the PL decay rate without Ag. Figure 15.6 shows that the wavelength dependence of the SP coupling rates is similar to that of $dk/d\omega$. The SP coupling becomes remarkable when the energy is near to the

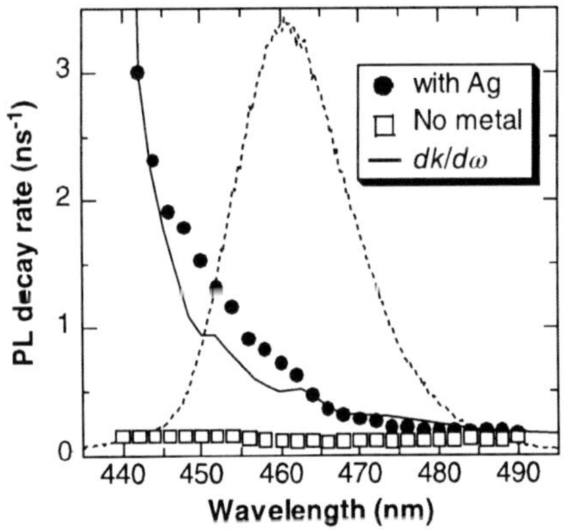


**Figure 15.6** The spontaneous emission rates of InGaN/GaN with/without silver layer plotted against wavelength. The solid line is dk/dω of the SPP mode at the silver/GaN interface obtained by the dispersion curve (Fir. 2b). The dashed line is the emission spectrum.

SP frequency described in Figure 15.2b as 2.84 eV (437 nm). At this shorter wavelength region, the SP coupling rate should be much faster than the radiative or nonradiative recombination rate of electron-hole pairs ($k_{SPC} \gg k_{rad} + k_{non}$) and the internal quantum efficiencies should be reached to almost 100% [11]. This suggests one of the most important advantages of the SP coupling technique to enhance the emission efficiencies. If we can control the SP frequency and obtain the best matching condition between the emission wavelength and the SP frequency, we can increase the internal quantum efficiencies to 100% at any wavelength. It would bring full color devices and natural white LEDs. Tuning of SP coupling should be available by choosing the appropriate metal, metal mixture alloys, multiple layers, or nanostructures.

Our proposed mechanism predicts several very important advantages of this method. We found that the SP coupling enhances internal quantum efficiencies and it reached almost 100% at the best matched wavelength. This suggests that this technique should be applicable for electrical pumping because the internal quantum efficiencies do not depend on the pumping method. So this should provide super bright plasmonic LEDs by electrical pumping[10, 11].

## 15.5 Applications for Organic Materials

As described above, high-efficiency emissions using plasmonics have been developed mainly for inorganic materials. The most important advantage of the SP coupling technique is that the technique can be applied not only to InGaN-based materials but also to various materials. Therefore, we have tried to use this technique for various organic light-emitting materials. For example, polymers, appropriately doped with dye molecules, emitting in the visible spectrum provide stable sources of light for displays and illumination sources at a significantly lower cost than inorganic semiconductors. Organic LEDs have become widely available and are used for replacing inorganic LEDs as they are less expensive and provide many opportunities with regard to structural placement. Despite the tremendous promise of efficient solid-state lighting offered by such organic light emitters, the road toward spectrally broad white light polymer emitters still holds many design challenges. Thus, it is of both commercial and scientific interest to improve the IQEs of the polymer dyes within such light emitters, as well as to increase the light extraction efficiencies from such organic films. Here, we focus on enhancing the light emission efficiency from organic thin films by using the SP coupling.

The experimental setup used to measure our samples is shown in Figure 15.7a. Dye polymer solution was prepared by dissolving common laser dye molecules of Coumarin 460 in chlorobenzene [12]. This laser dye emits blue light at 460 nm with UV excitation. Then, 2% polymethylmethacrylate (PMMA) was added to the mixture as a host matrix to obtain a 20 mM/L solution of the dye doped polymer solution. Only half of each substrate was metallized, enabling the rapid comparison between polymer emissions on top of metal layers with polymer deposited on quartz. After the metallization step, the dye doped PMMA layers were spun onto both gold and silver substrates to obtain layer thicknesses of ~200 nm. Figure 15.7b shows typical PL spectra of Coumarin 460 on Ag, Au, and bare quartz substrate. While the Au assisted in reflecting the pump laser, the surface plasmons did not seem to couple to the emission wavelength of Coumarin 460 to offer any measurable enhancement. However, we do observe an 11-fold enhancement of the emission light from the Coumarin doped PMMA on silver due to the coupling of the surface plasmons generated on the Ag film as the plasmon resonance frequency closely matches the emission frequency of the dye. Indeed, the dielectric constants for Ag match well with the emission wavelength of Coumarin 460, and if the data with the Coumarin 460 PL intensity normalized to 1. While reflection can be used to account for some of the increased brightness, only the SP coupling can explain the enhancement measured.

Likewise, we obtained obvious enhancements of both PL intensities and emission rates for three conjugated polymers: polyfluorenes (PF)-cyanophenylene(CNP) (1:1), PF-CNP (3:1), and polyfluorenes(PF)-triphenylamine(TPA)-quinoline(Q) [13]. These polymers have actually been used for organic LEDs as high-efficient light emitting materials.

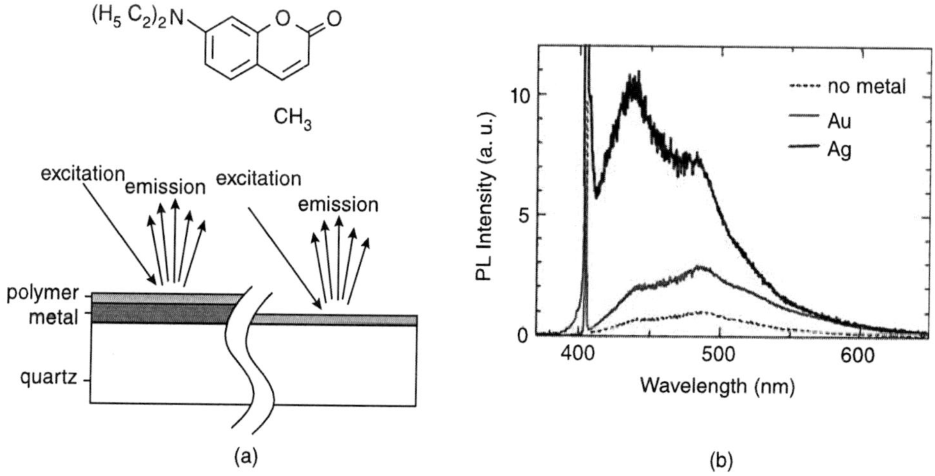


**Figure 15.7** (a) Sample structure of dye doped polymer with both pump light and emission light configurations. (b) PL spectra of Coumarin 460 on Ag, Au, and quartz. (b) The PL peak intensity of Coumarin 460 on quartz was normalized to 1.

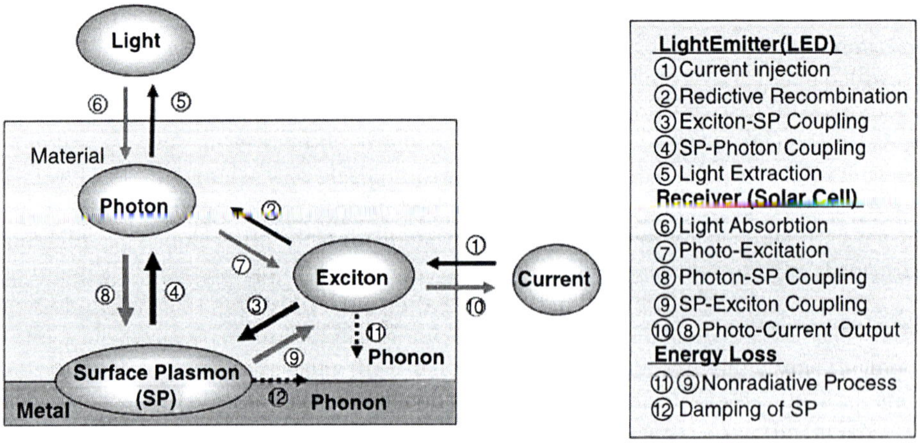


**Figure 15.8** Energy conversion schemes of the SP enhanced LEDs and solar cells.

## 15.6 Device Application for Light-Emitting Devices

Figure 15.8 shows the energy conversion scheme of the SP coupling and light emission [14]. The SP-exciton and SP-photon coupling processes provide new emission pathways. The high efficiency LEDs should be achievable if the new emission path through the SP coupling is much faster than the original emission path [(2) + (11) > (3), (4) > (12) in Figure 15.8]. In a similar way, plasmonics should also be able to improve high-efficiency solar cells, because the SP-exciton and SP-photon coupling processes are reversible. The sunlight can couple to the SP at the metal/dielectric interface and generate the excitons in the dielectric materials. This process should increase efficiencies of light absorption and photocurrent conversion if (8) > (7), (9) > (12) in Figure 15.8.

Possible device structures of plasmonic LEDs are shown in Figure 15.9 [10, 11]. Type A is the simplest structure using the usual LED structure with a p-n junction. The metal layer can be used both as an electrical contact and for exciting plasmons. The important point of this structure is that the distance between the metal surface and the InGaN QW must be very close to obtain a good SP coupling. Therefore, the p-type GaN layer must be thinner than 10 nm. As we mention above, the PL enhancement ratios become exponentially decayed with increasing of thickness of the GaN spacer layer. This feature makes the device application of the SP coupling so difficult. We already fabricated the structure of Type A but we were not able to obtain a huge enhancement of emission. There are two reasons; first, p-doping was very difficult into a 10 nm thick GaN layer. Second, we could not get a good ohmic contact because the p-GaN layer is too thin. Type B is the structure which has the metal nanoparticles or metal nanostructures on the top of the wafers. This structure is also very simple and has been fabricated and reported on by many groups. For example, Yeh et al. reported the SP coupling effect in an InGaN/GaN single-QW LED structure [15]. Their LED structure has a 10 nm p-type AlGaN current blocking layer and a 70 nm p-type GaN layer between the metal surface and the InGaN QW layer. The total distance is 80 nm, which is too far to obtain an effective SP coupling. For this reason, they obtained only 1.5-fold enhancement of the emission. Kwon, et al. also reported a plasmonic LED which has a similar structure of Type C [16]. They put

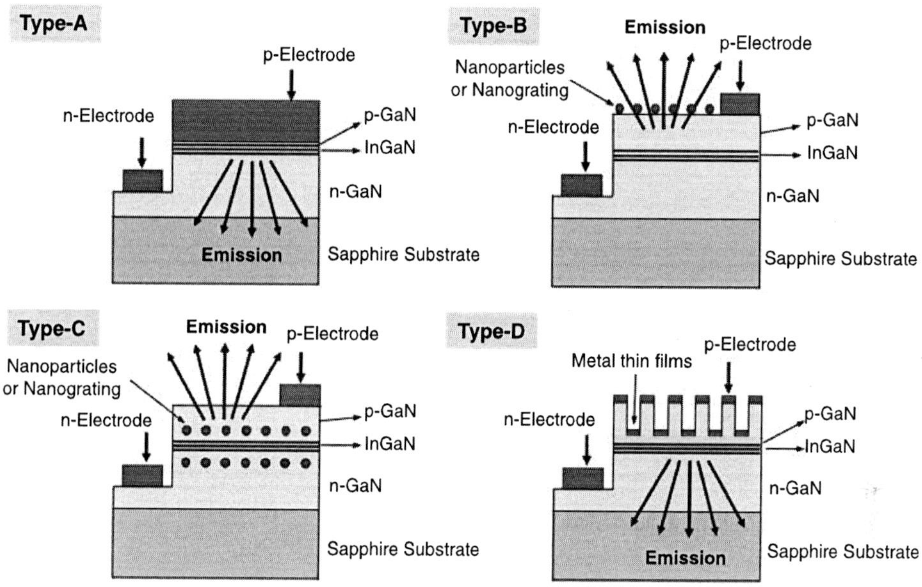


**Figure 15.9** Possible device structures for the plasmonic LED with electric pumping.

silver particles on the InGaN QW layer first, and overgrew a GaN layer above the Ag particles. However, a large number of Ag particles were gone by the high temperature of the crystal growth and only 3% of particles remained. Therefore, they obtained only 1.3-fold enhancement of the emission. Lu et al. fabricated the device structure of Type D by using the E-beam lithography and the dry etching process [17]. The initial thickness of the p-GaN layer was 150 nm and the etching depth was 110 nm. Therefore, the distance between the QW and metal interface was 40 nm, which should be enough to obtain the effective exciton-SP coupling. 2.8-fold enhancement was obtained by this structure with PL measurements; however, the electric pumping of this structure was difficult and has not yet succeeded. These tiny enhancement ratios should not be good enough for device application. Therefore, a highly efficient LED structure based on plasmonics has not yet been achieved.

As described above, the effective plasmonic LEDs have not yet been attained mainly due to technical difficulties. On the other hand, the organic devices are very easy to fabricate and should be more suitable for application to the plasmonic LED device structures. In 2005, which is the next year of our first report for the PL enhancement by plasmonics, Feng et al. reported enhancements of electroluminescence (EL) by using organic materials [18]. The 2D corrugated structures of a 80 nm thick tris-(8-hydroxyquinoline) aluminum (Alq) layer and an 80 nm thick N, N-diphenyl-N,N-bis(1-naphthyl)-(1,1'-biphenyl)-4,4'-diamine (NPB) layer were deposited on the grating structure of an indium tin oxide (ITO)/quarts substrates (Figure 15.10a). The observed emission intensity was enhanced by a factor of 4 compared with that of uncorrugated organic LEDs due to the increasing of the light extraction efficiency. In 2010, more significant increase in EL was achieved by Fujiki et al. through coupling with localized SP modes in a single layer of Au nanoparticles in the thin-film organic LEDs (Figure 15.10b) [19]. The device structure, with

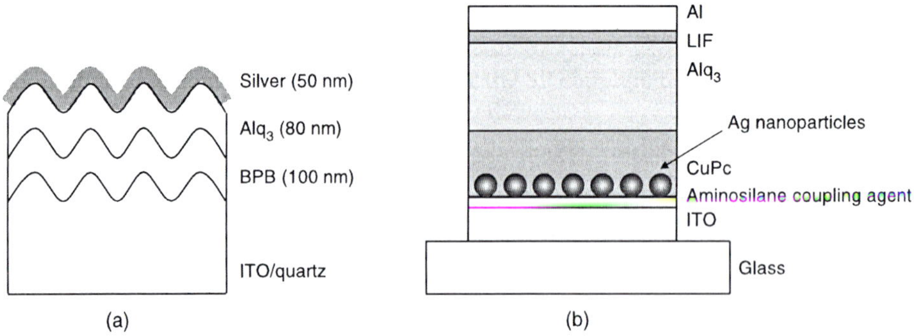


**Figure 15.10** Reported plasmonic organic LEDs driven by electrical pumping.

size-controlled Au particles embedded on ITO, can be used to realize the optimum distance for exciton-plasmon interactions by simply adjusting the thickness of the hole transport layer. A 20-fold increase in the molecular fluorescence was observed compared with that of a conventional diode structure. Moreover, we obtained a large PL enhancement for silicon nanocrystals in silicon dioxide media. Usually, the emission efficiencies of such indirect semiconductors are quite low, but by using the SP coupling, it is possible to increase these efficiencies up to values as large as those of direct compound semiconductors. We believe that the SP coupling technique would provide extremely bright silicon-based light-emitting devices.

## 15.7 Applications to High-Efficiency Solar Cells

Next, very important application of plasmonics is high-efficiency light-resaving devices, namely, photovoltaic devices [20]. The SP-exciton and SP-photon coupling processes are reversible processes as shown in Figure 15.8. Therefore, if the SP coupling increases light emission processes, it should also increase the light receiving processes. The sunlight can couple to the SP at the metal/dielectric interface and generate the excitons in the dielectric materials. The SP coupling make giant electric field at the metal surface by the light-antenna effect of the SP. Therefore, the excitation processes through the SP coupling should be much faster than the direct excitation processes as shown in Figure 15.8, and increase light absorption efficiencies.

Until now, several types of the plasmonic solar cells have been reported by using several structures as shown in Figure 15.11 [14]. The using of propagating modes of SPP must be useful to enhance the efficiency and making ultra-thin structures of solar cells. But the coupling between the light and the SPP is not easy. The wave vectors of the light and the SPP need to be matched in order to be coupled each other. Usually all-reflection setting with prism (Figure 15.11a) [21–23]or periodic nano-structures (Figure 15.11b) [24–26] are necessary to satisfy the matching condition of the wave vectors. These settings limit the plasmonic enhancement effect into some angle and wavelength of light. Figure 15.11c shows the structure by using the metal nanoparticles, which were simply dispersed on top of the solar cells. This structure is very easy to fabricate and useful to any type of solar cells such as silicon [27, 28], amorphous silicon [29], InGaAsP [30], GaAs [31], and GaN [32]. In this case, energy transfer from the localized SP mode to

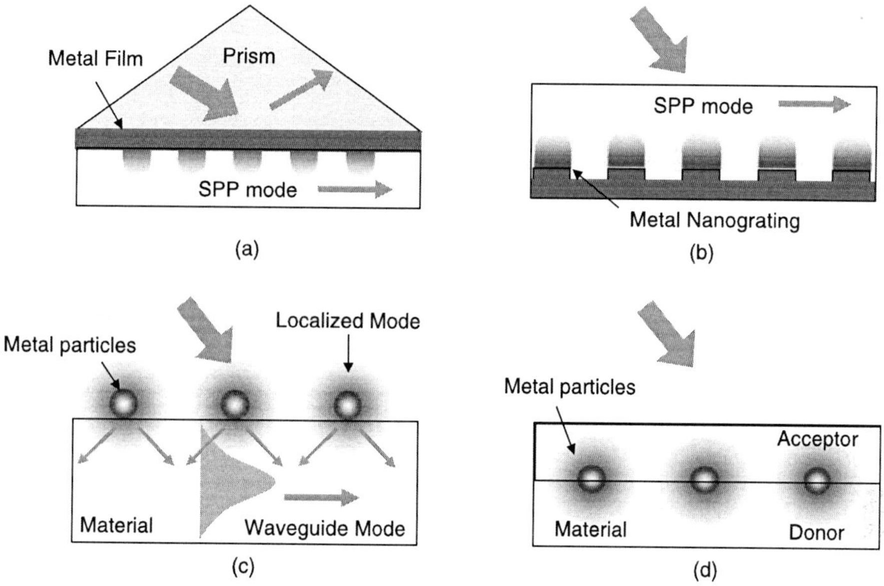


**Figure 15.11** Several types of the device structures of reported photovoltaic devices.

material is difficult because the electrical field is strongly confined around the metal nanoparticle. In many cases, the localized SP mode must couple to the waveguide mode in the materials. This fact limits the plasmonic enhancement effect, and the device structures must be thicker than the wavelength.

As well as the case of effective plasmonic LEDs, the effective exciton-SP coupling requires a very short distance between the metal surface and the exciton. Figure 15.11d shows the device structures which include nanoparticles inside. Such device structures are very difficult to fabricate using inorganic materials. Therefore, the several reports have been published for this structure by using organic thin films. For example, Rand et al. fabricate the tandem ultrathin-film organic photovoltaic cells which include Ag nanoparticles at the interface of the donor/accepter [33]. The organic layers consisted of the electron donor, copper phthalocyanine (CuPc), and the acceptor, 3,4,9,10-perylenetetracarboxylic bis-benzimidazole (PTCBI). The photocurrent conversion efficiency was increased from 1.1% to 2.4% by the SP coupling effect. Kim et al. employed the Ag nanoparticles into the organic thin-film solar cell with poly (3,4-ethylenedioxythiophene): poly(styrene-sulfonate) (PEDOT:PSS) as a hole transport layer and poly (3-hexylthiophene) (P3HT)/1-(3-methoxycarbonyl)-propyl-1-phenyl-(6,6) $C_{61}$ (PCBM) as the bulk heterojunction [34]. By incorporating plasmonic Ag nanoparticles on surface-modified transparent electrodes, overall power conversion efficiency was increased from 3.05% to 3.69%, mainly resulting from the improved photocurrent, and coupled to the radiative photon. We calculated the improved quantum efficiency and photocurrent of the CuPc/PTCBI solar-cell structure with Ag nanoparticles as shown in Figure 15.12a by using the combination of the FDTD and device simulations. The enhanced quantum efficiency and photocurrent were shown in Figures 15.12b and 15.12c, respectively. More than double enhancement, mainly by increasing the photocurrent, was clearly obtained. As described above, the SP-coupling

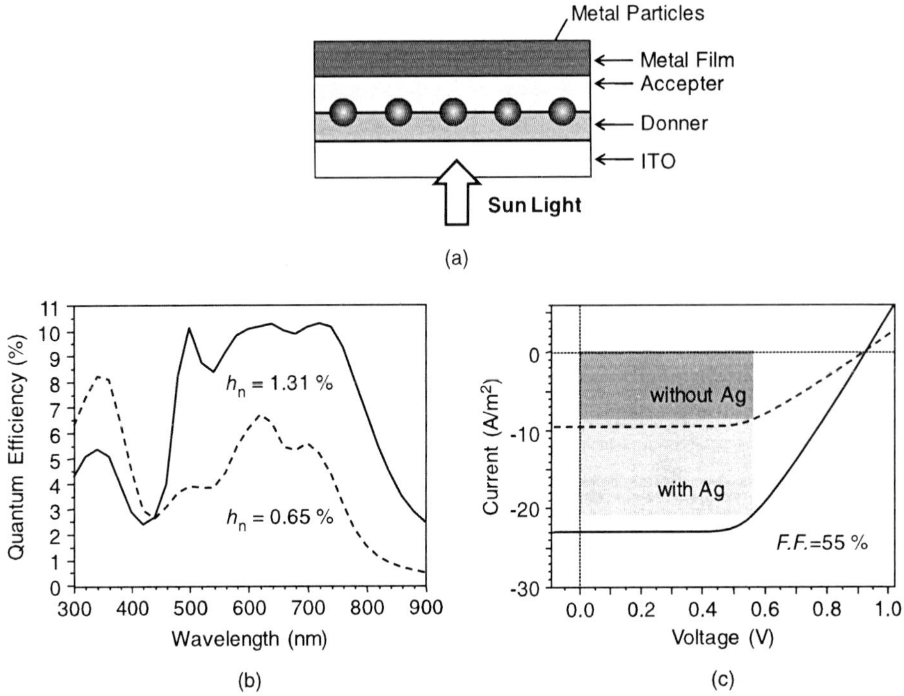


**Figure 15.12** Energy conversion schemes of the SP enhanced LEDs and solar cells.

technique has been expected to progress solar-cell technology. The SP-enhanced solar cell has been studied by several groups, however. such devices have not so far been used practically. We believe that further optimization of nanostructures and controlling the SP coupling would provide highly efficient and ultra-thin plasmonic solar cells.

## Acknowledgements

The author wishes to thank Prof. Y. Kawakami (Kyoto University), Prof. K. Tamada (Kyushu University) and Prof. A. Scherer (California Institute of Technology) for valuable discussions and support. This work was supported by JSPS KAKENHI (18H01903, 19H05627, 20H05622).

## References

1 Raether, H. (1988). *Surface Plasmons on Smooth and Rough Surfaces and on Gratings*. Berlin: Springer-Verlag.
2 Barnes, W.L., Dereux, A., and Ebbesen, T.W. (2003). *Nature* 424: 824.
3 Atwater, H.A. (2007). *Sci. Am.* 296: 56.
4 Nakamura, S., Senoh, M., and Mukai, T. (1993). *Jpn. J. Appl. Phys. Part 2 - Lett. Express Lett.* 32: L8.

5 Gontijo, I., Boroditsky, M., Yablonovitch, E. et al. (1999). *Phys. Rev. B* 60: 11564.
6 Neogi, A., Lee, C.-W., Everitt, H.O. et al. (2002). *Phys. Rev. B* 66: 153305.
7 Okamoto, K., Niki, I., Shvartser, A. et al. (2004). *Nat. Mater* 3: 601.
8 Okamoto, K., Niki, I., Scherer, A. et al. (2005). *Appl. Phys. Lett.* 87: 071102.
9 Xu, X., Funato, M., Kawakami, Y. et al. (2013). *Opt. Express* 21: 3145.
10 Okamoto, K. and Kawakami, Y. (2009). *IEEE J. Sel. Top. Quantum Electron.* 15: 1199.
11 Okamoto, K. and Kawakami, Y. (2010). *Phys. Status Solidi C* 7: 2582.
12 Neal, T.D., Okamoto, K., and Scherer, A. (2005). *Opt. Express* 13: 5522.
13 Neal, T.D., Okamoto, K., Scherer, A. et al. (2006). *Appl. Phys. Lett.* 89: 221106.
14 Okamoto, K. (2012). *Plasmonics for Green Technologies: Toward High-Efficiency LEDs and Solar Cells.* InTech: Advanced Photonic Sciences.
15 Yeh, D.-M., Huang, C.-F., Chen, C.-Y. et al. (2007). *Appl. Phys. Lett.* 91: 171103.
16 Kwon, M.-K., Kim, J.-Y., Kim, B.-H. et al. (2008). *Adv. Mater.* 20: 1253.
17 Lu, C.-H., Lan, C.-C., Lai, Y.-L. et al. (2001). *Adv. Funct. Mater.* 21: 4719.
18 Feng, J. and Okamoto, T. (2005). *Opt. Lett.* 30: 2302.
19 Fujiki, A., Uemura, T., Zettsu, N. et al. (2010). *Appl. Phys. Lett.* 96: 3.
20 Atwater, H.A. and Polman, A. (2010). *Nat. Mater.* 9: 205.
21 Hayashi, S., Kozaru, K., and Yamamoto, K. (1991). *Solid State Commun.* 79: 763.
22 Kume, T., Hayashi, S., Ohkuma, H., and Yamamoto, K. (1995). *Jpn. J. Appl. Phys.* 34: 6448.
23 Mapel, J.K., Singh, M., Baldo, M.A., and Celebi, K. (2007). *Appl. Phys. Lett.* 90: 121102.
24 Tvingstedt, K., Persson, N.K., Inganas, O. et al. *Appl. Phys. Lett.* 91: 113514.
25 Ferry, V.E., Sweatlock, L.A., Pacifici, D., and Atwater, H.A. (2008). *Nano Lett.* 8: 4391.
26 Ferry, V.E., Verschuuren, M.A., van Lare, M.C. et al. (2011). *Nano Lett.* 11: 4239.
27 Stuart, H.R. and Hall, D.G. (1996). *Appl. Phys. Lett.* 69: 2327.
28 Catchpole, K.R. and Pillai, S. (2006). *J. Lumin.* 121: 315.
29 Derkacs, D., Lim, S.H., Matheu, P. et al. (2006). *Appl. Phys. Lett.* 89: 093103.
30 Derkacs, D., Chen, W.V., Matheu, P.M. et al. (2008). *Appl. Phys. Lett.* 93: 091107.
31 Nakayama, K., Tanabe, K., and Atwater, H.A. (2008). *Appl. Phys. Lett.* 93: 121904.
32 Pryce, I.M., Koleske, D.D., Fischer, A.J., and Atwater, H.A. (2010). *Appl. Phys. Lett.* 96: 153501.
33 Rand, B.P., Peumans, P., and Forrest, S.R. (2004). *J. Appl. Phys.* 96: 7519.
34 Kim, S.-S., Na, S.-I., Jo, J. et al. (2008). *Appl. Phys. Lett.* 93: 073307.

# Index

## m



## n



## o



## p